WVU-PARKERSBURG LIBRARY
TS 156.8 .C62 1991
Comprehensive dictionary of me

3 8931 100036946 2

Ref.
TS
156
.8
.C62
1991

Comprehensive diction-
ary of measurement
and control

For Reference
Do Not Take
From the Library

D1154193

Comprehensive Dictionary of Measurement and Control

Second Edition

Reference Guides for Measurement and Control

W. H. Cubberly, Editor

Instrument Society of America

Ref.
TS
156
.8
.C62
1991

The Library
West Virginia University
at Parkersburg
Parkersburg, WV 26101

Copyright © Instrument Society of America 1991

All rights reserved

Printed in the United States of America

No part of this publication may be reproduced, stored in a
retrieval system, or transmitted, in any form or by any means,
electronic, mechanical, photocopying, recording or otherwise,
without the prior written permission of the publisher.

INSTRUMENT SOCIETY OF AMERICA
67 Alexander Drive
P.O. Box 12277
Research Triangle Park
North Carolina 27709

Library of Congress Cataloging-in-Publication Data

Comprehensive dictionary of instrumentation and
    control, 2nd ed.

    1. Process control—Dictionaries.   2. Engineering
instruments—Dictionaries.   I. Cubberly, W. H.
II. Instrument Society of America.
TS156.8.C62 1991   670.42-dc20   91-24460
ISBN 1-55617-320-2

11/94

ISA Reviewers

Mr. Stan Koloboff, Chevron Research & Technology Company

Mr. Rick Mergen, Lubrizol Corporation

Mr. Frank McGowan, Factory Mutual Research Corporation

Mr. Robert Hubby, Leeds & Northrup

Mr. Warren Weidman, Gilbert Commonwealth, Inc.

Mr. Harry Conner, Consultant

Table of Contents

How to Use
This Dictionary

Basic Format

The format for a defined entry provides the term in boldface and the definition in regular typeface. A term may have more than one definition, in which case the definition is preceded by a number. ISA Standards definitions always appear first. Abbreviations and acronyms are shown as such as well as spelled out, except in those cases where the abbreviation or acronym is so well known as to exclude the need for a spelled-out entry. Symbols are listed, starting with the most general and moving to the very specific.

Alphabetization

The terms are alphabetized on a letter-by-letter basis. Hyphens, commas, word spacing, etc., in a term are ignored in the sequencing of terms. To aid the user in finding words or abbreviations, the first and last term to appear on each page is shown in large type at the top of the page.

Definitions

A

a See ampere.

A See Ängstrom.

aberration Deviation from ideal behavior by a lens, optical system, or optical component. It exists in all optical systems and designers have to make trade-offs among the different types.

abort In data processing, the termination of a computer operation before its normal conclusion.

abrasion 1. Removal of surface material by sliding or rolling contact with hard particles of the same substance or another substance; the particles may be loose or may be part of another surface in contact with the first. 2. A surface blemish caused by roughening or scratching.

abrasive 1. Particulate matter, usually having sharp edges or points, that can be used to shape and finish workpieces in grinding, honing, lapping, polishing, blasting or tumbling processes; depending on the process, abrasives may be loose, formed into solid shapes, glued to paper or cloth, or suspended in a paste, slurry or air stream. 2. Any substance capable of removing material from a surface by abrasion. 3. A material formed into a solid mass, usually fired or sintered, and used to grind or polish workpieces; common forms are grinding wheels, abrasive discs, honing sticks, cones, and burrs.

ABS See Acrylonitrile Butadiene Styrene.

absolute accuracy error The deviation of the analog value at any code from its theoretical value after the full-scale range has been calibrated. Expressed in percent, ppm, or fractions of 1 LSB.

absolute address An address which indicates the exact storage location where the referenced operand is to be found or stored in the actual machine code address numbering system. Synonymous with specific address and actual address and related to absolute code.

absolute alarm An alarm caused by the detection of a variable which has exceeded a set of prescribed high- or low-limit conditions.

absolute altimeter See terrain clearance indicator.

absolute altitude Distance from an aircraft or spacecraft to the actual surface of a planet or natural satellite.

absolute code Coding that uses machine instructions with absolute addresses. Synonymous with specific code.

absolute encoder An electronic or electromechanical device which produces a unique digital output (in coded form) for each value of an analog or digital input; in an absolute rotary encoder, for instance, the position following any incremental movement can be determined directly, without reference to the starting position.

absolute feedback In numerical control, assignment of a unique value to each possible position of machine slide or actuating member.

absolute humidity The weight of water vapor in a gas water-vapor mixture per unit volume of space occupied, as, for example, grains or pounds per cubic foot.

absolute instrument An instrument that determines the value of a measured quantity in absolute units by making a simple physical measurement.

absolute measurement A measured value expressed in terms of fundamental standards of distance, mass and time.

absolute pressure 1. The pressure measured relative to zero pressure (vacuum). 2. The combined local pressure induced by some source and the atmospheric pressure at the location of the measurement. 3. Gage pressure plus barometric pressure in the same units.

absolute programming In numerical control, using a single point of reference for determining all positions and dimensions.

1

absolute stability   A linear system is absolutely stable if there exists a limiting value of the open-loop gain such that the system is stable for all lower values of that gain, and unstable for all higher values.

absolute value error   The magnitude of the error disregarding the algebraic sign or, if a vectorial error, disregarding its direction.

absolute viscosity   A measure of the internal shear properties of fluids expressed as the tangential force per unit area at either of two horizontal planes separated by one unit thickness of a given fluid, one of the planes being fixed and the other moving with unit velocity.

absorbance   An optical property expressed as $\log(1/T)$, where T is the transmittance.

absorptance   The fraction of the incident light absorbed.

absorption   The reduction in intensity of a beam of electromagnetic or particulate radiation as it passes through matter, chiefly due to interactions with atoms or electrons, or with their electric and magnetic fields.

absorption band   A region of the electromagnetic spectrum where a given substance exhibits a high absorption coefficient compared to adjacent regions of the spectrum.

absorption coefficient   An inherent material property expressed as the fractional loss in radiation intensity per unit mass or per unit thickness determined over an infinitesimal thickness of the given material at a fixed wavelength and band width.

absorption curve   A graph of the variation of transmitted radiation through a fixed sample while the wavelength material of a given thickness is changed at a uniform rate.

absorption dynamometer   A device for measuring mechanical force or power by converting the mechanical energy to heat in a friction mechanism or bank of electrical resistors.

absorption-emission pyrometer   An instrument for determining gas temperature by measuring the radiation emitted by a calibrated reference source both before and after the radiation passes through the gas, where it is partly absorbed.

absorption hygrometer   An instrument for determining water vapor content of the atmosphere by measuring the amount absorbed by a hygroscopic chemical.

absorption meter   An instrument for measuring the quantity of light transmitted through a transparent medium by means of a photocell or other light detecting device.

absorption spectroscopy   The study of the wavelengths of light absorbed by materials and the relative intensities at which different wavelengths are absorbed. This technique can be used to identify materials and measure their optical densities.

absorption tower   A vertical tube in which a gas rising through a falling stream of liquid droplets is partially absorbed by the liquid.

accelerated life test   A method of estimating reliability or durability of a product by subjecting it to operating conditions above its maximum ratings.

accelerating agent   1. A substance which increases a chemical reaction rate. 2. A chemical that hastens the curing of rubber, plastic, cement or adhesives, and may also improve their properties. Also known as accelerator.

accelerating electrode   An auxiliary electrode in an electron tube that is maintained at an applied potential to accelerate electrons in a beam.

acceleration   The time rate of change of velocity; the second derivative of a distance function with respect to time.

acceleration error   The maximum difference, at any measurand value within the specified range, between output readings taken with and without the application of specified constant acceleration along specified axes [S37.1].

acceleration limit   The maximum vibration and shock acceleration which the transducer can accept in either direction along its sensitive axis without permanent damage, usually stated as ± _____ g's. The acceleration limits are usually much wider than the acceleration range and thereby represent a measure of the overload capability of the transducer [RP37.2].

acceleration range   The range of accelerations over which the transducer has the specified linearity [RP37.2].

acceleration time   1. The span of time it takes a mechanical component of a computer to go from rest to running speed. 2. The measurement of time for any object to reach a predetermined speed.

accelerometer   1. An instrument for measuring acceleration or an accelerating force such as gravity; if it includes provisions for making a recorded output, it is called an accelerograph. 2.

A transducer used to measure linear or angular acceleration.

access Pertaining to the ability to place information into, or retrieve information from a storage device.

access direct storage (DSA) The procedure whereby data are transferred to or from storage essentially coincident with normal computer operation, without disturbing the central processing unit registers [RP55.1].

accessible A system feature that is viewable by and interactive with the operator, and allows the operator to perform user permissible control actions, e.g. set point changes, auto-manual transfers, or on-off action [S5.3].

accessible area An area routinely or periodically entered by plant personnel in the performance of routine functions during normal plant operation and in accordance with applicable health physics procedures [S67.03].

accessible isolation valve The isolation valve nearest the measured process on an instrument sensing line which is available to personnel during normal plant operation. The root valve may or may not perform the function of the accessible isolation valve, dependent on its location [S67.02].

accessible part A part that can be touched during normal use or operator servicing [S82.01].

accessible terminal A node in an electronic network that is configured to allow it to be connected to an external circuit.

access method Any of the data-management techniques available to the user for transferring data between main storage and an input/output device.

access privilege The right or permission to access (read or write) a file granted by the processor following a request for such permission [S61.2].

access procedures The procedure by which the devices attached to the network gains access to the medium. The access procedure typically includes provisions to guarantee fairness in sharing the network bandwidth between attached devices. The most common access procedures for LANs are CSMA/CD, Token Bus, Token Ring, and Slotted Ring. See MAC.

access random 1. Pertaining to the process of obtaining data from, or placing data into, storage where the time required for such access is independent of the location of the data most re-cently obtained or placed in storage. 2. Pertaining to a storage device in which the access time is effectively independent of the location of the data. [RP55.1].

access, serial Pertaining to the process of obtaining data from, or placing data into, storage when there is a sequential relation governing the access time to successive storage locations [RP55.1].

access time 1. The interval between a request for stored information and the delivery of the information; often used as a reference to the speed of memory. 2. The time interval that is characteristic of a storage unit; a measure of the time required to locate information in a storage position and make it available for processing or to return information from the processing unit to a storage location.

Access Unit Interface The optional interface between a data station using an IEEE 802.3 LAN and a transceiver or modem. The AUI permits transparent connection of a data station to either baseband or broadband media.

Accredited Standard Committee A standard committee accredited to ANSI.

accumulator 1. The register and associated equipment in the arithmetic unit of the computer in which arithmetical and logical operations are performed. 2. A unit in the digital computer where numbers are totaled, i.e., accumulated. Often the accumulator stores one operand and upon receipt of any second operand, it forms and stores the result of performing the indicated operations on the first and second operands. Related to adder. 3. A pressure vessel containing water and steam, which is used to store the heat of steam for use at a later period and at some lower pressure. 4. A relatively large-volume chamber or other hydraulic device which receives fluid under low hydraulic power, stores it, and then discharges it at high hydraulic power, after which it is ready to repeat the cycle. 5. A chamber or vessel for storing low-side liquid refrigerant in a refrigeration system. 6. An accumulator is also referred to as a receiver, a reflux receiver or a reflux drum.

accuracy 1. The ratio of the error to the full-scale output or the ratio of the error to the output, as specified, expressed in percent [S37.1]. 2. In process instrumentation, degree of conformity of an indicated value to a recognized accepted standard value or ideal value [S51.1]. 3. The degree to which an indicated value matches the actual value of a measured variable. 4. Quantitatively, the difference between the measured value and the most probable value for the same quantity, when the latter is determined from all available data, critically adjusted for sources of error. 5. In process instrumentation, degree of conformity of an indicated value to a recognized accepted standard value, or ideal value. 6. The deviation, or error, by which an actual output varies from an expected ideal or absolute output. Each element in a measurement system contributes to errors, which should be separately specified if they significantly contribute to the degradation of total system accuracy. 7. In analog-to-digital converter, accuracy is tied to resolution, a 13-bit A/D, as used in the controller for example, can resolve to one part in 2^{13} or 8192, so best accuracy as a percentage of full scale range is theoretically 1/8192, or about 0.0125 percent.

accuracy (data processing) The degree of freedom from error, that is, the degree of conformity to truth or to a rule. Accuracy is contrasted with precision, e.g., four-place numerals are less precise than six-place numerals, nevertheless a properly computed four-place numeral might be more accurate than an improperly computed six-place numeral [RP55.1].

accuracy, mean (data processing) Mean accuracy is precisely defined as (100-E) % F.R. Where the mean error E is expressed as a percentage of full range (F.R.). It is common practice, however, to equate mean accuracy with the value of the mean error. That is, mean accuracy is commonly stated as 0.1% F.R. Whereas a more precise and acceptable statement is that mean accuracy is 99.9% F.R. [RP55.1].

accuracy, measured The maximum positive and negative deviation observed in testing a device under specified conditions and by a specified procedure. Note 1: It is usually measured as an inaccuracy and expressed as accuracy. 2: It is typically expressed in terms of the measured variable, percent of span, percent of upper range-value, percent of scale length or percent of actual output reading [S51.1].

accuracy rating In process instrumentation, a number or quantity that defines a limit that errors will not exceed when a device is used under specified operating conditions. Note 1: When operating conditions are not specified, reference operating conditions shall be assumed. 2: As a performance specification, accuracy (or reference accuracy) shall be assumed to mean accuracy rating of the device, when used at reference operating conditions. 3: Accuracy rating includes the combined effects of conformity, hysteresis, dead band and repeatability errors. The units being used are to be stated explicitly. It is preferred that a ± sign precede the number or quantity. The absence of a sign indicates a + and a – sign. Accuracy rating can be expressed in a number of forms. The following five examples are typical: (a) accuracy rating expressed in terms of the measured variable. Typical expression: The accuracy rating is ± 1 °C, or ± 2 °F. (b) accuracy rating expressed in percent of span. Typical expression: The accuracy rating is ±0.5% of span. (This percentage is calculated using scale units such as degrees F, psig, etc.). (c) accuracy rating expressed in percent of the upper range-value. Typical expression: The accuracy rating is ±0.5% of upper-range value. (This percentage is calculated using scale units such as kPa, degrees F, etc.). (d) accuracy rating expressed in percent of scale length. Typical expression: The accuracy rating is ±0.5% of scale length. (e) accuracy rating expressed in percent of actual output reading. Typical expression: The accuracy rating is ±1% of actual output reading. [S51.1].

accuracy, reference See accuracy, rating.

accuracy, total (data processing) Total accuracy is precisely defined as (100-EMAX) % F.S. Where the maximum error EMAX is expressed as a percentage of full scale value. It is a measurement of the worst case effect of all the errors present in the analog subsystem [RP55.1].

achromatic Optical elements which are designed to refract light of different wavelengths at the same angle. Typically achromatic lenses are made of two or more components of different refractive index, and are designed for use at visible wavelengths only.

acid cleaning The process of cleaning the interior surfaces of steam generating units by an inhibitor to prevent corrosion, and subsequently draining, washing and neutralizing the acid by a further wash of alkaline water.

acidity Represents the amount of free carbon dioxide mineral acids and salts (especially sulphates of iron and aluminum) which hydrolize to give hydrogen ions in water and is reported as milliequivalents per liter of acid, or p.p.m. acidity as calcium carbonate, or pH the measure of hydrogen ions concentration.

acid pickle Industrial waste consisting of spent liquor from an acidic process for cleaning metal surfaces.

acid-resistant Able to withstand chemical attack by strongly acidic solutions.

acid sludge Oil refinery waste fuel from acid treatment of unrefined petroleum.

acid wash A chemical solution containing phosphoric acid which is used to neutralize residues from alkaline cleaners and to simultaneously produce a phosphate coating that protects a surface of metal from rusting and prepares it for painting.

a-c input module I/O module that converts process switched a-c to logic levels for use in the PC.

ACK See acknowledge.

acknowledge 1. The sequence action that indicates recognition of a new alarm [S18.1]. 2. A message sent between peer entities to indicate that data was properly received.

Acme screw thread A type of power-transmission thread made in four series—29° general purpose, 29° stub, 60° stub and 10° modified square; the number of threads per inch is not standardized according to shank diameter.

acoustic Related to sound.

acoustical ohm The unit of measure for acoustic resistance, reactance or impedance; it equals unity when a sound pressure of one microbar produces a volume velocity of one cubic centimetre per second.

acoustic compliance The reciprocal of acoustic stiffness.

acoustic coupler A type of communications device that converts digital signals into audio tones that can be transmitted by telephone.

acoustic dispersion Separation of a complex sound wave into its various frequency components, usually due to variation of wave velocity in the medium with sound frequency; usually expressed in terms of the rate of change of velocity with frequency.

acoustic generator A transducer for converting electrical, mechanical or some other form of energy into sound waves.

acoustic holography A technique for detecting flaws or regions of inhomogeneity in a part by subjecting it to ultrasonic energy, producing an interference pattern on the free surface of water

in an immersion tank, and reading the interference pattern by laser holography to produce an image of the test object.

acoustic impedance The complex quotient obtained by dividing sound pressure on a surface by the flux through the surface.

acoustic inertance A property related to the kinetic energy of a sound medium which equals $Z_a/2\pi f$, where Z_a is the acoustic reactance and f is sound frequency; the usual units of measure are g/cm4. Also known as acoustic mass.

acoustic interferometer An instrument for measuring either the velocity or frequency of sound pressure in a standing wave established in a liquid or gas medium between a sound source and reflector as the reflector is moved or the frequency is varied.

acoustic radiometer An instrument that measures sound intensity by determining unidirectional steady-state pressure when the sound wave is reflected or absorbed at a boundary.

acoustic reactance The imaginary component of acoustic impedance.

acoustic resistance The real component of acoustic impedance.

acoustics 1. The technology associated with the production, transmission and utilization of sound, and the science associated with sound and its effects. 2. The architectural quality of a room—especially a concert hall, theater or auditorium—that influences the ability of a listener to hear sound clearly at any location.

acoustic sensitivity The output of a transducer (not due to rigid body motion) in response to a specified acoustical environment. This is sometimes expressed as the acceleration in g rms sufficient to produce the same output as induced by a specified sound pressure level spectrum having an over-all value of 140 db referred to 0.0002 dyne per sq. cm. rms. [RP37.2].

acoustic signature In sonar applications, the profile that is characteristic of a particular undersea object (or class of objects)—for example, the profile of a school of fish or a sea-bottom formation.

acoustic spectrometer An instrument for analyzing a complex sound wave by determining the volume (intensity) of sound-wave components having different frequencies.

acoustic stiffness A property related to the potential energy of a medium or its boundaries which equals $2\pi f Z_a$, where Z_a is the acoustic

5

reactance and *f* is sound frequency; the usual units of measure are dyne/cm5.

acousto-optic An interaction between an acoustic wave and a lightwave passing through the same material. Acousto-optic devices can serve for beam deflection, modulation, signal processing, and Q switching.

acousto-optic glass Glass with a composition designed to maximize the acousto-optic effect.

a-c output module I/O module that converts PC logic levels to output switch action for a-c load control.

acronym A word made up of the initial letters of a long or complex technical term, e.g. RAM is the acronym for Random Access Memory.

Acrylonitrile Butadiene Styrene A type of plastic material.

actinicity The ability of radiation to induce chemical change.

actinometer 1. An instrument for measuring the actinic quality of radiation—that is, its relative ability to induce chemical change. 2. An instrument for measuring the flux density of solar radiation.

action, air-to-close See fail-open.

action, air-to-open See fail-close.

activation analysis A method of determining composition, especially the concentration of trace elements, by bombarding the composite substance with neutrons and measuring the wavelengths and intensities of characteristic gamma rays emitted from activated nuclides.

active alarm point See alarm point.

active medium The material in a laser which produces the amplified stimulated emission. The name of the laser identifies the active medium.

active transducer A transducer whose output waves are produced by power derived from a source other than any of the actuating waves, but whose output power is controlled by the actuating waves.

activity Ratio of escaping tendency of the component in solution to that at a standard state. The ion concentration multiplied by an activity coefficient is equal to the ion activity.

actual address See absolute address.

actual flow The actual volume of liquid passing through the flowmeter in a unit time as computed by applying all necessary corrections for the effects of temperature, pressure, air buoyancy, etc. to the corresponding readings indicated by the calibrator [RP31.1].

actuate To put into action or motion.

actuating error signal See signal, actuating error.

actuation signal The setpoint minus the controlled variable at a given instant. Same as error.

actuator 1. A fluid powered or electrically powered device which supplies force and motion to a valve closure member [S75.05]. 2. A part of the final control element that translates the control signal into action of the final control device in the process. Typical examples are motors, solenoids, cylinders, etc. 3. A device responsible for actuating a mechanical device such as a control valve. 4. A device that actuates.

actuator, bellows type A fluid powered device in which the fluid acts upon a flexible convoluted component, the bellows.

actuator, diaphragm A fluid powered device in which the fluid acts upon a flexible component, the diaphragm.

actuator, double acting An actuator in which power is supplied in either direction.

actuator effective area The net area of piston, the bellows, vane or diaphragm acted on by fluid pressure to generate actuator output thrust. It may vary with relative stroke position depending upon the actuator design [S75.05].

actuator, electric A device which converts electrical energy into motion.

actuator, electro-hydraulic type A device which converts electrical energy to hydraulic pressure and into motion.

actuator, electro-mechanical type A device which converts electrical energy into motion.

actuator environment The temperature, pressure, humidity, radioactivity and corrosiveness of the atmosphere surrounding the actuator. Also, the mechanical and seismic vibration transmitted to the actuator through the piping or heat radiated toward the actuator from the valve body [R75.05].

actuator, hydraulic A fluid device which converts the energy of an incompressible fluid into motion.

actuator, piston type A fluid powered device in which the fluid acts upon a movable piston, to provide motion to the actuator stem.

actuator, pneumatic A device which converts

the energy of a compressible fluid, usually air, into motion.

actuator, single acting An actuator in which the power supply acts in only one direction, e.g., a spring diaphragm actuator.

actuator travel time See stroke time.

actuator, vane type A fluid powered device in which the fluid acts upon a pivoted member, the vane, to provide rotary motion.

ADA A PASCAL-based, real-time systems programming language developed by CII-Honeywell Bull for the United States Department of Defense.

adapting See self-adapting.

adaptive control A control system which adjusts its response to its inputs based on its previous experience. Automatic means are used to change the type or influence (or both) of control parameters in such a way as to improve the performance of the control system. See control, adaptive.

adaptive gain control A control technique which changes a feedback controller's gain based on measured process variables or controller setpoints.

adaptive optics Optical components which can be made to change the way in which they reflect or refract light. In practice, the term usually means mirrors with surface shapes that can be adjusted.

adaptive tuning 1. In a control system, a way to change control parameters according to current process conditions. 2. The identification of process gains and time that can be used to improve the response of a control loop.

adaptor bushing The part which attaches a close coupled diaphragm actuator to the bonnet of the diaphragm valve body [S75.05].

ADC See analog-to-digital converter.

A-D converter (ADC) A hardware device that converts analog data into digital form; also called an encoder.

ADD 1. See OR and false add. 2. See sum.

adder A device which forms, as output, the sum of two or more numbers presented as inputs. Often no data-retention feature is included, i.e., the output signal remains only as long as the input signals are present. Related to accumulator.

adder-subtractor A device whose output is a

representation of either the arithmetic sum or difference, or both, of the quantities represented by its operand inputs.

address 1. An identification, represented by a name, label or number, of a register or location in storage. Addresses also are part of an instruction word along with commands, tags, and other symbols. 2. The part of an instruction, which specifies an operand for the instruction.

address bus The highway linking subcomponents of the microcomputer system along which address data is transferred.

address field That part of an instruction or word containing an address or operand.

address format The arrangement of the address parts of an instruction.

addressing The means whereby the originator or control station selects the unit to which it is going to send a message.

addressing mode Method for addressing a location used for data storage.

address modification The hardware action of computing an instruction's effective operand address by some sequence of the following operations as prescribed within the instruction: a) Indexing, adding an index to the address; b) Indirect addressing, using the intermediate computed address to obtain another address from memory.

address register A register in which an address is stored.

add time The time required for one addition, not including the time required to get and return the quantities from storage.

adhesion 1. A bonding between two surfaces, usually applied to localized welding at high points under substantial contact pressures. 2. Bonding between two surfaces, assisted by an adhesive substance.

adhesive Any substance capable of bonding two surfaces together.

adhesive bonding A commercial process for fastening parts together in an assembly using only glue, cement, resin or other adhesive.

adhesive strength The strength of an adhesively bonded joint, usually measured in tension (perpendicular to the plane of the bonded joint) or in shear (parallel to the plane of the joint).

adiabatic Referring to a process which takes place without any exchange of heat between the process system and another system or its surroundings.

adiabatic curing Curing concrete or mortar under conditions where heat is neither gained nor lost.

adiabatic engine Any heat engine that produces power without a gain or loss of heat.

adiabatic temperature The theoretical temperature that would be attained by the products of combustion provided the entire chemical energy of the fuel, the sensible heat content of the fuel, and combustion air above the datum temperature were transferred to the products of combustion. This assumes: a) combustion is complete; b) there is no heat loss; c) there is no dissociation of the gaseous compounds formed, and d) inert gases play no part in the reaction.

adjacent channel In FM/FM telemetry, the modulated signal bandwidth immediately below or above the channel of interest.

adjacent equipment The auxiliary equipment which may be located adjacent to the valve or actuator [S75.05].

adjustment The process of altering the value of some circuit element or some component of the mechanism of an instrument, controller or auxiliary device to bring the indication to a desired value, usually a value corresponding to an independently determined value of the measured variable within a specified tolerance.

adjustment span Means provided in an instrument to change the slope of the input-output curve [S51.1]. See span shift.

adjustment, zero Means provided in an instrument to produce a parallel shift of the input-output curve. See zero shift [S51.1].

adjust, span Means provided in an instrument to change the slope of the input-output curve. See span shift.

adsorption The concentration of molecules of one or more specific elements or compounds at a phase boundary, usually at a solid surface bounding a liquid or gaseous medium containing the specific element or compound.

aeolight A type of glow lamp whose intensity of light output varies with an applied signal voltage; its construction employs a cold cathode and an envelope filled with a mixture of gases.

aerator Any device for injecting air into a material or process stream.

aerodynamics A branch of mechanics that deals with the motion of gases, such as air, and with the forces acting on solids in relative motion with respect to a gas.

aerograph Any self-recording instrument carried aloft to take meteorological data.

aerometer An instrument for determining the density of air or other gases.

aerosol A dispersion of fine liquid or solid particles in a gas—for instance, both smoke and fog are aerosols.

AFC See automatic frequency control.

afterglow Luminosity which persists in a gas after an electrical discharge passes through it; the phenomenon is sometimes utilized in flow measurement.

AGC See automatic gain control.

age hardening Raising the strength and hardness of an alloy by heating a supersaturated solid solution at a relatively low temperature to induce precipitation of a finely dispersed second phase. Also known as aging; precipitation hardening.

agglomeration Any process for converting a mass of relatively fine solid material into a mass of larger lumps.

aggregate Natural sand, gravel and crushed stone that is mixed with cement to make mortar or concrete.

aging 1. Alteration of the characteristics of a device due to use. 2. Operating a product before shipping it to stabilize component functions or detect early failures. 3. Any time-dependent change in properties of a material, but especially age hardening at room or slightly elevated temperatures. 4. Curing or stabilizing parts or materials by long-term storage outdoors or under closely controlled storage conditions.

agitator A device for mixing, stirring or shaking liquids or liquid-solid mixtures to keep them in motion.

air 1. Air implies use of any suitable, and normally clean, dry, safe gas [RP60.9]. 2. The mixture of oxygen, nitrogen and other gases, which with varying amounts of water vapor, forms the atmosphere of the earth.

air atomizing oil burner A burner for firing oil in which the oil is atomized by compressed air which is forced into and through one or more streams of oil breaking the oil into a fine spray.

air bearing A device which lubricates motion with flowing air. A linear air bearing, floats a table on air as it travels a straight line.

air bind An air pocket in a pump, conduit or piping system that prevents liquid from flowing past it. Also called a liquid trap.

air binding The inclusion of air in a space hindering the flow of some other gas or liquid.

air blast The flow of air at a high velocity, usually for a short period.

airborne Carried in the atmosphere—either by being transported in an aircraft or by being dispersed in the atmosphere.

air-bubbler liquid-level detector A device for indirectly measuring the level of liquid in a vessel—especially a corrosive liquid, viscous liquid or liquid containing suspended solids; it consists of a standpipe open at the bottom and closed at the top, which is connected to an air supply whose pressure is maintained slightly above maximum head of liquid in the vessel; air bubbles out of the bottom of the pipe, maintaining the internal pressure equal to the head of liquid in the vessel, pressure being measured by a simple gage or transducer. Also known as purge-type liquid-level detector.

air-bubbler specific-gravity meter Any of several devices that measure specific gravity by determining differential pressure between two air-purged bubbler columns; the devices ordinarily use either of two principles for determining specific gravity— comparison of sample density with density of a known liquid, or comparison of pressure between two bubbler columns immersed at different depths in the process liquid.

air buoyancy The lifting effect or buoyancy of the ambient air which acts during a "weighing" procedure with open gravimetric calibrations. This is caused by displacement of air from the measuring vessel during the calibration run. The standard air (50% R.H.) for correcting to "weights" in vacuum has a density of 1.217 kg/m³ at 288.7 K and 1.013 250E+05 Pa. When "weighings" are made against "weights", the buoyancy force on these must also be considered. For brass "weights" the net effect of air buoyancy in air at standard conditions is about 0.015% [RP31.1].

air compressor A machine that raises the pressure of air above atmospheric pressure and normally delivers it to an accumulator or distribution system.

air condenser 1. A heat exchanger for converting steam to water where the heat-transfer fluid is air. Also known as air-cooled condenser. 2. A device for removing oil or water vapors from a compressed-air line.

air conditioned area See area, air conditioned.

air conditioning Controlling the atmospheric environment in a confined space by measuring and continually adjusting factors such as temperature, humidity, air motion, and concentrations of dust, gases, odors, pollen or microorganisms.

air consumption The maximum rate at which air is consumed by a device within its operating range during steady-state signal conditions. It is usually expressed in cubic feet per minute (ft³/min) or cubic meters per hour (m³/h) at a standard (or normal) specified temperature and pressure [S51.1].

air-cooled engine An engine, such as an internal combustion engine, whose waste heat is removed directly by a flowing stream of air—either a stream blown across the engine's external surfaces, or one blown through internal cooling passages.

air-cooled heat exchanger A device for removing heat from a process fluid by passing it through a bank of finned tubes that are cooled by blowing or drawing a stream of air across the tube exteriors.

air curtain A stream of high-velocity conditioned air directed downward across an opening such as a door or window to exclude insects and exterior drafts, prevent heat transfer through the opening, and permit the interior space to be air conditioned.

air cushion 1. A mechanical device that uses trapped air to absorb shocks or arrest motion without shock. 2. The partly confined stream of low-pressure, low-velocity air that supports a vehicle known as an air-cushion vehicle, ground-effect machine or hovercraft and allows it to travel equally well over water, ice, marshland, or relatively level ground.

air cylinder A cylindrical body for storing compressed air, for compressing air with a piston, or for driving a piston with compressed air.

air deficiency Insufficient air, in an air-fuel mixture, to supply the oxygen theoretically required for complete oxidation of the fuel.

air dry 1. Air with which no water vapor is mixed. This term is used comparatively, since in nature there is always some water vapor included in air, and such water vapor being a gas, is dry. 2. A papermaking term used to describe "dry" pulp containing about 10% moisture.

air ejector   A device for removing air or noncondensible gases from a confined space, such as the shell of a steam condenser, by eduction using a fluid jet.

air entrainment   Artificial infusion of a semisolid mass such as concrete or a dense slurry with minute bubbles of air, especially by mechanical agitation.

air filter   A device for removing solid particles such as dust or pollen from a stream of air, especially by causing the airstream to pass through layered porous material such as cloth, paper or screening.

airfoil-vane fan   A device for creating a stream of moving air by drawing it into a fan casing near the hub and propelling it centrifugally with a rotor whose vanes are curved backward from the direction of rotation.

air-free   The descriptive characteristic of a substance from which air has been removed.

air-fuel ratio   The ratio of the weight, or volume, of air to fuel.

air furnace   Any furnace whose combustion air is supplied by natural draft, or whose internal atmosphere is predominantly heated air.

air gage   1. A device for measuring air pressure. 2. A device for precisely measuring physical dimensions by measuring the pressure or flow of air from a nozzle against a workpiece surface and relating the measurement to distance from the nozzle to the workpiece.

air gap   The space between two ferromagnetic elements of a magnetic circuit.

air-handling unit   An assembly of air-conditioning equipment, usually confined within a single enclosure, which treats air prior to distribution and provides the means of propelling the treated air through the distributing ducts.

air-hardening steel   A type of tool steel containing sufficient alloying elements to permit it to harden fully on cooling in air from a temperature above its transformation temperature. Also known as self-hardening steel.

air hoist   A lifting or hauling tackle whose power is provided by air-driven pistons (for reciprocating motion) or air motors (for rotary motion).

air infiltration   The leakage of air into a setting or duct.

air knife   A device that uses a thin, flat jet of air to remove excess coating material from sheet stock such as paper.

air lance   A device for directing a high-velocity stream of pressurized air into a process vessel, or against a surface such as a boiler wall to remove unwanted deposits.

air lock   1. An intermediate chamber between an environmentally controlled confined space and the outside atmosphere that provides for entry of personnel and materials by sealing a door between the chamber and the confined space, opening a door to the outside to admit personnel or materials, closing and sealing that door, changing environmental conditions in the chamber to match those in the confined space, then opening an interior door to permit entry into the confined space; the process is reversed when exiting the confined space. 2. See air bind.

air meter   A device for measuring the flow of air or other gas and expressing it as weight or volume per unit time.

air moisture   The water vapor suspended in the air.

air monitor   A warning device that detects airborne radioactivity or chemical contamination and sounds an alarm when the radiation, gas or vapor level exceeds a preset value.

air motor   An engine that produces rotary motion using compressed air or other gas as the working fluid.

air nozzle   An air port having direction and appreciable length for directing an air stream.

air permeability   A method of measuring the fineness of powdered materials, such as portland cement, by determining the ease with which air passes through a defined mass or volume.

air port   An opening through which air passes.

air-position indicator   An aircraft navigation device which integrates headings and speeds to give a continuous, dead-reckoned indication of position with respect to the surrounding air mass.

air preheater   A heat exchanger for transferring some of the waste heat in flue gases from a boiler or furnace to incoming air, which increases the efficiency of combustion.

air-puff blower   A soot blower automatically controlled to deliver intermittently jets or puffs of compressed air for removing ash, refuse, or soot from heat absorbing surfaces.

air purge   The removal of undesired matter by replacement with air.

air purging Removing airborne contaminants, gases or odors from a confined space by introducing fresh, clean air.

air regulator A device for controlling airflow—for example, a damper to control flow of air through a furnace, or a register to control flow of heated air into a room.

air reheater A device in a forced-air heating system that adds heat to air circulating in the system.

air resistance The opposition offered to the passage of air through any flow path.

air, saturated Air which contains the maximum amount of the vapor of water or other compound that it can hold at its temperature and pressure.

air separator A device for separating materials of different density, or particles of different sizes, by means of a flowing current of air.

air set A regulator which is used to control the supply pressure to the valve actuator and its auxiliaries [S75.05].

airspeed Speed of an airborne object with respect to the surrounding air mass; in calm air, airspeed is equal to ground speed; true airspeed is a calibrated airspeed that has been corrected for pressure and temperature effects due to altitude, and for compressibility effects at high airspeeds.

air spring A device commonly used instead of a mechanical spring in heavy vehicles to support the vehicle's body on its running gear; the energy-storage element is an air-filled container with an internal elastomeric bellows or diaphragm.

air supply (AS) 1. The supply of air used in pneumatic instrumentation as a power supply. 2. Plant air supply (PA). 3. Instrument air (IA). 4. The energy supply for pneumatic instrumentation.

air thermometer A device for measuring temperature in a confined space by detecting variations in pressure or volume of air in a bulb inside the space.

airtight Sealed to prevent passage of air or other gas; impervious to leakage of gases across a boundary.

air vent A valve opening in the top of the highest drum of a boiler or pressure vessel for venting air.

air vessel An enclosed chamber, partly filled with pressurized air, that is connected to a piping system to counteract water hammer or promote uniform flow of liquid.

airy disk The central bright spot produced by a theoretically perfect circular lens or mirror. The spot is surrounded by a series of dark and light rings, produced by diffraction effects.

alarm 1. A device or function that signals the existence of an abnormal condition by means of an audible or visible discrete change, or both, intended to attract attention [S5.1]. 2. An abnormal process condition. 3. The sequence state when an abnormal process condition occurs. 4. A device that calls attention to the existence of an abnormal process condition [S18.1]. See annunciator. 5. An audible, visual, or physical presentation designed to alert the instrument user that a specific level of hydrogen sulfide concentration has been reached or exceeded [S12.15]. 6. An instrument, such as a bell, light, printer, or buzzer, which indicates when the value of a variable is out of limits.

alarm extensions, electrically operated Usually a highly sensitive induction type device for signaling high or low flows or deviations from any set flow. The device consists of a sensing coil positioned around the extension tube of the rotameter. Movement of the metering float into the field of the coil causes a low level signal change which is usually amplified to a level suitable for performing annunciator or control functions [RP16.4].

alarm extensions, magnetically actuated A device attached to the meter body which contains an electrical switch and which is magnetically actuated by the metering float extension to signal a high or low flow. The switch is adjustable with respect to the float position over a range equal to the travel of the metering float. Standard switch ratings are usually 0.3 amperes for 110 volt, 60 cycle AC supply (five amperes or more if relays are used) [RP16.4].

alarm, maintained An alarm that returns to normal after being acknowledged [S18.1].

alarm module (point or sequence module) A plug-in assembly containing the sequence logic circuit. Some alarm modules also contain visual display lamps or lamps and windows [S18.1].

alarm, momentary An alarm that returns to normal before being acknowledged [S18.1].

alarm-only instrument An instrument providing an alarm(s), but which does not have an integral meter or other readout device indicating

11

current hydrogen sulfide concentration levels [S12.15].

alarm point The sequence logic circuit, visual display, auxiliary devices, and internal wiring related to one visual display [S18.1].

alarm point, active An alarm point that is wired internally and completely equipped. The window is labeled to identify a specific monitored variable [S18.1].

alarm point, future (blank) An alarm point that is wired internally and equipped except for the plug-in alarm module. The window is not labeled to identify a monitored variable [S18.1].

alarm point, spare An alarm point that is wired internally and completely equipped. The window is not labeled to identify a monitored variable [S18.1].

alarm set point The selected gas concentration level(s) at which an indication, alarm, or other output function is initiated [S12.13] [S12.15].

alarm severity A selection of levels of priority for the alarming of each input, output, or rate of change.

alarm system An integrated combination of detecting instruments and visible or audible warning devices that actuates when an environmental condition or process variable exceeds some predetermined value.

alarm valve A device that detects water flow and sounds an alarm when an automatic sprinkler system is activated.

alert See process condition; see also sequence state.

alert box In data processing, a window that appears on a computer screen to alert the user of an error condition.

algebraic adder An electronic or mechanical device that can automatically find the algebraic sum of two quantities.

ALGOL See Algorithmic-Oriented Language.

algorithm 1. A prescribed set of well-defined rules or processes for the solution of a problem in a finite number of steps, for example, full statement of an arithmetic procedure for evaluating sin X to a stated precision. 2. Detailed procedures for giving instructions to a computer. 3. Contrast with heuristic and stochastic. 4. A recursive computational procedure. 5. A step-by-step procedure for solving a problem or accomplishing an end. 6. Sometimes used in reference to a firmware or a software program.

algorithmic language A language designed for expressing algorithms.

Algorithmic-Oriented Language An international procedure-oriented language.

alias When varying signals are sampled at equally spaced intervals, two frequencies are considered to be aliases of one another if they cannot be distinguished from each other by an analysis of their equally spaced values.

aliasing A peculiar problem in data sampling, where data are not sampled enough times per cycle, and the sampled data cannot be reconstructed.

aliasing error An inherent error in time-shared telemetry systems where improper filtering is employed prior to sampling.

alidade 1. An instrument used in the plane-table method of topographic surveying and mapping. 2. Any sighting device for making angular measurements.

alkaline cleaner An alkali-based aqueous solution for removing soil from metal surfaces.

alkalinity Represents the amount of carbonates, bicarbonates, hydroxides and silicates or phosphates in the water and is reported as grains per gallon, or p.p.m., as calcium carbonate.

Allen screw A screw or bolt that has a hexagonal socket in its head, and is turned by inserting a straight or bent hexagonal rod into the socket.

alligatoring 1. Cracking in a film of paint or varnish characterized by broad, deep cracks extending through one or more coats. Also known as crocodiling. 2. Surface roughening of very coarse-grained sheet metal during forming. 3. Longitudinal splitting of flat slabs in a plane parallel to the rolled surface that occurs during hot rolling. Also called fishmouthing.

allobar A form of an element having a distribution of isotopes that is different from the distribution in the naturally occurring form; thus an allobar has a different apparent atomic weight than the naturally occurring form of the element.

allowable response time The limiting response time established in the safety analysis and documented in the plant's technical specifications [S67.06].

allowable value The limiting value that the trip setpoint can have when tested periodically, beyond which the instrument channel is declared inoperable and corrective action must be taken [S67.04].

allowable working pressure See design pressure.

allowance Specified difference in limiting sizes—either minimum clearance or maximum interference between mating parts—computed mathematically from the specified dimensions and tolerances of both parts.

alloy A solid material having metallic properties and composed of two or more chemical elements.

alloy steel An alloy of iron and carbon which also contains one or more additional elements intentionally added to increase hardenability or to enhance other properties.

all-pass network A network designed to introduce phase shift or delay into an electronic signal without appreciably reducing amplitude at any frequency.

Alnico Any of a series of commercial iron-base permanent magnet alloys containing varying amounts of aluminum, nickel and cobalt as the chief alloying elements; the Alnicos are characterized by their ability to produce a strong magnetic field for a relatively small magnet mass, and to retain their magnetism, with relatively insignificant loss in field strength when the magnetizing field is removed.

alphabet The specific character set used by a computer.

alphabetic word 1. A word consisting solely of letters. 2. A word consisting of characters from the same alphabet.

alpha counter 1. A system for detecting and counting energetic alpha particles; it consists of an alpha counter tube, amplifier, pulse-height discriminator, scaler and recording or indicating mechanism. 2. An alpha counter tube and necessary auxiliary circuits alone. 3. A term sometimes loosely used to describe just the alpha counter tube or chamber itself.

alpha emitter A radionuclide that disintegrates by emitting an alpha particle from its nucleus.

alphanumeric Pertaining to a character set that contains both letters and numerals, and usually other characters. Synonymous with alphameric [RP55.1].

alpha particle A positively charged, energetic atomic particle consisting of two protons and two neutrons, identical in all measured properties with the nucleus of a helium atom; it may be produced by radioactive decay of certain nuclides or by stripping a helium atom of its electrons.

alpha-ray spectrometer An instrument used to determine the energy distribution in a beam of alpha particles.

altazimuth A sighting instrument having both horizontal and vertical graduated circles so that both azimuth and declination can be determined from a single reading. Also known as astronomical theodolite; universal instrument.

alteration switch A manual switch on the computer console or a program-simulated switch which can be set on or off to control coded machine instructions.

alternate code complement In a frame synchronization scheme, a frame synchronization pattern is complemented on alternate frames to give better synchronization.

alternate immersion test A type of accelerated corrosion test in which a test specimen is repeatedly immersed in a corrosive medium, then withdrawn and allowed to drain and dry.

alternating-current bridge A bridge circuit that utilizes an a-c signal source and a-c null detector; generally, both in-phase (resistive) and quadrature (reactive) balance conditions must be established to balance the bridge. Some bridges require only one balance (resistive or reactive) and use a phase-sensitive detector.

altigraph A recording pressure altimeter.

altimeter An instrument for determining height of an object above a fixed level or reference plane—sea level, for example; the aneroid altimeter and the radio altimeter are the most common types.

altitude 1. The vertical distance above a stated reference level. Unless otherwise specified, this reference is mean sea level [S37.1]. 2. Height above a specified reference plane, such as average sea level, usually given as a distance measurement in feet or metres regardless of the method of measurement.

altitude signals Reflected radio signals returned to an airborne electronic device from the land or sea surface directly underneath the vehicle.

ALU See arithmetic and logical unit.

alum A general name for a class of double sulfates containing aluminum and another cation such as potassium, ammonium or iron.

alumina The oxide of aluminum—Al_2O_3.

aluminizing 1. Applying a thin film of aluminum to a material such as glass. 2. Forming a protective coating on metal by depositing aluminum on the surface, or reacting surface material with an aluminum compound, and diffusing the aluminum into the surface layer at elevated temperature.

aluminum A soft, white metal that in pure form exhibits excellent electrical conductivity and oxidation resistance; it is the base metal for an extensive series of lightweight structural alloys used in such diverse applications as aircraft frames and skin panels, automotive body panels and trim, lawn furniture, ladders, and domestic cookware.

amalgamation 1. Forming an alloy of any metal with mercury. 2. A process for separating a metal from its ore by extracting it with mercury in the form of an amalgam; the process was formerly used to recover gold and silver, which are now extracted chiefly with the cyanide process.

ambient A surrounding or prevailing condition, especially one that is not affected by a body or process contained in it.

ambient air 1. Air to which the sensing element is normally exposed [S12.15]. 2. The air that surrounds the equipment. The standard ambient air for performance calculations is air at 80 °F, 60% relative humidity, and a barometric pressure of 29.921 in. Hg., giving a specific humidity of 0.013 lb of water vapor per lb of air.

ambient conditions 1. The conditions (pressure, temperature, etc.,) of the medium surrounding the case of the transducer [S37.1]. 2. The environment of an enclosure (room, cabinet, etc.) surrounding a given device or equipment.

ambient pressure See pressure, ambient.

ambient pressure effects The change in sensitivity and the change in zero-measurand output due to subjecting the transducer to a specified ambient pressure change [S37.8].

ambient pressure error The maximum change in output, at any measured valued within the specified range, when the ambient pressure is changed between specified values [S37.1].

ambient temperature 1. The temperature of the atmosphere encompassing the area of the entire instrument air system installation, including the compressor, piping, dryer, and the instruments [S7.3]. 2. The temperature of the air surrounding a particular location. See also temperature, ambient.

American standard pipe thread A series of specified sizes for tapered, straight and dryseal pipe threads established as a standard in the United States. Also known as Briggs pipe thread.

American standard screw thread A series of specified sizes for threaded fasteners, such as bolts, nuts and machine screws, established as a standard in the United States.

Amici prism Also known as a roof prism. A right-angle prism in which the hypotenuse has been replaced by a roof, where two flat faces meet at a 90° angle. The prism performs image erection while deflecting the light by 90°; it is the same as rotating the image by 180°—reversing it left to right and at the same time inverting it top to bottom.

ammeter An instrument for determining the magnitude of an electric current.

ammonia A pungent, colorless, gaseous compound of hydrogen and nitrogen—NH_3; it is readily soluble in water, where it reacts to form the base, ammonium hydroxide.

amorphous film A film of material deposited on a substrate for corrosion protection, insulation, conductive properties or a variety of other purposes. It is non-crystalline and can be deposited by evaporation, chemical deposition or by condensation. The method employed would be dictated by its composition and ultimate use.

amp or ampere Metric unit for electric current produced by one volt acting through a resistance of one ohm. Also the current that will deposit silver at the rate of 0.001118 grams per second, with the current flowing at 1 coulomb per second.

ampere-hour A quantity of electricity equal to the amount of electrical energy passing a given point when a current of one ampere flows for one hour.

ampere-hour meter An integrating meter that measures electric current flowing in a circuit and indicates the integral of current with respect to time.

ampere per metre The SI unit of magnetic field strength; it equals the field strength developed in the interior of an elongated, uniformly wound coil excited with a linear current density in the winding of one ampere per metre of axial distance.

amplification 1. Increasing the amplitude of a signal by using a signal input to control the amplitude of a second signal supplied from another source. 2. The ratio of the output-signal amplitude from an amplifier circuit to the input-signal amplitude from the control network, both expressed in the same units.

amplification factor The μ factor for plate and control electrodes of an electron tube when the plate current is held constant.

amplification factor at resonant frequency The ratio of the maximum sensitivity of a transducer (at its resonant frequency) to its reference sensitivity. Amplification factor at resonant frequency is sometimes referred to as "Q" [RP37.2].

amplifier 1. In process instrumentation a device that enables an input signal to control power from a source independent of the signal and thus be capable of delivering an output that bears some relationship to and is generally greater than, the input signal [S51.1]. 2. Any device that can increase the magnitude of a physical quantity, such as mechanical force or electric current, without significant distortion of the wave shape of any variation with time associated with the quantity. 3. A component used in electronic equipment to raise the level of an input signal so that the corresponding output signal has sufficient power to drive an output device such as a recorder or loudspeaker.

amplifier (laser) A laser amplifier is a device which amplifies the light produced by an external laser, but lacks the mirrors needed to sustain oscillation and independently produce a laser beam.

amplitude A measure of the departure of a phenomenon from any given reference.

amplitude distortion A condition in an amplifier or other device when the amplitude of the output signal is not an exact linear function of the input (control) signal.

amplitude-frequency response See frequency response.

amplitude linearity, shock Closeness of sensitivity to reference sensitivity over a stated range of acceleration amplitudes, under shock conditions, usually specified as "within ± _____ percent for acceleration rise times longer than _____ microseconds." [RP37.2].

amplitude linearity, vibration Closeness of sensitivity over a stated range of acceleration amplitudes, at a stated fixed frequency, usually specified as "within ± _____ percent." [RP37.2].

amplitude modulation (AM) The process (or the results of the process) of varying the amplitude of the carrier in synchronism with and in proportion to the variation in the modulating signal.

amplitude noise Random fluctuations in the output of a light source or signal from other generating or detecting means.

amplitude response A measure of the time taken for a defined change of amplitude.

AM rejection The removal of unwanted amplitude modulation of a signal; usually performed by using signal clipping or limiting circuitry.

A/M station (automatic/manual station) In control systems, a device which enables the process operator to manually position one or more valves. A single-loop station enables manual positioning of a single valve; a shared station enables control of multiple valves; and a cascade station provides control of paired loops.

analog 1. Pertaining to data in the form of continuously variable physical quantities. Contrast with digital [RP55.1]. 2. The representation of numerical quantities by means of physical variables, such as translation, rotation, voltage, or resistance. 3. Contrasted with digital. A waveform is analog if it is continuous and varies over an arbitrary range.

analog back-up An alternate method of process control by conventional analog instrumentation in the event of a failure in the computer system.

analog channel A channel on which the information transmitted can take any value between the limits defined by the channel. Voice channels are analog channels.

analog computer 1. A computer in which analog representation of data is mainly used. 2. A computer that operates on analog data by performing physical processes on these data. Contrast with digital computer.

analog control Implementation of automatic control loops with analog (pneumatic or electronic) equipment. Contrast with direct digital control.

analog control system Classically, a system that consists of electronic or pneumatic single-loop analog controllers, in which each loop is controlled by a single, manually-adjusted device.

analog data   Data represented in a continuous form, as contrasted with digital data represented in a discrete, discontinuous form. Analog data are usually represented by means of physical variables, such as voltage, resistance, rotation, etc.

analog dc current signal   A signal used for transmission which varies in a continuous manner according to one or several physical quantities [S50.1].

analog device   A mechanism which represents numbers by physical quantities, e.g., by lengths, as in a slide rule, or by voltage or currents as in a differential analyzer or a computer of the analog type.

analog electronic controller   Any of several adaptations of analog computers to perform control functions; they may produce an output signal directly related to the difference between a measured value and a predetermined setpoint, or they may produce an output signal modified by rate-of-change or other feedback signals.

analog hardware description language (AHDL)   A modeling language capable of representing both the structural and behavioral properties of analog circuits. Structural refers to the connectivity or net-list properties of a circuit; behavioral refers to the mathematical equations for individual components.

analog input   1. A continuous variable input. 2. A termination panel used to connect field wiring from the input device. See input, analog.

analog input module   I/O module that converts a process voltage or current signal into a multiple-bit form for use in the PC. The signal is the analog of some process variable.

analog input point   An alarm point for use with an analog monitored variable signal, usually current or voltage. The logic circuit initiates an alarm when the analog signal is above or below a set point [S18.1].

analog output   1. Transducer output which is a continuous function of the measurand, except as modified by the resolution of the transducer [S37.1]. 2. A continuously variable output (generally 4-20 ma or 3-15 psi). See also output, analog.

analog output module   I/O module that converts a multiple-bit number calculated in the PC to a voltage or current output signal for use in control.

analog signal   An analog signal is a continuously variable representation of a physical quantity, property, or condition such as pressure, flow, temperature, etc. The signal may be transmitted as pneumatic, mechanical, or electrical energy. See signal, analog.

analog simulation   The calculation of the time or frequency domain response of electrical circuits to input stimulus. It assembles and solves a set of simultaneous equations associated with circuit topology.

analog-to-digital (A/D)   1. A device, or sub-system, that changes real-world analog data (as from transducers) to a form compatible with binary (digital) processing, as done in a microprocessor. 2. The conversion of analog data to digital data. See analog-to-digital converter.

analog-to-digital converter (ADC)   Any unit or device used to convert analog information to approximate corresponding digital information. See converter, analog to digital.

analysis   Quantitative determination of the constituent parts.

analysis, ultimate   Chemical analysis of solid, liquid or gaseous fuels. In the case of coal or coke, determination of carbon, hydrogen, sulfur, nitrogen, oxygen, and ash.

analytical balance   Any weighing device having a sensitivity of at least 0.1 mg.

analytical curve   A graphical representation of some function of relative intensity in spectroscopic analysis plotted against some function of concentration.

analytical gap   The separation between the source electrodes in a spectrograph.

analytical limit   Limit of a measured or calculated variable established by the safety analysis to ensure that a safety limit is not exceeded [S67.04].

analytical line   The spectral line of an element used to determine its concentration in spectroscopic analysis.

analytical scale   In spectroscopic analysis, the scale that results when an analytical curve is projected onto the intensity axis; it is often used in lieu of an analytical curve to permit direct reading of spectral intensity as element concentration.

analyzer   1. Any of several types of test instruments, ordinarily one that can measure several different variables either simultaneously or sequentially. 2. In an absorption refrigeration system, the component that allows the mixture of water and ammonia vapors leaving the generator to come in contact with the relatively cool ammonia solution entering the generator where the mixture loses some of its vapor content.

AND A logic operator having the property that if P is an expression, Q is an expression, R is an expression..., then the AND of P, Q, R...is true if all expressions are true, false if any expression is false.

Anderson bridge A type of a-c bridge especially suited to measuring the characteristics of extremely low-Q coils.

AND gate A basic electronic circuit used in microprocessor systems. A logical 1 value on output is produced only if all of the inputs have logical 1 values.

anechoic chamber 1. A test room having all surfaces lined with a sound-absorbing material. Also known as dead room. 2. A room lined with a material that absorbs radio waves of a particular frequency or band of frequencies; it is used chiefly for tests at microwave frequencies, such as a radar-beam cross section.

anemobiagraph A recording pressure-tube anemometer, such as a Dines anemometer, in which springs are used to make the output from the float manometer linear with wind speed.

anemoclinometer An instrument for determining the inclination of the wind to a horizontal plane.

anemometer A device for measuring wind speed; if it produces a recorded output, it is known as an anemograph.

anemoscope A device for indicating wind direction.

aneroid Describing a device or system that does not contain or use liquid.

angle beam In ultrasonic testing, a longitudinal wave from an ultrasonic search unit that enters the test surface at an acute angle.

angle modulation A type of modulation involving the variation of carrier-wave angle in accordance with some characteristic of a modulating wave; angle modulation can take the form of either phase modulation or frequency modulation.

angle of elevation The angle between a horizontal plane and the line of sight to an object lying above the plane of the observer.

angle of extinction The phase angle of the stopping instant of anode current flow in a gas tube with respect to the starting instant of the corresponding half cycle of anode voltage.

angle of ignition The phase angle of the starting instant of anode current flow in a gas tube with respect to the starting instant of anode current flow in a gas tube with respect to the starting instant of the corresponding positive half cycle of anode voltage.

angle of incidence The angle between the direction of propagation of a ray of incident radiation and a normal to the surface it strikes; for a reflected wave, the angle of reflection and the angle of incidence are equal.

angle of repose A characteristic of bulk solids equal to the maximum angle with the horizontal at which an object on an inclined plane will retain its position without tending to slide; the tangent of the angle of repose equals the coefficient of static friction.

angle valve A valve design in which one port is colinear with the valve stem or actuator, and the other port is at right angles to the valve stem.

Ängstrom A unit of length defined as $1/6438.4696$ of the wavelength of the red line in the Cd spectrum; it equals almost exactly 10^{-10} metre; this unit was once used almost exclusively for expressing wavelengths of light and x-rays, but it has now been largely replaced by the SI unit nanometre, or 10^{-9} metre.

angular accelerometer A device for measuring the rate of change of angular velocity between two objects.

angular frequency A frequency expressed in radians per second; it equals 2π times the frequency in Hz.

angular momentum The product of a body's moment of inertia and its angular velocity.

angular momentum flowmeter A device for determining mass flow rate in which an impeller turning at constant speed imparts angular momentum to a stream of fluid passing through the meter; a restrained turbine located just downstream of the impeller removes the angular momentum, and the reaction torque is taken as the meter output; under proper calibration conditions, the reaction torque is directly proportional to mass flow rate. Also called an axial flowmeter.

angular velocity Rate of motion along a circular path, measured in terms of angle traversed per unit time.

anhydrous Describing a chemical or other solid

substance whose water of crystallization has been removed.

aniline (phenylamine) A substance produced from coal tar the indigo plant that is used to make inks, dyes, and plastics.

anisotropic Exhibiting different properties when characteristics are measured along different directions or axes.

anisotropy Exhibiting different properties or other characteristics—strength or coefficient of thermal expansion, for instance—in different directions with respect to a given reference, such as a specific lattice direction in a crystalline substance.

annealing Treating metals, alloys or glass by heating and controlled slow cooling, primarily to soften them and remove residual internal stress but sometimes to simultaneously produce desired changes in other properties or in microstructure.

annotate To add explanatory text to computer programming or any other instructions.

annular nozzle A nozzle whose inlet opening is ring shaped rather than an open circle.

annulus 1. Any ring-shaped cavity or opening. 2. A plate that protects or covers a machine.

annunciator 1. A device or group of devices that call attention to changes in process conditions that have occurred. An annunciator usually calls attention to abnormal process conditions, but may be used also to show normal process status. Usually included are sequence logic circuits, labeled visual displays, audible devices, and manually operated pushbuttons [S18.1]. 2. An electromagnetic, electronic or pneumatic signaling device that either displays or removes a signal light, metal flag or similar indicator, or sounds an alarm, or both, when occurrence of a specific event is detected; in most cases, the display or alarm is single-acting, and must be reset after being tripped before it can indicate another occurrence of the event.

anode 1. The metal plate or surface that acts as an electron donor in an electrochemical circuit; metal ions go into solution in an electrolyte at the anode during electroplating or electrochemical corrosion. 2. The negative electrode in a storage battery, or the positive electrode in an electrochemical cell. 3. The positive electrode in an x-ray tube or vacuum tube, where electrons leave the interelectrode space.

anode circuit A circuit which includes the anode-cathode path of an electron tube connected in series with other circuit elements.

anode supply voltage The voltage across the terminals of an electric power source connected in series in the anode circuit.

anodic coating An oxide film produced on a metal by treating in an electrolytic cell with the metal as the cell anode.

anodic protection Reducing the corrosion rate of a metal that exhibits active-passive behavior by imposing an external electrical potential on a part.

anodize To form a protective passive film (conversion coating) on a metal part, such as a film of Al_2O_3 on aluminum, by making the part an anode in an electrolytic cell and passing a controlled electric current through the cell.

anodizing A method of producing film on a metal surface which is particularly well suited for aluminum.

anomalous dispersion Inversion of the derivative of refractive index with respect to wavelength in the vicinity of an absorption band.

ANSI screen control An ANSI standard that specifies a specific set of character sequences which instruct the computer to perform certain actions on the computer screen.

ANSI X3J3 ANSI PL/I Language Standardization Committee.

antenna A device for sending or receiving radio waves, but not including the means of connecting the device to a transmitter or receiver. See also dipole antenna; horn antenna.

antenna array A single mounting containing two or more individual antennas coupled together to give specific directional characteristics.

anti-cavitation trim A combination of plug and seat ring or plug and cage that by its geometry permits non-cavitating operation or reduces the tendency to cavitate, thereby minimizing damage to the valve parts, and the downstream piping [S75.05].

anticoincidence circuit A circuit with two inputs and one output, which produces an output pulse only if either input terminal receives a pulse within a specified time interval but does not produce a pulse if both input terminals receive a pulse within that interval.

anticorrosive Describing a substance, such as paint or grease, that contains a chemical which counteracts corrosion or produces a corrosion-resistant film by reacting with the underlying surface.

antifriction Describing a device, such as a bearing or other mechanism, that employs rolling contact with another part rather than sliding contact.

antimagnetic Describing a device which is made of nonmagnetic materials or employs magnetic shielding to avoid being influenced by magnetic fields during operation.

antinodes The points, lines or surfaces in a medium containing a standing wave where some characteristic of the wave field is at maximum amplitude. Also known as loops.

anti-noise trim A combination of plug and seat ring or plug and cage that by its geometry reduces the noise generated by fluid flowing through the valve [S75.05].

antireflective coating A coating designed to suppress reflections from an optical surface.

antiresonance A condition existing between an externally excited system and the external sinusoidal excitation when any small increase or decrease in the frequency of the excitation signal causes the peak-to-peak amplitude of a specified response to increase.

antiresonant Describing an electric, acoustic or other dynamic system whose impedance is very high, approaching infinity.

antiresonant frequency A frequency at which antiresonance exists between a system and its external sinusoidal excitation.

antiskid Describing a material, surface or coating which has been roughened or which contains abrasive particles to increase the coefficient of friction and prevent sliding or slipping. Also known as antislip.

antisurge control Control by which the unstable operating mode of compressors known as "surge" is avoided.

anvil 1. The part of a machine that absorbs the energy of a sharp blow. 2. A heavy block made of wrought iron, cast iron or steel and used to support metal being smith forged. 3. The base of a forging press or drop hammer that supports the die bed and lower die. 4. The stationary contact of a micrometer caliper or similar gaging device.

APD See avalanche photodiode.

aperiodic Varying in a manner that is not periodically repeated.

aperiodically damped Reaching a constant value or steady state of change without introducing oscillation.

aperture A hole in a surface through which light is transmitted. Apertures are sometimes called spatial filters, a more descriptive term when placed in the Fourier (focal) plane.

aperture time The time required, in a sample-and-hold circuit, for the switch to open after the "hold" command has been given.

APL A Programming Language. A powerful systems programming language developed by the International Business Machines Corporation.

apparent density The density of loose or compacted particulate matter determined by dividing actual weight by volume occupied; apparent density is always less than true density of a material comprising the particulate matter because volume occupied includes the space devoted to pores or cavities between particles.

apparent flow The uncorrected volume flow as indicated by the calibrator [RP31.1].

apparent viscosity The resistance to continuous deformation (viscosity) in a non-Newtonian fluid subjected to shear stress.

appearance potential The minimum electron-beam energy required to produce ions of a particular type in the ion source of a mass spectrometer.

application The system or problem to which a computer is applied. Reference is often made to computation, data processing, and control as the three categories of application.

application layer Layer of 7 of the OSI.

application program A program that performs a task specific to a particular end-user's needs; generally, an application program is any program written on a program development operating system that is not part of a basic operating system.

applications software Programs which are unique to a specific process control system installation or other specific installations, rather than general purpose and of broad applicability.

applied load 1. Weight carried or force sustained by a structural member in service, in most cases the load includes the weight of the member itself. 2. Material carried by the load-receiving member of a weighing scale, not including any load necessary to bring the scale into initial balance.

applied shock Any rapidly applied load or other form of excitation that produces shock motion within a system.

approach idler The last idler passed before the material on a belt reaches the weighbridge [RP74.01].

approved Acceptable to the authority having jurisdiction. The term is considered synonymous with "listed" and "certified" [RP12.6] [S12.15].

APT See automatically programmed tools.

arbitration bar A test bar cast from molten metal at the same time as a lot of castings; it is used to determine mechanical properties in a standard tensile test, which are then evaluated to determine acceptability of the lot of castings.

arbor 1. In machine grinding, the spindle for mounting and driving the grinding wheel. 2. In machine cutting, such as milling, the shaft for holding and driving a rotating cutter. 3. Generically, the principal spindle or axis of a rotating machine which transmits power and motion to other parts. 4. In metal founding, a bar, rod or other support embedded in a sand core to keep it from collapsing during pouring.

arbor press A mechanical or hydraulic machine for forcing arbors, mandrels, bushings, shafts or pins into or out of drilled or bored holes.

arc 1. A segment of the circumference of a circle. 2. The graduated scale on an instrument for measuring angles. 3. A discharge of electricity across a gap between electrical conductors.

architecture The structure, functional and performance characteristics of a system, specified in an implementation independent way.

archival file In data processing, storage of seldom used data that must be retained for several years.

arc lamp A high intensity lamp in which a direct current electric discharge produces light that is continuous, as opposed to a flashlamp, which produces pulsed light.

arc line A spectral line in spectroscopy.

arc melting Raising the temperature of a metal to its melting point using heat generated by an electric arc; usually refers to melting in a specially designed furnace to refine a metal, produce an alloy or prepare a metallic material for casting.

arc strike See strike.

arc welding A group of welding processes which produce coalescence of metals by heating them with an arc, with or without pressure or the use of filler metal.

area, air conditioned A location with temperature at a nominal value maintained constant within narrow tolerance at some point in a specified band of typical comfortable room temperature. Humidity is maintained within a narrow specified band [S51.1].

area classification The classification of hazardous (classified) locations by Class I, II or III depending upon the presence of flammable gases or vapors, flammable liquids, combustible dust, or ignitible fibers or flyings and by Division 1 or 2 depending upon the existence of these materials to exist in an ignitible concentration under normal or abnormal conditions.

area classification (Class) 1. Class I locations are those in which flammable gases or vapors are or may be present in the air in quantities sufficient to produce explosive or ignitable mixtures. 2. Class II locations are those that are hazardous because of the presence of combustible dust. 3. Class III locations are those that are hazardous because of the presence of easily ignitable fibers or flyings, but in which such fibers or flyings are not likely to be in suspension in the air in quantities sufficient to produce ignitable mixtures.

area classification (Division) 1. Division I (hazardous). Where concentrations of flammable gases or vapors exist (a) continuously or periodically during normal operations; (b) frequently during repair or maintenance or because of leakage; or (c) due to equipment breakdown or faulty operation which could cause simultaneous failure of electrical equipment. 2. Division 2 (normally non-hazardous). Locations in which the atmosphere is normally non-hazardous and may become hazardous only through the failure of the ventilating system, opening of pipe lines, or other unusual situations.

area classification (Group) Identified groups of chemicals and compounds whose air mixtures have similar ease of ignition and explosive characteristics, for the purpose of testing, approval, and area classification. Group A: atmospheres containing acetylene. Group B: atmosphere containing butadiene, ethylene oxide, propylene oxide, acrolein, or hydrogen (or gases or vapors equivalent in hazard to hydrogen). Group C: atmospheres such as cyclopropane, ethyl ether, ethylene, or gases or vapors of equivalent hazard. Group D: atmospheres such as acetone, alcohol ammonia, benzene, benzol, butane, gasoline, hexane, lacquer solvent vapors, naphtha, natural gas, propane, or gases or vapors of equivalent hazard. Group E: atmospheres containing metal dusts. Group G: atmospheres containing combustible dusts having resistivity of $> 10^5$ ohm — cm^2.

area, control room A location with heat and/or cooling facilities. Conditions are maintained within specified limits. Provisions for automatically maintaining constant temperature and humidity may or may not be provided [S51.1].

area, environmental A basic qualified location in a plant with specified environmental conditions dependent on severity. Environmental areas include: air conditioned areas; control room areas, heated and/or cooled; sheltered areas (process facilities); outdoor areas (remote field sites) [S51.1].

area meter A device for measuring the flow of fluid through a passage of fixed cross-sectional area, usually through use of a weighted piston or float supported by the flowing fluid.

area, outdoor A location in which equipment is exposed to outdoor ambient conditions; including temperature, humidity, direct sunshine, wind and precipitation [S51.1].

area, sheltered An industrial process location, area, storage or transportation facility, with protection against direct exposure to the elements, such as direct sunlight, rain or other precipitation or full wind pressure. Minimum and maximum temperatures and humidity may be the same as outdoors. Condensation can occur. Ventilation, if any, is by natural means [S51.1].

argentometer A hydrometer used to find the concentration of a silver salt in water solution.

argument 1. An independent variable, e.g., in looking up a quantity in a table, the number or any of the numbers which identifies the location of the desired value, or in a mathematical function the variable which, when a certain value is substituted for it, determines the value of the function. 2. An operand in an operation on one or more variables. See also parameter.

arithmetic ability The capability of performing (at least) addition and subtraction in the PC.

arithmetic and logical unit A component of the central processing unit in a computer where

data items are compared, arithmetic operations performed and logical operations executed.

arithmetic check See mathematical check.

arithmetic element The portion of a mechanical calculator or electronic computer that performs arithmetic operations.

arithmetic expression An expression containing any combination of data names, numeric literals, and named constants, joined by one or more arithmetic operators in such a way that the expression as a whole can be reduced to a single numeric flue.

arithmetic operation A computer operation in which the ordinary elementary arithmetic operations are performed on numerical quantities. Contrasted with logical operation.

arithmetic operator Any of the operators, +, and, -, or the infix operators, +, -, *, /, and **.

arithmetic unit The unit of a computing system that contains the circuits that perform arithmetic operations.

arm Allows a hardware interrupt to be recognized and remembered. Contrasted with disarm; see enable.

armature 1. The core and windings of the rotor in an electric motor or generator. 2. The portion of the moving element of an instrument which is acted upon by magnetic flux to produce torque.

armored meter tube Variable area meter tube (rotometer) of all metal construction utilizing magnetic coupling between the float and an external follower.

array 1. An arrangement of elements in one or more dimensions. See also matrix and vector. 2. In a computer program, a numbered, ordered collection of elements, all of which have identical data attributes. 3. A group of detecting elements usually arranged in a straight line (linear array) or in two dimensional matrix (imaging array). 4. A series of data samples, all from the same measurement point. Typically, an array is assembled at the telemetry ground station for frequency analysis.

array dimension The number of subscripts needed to identify an element in the array.

array process A hardware device that processes data arrays; Fast Fourier transforms (FFT) and power-spectral density (PSD) are typical processes.

array processor The capability of a computer to operate at a variety of data locations at the same time.

arrester A device to impede the flow of large dust particles or sparks from a stack, usually screening at the top.

arrow keys Keys on a computer keyboard that will move the cursor.

articulated arms (waveguides) A beam-direction arrangement in which light passes through a series of jointed pipes containing optics.

articulated structure A structure—either stationary or movable, such as a motor vehicle or train—which is permanently or semipermanently connected so that different sections of the structure can move relative to the others, usually involving pinned or sliding joints.

artificial aging Heat treating a metal at a moderately elevated temperature to hasten age hardening.

artificial intelligence (AI) The use of computers to simulate the way the human mind operates, as learning or adaptation.

artificial language A language specifically designed for ease of communication in a particular area of endeavor, but one that is not yet natural to that area. This is contrasted with a natural language which has evolved through long usage.

artificial radioactivity Radioactivity induced by bombarding a material with a beam of energetic particles or with electromagnetic radiation.

artificial weathering Producing controlled changes in materials, such as surface appearance, under laboratory conditions that simulate outdoor exposure.

asbestos A fibrous variety of the mineral horneblend, used extensively for its fire-resistant qualities to make insulation and fire barriers; it has been stitched, bonded or woven into blankets, mixed with portland cement and water to make sheet roofing, wall cladding, drainage tiles and corrosion-resistant pipe, and combined with binders such as asphalt or bentonite to make asbestos felt or plaster.

ASC See Accredited Standard Committee.

ASCII 1. A widely-used code (American Standard Code for Information Interchange) in which alphanumeries, punctuation marks, and certain special machine characters are represented by unique, 7-bit, binary numbers; 128 different binary combinations are possible ($2^7 = 128$), thus 128 characters may be represented. 2. A protocol.

ASCII file A text file that uses the ASCII character set.

as-fabricated Describing the condition of a structure or material after assembly, and without any conditioning treatment such as a stress-relieving heat treatment; specific terms such as as-welded, as-brazed or as-polished are used to designate the nature of the final step in fabrication.

as-fired fuel Fuel in the condition as fed to the fuel burning equipment.

ash The noncombustible inorganic matter in the fuel.

ash content The incombustible residue remaining after burning a combustible material completely.

ash-free basis The method of reporting fuel analysis whereby ash is deducted and other constituents are recalculated to total 100%.

ash pit A pit or hopper located below a furnace where refuse is accumulated and from which it is removed at intervals.

ASN.1 Abstract Syntax Notation One. An ISO standard (DIS 8824) that specifies a canonical method of data encoding. This standard is an extension of CCITT standard X.409.

aspect ratio 1. The ratio of a symbol's height to its width [S5.5]. 2. The ratio of frame width to height for a television picture, which is 4:3 in the United States, Canada and United Kingdom. 3. In any rectangular structure, such as the cross section of a duct or tubular beam, the ratio of the longer dimension to the shorter. 4. A ratio used in calculating resistance to flow in a rectangular elbow and is the ratio of width to depth.

asphalt A brown to black bituminous solid that melts on heating and is impervious to water but soluble in gasoline; it is used extensively in paving and roofing applications, and in paints and varnishes; it occurs naturally in certain oil-bearing rocks, and can be made by pyrolysis of coal tar, lignite tar and certain petroleums.

aspheric For optical elements, surfaces are aspheric if they are not spherical or flat. Lenses with aspheric surfaces are sometimes called aspheres.

aspirating burner A burner in which the fuel in a gaseous or finely divided form is burned in suspension, the air for combustion being supplied by bringing into contact with the fuel, air drawn through one or more openings by the lower static pressure created by the velocity of the fuel stream.

aspiration Using a vacuum to draw up gas or granular material, often by passing a stream of water across the end of an open tube, or through the run of a tee joint, where the open tube or branch pipe extends into a reservoir containing the gas or granular material.

ASR See Automatic Send/Receive.

as-received fuel Fuel in the condition as received at the plant.

assemble To prepare a machine-language program from a symbolic language program by substituting absolute code for symbolic operation codes and absolute or relocatable addresses for symbolic addresses.

assembler A program that translates symbolic source code into machine instructions by replacing symbolic operation codes with binary operation codes, and symbolic addresses with absolute or relocatable addresses.

assembly 1. A unit constructed of many parts or components, and which functions in service as a single device, mechanism or structure. 2. A mid-level computer language.

assembly language A computer programming language, similar to computer language, in which the instructions usually have a one-to-one correspondence with computer instructions in machine language, and which utilizes mnemonics for representing instructions.

assembly list A printed list which is the by-product of an assembly procedure. It lists in logical instruction sequence all details of a routine showing the coded and symbolic notation next to the actual notations established by the assembly procedure. This listing is highly useful in the debugging of a routine.

assembly program See assembly system.

assembly system An assembly system comprises two elements, a symbolic language and an assembly program that translates source programs written in the symbolic language into machine language.

assign To designate a part of a system for a specific purpose.

assignable A term applied to a feature permitting the channeling (or directing) of a signal from one device to another without the need for switching, patching, or changes in wiring [S5.1] [S5.3].

assignment statement A program statement that calculates the value of an expression and assigns it a name.

associated apparatus Apparatus in which the circuits are not necessarily intrinsically safe themselves, but which affect the energy in the intrinsically safe circuits and are relied upon to maintain intrinsic safety. Associated electrical apparatus may be either: (a) electrical apparatus that has an alternative type of protection for use in the appropriate hazardous (classified) location, or (b) electrical apparatus not protected shall not be used within a hazardous (classified location) [RP12.6].

associated electrical apparatus Electrical apparatus in which the circuits are not all intrinsically safe but which contains circuits that can affect the safety of intrinsically safe circuits connected to it.

association The combining of ions into larger ion clusters in concentrated solutions.

associative storage A storage device in which storage locations are identified by their contents, not by names or positions. Synonymous with content-addressed storage, contrast with parallel search storage.

astatic Without polarity; independent of the earth's magnetic field.

astigmatism A defect in an optical element that causes rays from a single point in the outer portion of a field of view to fall on different points in the focused image.

astrodynamics Practical application of fundamental science to the problem of planning and controlling the trajectories of space vehicles.

astrolabe An instrument formerly used to find the altitudes of celestial bodies; a predecessor of the sextant.

astronomical theodolite See altazimuth.

asymmetric rotor A rotating machine element whose axis of rotation is not the same as its axis of symmetry.

asymmetry potential The difference in potential between the inside and outside pH sensitive glass layers when they are both in contact with 7 pH solutions. It is caused by deterioration of the pH sensitive glass layers or contamination of the internal fill of the measurement electrode.

asynchronous 1. A mode of operation in which an operation is started by a signal before the operation on which this operation depends is completed. When referring to hardware devices, it is the method in which each character is sent with its own synchronizing information. The hardware operations are scheduled by "ready" and "done" signals rather than by time intervals. This implies that a second operation can begin before the first operation is completed. 2. Not synchronous with the line frequency as applied to rotating a.e. machinery.

asynchronous communication Often called start/stop transmission, a way of transmitting data in which each character is preceded by a start bit and followed by a stop bit.

asynchronous transmission 1. Transmission in which information character, or sometimes each word or small block, is individually synchronized, usually by the use of start and stop elements. The gap between each character (or word) is not of a necessarily fixed length. (Compare with synchronous transmission.) 2. Asynchronous transmission is called start-stop transmission. 3. Data transmission mode in which the timing is self-determined and not controlled by an external clock.

atmometer A generic name for any instrument that measures evaporation rates; also known as atmidometer; evaporimeter; evaporation gage.

atmospheric air Air under the prevailing atmospheric conditions.

atmospheric communication Sending signals in the form of modulated light through the atmosphere, without the use of fiber optics to contain and direct the beam.

atmospheric corrosion Corrosion that occurs naturally due to exposure to climatic conditions; corrosion rates vary depending on specific global location because of variations in average temperature, humidity, rainfall, airborne substances such as sea spray, dust and pollen, and airborne pollutants such as sulfur dioxide, chlorine compounds, fly ash and other combustion products.

atmospheric monochromator A monochromator in which the optical path is through air. This is the standard type used for visible and infrared wavelengths transmitted by air.

atmospheric pressure The barometric reading of pressure exerted by the atmosphere. At sea level 14.7 lb per sq in. or 29.92 in. of mercury.

atomic mass unit A unit for expressing atomic weights and other small masses; it equals, exactly, 1/12 the mass of the carbon-12 nuclide.

atomic number An integer that designates the position of an element in the periodic table of the elements; it equals the number of protons in the nucleus and the number of electrons in the electrically neutral atom.

atomic weight The weight of a single atom of any given chemical element; it is usually taken as the weighted average of the weights of the naturally occurring nuclides, expressed in atomic mass units.

atomization Mechanically producing fine droplets or mist from a bulk liquid or molten substance.

atomizer A device by means of which a liquid is reduced to a very fine spray.

atom probe An instrument, consisting of a field-ion microscope with a probe hole in its screen that opens into a mass spectrometer, used to identify a single atom or molecule on a metal surface.

attached equipment The auxiliary equipment which must be located on the valve or actuator [S75.05].

attachment plug A connecting device for a flexible cord that, by insertion into a receptacle, establishes supply-circuit connections between the flexible cord and the receptacle [S82.01].

attemperation Regulating the temperature of a substance—for instance, passing superheated steam through a heat exchanger or injecting water mist into it to regulate final steam temperature.

attenuate To weaken or make thinner—for example, to reduce the intensity of sound or ultrasonic waves by passing them through an absorbing medium.

attenuation 1. A decrease in signal magnitude between two points, or between two frequencies. 2. The reciprocal of gain, when the gain is less than one. It may be expressed as a dimensionless ratio, scalar ratio, or in decibels as 20 times the log10 of that ratio [S51.1] [RP55.1]. 3. The loss of amplitude in a signal as it is transmitted through a conductor. See gain.

attenuator 1. An optical device which reduces the intensity of a beam of light passing through it. 2. An electrical component that reduces the amplitude of a signal in a controlled manner.

attitude 1. The relative orientation of a vehicle or object represented by its angles of inclination to three orthogonal reference axes [S37.1]. 2. The position of an object in space determined by the angles between its axes and a selected set of planes.

attitude error The error due to the orientation of the transducer relative to the direction in which gravity acts upon the transducer [S37.1]. See acceleration error.

attribute sampling A type of sampling inspection in which an entire production lot is accepted or rejected depending on the number of items in a statistical sample that have at least one characteristic (attribute) that does not meet specifications.

auctioneering device See signal selector.

audible device A device that calls attention by sound to the occurrence of abnormal process conditions. An audible device may also call attention to return to normal conditions [S18.1].

audible device follower See auxiliary output.

audio Pertaining to audible sound—usually taken as sound frequencies in the range 20 to 20,000 Hz.

audiometer An instrument used to measure the ability of people to hear sounds; it consists of an oscillator, amplifier and attenuator, and may be adapted to generate pure tones, speech or bone-conducted vibrations.

auger 1. A woodboring tool consisting of a shank with a T-shaped handle. 2. A feeding device consisting primarily of a set of spiral blades mounted on a central shaft or fastened together to make a spiral rotating assembly; it may rotate in a tube, trough or housing to move powdered, granular or semisolid material axially; in some applications, the auger may be constructed of two counterspiraled augers, which feed material toward the midpoint or outward from the midpoint of the axis depending on the direction of rotation.

AUI See Access Unit Interface.

austenitic stainless steel An alloy of iron containing at least 12% Cr plus sufficient Ni (or in some specialty stainless steels, Mn) to stabilize the face-centered cubic crystal structure of iron at room temperature.

autoclave An airtight vessel for heating its contents and sometimes agitating them; it usually uses high pressure steam to perform processing, sterilizing or cooking steps using moist or dry heat.

autocollimator A telescopic sight including a light source and a partially reflecting mirror, focused to infinity, for use in measuring small angular motion and checking alignment.

AUTOEXEC The name of the file in MS-DOS that has commands to be executed when the computer is booted. Usually named AUTOEXEC.BAT.

autoignition temperature (AIT) The temperature at which a flammable mixture ignites spontaneously, without exposure to an external spark or flame.

auto-manual station Synonym for control station [S5.1].

automate 1. To apply the principles of automation. 2. To operate or control by automation. 3. To install automatic procedures, as for manufacturing, servicing, etc.

automatic 1. Having the power of self-motion; self-moving; or self-acting; an automatic device. 2. A machine that operates automatically. 3. Pertaining to a process or device that, under specified conditions, functions without intervention by a human operator.

automatically programmed tools A numerical language.

automatic control The type of control in which there is no direct action of man on the controlling device.

automatic control engineering The branch of science and technology which deals with the design and use of automatic control devices and systems.

automatic controller Any device which measures the value of a process variable and generates a signal or some controlling action to maintain the value in correspondence with a reference value, or setpoint.

automatic control panel A panel of indicator lights and switches on which are displayed an indication of process conditions, and from which an operator can control the operation of the process.

automatic control system See control system, automatic.

automatic error correction A technique usually requiring the use of special codes or automatic retransmission, which detects and corrects errors occurring in transmission. The degree of correction depends upon coding and equipment configuration.

automatic frequency control A device or circuit designed to maintain the frequency of an oscillator within a preselected band of frequencies. In an FM radio receiver, the circuitry that senses frequency drift and automatically controls an internal oscillator to compensate for the drift.

automatic gain control An auxiliary circuit that adjusts gain of the main circuit in a predetermined manner when the value of a selected input signal varies.

automatic lighter A means for starting ignition of fuel without manual intervention. Usually applied to liquid, gaseous or pulverized fuel.

automatic/manual station A device which enables an operator to select an automatic signal or a manual signal as the input to a controlling element. The automatic signal is normally the output of a controller while the manual signal is the output of a manually operated device [S51.1].

automatic pilot An automatic control system adapted for maintaining an aircraft in stable level flight or for executing selected maneuvers.

automatic reset See reset.

automatic send/receive (ASR) A teletypewriter unit with keyboard, printer, paper tape, reader/transmitter, and paper tape punch. This combination of units may be based on-line or off-line and, in some cases, on-line and off-line simultaneously.

automatic utility translator (AUTRAN) A process control language and system offered by Control Data Corporation.

automatic zero- and full-scale calibration Zero and sensitivity stabilization by servos for comparison of demodulated zero- and full-scale signals with zero- and full-scale references.

automation 1. The implementation of processes by automatic means. 2. The theory, art, or technique of making a process more automatic. 3. The investigation, design, development, and application of methods of rendering processes automatic, self-moving, or self-controlling. 4. The conversion of a procedure, a process, or equipment to automatic operation.

automotive engineering A branch of mechanical engineering that deals with design and construction of landgoing vehicles, especially self-propelled highway vehicles such as automobiles, trucks and buses.

autoradiography A technique for producing a radiographic image using ionizing radiation produced by radioactive decay of atoms within the test object itself.

Autosyn A trade name for a type of synchro.

auto-tracking antenna A receiving antenna which always points to the transmitting site, automatically tracking all movements of the vehicle being telemetered.

autotransformer A type of transformer in which certain portions of the windings are shared by the primary and secondary circuits.

auto-zero logic module A component of a digital controller whose function is primarily to establish an arbitrary zero-reference value for each individual measurement.

AUTRAN See automatic utility translator.

auxiliary contact See auxiliary output.

auxiliary device 1. Generally, any device which is separate from a main device but which is necessary or desirable for effective operation of the system. 2. Specifically, any device used in conjunction with an instrument to extend its range, increase its accuracy, otherwise assist in making a measurement, or perform a function not directly involved in making the measurement.

auxiliary means A device or subsystem, usually placed ahead of the primary detector, which alters the magnitude of the measured quantity to make it more suitable for the primary detector without changing the nature of the measured quantity.

auxiliary output, audible device follower (horn relay contact) An auxiliary output that operates while the common alarm audible device operates [S18.1].

auxiliary output (auxiliary contact) 1. An output signal operated by a single alarm point or group of points for use with a remote device [S18.1]. 2. A secondary output.

auxiliary output, field contact follower An auxiliary output that operates while the field contact indicates an abnormal process condition [S18.1].

auxiliary output, lamp follower An auxiliary output that operates while the visual display lamps indicate an alarm, silenced, or acknowledged state [S18.1].

auxiliary output, reflash An auxiliary output that operates when any one of a group of alarm points indicates an abnormal process condition. The output usually returns to normal briefly when each alarm point changes to an abnormal process condition and returns to normal when all alarm points in the group indicate normal process conditions [S18.1].

auxiliary panel 1. A panel which is not in the main control room. The front of an auxiliary panel is normally accessible to an operator but the rear is normally accessible only by maintenance personnel. 2. Located at an auxiliary location.

auxiliary storage A storage device in addition to the main storage of a computer, e.g., magnetic tape, disk, magnetic drum or core. Auxiliary storage usually holds much larger amounts of information than the main storage, and the information is accessible less rapidly. Contrasted with main storage.

availability The number of hours in the reporting period less the total downtime for the reporting period divided by the number of hours in the reporting period (expressed in percent).

availability factor The fraction of the time during which the unit is in operable condition.

available draft The draft which may be utilized to cause the flow of air for combustion or the flow of products of combustion.

available energy Energy that theoretically can be converted to mechanical power.

available heat In a thermodynamic working fluid, the amount of heat that could be transformed into mechanical work under ideal conditions by reducing the temperature of the working fluid to the lowest temperature available for heat discard.

available power An attribute of a linear source of electric power defined as $V_{rms}/4R$, where V_{rms} is the open circuit rms voltage of the power source and R is the resistive component of the internal impedance of the power source.

available power gain An attribute of a linear transducer defined as the ratio of power available from the output terminals of the transducer to the power available from the input circuit under specified conditions of input termination.

available work The capacity of a fluid or body to do work if applied to an ideal engine.

avalanche Production of a large number of ions by cascade action in which a single charged particle, accelerated by a strong electric field, collides with neutral gas molecules and ionizes them.

avalanche photodiode (APD) A photodiode designed to take advantage of avalanche multiplication of photocurrent. As the reverse-bias voltage approaches the breakdown voltage, hole-electron pairs created by absorbed photons acquire sufficient energy to create additional hole-electron pairs when they collide with substrate atoms, producing a multiplication effect.

average outgoing quality limit The average percent of defective units that remain undetected in all lots that pass final inspection; it is a measure of the ability of sampling inspection to limit the probability of shipping defective product; here, a defective unit is considered to be one containing at least one attribute that does not meet specifications.

average-position action A type of control-system action in which the final control element is positioned in either of two fixed positions, the average time at each position being determined from some function of the measured value of the controlled variable.

averaging pitot tube An adaptation of the pitot tube in which a multiple-ported pitot tube spans the process tube; total pressure is measured as a composite of the pressures on several ports facing upstream while static pressure is measured using one or more ports facing downstream; the device works best for clean liquids, vapors and gases, but can be used for streams containing suspended solids or viscous contaminants if the purging flow is supplied to the measuring tube.

average resolution The reciprocal of the total number of output steps over the unit range multiplied by 100 and expressed in % VR [S37.6] [S37.12].

axial fan Consists of a propeller or disc type of wheel within a cylinder discharging the air parallel to the axis of the wheel.

axial-flow Describing a machine such as a pump or compressor in which the general direction of fluid flow is parallel to the axis of its rotating shaft.

axial hydraulic thrust In single-stage and multiple-stage pumps, the axial component of the summation of all unbalanced impeller forces.

axial runout For a rotating member, the total amount that a specific surface deviates from a plane perpendicular to the axis of rotation in one complete revolution; it is usually expressed in 0.001 in., or some other suitable unit of measure, taken at a specific radial distance from the axis of rotation.

axle A rod, shaft or other supporting member that carries wheels and either transmits rotating motion to the wheels or allows the wheels to rotate freely about it.

azeotrope A mixture whose evolved vapor composition is the same as the liquid it comes from. This phenomenon occurs at one fixed composition for a given system. At either side of the azeotropic point, the vapors will have different compositions from that of the liquid they evolved from. Such mixtures act as pure substances in distillation and thus are inseparable by standard distillation methods. Azeotropic distillation is necessary to separate such a mixture.

azeotropic distillation A distillation technique in which one of the product streams is an azeotrope. It is sometimes used to separate two components by adding a third, which forms an azeotrope with one of the original two components.

azimuth angle An angular measurement in a horizontal plane about some arbitrary center point using true North or some other arbitrary direction as a reference direction (0°).

azimuth circle A ring scale graduated from 0° to 360°, and used with a compass, radar plan position indicator, direction finder or other device to indicate compass direction, relative bearing or azimuth angle.

B

babbitt Any of the white alloys composed principally of lead or tin which are used extensively to make linings for sliding bearings.

babble The composite signal resulting from cross talk among a large number of interfering channels.

backbone The trunk media of a multi-media LAN separated into sections by bridges, routers, or gateways.

back draft A reverse taper on the sidewall of a casting mold or forging die that prevents a casting pattern or forged part from being removed from the cavity.

back face The machined surface on the side of a through-bolted flange, opposite the gasket face, that is provided for nut seating [S75.05].

backflush The injecting of a fluid in a reverse flow manner to remove sample-line fluid or obstructions [S67.10].

background 1. The field that information is displayed upon for contrast [S5.5]. 2. In radiation counting, a low-level signal caused by radiation from sources other than the source of radiation being measured.

background discrimination The ability of a measuring instrument or detection circuit to distinguish an input signal from electronic noise or other background signals.

background noise Undesired signals or other stimuli that are always present in a transducer output or electronic circuit, regardless of whether a desired signal or stimulus is also present.

background program A program of the lowest urgency with regard to time and which may be preempted by a program of higher urgency and priority. Contrast to foreground program.

backhand welding Laying down a weld bead with the back of the welder's principal hand (the one holding the torch or welding electrode) facing the direction of welding; in torch welding, this directs the flame backward against the weld bead to provide postheating.

backing pump In a vacuum system using two pumps, the pump discharging directly to the atmosphere which reduces system pressure to an intermediate value, usually 10^{-2} to 10^{-5} psia. Also known as fore pump.

backing ring A ring of steel or other material placed behind the welding groove when joining tubes or pipes by welding, to confine the weld metal.

backing strip A piece of metal, asbestos or other nonflammable material placed behind a joint prior to welding to enhance weld quality.

backlash 1. In process instrumentation a relative movement between interacting mechanical parts resulting from looseness, when motion is reversed [S51.1]. 2. In a mechanical linkage or gear train, the amount by which the driving shaft must rotate, when reversing direction, in order to merely take up looseness in the linkage or gear train before it begins to transmit motion in the reverse direction. 3. The difference in actual values of a controlled variable when a control dial is brought to the same indicated position from opposite rotational directions.

back pressure The absolute pressure level as measured four pipe-diameters downstream from the turbine flowmeter under operating conditions, expressed in pascals [RP31.1].

backscattering The scattering of light in the direction opposite to the original one in which it was traveling.

back seat A seating surface in the bonnet that mates with the closure or valve stem in the extreme open position to provide pressure isolation of the stem seal [S75.05].

backstep sequence A method of laying down a weld bead in which a segment is welded in one direction, then the torch is moved in the opposite direction a distance approximately twice the length of the first segment, and another segment is welded back toward the first; thus the general direction of progress along the joint is opposite to the direction of welding individual segments, with the end point of each segment coinciding with the starting point of the preceding segment.

backtracking A technique used to synchronize mixed-signal simulation systems where an analog simulator is required to back up to a previous time point in order to process a signal originating in the digital simulator.

back-up 1. Equipment which is available to complete an operation in the event that the primary equipment fails. 2. A copy of a computer diskette which protects against destruction or loss of the original.

backup copy In data processing, a copy of data or a program that can be used if the original copy is lost or destroyed.

baffle 1. A plate or vane, plain or perforated, used to regulate or direct the flow of fluid. 2. A cabinet or partition used with a loudspeaker produced simultaneously by the front and rear surfaces of the diaphragm.

baffle-nozzle amplifier A device for converting mechanical motion to a pneumatic signal which consists of a supply tube ending in a small nozzle and a movable baffle plate attached to a mechanical arm; the supply tube has a restriction a short distance before the nozzle, so that as the baffle plate moves closer to the nozzle opening the pressure rises in the section of the supply tube between the restriction and the nozzle; arm motion and nozzle clearance are small—on the order of 0.2 mm or less; a baffle-nozzle amplifier serves as the primary detector in almost all pneumatic transmitters and controllers. Often referred to as a flapper-nozzle amplifier because the baffle plate is mounted on a pivoting arm.

baffle plate A tray or partition, solid or perforated, positioned in the flowpath through a process vessel so as to cause the process stream to flow in a certain direction, to reverse its direction of flow, or to slow its velocity.

baffle-type collector A device in gas paths utilizing baffles so arranged as to deflect dust particles out of the gas stream.

bag A deep bulge in the shell or of a furnace or fire-tube boiler.

bag filter A device containing one or more cloth bags for recovering particles from the dust laden gas or air which is blown through it.

bag-type collector A filter wherein the cloth filtering medium is made in the form of cylindrical bags.

bakeout Heating the surfaces of a vacuum system during evacuation to degas them and aid the process of reaching a stable final vacuum level.

balance 1. Generically, a state of equilibrium—static, as when forces on a body exactly counteract each other, or dynamic, as when material flowing into and out of a pipeline or process has reached steady state and there is no discernible rate of change in process variables. 2. An instrument for making precise measurements of mass or weight.

balanced (to ground) See unbalanced (to ground).

balanced trim An arrangement of ports and plug or combination of plug, cage, seals and ports that tends to equalize the pressure above and below the valve plug to minimize the net static and dynamic fluid flow forces acting along the axis of the stem of a globe valve [S75.05].

balance weight A mass positioned on the balance arms of a weighing device so that the arms can be brought to a predetermined position (null position) for all conditions of use.

ball A spherically shaped part which uses a portion of a spherical surface or an internal path to modify flow rate with rotary motion [S75.05].

ballast 1. Relatively dense material placed in the keel of a ship or gondola of a lighter-than-air craft to increase stability or control buoyancy, or both. 2. Crushed stone placed along a railroad bed to help support the ties.

ball bearing A type of antifriction bearing in which the load is borne on a series of hard spherical elements (balls) confined between inner and outer retaining rings (races).

ball burnishing 1. Producing a smooth, dimensionally precise hole by forcing a slightly oversize tungsten-carbide ball through a slightly undersize hole at high speed. 2. A method of producing a lustrous finish on small parts by tumbling them in a wood-lined barrel with burnishing soap, water, and hardened steel balls.

ball bushing A variation of ball bearing that permits axial motion of a shaft instead of rotating motion.

ball check valve A valve that permits flow in one direction only by lifting a spring-loaded ball off its seat when a pressure differential acts in that direction and by forcing the ball more tightly against the seat when a pressure differential acts in the opposite flow direction.

ball-float liquid-level meter A device consisting of a hollow or low-density float attached by means of a linkage to a pointer; in operation, the float rises and falls with the level of liquid in a tank, while the pointer indicates position of the float on a scale outside the tank.

ball, full A closure component that has a complete spherical surface with a flow passage through it.

balloon 1. The circular symbol used to denote and identify the purpose of an instrument or function. It may contain a tag number. 2. Synonym for bubble. See also bubble.

ball, segmented A closure component that is a segment of a spherical surface which may have one edge contoured to yield a desired flow characteristic.

ball sizing See ball burnishing.

ball-type viscometer An apparatus for determining viscosity, especially of high-viscosity oils and other fluids, in which the time required for a ball to fall through liquid confined in a tube is measured.

ball valve A type of shutoff valve consisting of a solid ball with a diametral hole through it which can be rotated within a spherical seat about an axis perpendicular to the axis of the hole; to permit flow, the ball is rotated so that the hole lines up with inlet and outlet ports of the valve, whereas to shut off flow, the ball is rotated so that the hole does not line up with the ports.

Banbury mixer A heavy-duty batch mixer with two counterrotating rotors; it is designed for blending doughy material such as uncured rubber and plastics.

band 1.The gamut or range of frequencies. 2. The frequency spectrum between two defined limits. 3. Frequencies that are within two definite limits and used for a different purpose. 4. A group of channels; see channel. 4. A group of recording tracks on a computer magnetic disk or drum.

band brake A device for stopping or slowing rotational motion by increasing the tension in a flexible band to tighten it around a drum that is attached to the rotating member.

band-elimination filter A wave filter having a single attenuation band whose critical and cut-off frequencies are finite, nonzero values.

bandpass filter A process or device in which all signals outside a selected band are strongly attenuated, while the signal components lying within the band are passed with a minimum of change.

B and S gage Brown and Sharp gage; see AWG in acronym appendix.

band spectrum A spectral distribution of light or other complex wave in which the wave components can be separated into a series of discrete bands of wavelengths. See also continuous spectrum.

bandwidth 1. The difference, expressed in hertz, between the two boundaries of a frequency range. 2. A group of consecutive frequencies constituting a band that exists between limits of stated frequency attenuation. A band is normally defined as more than 3.0 decibels greater than the mean attenuation across the band. 3. A group of consecutive frequencies constituting a band that exists between limits of stated frequency delay. 4. The range of frequencies that can be transmitted in an electronic system.

bank switching A method of equipping a computer with greater memory by giving the same address to added memory chips.

bar A solid elongated piece of metal, usually having a simple cross section and usually produced by hot rolling or extrusion, which may or may not be followed by cold drawing.

BAR One atmosphere.

bar code A pattern of narrow and wide bars that can be scanned and interpreted into alpha and numeric characters.

bar-code scanner A type of optical scanner developed to read the 12-character Universal Product Code used to identify groceries and other products.

bark A decarburized layer on steel, just beneath oxide scale formed by heating the steel in air.

Barkometer scale A specific gravity scale used primarily in the tanning industry, in which specific gravity of a water solution is determined from the formula: sp gr=1.000 + or –0.001n where n is degrees Barkometer; on this scale, water has a specific gravity of zero Barkometer.

barn A unit of nuclear cross section where the probability of a specific nuclear interaction, such as neutron capture, is expressed as an apparent area; in this context, one barn equals 10^{-28} m^2.

barometer An absolute pressure gage for determining atmospheric pressure; if it is a recording instrument, it is known as a barograph.

barometric hypsometry Determining elevation above some arbitrary reference plane (usually sea level) through the use of mercury or aneroid barometers.

barometric pressure Atmospheric pressure as determined by a barometer usually expressed in inches of mercury.

barometry The study of atmospheric pressure measurement; in particular, determining errors in barometric instrument readings and correcting them.

barostat A device for maintaining constant pressure within a chamber.

barothermograph An instrument for automatically recording both atmospheric temperature and pressure.

barothermohygrograph An instrument for automatically recording atmospheric pressure, temperature and humidity on the same chart.

barrel A unit of volume; for petroleum, it equals 9702 in^3; for fruits, vegetables, other dry commodities and some liquids, a different standard barrel is used.

barrel finishing Producing a lustrous surface finish on metal parts by tumbling them in bulk in a barrel partly filled with an abrasive slurry; similar processes are used for cleaning and electroplating using detergent solutions or electrolytes instead of an abrasive slurry.

base 1. The foundation or support upon which a machine or instrument rests. 2. The fundamental number of characters available for use in each digital position in a numbering system. 3. A chemical substance that hydrolyzes to yield OH$^-$ ions. 4. A reference value. 5. A number that is multiplied by itself as many times as indicated by an exponent. 6. See radix number.

base address 1. A number that appears as an address in a computer instruction but serves as the base, index, initial, or starting point for subsequent addresses to be modified; synonymous with presumptive address and reference address. 2. A number used in symbolic coding in conjunction with a relative address; an address used as the basis for computing the value of some other relative address.

baseband 1. A single channel signaling technique in which the digital signal is encoded and impressed on the physical medium. 2. The frequencies starting at or near d-c.

baseline 1. Generally, a reference set of data against which operating data or test results are compared to determine such characteristics as operating efficiency or system degradation with time. 2. In navigation, the geodesic line between two stations operating in conjunction with each other.

base load Base load is the term applied to that portion of a station or boiler load that is practically constant for long periods.

base metal 1. The metallic element present in greatest proportion in an alloy. 2. The type of metal to be welded, brazed, cut or soldered. 3. In the welded joint, metal that was not melted during welding. 4. Any metal that will oxidize in air or that will form metallic ions in an aqueous solution. 5. Metal to which a plated, sprayed or conversion coating is applied. Also known as basis metal.

base number Same as radix number.

BASIC See Beginner's All-purpose Symbolic Instruction Code.

basic element A single component or subsystem that performs one necessary and distinct function in a measurement sequence; to be considered a basic element, the component must perform one and only one of the smallest steps into which the measuring sequence can be conveniently divided.

basic frequency In a waveform made up of several sinusoidal components of different frequencies, the single component having the largest amplitude or having some other characteristic that makes it the principal component of the composite wave.

basic input output system (BIOS) That part of a computer operating system that handles input and output.

basic recipe A generic, transportable recipe consisting of header information, equipment requirements, formula and procedure.

basis weight For paper and certain other sheet products, the weight per unit area.

batch 1. The quantity of material required for or produced by a production operation at a single time. 2. An amount of material that undergoes some unit chemical process or physical mixing operation to make the final product homogeneous or uniform. 3. A group of similar computer transactions joined together for processing as a single unit.

batch distillation A distillation process in which a fixed amount of a mixture is charged, followed by an increase in temperature to boil off the volatile components. This process differs from continuous distillation, in which the feed is charged continuously.

batch mixer A type of mixer in which starting ingredients are fed all at once and the mixture removed all at once at some later time. Contrast with continuous mixer.

batch process A process that manufactures a finite quantity of material by subjecting measured quantities of raw materials to a time-sequential order of processing actions using one or more pieces of equipment.

batch processing 1. Pertaining to the technique of executing a set of programs such that each is completed before the next program of the set is started. 2. Loosely, the execution of programs serially.

bat file A file name ending in .bat which contains a list of commands most often used to initiate a computer program.

bathochrome An agent or chemical group that causes the absorption band of a solution to shift to lower frequencies.

bathometer An instrument for measuring depth in the ocean or other body of water.

bathyclinograph An instrument for measuring vertical ocean currents.

bathyconductograph An instrument for measuring electrical conductivity of sea water as it is towed at various depths behind a moving ship.

bathymetry Application of scientific principles to the measurement of ocean depths.

bathythermograph An instrument for recording sea temperature versus depth (pressure) as it is towed behind a moving ship. Also known as bathythermosphere.

battery setting Describes a setting of two or more boilers with common division walls.

baud 1. A unit of signaling speed equal to the number of code elements per second. (This is applied only to the actual signals on a communication line) 2. If each signal event represents only one bit condition, baud is the same as bits per second. 3. When each signal event represents other than the logical state of only one bit, used for data entry only in the simplest of sys-

tems. 4. The unit of signal speed equal to twice the number of Morse code dots continuously sent per second; clarified by rate, bit and capacity, and channel.

Baudot code A three-part teletype code consisting of a start pulse (always a space), five data pulses, and a stop pulse (1.42 times the length of the other pulses) for each character transmitted; various combinations of data pulses are used to designate letters of the alphabet, numerals 0 to 9, and certain standard symbols.

baud rate Any of the standard transmission rates for sending or receiving binary coded data; standard rates are generally between 50 and 19,200 bauds.

Baumé scale Either of two specific gravity scales devised by French chemist Antoine Baumé in 1768 and often used to express the specific gravity of acids, syrups and other liquids; for light liquids the scale is determined from the formula: $°Bé=(140/sp\ gr)-130$. For heavy liquids it is determined from: $°Bé=145-(145/sp\ gr)$. $60°F$ is the standard temperature used.

Bauschinger effect The phenomenon wherein plastic deformation of a metal raises its tensile yield strength but decreases its compressive yield strength.

BCD See binary coded decimal.

BCOMP See buffer complete.

BDC See buffered data channel.

bead 1. A rolled or folded seam along the edge of metal sheet. 2. A projecting band or rim. 3. A drop of precious metal produced during cupellation in fire assaying. 4. An elongated seam produced by welding in a single pass.

beaded tube end The rounded exposed end of a rolled tube when the tube metal is formed over against the sheet in which the tube is rolled.

beam 1. An elongated structural member that carries lateral loads or bending moments. 2. A confined or unidirectional ray of light, sound, electromagnetic radiation or vibrational energy, usually of relatively small cross section.

beam divergence The increase in beam diameter with increase in distance from a laser's exit aperture. Divergence, expressed in milliradians, is measured at specified points across the beam's diameter.

beam expander An optical system which expands a narrow beam to a larger diameter, ideally without changing the divergence of the beam.

beam integrator A device which integrates the energy in a light beam to make it uniform across the beam cross section.

beam splitter A device which separates a light beam into two beams. Some types affect polarization of the beam.

beam spread The angle of divergence of an acoustic or electromagnetic beam from its central axis as it travels through a material.

bearing 1. A machine part that supports another machine part while the latter undergoes rotating, sliding or oscillating motion. 2. That portion of a beam, truss or other structural member which rests on the supports. 3. The angle in a horizontal plane between the line of sight to a distant object and some absolute or relative reference direction.

bearing circle A ring-shaped device that fits over a compass or compass repeater to facilitate taking compass bearings.

beat-frequency oscillator An electrical oscillator which generates a frequency which in turn is beat against another frequency to generate a third usually audible frequency. Generally used in communications receivers to provide an audible signal for CW reception or to reinsert a carrier for reception of single side band signals.

beating A resultant pulsating waveform sometimes produced when two or more periodic quantities of different frequencies combine.

beat note The wave of different frequency resulting when two sinusoidal waves whose frequencies differ from each other are supplied to a nonlinear device.

beats Periodic pulsations in amplitude that are created when a wave of one frequency is combined with a wave of a different frequency.

bed 1. The part of a machine having precisely machined ways or bearing surfaces for supporting and aligning other parts such as toolholders or dies. 2. A perforated floor, lining or support structure, often covered with a layer of granular material, in a furnace, chemical processing tank or filtration tank.

Beer's law The law relating the absorption coefficient to the molar density.

Beginner's All-purpose Symbolic Instruction Code A widely used computer language for personal computers.

behavioral modeling Modeling a device or component directly in terms of its underlying mathematical equations.

behind the panel 1. A term applied to a location that is within an area that contains (a) the instrument panel, (b) its associated rack-mounted hardware, or (c) is enclosed within the panel. 2. Behind the panel devices are not accessible for the operator's normal use. 3. Behind the panel devices are not designated as local or front-of-panel-mounted. 4. In a very broad sense, "behind the panel" is equivalent to "not normally accessible to the operator." [S5.1]

bel A dimensionless unit for expressing the ratio of two power levels; the value in bels equals log (P_2/P_1), where P_1 and P_2 are the two power levels.

belled tube end See flared tube-end.

Belleville washer See disk spring.

bellows 1. A pressure sensing element of generally cylindrical shape whose walls contain deep convolutions, and for which the length changes when a pressure differential is applied [S37.1]. 2. An enclosed chamber with pleated or corrugated walls so that its interior volume may be varied, either to alternately draw in and expel a gas or other fluid, or to expand and contract in response to variations in internal pressure. 3. A pressure transducer that converts pressure into a nearly linear displacement.

bellows expansion joint A type of coupling between two pieces of pipe that uses a flexible metal bellows to prevent leakage while allowing limited linear movement, such as to accommodate thermal expansion and contraction.

bellows gage A pressure-measuring device in which variations in internal pressure within a flexible bellows causes movement of an end plate against spring force; the position of the end plate is directly related to bellows internal pressure.

bellows meter A differential pressure measuring instrument having a measuring element of opposed metal bellows, the motion of which positions the output actuator.

bellows seal 1. A multi-convolution type element used as a protective barrier between the instrument and the process fluid. 2. A seal in the shape of a bellows used to prevent air or gas leakage.

bellows sealed valve A valve utilizing a bellows to replace the conventional packing gland. One end of the bellows is welded to the rising stem; the other is sealed against the valve body.

bellows stem seal A thin wall, convoluted, flexible component which makes a seal between the stem and bonnet or body and allows stem motion while maintaining a hermetic seal.

bellows type valve A fluid powered device in which the fluid acts upon a flexible convoluted member, the bellows, to provide linear motion to the actuator stem [S75.05].

bell-type manometer A gage for measuring differential pressure which consists essentially of a cup inverted in a container of liquid; pressure from one source is fed to the inside of the cup while pressure from a second source is applied to the exterior of the cup; pressure difference is indicated by the position of the cup in relation to the liquid level.

belt conveyor An endless fabric, rubber, plastic, leather, or metal belt operating over suitable drive, tail-end, and bend terminals; belt idlers; or slider beds for handling bulk materials, packages, or objects placed directly upon the belt [RP74.01].

belt conveyor scale A device installed on a belt-conveyor structure that continuously weighs the material being conveyed [RP74.01].

belt-speed sensor A device that generates a signal as a function of belt speed [RP74.01].

belt-speed transmitter A device that transmits a belt speed signal to a receiver [RP74.01].

bench check A laboratory-type test of an assembly, component or subassembly to verify its function or identify a source of malfunction, often done with the unit removed from its housing or system for service or repair. Also known as bench test.

bench mark A natural or artificial object having a specific point marked to identify a reference location, such as a reference elevation.

benchmark program A routine used to determine the performance of a computer or software.

bench (optical) A mounting surface for optical components.

bench set The calibration of the actuator spring range of a control valve, to account for the in-service process forces [S75.05].

bench top equipment Equipment designed to be used on and supported by a bench, table, stand, etc., but is neither fixed nor portable as determined by the following: (a) It has at least one handle and the weight exceeds 20 kilograms (44 pounds), or (b) It has no handle and the weight exceeds 5 kilograms (11 pounds), or (c) It is not mobile (does not have casters, wheels,

rollers, etc., nor is it provided with a cart) [S82.01].

bending Applying mechanical force or pressure to form a metal part by plastic deformation around an axis lying parallel to the metal surface; commonly used to produce angular, curved or flanged parts from sheet metal, rod or wire.

bend loss Attenuation caused by high-order modes radiating from the side of a fiber. The two common types of bend losses are: a) those occurring when the fiber is curved around a restrictive radius of curvature and b) microbends caused by small distortions of the fiber imposed by externally induced perturbations, such as poor cabling techniques.

bend pulley Any pulley used to change the direction of travel of a belt [RP74.01].

bend test A ductility test in which a metal specimen is bent through a specified arc around a support of known radius; used primarily to evaluate inherent formability of metal sheet, rod or wire, or to evaluate weld quality produced with specific materials, joint design and welding technique.

bent tube boiler A water tube boiler consisting of two or more drums connected by tubes, practically all of which are bent near the ends to permit attachment to the drum shell on radial lines.

BER See bit error rate.

Bernoulli coefficient In any stream, if the area is changed, as by a reducer, there is a change in the velocity and a corresponding change in the static pressure, or "head." This pressure change is measured in units of velocity head. The dimensionless coefficient used for this purpose is the Bernoulli coefficient K_8.

bessel The filter characteristic in which phase-linearity across the pass band, rather than amplitude linearity, is emphasized; known also as constant-delay.

best straight line A line midway between the two parallel straight lines closest together and enclosing all output vs. measurand values on a calibration curve [S37.1].

best-straight-line linearity Also called independent linearity; an average of the deviation of all calibration points.

beta emitter A radioactive nuclide that disintegrates by emitting a beta particle.

beta particle An electron or positron emitted from the nucleus of a radioactive nuclide.

beta ratio The ratio of the diameter of the constriction to the pipe diameter, $\beta = D_{const}/D_{pipe}$.

beta ray A stream of beta particles.

beta-ray spectrometer An instrument used to measure the energy distribution in a stream of beta particles or secondary electrons.

beta test The second stage of testing a new software program.

betatron A large particle accelerator used to impart energy to a stream of electrons by means of magnetic induction.

bevel gear One of a pair of gears whose teeth run parallel to a conical surface so that they can transmit power and motion between two shafts whose axes intersect.

bezel A ring-shaped member surrounding a cover glass, window, cathode-ray tube face or similar area to protect its edges and often to also provide a decorative appearance.

B-H meter An instrument used to determine the intrinsic hysteresis loop of a magnetic material.

bias 1. The departure from a reference value of the average of a set of values; thus, a measurement of the amount of unbalance of a set of measurements or conditions; that is...error having an average value that is non-zero. 2. The average d-c voltage or current maintained between a control electrode and the common electrode in a transistor or vacuum tube.

bias (tape) The sine wave, typically ten times the amplitude and 3.5 times the top frequency, applied to tape recording heads with a signal in order to eliminate most signal distortion.

BICEPS A General Electric process-oriented language.

bidirectional load cell A column-type strain-gage load cell with female or male fittings at both ends for attaching load hardware; it can be used to measure either tension or compression loading. Also known as universal load cell.

bi-directional printer An electronic printer capable of printing either forward or backward.

bidirectional pulse A wave pulse in which intended deviations from the normally constant values occur in two opposing directions.

Bielby layer An amorphous layer at the surface of mechanically polished metal.

bilateral tolerance The amount of allowable variation about a given dimension, usually expressed as plus-or-minus a specific fraction or decimal.

bilateral transducer A transducer that can transmit signals simultaneously in both directions between two or more terminations.

billet 1. A semifinished primary mill product ordinarily produced by hot rolling metal ingot to a cylinder or prism of simple cross-sectional shape and limited cross-sectional area. 2. A general term for the starting stock used to make forgings and extrusions.

bimetal A bonded laminate consisting of two strips of dissimilar metals; the bond is usually a stable metallic bond produced by corolling or diffusion bonding; the composite material is used most often as an element for detecting temperature changes by means of differential thermal expansion in the two layers.

bimetallic corrosion A type of accelerated corrosion induced by differences in galvanic potential between dissimilar metals immersed in the same liquid medium (electrolyte) and also in electrical contact with each other.

bimetallic thermometer element A temperature-sensitive strip of metal (or other configuration) made by bonding or mechanically joining two dissimilar strips of metal together in such a manner that small changes in temperature will cause the composite assembly to distort elastically, and produce a predictable deflection; the element is designed to take advantage of the fact that different metals have different coefficients of thermal expansion.

bin activator A vibratory device sometimes installed in the discharge path of a mass-flow bin or storage hopper to promote steady discharge of dry granular material.

binary 1. A term applied to a signal or device that has only two discrete positions or states. When used in its simplest form, as in "binary signal" (as opposed to "analog signal"), the term denotes an "on-off" or "high-low" state, i.e., one which does not represent continuously varying quantities [S5.1]. 2. Pertaining to the characteristic or property involving a selection, choice, or condition in which there are two possibilities [RP55.1]. 3. Pertaining to the numeration system with a radix of two [RP55.1]. 4. A computer numbering system that uses two as its base rather than ten. The binary system uses only 0 and 1 in its written form. 5. A device that uses only two states or levels to perform its functions, such as a computer.

binary alloy A metallic material composed of

only two chemical elements (neglecting minor impurities), at least one of which is a metal.

binary cell An information-storage element that can assume either of two stable conditions, and no others.

binary code A code that uses two distinct characters, usually 0 and 1.

binary coded decimal (BCD) Describing a decimal notation in which the individual decimal digits are represented by a group of binary bits, e.g., in the 8-4-2-1 coded decimal notation each decimal digit is represented by a group of four binary bits. The number twelve is represented as 0001 0010 for 1 and 2, respectively, whereas in binary notation it is represented as 1100. Related to binary.

binary-coded decimal system A system of number representation in which each digit in a decimal number is expressed as a binary number.

binary counter 1. A counter which counts according to the binary number system. 2. A counter whose basic counting elements are capable of assuming one of two stable states.

binary digit 1. In binary notation, either of the characters 0 or 1. 2. Same as bit. 3. See equivalent binary digits.

binary distillation A distillation process that separates only two components.

binary file An electronic term for a file that is not a text file.

binary notation A numbering system using the digits 0 and 1 with a base of 2.

binary number A number composed of the characters 0 and 1, in which each character represents a power of two. The number 2 is 10; the number 12 is 1100; the number 31 is 11111, etc.

binary point The radix point in a binary number system.

binary scaler A signal-modifying device (scaler) with a scaling factor of 2.

binary synchronous A procedure for connecting many terminals that share a single link.

binary synchronous communications A communications procedure using special characters for control of synchronized transmission

binary unit 1. A binary digit. 2. A unit of information content, equal to one binary decision, or the designation of one of two possible and equally likely values or states of anything used to store or convey information. 3. See check bit and parity bit. 4. Same as bit.

binary word A group of binary digits with place values in increasing powers of two.

binder 1. In metal founding, a material other than water added to foundry sand to make the particles stick together. 2. In powder metallurgy, a substance added to the powder to increase green strength of the compact, or a material (usually of relatively low melting point) added to a powder mixture to bond particles together during sintering that otherwise would not bond into a strong sintered body.

Bingham body A non-Newtonian substance that exhibits true plastic behavior—that is, it flows when subjected to a continually increasing shear stress only after a definite yield point has been exceeded.

Bingham viscometer A time-of-discharge device for measuring fluid viscosity in which the fluid is discharged through a capillary tube instead of an orifice or nozzle.

bioinstrumentation Instruments that can be attached to humans or animals to record biological parameters, such as pulse rate, breathing rate or body temperature.

biological corrosion Deterioration of metal surfaces due to the presence of plant or animal life; deterioration may be caused by chemicals excreted by the life form, or by concentration cells such as those under a barnacle, or by other interactions.

biomedical engineering The application of engineering principles to the solution of medical problems, including the design and fabrication of prostheses, diagnostic instrumentation and surgical tools.

bi-phase A method of bit encoding for serial data transmission or recording whereby there is a signal transition every bit period.

bipolar technology Technology that uses two different polarity electrical signals to represent logic states of 1 and 0.

bipolar transistor A transistor created by placing a layer of P- or N-type semiconductors between two regions of an opposite type of semiconductor.

biquinary code A method of coding decimal digits in which each numeral is coded in two parts—the first being either 0 or 5, and the second any value from 0 to 4; the digit equals the sum of the two parts.

birefringent element A device that has a refractive index which is different for lightwaves of different orthogonal polarizations. Because of this difference, light of the two orthogonal polarizations travels at different speeds and is refracted slightly differently.

Birmingham wire gage A system of standard sizes used in the United States for brass wire, and for strip, bands, hoops and wire made of ferrous and nonferrous metals; the decimal equivalent of standard Bwg sizes is generally larger than for the same gage number in both the American wire gage and U.S. steel wire gage systems.

biscuit 1. A piece of pottery that has been fired but not glazed. 2. An upset blank for drop forging. 3. A small cake of primary metal, generally one produced by bomb reduction or a similar process.

bistable The capability of assuming either of two stable states, hence of storing one bit of information.

BISYNC See binary synchronous communications.

bit 1. A cutting tool for drilling or boring. 2. The blade of a cutting tool such as a plane or ax. 3. A removable tooth of a saw, milling cutter or carbide-tipped cutting tool. 4. The heated tip of a soldering iron. 5. An abbreviation of binary digit. 6. A single character in a binary number. 7. A single pulse in a group of pulses. 8. A unit of information capacity of a storage device. The capacity in bits is the logarithm to the base two of the number of possible states of the device. Related to storage capacity. 9. The smallest unit of information that can be recognized by a computer.

bit density A measure of the number of bits recorded per unit of length or area.

bit error rate The ratio of bits received in error to bits sent.

bit error rate tester A system which measures the fraction of bits transmitted incorrectly by a digital communication system.

bit map A table that describes the state of each member of a related set; bit map is most often used to describe the allocation of storage space; each bit in the table indicates whether a particular block in the storage medium is occupied or free.

bit pattern A combination of n binary digits to represent 2 to the n possible choices, e.g., a 3-bit pattern represents 8 possible combinations.

bit rate 1. The speed at which bits are transmitted, usually expressed in bits per second. (Com-

pare with baud). 2. The rate at which binary digits, or pulses representing them, pass a given point on a communications line or channel. Clarified by baud.

bits per second In a serial transmission, the instantaneous bit speed within one character, as transmitted by a machine or a channel. See baud.

bit stream A binary signal without regard to grouping by character.

bit string A string of binary digits in which each bit position is considered as an independent unit.

bit synchronizer A hardware device that establishes a series of clock pulses in synchronism with an incoming bit stream and identifies each bit.

bituminous Describing a substance that contains organic matter, mostly in the form of tarry hydrocarbons (described as bitumen).

black body 1. A physical object that absorbs incident radiation, regardless of spectral character or directional preference of the incident radiation; a perfect black body is most closely approximated by a hollow sphere with a small hole in its wall—the plane of the hole being the black body; a perfect black body is used as an ideal reference concept in the study of radiant energy. 2. Denotes a perfectly absorbing object, none of the incident energy is reflected. It radiates (perfectly) at a rate expressed by the Stefan-Boltzmann Law; the spectral distribution of radiation is expressed by Planck's radiation formula. When in thermal equilibrium, a black body absorbs and radiates at the same rate.

blackbody temperature The true temperature of a blackbody source. When used to calibrate a radiation pyrometer, the radiation pyrometer will measure brightness temperature of sources other than the blackbody. To obtain true temperature of non-blackbodies using a radiation pyrometer, multiply the brightness temperature by the emissivity of the observed source.

black box A generic term used to describe an unspecified device which performs a special function or in which known inputs produce known outputs in a fixed relationship.

black-bulb thermometer A thermometer whose sensitive element is covered with lampblack to make it approximate a black body.

black liquor 1. The solution remaining after cooking pulpwood in the soda or sulfite papermaking process. 2. A black, iron-acetate solution containing 5 to 5.5% Fe, and sometimes tannin or copperas, used in dyes and printing inks.

blade-type consistency sensor A pneumatic device for determining changes in consistency of a flowing non-Newtonian substance such as a slurry; it senses the force required for a shaped blade to shear through the flowing stock, and transmits a pneumatic output signal proportional to changes in consistency; its normal operating range is 1.75 to 6.0% suspended solids, with a sensitivity of -0.02% in many applications.

blank In computer programming, the character used to represent a space.

blank alarm point See alarm point.

blanking 1. Inserting a solid disc at a pipe joint or union to close off flow during maintenance, repair or testing. 2. Using a punch and die to cut a shaped piece from sheet metal or plastic for use in a subsequent forming operation. 3. Using a punch and die to make a semifinished powder-metal compact.

blast furnace gas Lean combustible by-product gas resulting from burning coke with a deficiency of air in a blast furnace.

blasting 1. Detonating an explosive. 2. Using abrasive grit, sand or shot carried in a strong stream of air or other medium to remove soil or scale from a surface.

bleeding 1. Allowing a fluid to drain or escape to the atmosphere through a small valve or cock; used to provide controlled slow reduction of slight overpressure, to withdraw a sample for analysis, to drain condensation from compressed air lines, or to reduce the airspace above the liquid level in a pressurized tank. 2. Withdrawing steam from an intermediate stage of a turbine to heat a process fluid or boiler feedwater. 3. Natural separation of liquid from a semisolid mixture—such as oil from a lubricating grease or water from freshly poured concrete.

bleeding cycle A type of steam cycle where steam is withdrawn from the turbine at one or more intermediate stages and used to heat feedwater before it enters the boiler.

blend 1. To mix ingredients so that they are indistinguishable from each other in the mixture. 2. To produce a smooth transition between two intersecting surfaces, such as at the edges of a radiused fillet between a shaft and an integral flange or collar.

blind hole A hole in a piece of material that does not completely penetrate to the back surface.

blind nipple A short piece of pipe or tubing with one end closed and sealed.

blind pressure transmitter A pressure transmitter not having an integral readout device.

blinking A periodic change of hue, saturation, or intensity of a video display unit pixel, character, or graphic symbol [S5.5].

blip Any erratic signal on a computer screen.

blister 1. A small area on the surface of metal or plastic where a thin layer of the material has been separated from underlying material and is raised due to gas trapped between the layers, yet remains attached around the edges of the raised area. 2. An enclosed macroscopic cavity in a glaze or other fired ceramic coating. 3. A raised area where a paint, electroplate or other coating has become detached from the substrate due to accumulation of gas or moisture at the coating-substrate interface.

block 1. A set of things, such as words, characters, or digits, handled as a unit. 2. A collection of contiguous records recorded as a unit, blocks are separated by interblock gaps, and each block may contain one or more records. 3. In data communication, a group of contiguous characters formed for transmission purposes. The groups are separated by interblock characters. 4. A group of physically adjacent words or bytes of a specified size particular to a device. The smallest system-addressable segment on a mass-storage device in reference to I/O. See also cylinder block; block-and-tackle.

block-and-tackle A hoisting gear consisting of a rope or cable and one or more independently rotating frictionless pulleys. Also known as block and fall.

block, data A set of associated characters or words handled as a unit.

block diagram 1. A graphical representation of the hardware in a computer system. The primary purpose of a block diagram is to indicate the paths along which information or control flows between the various parts of a computer system. It should not be confused with the term flow chart. 2. A coarser and less symbolic representation than a flow chart. 3. A graphical representation of a computer program. 4. Used to provide a simple pictorial representation of a control system. Block diagrams have two basic symbols, the circle and the function block. The arrows entering and leaving the circle represent the flow of information and the head of each arrow has an algebraic sign associated with it, either plus or minus. 5. Block diagrams show the graphical representation of the hardware in a system. The primary purpose of a block diagram is to indicate the paths along which information or control flows between various parts of the system.

blocked impedance Of an electro-mechanical transducer; the electrical impedance at the input terminals when the mechanical system is "blocked", or prevented from moving.

blockend interrupt (BIN) A signal in TELE-VENT that indicates that a buffer is completely filled with data.

blocker-type forging A shape forging designed for easy forging and extraction from the die through the use of generous radii, large draft angles, smooth contours and generous machining allowances; used as a preliminary stage in multiple-die forging or when machining to final shape is less costly than forging to final shape.

blocking 1. Producing a semifinished forging of approximate shape suitable for further forging or machining to final size and shape. 2. Reducing the oxygen content of the bath in an open-hearth furnace. 3. Undesired adhesion between plastics surfaces during storage or use. 4. Of computer records, see grouping.

block sequence A welding sequence in which separated lengths of a continuous multiple-pass weld are built up to full cross section before gaps between the segments are filled in. Compare with cascade sequence.

block switching A two-level multiplexing technique used in data transmission, whereby one level selects the input channel to be transmitted and the second level selects the group of first-level input channels to be addressed; the chief advantage of block switching is reduction of leakage currents from "off" channels which interfere with data signals being transmitted. Also known as submultiplexing.

bloom 1. A semifinished metal bar of large cross section (usually a square or rectangle exceeding 36 sq in.) hot rolled or sometimes forged from ingot. 2. Visible fluorescence on the surface of lubricating oil or an electroplating bath. 3. A bluish fluorescent cast to a painted surface caused by a thin film of smoke, dust or oil. 4. A loose, flower-like corrosion product formed when certain nonferrous metals are exposed in a moist environ-

ment. 5. To apply an antireflection coating to glass. 6. To hammer or roll metal to brighten its surface.

blowback The difference between the pressure at which a safety valve opens and at which it closes, usually about 3% of the pressure at which valve opens.

blowby Leakage of fluid through the clearance between a piston and its cylinder during operation.

blowdown 1. In a safety valve, the difference between opening and closing pressures. 2. In a steam boiler, the practice of periodically opening valves attached to the bottom of steam drums and water drums, during boiler operation, to drain off accumulations of sediment.

blow down valve A valve generally used to continuously regulate concentration of solids in the boiler, not a drain valve.

blower A fan used to force air under pressure.

blowhole A pocket of air or gas trapped during solidification of a cast metal.

blow-off valve A specially designed, manually operated, valve connected to the boiler for the purpose of reducing the concentration of solids in the boiler or for draining purposes.

blowout disk See rupture disk device.

blowtorch action Impingement of a localized jet of hot gas on a surface.

blue brittleness In some steels, loss of ductility associated with tempering or service temperatures in the blue heat range, 400 to 600 °F.

blue vitriol A solution of copper sulfate sometimes applied to metal surfaces to make scribed layout lines more visible.

bluing Also spelled blueing. 1. Forming a bluish oxide film on steel by exposing it to steam, air or other agents at a suitable temperature, thus giving scale-free surfaces an attractive appearance and improved corrosion resistance. 2. Heating formed springs after fabrication to improve their properties and reduce residual stress. 3. A thin blue oxide formed on polished metal surfaces when exposed briefly to air at high temperatures.

BNI See Bureau d'Orientation de la Normalisatin en Informatique.

board In computers, a flat sheet in which integrated circuits are mounted. See panel.

Bode diagram In process instrumentation, a plot of log gain (magnitude ratio) and phase angle values on a log frequency base for a transfer function [S51.1].

Bode plot A graph of transfer function versus frequency wherein the gain (often in decibels) and phase (in degrees) are plotted against the frequency on log scale. Also called bode diagram.

body The part of the valve which is the main pressure boundary. The body also provides the pipe connecting ends, the fluid flow passageway, and may support the seating surfaces and the valve closure member [S75.05].

body cavity The internal chamber of the valve body including the bonnet zone and excluding the body ends [S75.05].

body, encapsulated A body with all surfaces covered by a continuous surface layer of a different material, usually an elastomeric or polymeric material.

body, split A valve body design in which trim is secured between two segments of a valve body.

body, wafer A thin annular section body whose end surfaces are located and clamped between the piping flanges by bolts extending from flange to flange.

body, wafer, lugged A thin annular section body whose end surfaces mount between the pipeline flanges, or may be attached to the end of a pipeline without any additional flange or retaining parts, using either through-bolting and/or tapped holes.

body, weir type A body having a raised contour contacted by a diaphragm to shut off fluid flow.

bogie Also spelled bogey; bogy. 1. A type of aircraft landing gear consisting of two sets of wheels in tandem with a central strut. 2. A supporting and aligning idler wheel or roller on the inside of an endless track. 3. A swivel-mounted axle or truck that supports a railroad car, the leading end of a locomotive, or the end of a vehicle such as a gun carriage. 4. The drive-wheel assembly and supporting frame for the two rear axles of a three-axle motor truck, mounted so that the wheels are kept in contact with the road surface, especially around curves and over rough roads.

boiler The entire vessel in which steam or other vapor is generated for use external to itself, including the furnace, consisting of the following: waterwall tubes; the firebox area, including burners and dampers; the convection area, consisting of any superheater, reheater, and/or economizer sections, as well as drums and headers [S77.42].

boiler horsepower The evaporation of 34 1/2 lbs of water per hour from a temperature of 212 °F into dry saturated steam at the same temperature. Equivalent to 33,475 Btu.

boiler water A term construed to mean a representative sample of the circulating boiler water, after the generated steam has been separated and before the incoming feed water or added chemical becomes mixed with it so that its composition is affected.

boiling The conversion of a liquid into vapor with the formation of bubbles.

boiling out The boiling of a highly alkaline water in boiler pressure parts for the removal of oils, greases, etc.

boiling water reactor (BWR) A nuclear steam supply system in which process steam is generated in the reactor vessel [S67.03].

boilup Vapors that are generated in the column reboiler.

bolometer A sensitive infrared detector whose operation is based on a change in temperature induced by absorbing infrared radiation. It is made of two thin, blackened gratings of platinum, one illuminated and the other kept in the dark. The absorption of heat changes the electrical resistance which is detected by comparing the resistances of the two gratings in an electrical circuit.

bolster A steel block or plate used to support dies and attach them to a press bed; in drop forging, a bolster is also used to attach dies to the ram.

bolt A threaded fastener consisting of a rod, usually made of metal, having threads at one end and an integral round, square or hexagonal head at the other end; short bolts usually have threads running the entire length below the head, and longer bolts often have an unthreaded shank between the head and threaded end.

bolted joint An assembly of two or more parts held together by a bolt and nut, with or without washers, or by a bolt that threads into a tapped hole in one of the parts.

bolting 1. A collective term for threaded fasteners, especially bolts, nuts, screws and studs. 2. Assembling parts together using threaded fasteners.

bomb calorimeter An apparatus for measuring the quantity of heat released by a chemical reaction; it consists of a strong-walled metal container (bomb) immersed in about 2.5 litres of water in an insulated container; a sample is sealed in the bomb, the bomb immersed, the sample ignited (or a reaction started) by remote control, and the heat released measured by observing the rise in temperature of the water bath.

bond 1. A wire rope that attaches a load to a crane hook. 2. Adhesion between cement or mortar and masonry. 3. In an adhesive bonded or diffusion bonded joint, the junction between faying surfaces. 4. In welding, brazing or soldering, the junction between assembled parts; where filler metal is used, it is the junction between fused metal and heat-affected base metal. 5. In grinding wheels and other rigid abrasives, the material that holds abrasive grains together. 6. Material added to molding sand to hold the grains together. 7. The junction between base metal and cladding in a clad metal product.

bondable Designed to be permanently mounted to a surface by means of adhesives [S37.1].

bonded Permanently attached over the length and width of the active element [S37.1].

bonded liner 1. A liner vulcanized or cemented to the body bore [S75.05]. 2. In a butterfly valve body, a liner vulcanized or cemented to the body bore.

bonded strain gage A device for measuring strain which consists of a fine-wire resistance element, usually in zigzag form, embedded in nonconductive backing material such as impregnated paper or plastic, which is cemented to the test surface or sensing element.

bonded transducer A pressure sensor that uses a bonded strain gage to generate the output signal.

bonding An electrically conductive connection between metallic parts of the equipment that are required to be grounded, and some other part of the equipment to which a grounding conductor is connected [S82.01].

bone dry A papermaking term used to describe pulp fibers or paper from which all water has been removed. Also known as oven dry; moisture free.

bonnet That portion of the valve pressure retaining boundary which may guide the stem and contains the packing box and stem seal. It may also provide the principal opening to the body cavity for assembly of internal parts or be an integral part of the valve body. It may also provide the attachment of the actuator to the valve body [S75.05].

bonnet bolting A means of fastening the bonnet to the body. It may consist of studs with nuts for a flanged bonnet joint, studs threaded into the bonnet neck of the body, or bolts through the bonnet flange [S75.05].

bonnet gasket A deformable sealing element between the mating surfaces of the body and bonnet. It may be deformed by compressive stress or energized by fluid pressure within the valve body [S75.05].

bonnetless Gate valve which has packing between the gate and body, such that the gate extends outside the pressure boundary in the open position [S75.05].

bonnet, seal-welded A bonnet welded to a body to provide a zero leakage joint.

bonnetted Gate valve having a bonnet which enclosed the gate within the pressure boundary when in the open position. Packing is provided at the stem [S75.05].

bonnet types Typical bonnets are bolted, threaded, or welded to or integral with the body [S75.05].

Boolean Pertaining to logic quantities.

Boolean add See OR.

Boolean algebra A process of reasoning, or a deductive system of theorems using a symbolic logic, and dealing with classes, propositions, or on-off circuit elements. It employs symbols to represent operators such as and, or, not, except, if, then, etc., to permit mathematical calculation. Named after George Boole, famous English mathematician.

Boolean expression A quantity expressed as the result of Boolean operations such as and, or, and not upon Boolean variables.

Boolean functions A system of mathematical logic often executed in circuits to provide digital computations such as "OR", "AND", "NOR", "NOT", etc.

Boolean operator A logic operator each of whose operands and whose result have one of two values.

Boolean variable See logical variable.

booster fan A device for increasing the pressure or flow of a gas.

booster relay A volume or pressure amplifying pneumatic relay that is used to reduce the time lag in pneumatic circuits by reproducing pneumatic signals with high volume and/or high pressure outputs.

boot A computer routine in which a few instructions are loaded which then cause the rest of the system to be loaded.

bootstrap A technique for loading the first few instructions of a routine into storage, then using these instructions to bring in the rest of the routine. This usually involves either the entering of a few instructions manually or the use of a special key on the console.

bootstrap loader A routine whose first instruction is sufficient to load the remainder of itself into memory from an input device; normally used to start a complete system of programs.

bore 1. The inner cavity in a pipe or tube. 2. The diameter of the cylinder of a piston-cylinder device such as a reciprocating compressor, engine or pump, or a hydraulic or pneumatic power cylinder. 3. To penetrate or pierce a workpiece with a rotating cutting tool. 4. To increase the size of an existing hole, generally with a single-point cutting tool, while either the work or the cutting tool rotates about the central axis of the hole. 5. The inner surface of a gun tube. 6. The central hole in a laser or other type of tube (a capillary, waveguide, or hole in a micro-channel plate).

bore Reynolds number Calculated Reynolds number including R_d using V_{bore}, P_{bore}, μ_{bore}, d_{bore}; also $R_d R_D / \beta$.

borescope A straight-tube telescope incorporating mirrors or prisms that is used to visually inspect the inner surfaces of pipes or gun tubes.

boresighting Initially aligning a gun, directional antenna or other device by optical means or by observing a return signal from a fixed target at a known location; the term is derived from an early military practice of looking down the bore of an artillery piece to obtain an initial line of sight to a target.

boron counter tube A type of radiation counter tube used to detect slow neutrons; the tube has electrodes coated with a boron compound, and it also may be filled with BF_3; a slow neutron is easily absorbed by a B^{10} nucleus, with subsequent emission of an alpha particle.

borosilicate glass A type of heat-resisting glass that contains at least 5% boric acid.

bort Industrial diamonds or diamond fragments.

boss 1. A localized projection on a valve surface provided for various purposes, such as attachment of drain connections, or other accessories [S75.05]. 2. A raised portion of metal or small area and limited thickness on flat or curved metal surfaces. 3. A short projecting section of a casting, forging or molded plastics part, often cylindrical in shape, used to add strength or to provide for alignment or fastening of assembled parts.

bottom contraction The vertical distance from the crest to the floor of the weir box or channel bed.

bottom dead center The position of a piston and its connecting rod when at the extreme downstroke position.

bottom flange A part which closes a valve body opening opposite the bonnet opening. It may include a guide bushing and/or serve to allow reversal of the valve action. In three-way valves it may provide the lower flow connection and its seat [S75.05].

bottoming drill A flat-end twist drill that converts the conical bottom of a blind hole into a cylinder.

bottoming tap A tap designed for cutting full threads all the way to the bottom of a blind hole; usually used to finish the bottom of a hole tapped with a regular, tapered-end tap.

bottoms The higher boiling product streams usually taken from the bottom of a distillation column—sometimes from the reboiler and sometimes from a separate surge vessel.

boundary layer In a flowing fluid, a low-velocity region along a tube wall or other boundary surface.

boundary lubrication A condition in sliding contact when contact pressures are high enough and sliding velocities low enough that hydrodynamic lubrication is completely absent; mating surfaces slide across each other on a multimolecular layer of lubricant, often with some solid-to-solid surface contact; for liquid lubricants, a bearing-characteristic (Sommerfield) number of 0.01 is considered to be the upper limit of boundary lubrication.

bound water In a moist solid to be dried, that portion of the water content which is chemically combined with the solid matter.

Bourdon tube 1. A pressure sensing element consisting of a twisted or curved tube of noncircular cross section which tends to be straightened by the application of internal pressure [S37.1]. 2. A flattened tube, twisted or curved, and closed at one end, which is used as the pressure-sensing element in a mechanical pressure gage or recorder; a process stream pressure is routed to the open end of the tube, and the tube flexes or untwists in relation to the internal pressure, with the change in shape of the tube being used to operate a mechanical pointer or pen positioner. Also known as Bourdon element; Bourdon pressure gage.

box A flow-chart symbol.

boxcar averager A signal processing instrument which averages selected portions of repetitive signals to improve signal quality. Sometimes called a gated integrator because it passes or gates portions of the signal, then integrates them.

box girder A hollow girder or beam, usually having a spare or rectangular cross section. Also known as box beam.

box header boiler A horizontal boiler of the longitudinal or cross drum type consisting of a front and rear inclined rectangular header connected by tubes.

box wrench A closed-end wrench designed to fit a single size and shape nut; different wrench ends are needed for different nut sizes and different nut shapes. Also known as box end wrench.

B power supply An electrical power supply connected in the plate circuit of a vacuum tube electronic device.

BPS See bits per second. Also see baud.

Bragg law A principle describing the apparent reflection of x-rays (and DeBroglie waves associated with certain particulate beams) from atomic planes in crystals; maximum reflected intensity occurs along the family of directions defined by $\Theta = \arcsin \lambda\, n\, 1/2d$, where Θ is the Bragg angle (angle of reflection and of incidence), n is an integer, λ is the wavelength of monochromatic radiation reflected from the crystal, and d is the interplanar spacing of the reflecting parallel planes in the crystal.

brake A machine element for applying frictional force to slow or stop relative motion.

brake drum See drum.

brake horsepower The mechanical power developed by an engine as measured by absorbing engine output with a friction brake or dynamometer applied to the engine's shaft or flywheel.

brake lining A material having high coefficient of friction that is used as the principal friction element in a mechanical brake; it usually is made of fabric or molded asbestos, and usually

can be readily replaced to extend the brake's service life and restore braking efficiency.

brake shoe See shoe.

Brale A 120° conical diamond indenter used in Rockwell hardness testing of relatively hard metals.

branch The selection of one of two or more possible paths in the control of flow based on some criterion. The instructions which mechanize this concept are sometimes called branch instructions; however the terms transfer of control and jump are more widely used. Related to conditional transfer.

branch circuit That portion of permanently installed wiring between the final overcurrent protective device and the attachment-plug receptacle or outlet, or point of connection to the fixed equipment [S82.01].

branch instruction An instruction that performs a branch.

branchpoint A point in a routine where one of two or more choices is selected under control of the routine. See conditional transfer.

brass Any of the many alloys based on the binary system copper-zinc; most brasses contain no more than 40 wt% zinc.

braze welding A joining process similar to brazing but in which the filler metal is not distributed in the joint by capillary action.

brazing A method for joining metals using heat and a filler metal whose melting temperature is above 850 °F but below the melting temperature of the base metals; filler metal is distributed in the joint by capillary action.

breadboard model A prototype or uncased assembly of an instrument or electronic device whose parts are laid out on a flat surface and connected together to demonstrate or check its operation.

break 1. An interruption in computer processing. 2. To interrupt the sending end and take control of the circuit of the receiving end.

breakdown 1. Initial hot working of ingot-cast or slab-cast metal to reduce its size prior to final working to finished size. 2. A preliminary press-forging operation.

breakdown voltage rating The dc or sinusoidal ac voltage which can be applied across specified insulated portions of a transducer without causing arcing or conduction above a specified current value across the insulating material.

Note: Time duration of application, ambient conditions, and ac frequency must be specified [S37.1].

break point 1. The junction of the extension of two confluent straight-line segments of a plotted curve. In the asymptotic approximation of a log-gain vs. log-frequency relation in a Bode diagram, the value of the abscissa is called the corner frequency [S51.1]. 2. A location at which program operation is suspended so that partial results can be examined; a preset point in a program where control passes to a debugging routine.

breakpoint instruction 1. An instruction which will cause a computer to stop or to transfer control, in some standard fashion, to a supervisory routine which can monitor the progress of the interrupted program. 2. An instruction which, if some specified switch is set, will cause the computer to stop or take other special action.

breaks Creases or ridges, usually appearing in aged sheet or strip, where the yield point has been locally exceeded; depending on the origin of the break it may be termed a coil break, cross break, edge break or sticker break.

breeching A duct for the transport of the products of combustion between parts of a steam generating unit or to the stack.

Bremsstrahlung X-rays having a broad spectrum of wavelengths, which are formed due to deceleration of a beam of energetic electrons as they penetrate a target. Also known as white radiation.

Brewster-angle window A window inserted into an optical path at Brewster's angle—the angle at which unpolarized light must be incident upon a nonmetallic surface for the reflected radiation to acquire maximum plane polarization. At Brewsters angle, the reflected plane polarized beam and the refracted beam through the window are at 90°.

bridge 1. A network device that interconnects two local area networks that use the same LLC but may use different MACs. A bridge requires only OSI Level 1 and 2 protocols. See gateway and router. 2. The strain-to-voltage converter in many measurement systems (actually, a Wheatstone bridge).

bridge amplifier A type of amplifier circuit used extensively in instrumentation to provide gains up to 1000 at bandwidths up to 50 kHz; it is generally configured as a direct-coupled amplifier constructed of four subamplifiers and suitable fixed resistances.

bridge circuit An electronic network in which an input voltage is applied across two parallel elements and the output voltage—to an indicating device or load—is taken across two intermediate points on the parallel elements.

bridged-T network A T-network having a fourth branch connected in parallel with the two series branches of the T, the fourth branch termination at one input and one output terminal.

bridgewall A wall in a furnace over which the products of combustion pass.

bridging 1. Premature solidification of metal across a mold section before adjacent metal solidifies. 2. Welding or mechanical jamming of the charge in a downfeed furnace. 3. Forming an arched cavity in a powder metal compact. 4. Forming an unintended solder connection between two or more conductors, either a secure connection or merely an undesired electrical path without mechanical strength. Also known as crossed joint; solder short.

Briggs pipe thread See American standard pipe thread.

bright dipping Producing a bright surface on metal, such as by immersion in an acid bath.

brightness A term used in nonquantitative statements with reference to sensations and perceptions of light, in quantified statements with reference to the description of brightness by photometric units.

brightness temperature The temperature of any nonblackbody as determined using an optical pyrometer calibrated to give the true temperature of a blackbody; it is always less than the true temperature of the nonblackbody.

bright plating Electroplating to yield a highly reflective coated surface.

bright switch A solid-state switch consisting of two bipolar transistors connected in an inverted configuration to achieve a low offset voltage; used in only limited applications today.

Brinell test A standard bulk hardness test in which a 10-mm-diameter ball is pressed into the surface of a test piece and a hardness number determined by dividing applied load in kg by area of the circular impression in sq mm.

briquetting Producing relatively small lumps or block of compressed granular material, often incorporating a binder to help hold the particles together.

British thermal unit The mean British thermal unit is 1/180 of the heat required to raise the temperature of 1 lb of water from 32 °F to 212 °F at a constant atmospheric pressure. It is about equal to the quantity of heat required to raise 1 lb of water 1 °F. A Btu is essentially 252 calories.

brittle fracture Separation of solid material with little or no evidence of macroscopic plastic deformation, usually by rapid crack propagation involving less energy than for ductile fracture of a similar structure.

brittleness The tendency of a material to fracture without apparent plastic deformation.

Brix scale A specific gravity scale used almost exclusively in sugar refining; the degrees Brix represent the weight percent pure sucrose in water solution at 17.5 °C.

broaching Cutting a finished hole or contour in solid material by axially pulling or pushing a bar-shaped, toothed, tapered cutting tool across a workpiece surface or through a pilot hole.

broadband A medium based on CATV technology where multiple signals are frequency division multiplexed. Due to the use of CATV technology, a broadband cable is unidirectional (within any given block of frequencies). As a result, two types of broadband systems are in common use, single cable and dual cable. In a single cable system, stations transmit and receive on the same cable but at different frequencies. The station transmits on one frequency, travels down the network to the head end, gets translated into a different frequency, and sent back down the network where it is received by all stations. In a dual cable system, the stations transmit and receive at the same frequency but on different cables. The end of the transmit cable is connected to the beginning of the receive cable, forming a double loop through the plant.

broadband pyrometer See wideband radiation thermometer.

broadband transmission (fiber optic) Transmission of signals with a large bandwidth, such as video transmission or higher.

broadcast 1. The simultaneous dissemination of information to one or more stations, one-way, with no acknowledgment of receipt. 2. A message addressed to all stations connected to a LAN.

bronze 1. A copper-rich alloy of copper and tin, with or without small amounts of additional alloying elements. 2. By extension, certain copper-base alloys containing less tin than other elements, such as manganese bronze and leaded tin bronze, and certain other copper-base alloys that do not contain tin, such as aluminum bronze, beryllium bronze and silicon bronze. 3. Trade names for certain copper-zinc alloys (brasses), such as architectural bronze (Cu-40Zn-3Pb) and commercial bronze (Cu-10Zn).

brush plating An electroplating process in which the surface to be plated is not immersed, but rather rubbed with an electrode containing an absorbent pad or brush which holds (or is fed) a concentrated electrolyte solution or gel.

Btu See British thermal unit.

bubble 1. The circular symbol used to denote and identify the purpose of an instrument or function. It may contain a tag number. Synonym for balloon [S5.1]. 2. A small volume of steam enclosed within a surface film of water from which it was generated. 3. Any small volume of gas or vapor surrounded by liquid; surface-tension effects tend to make all bubbles spherical unless they are acted upon by outside forces.

bubblegas A gas selected to bubble from the end of a tube immersed in liquid for level measurement from the hydrostatic back pressure created in the tube.

bubble memory See magnetic bubble memory.

bubble point The temperature at which a liquid mixture begins to boil and evolve vapors.

bubbler-type specific-gravity meter See air-bubbler specific-gravity meter.

bubble sort In data processing, a method of arranging a group of numbers in some order.

bubble tight A non-standard term used to refer to control valve seat leakage. Refer to ANSI/FCI 70-2 for specification of seat leakage classifications.

bubble tube A length of pipe or tubing placed in a vessel at a specified depth to transport a gas injected into the liquid to measure level from hydrostatic back pressure in the tube.

bubble-type viscometer A device similar to a ball-type viscometer, except viscosity is determined from timed rise of a standard-size bubble through the sample liquid instead of timed fall of a ball.

buckle 1. Localized waviness in a metal bar or

sheet, usually transverse to the direction of rolling. 2. An indentation in a casting due to expansion of molding sand into the cavity.

Buckley gage A device that measures very low gas pressure by sensing the amount of ionization produced by a prescribed electric current.

buckling Producing a lateral bulge, bend, bow, kink or wavy condition in a beam, bar, column, plate or sheet by applying compressive loading.

buckstay A structural member placed against a furnace or boiler wall to restrain the motion of the wall.

buffer 1. An internal portion of a data processing system serving as intermediate storage between two storage or data-handling systems with different access times or formats; usually to connect an input or output device with the main or internal high-speed storage. Clarified by storage buffer. 2. An isolating component designed to eliminate the reaction of a driven circuit on the circuits driving it, e.g., a buffer amplifier.

buffer complete (BCOMP) In TELEVENT, the signal that indicates when the computer buffer is complete.

buffered computer A computing system with a storage device which permits input and output data to be stored temporarily in order to match the slow speeds of input and output devices. Thus, simultaneous input-output and computer operations are possible. A data transmission trap is essential for effective use of buffering since it obviates frequent testing for the availability of a data channel.

buffered data channel (BDC) A device that provides high-speed parallel data interfaces into and out of the computer memory.

buffered I/O channel A computer I/O channel that controls the movement of data between an external device and memory, under the control of self-contained registers (i.e., independent of the operating program). See BDC.

buffing Producing a very smooth and bright surface by rubbing it with a soft wheel, belt or cloth impregnated with fine abrasive such as jeweler's rouge.

bug An error, defect or malfunction in a computer program.

buildup 1. Excessive electrodeposition on areas of high current density, such as at corners and edges. 2. Small amounts of work metal that ad-

here to the cutting edge of a tool and reduce its cutting efficiency. 3. Deposition of metal by electrodeposition or spraying to restore required dimensions of worn or undersize machine parts.

bulge A local distortion of swelling outward caused by internal pressure on a tube wall or boiler shell caused by overheating. Also applied to similar distortion of a cylindrical furnace due to external pressure when overheated provided the distortion is of a degree that can be driven back.

bulk density Mass per unit volume of a bulk material, averaged over a relatively large number of samples.

bulk memory See secondary storage.

bulk modulus An elastic modulus determined by dividing hydrostatic stress by the associated volumetric strain (usually computed as the fractional change in volume).

bulk storage A hardware device in a computer system that supplements computer memory; typically, a magnetic tape or disk.

bulk storage memory Any non-programmed large memory. For example, discs, drums, or magnetic tape units.

bull block A machine with a power-driven rotating drum for pulling wire through a drawing die.

bull gear A bull wheel with gear teeth around its periphery.

bullion 1. A semirefined alloy containing enough precious metal to make its recovery economically feasible. 2. Refined gold or silver, ready for coining.

bull wheel 1. The main wheel or gear of a machine, usually the largest and strongest. 2. A cylinder with a rope wound around it for lifting or hauling.

bump A raised or flattened portion of a boiler drum head or shell formed by fabrication, generally used for nozzle or pipe attachments.

bumpless transfer Change from manual mode to automatic mode of control, or vice versa, without change in control signal to the process.

Buna-N A nitrile synthetic rubber known for resistance to oils and solvents.

bundle (fiber optic) A group of fibers packaged together which collectively transmit light, in a coherent bundle, the end fibers are in a fixed relationship to each other and can transmit an image.

bunker C oil Residual fuel oil of high viscosity

commonly used in marine and stationary steam power plants. (No. 6 fuel oil).

buoyancy The tendency of a fluid to lift any object submerged in the body of the fluid; the amount of force applied to the body equals the product of fluid density and volume of fluid displaced.

buoyancy displacers The technique of measuring liquid level by measuring the buoyant force on a partially immersed volumetric displacing device.

buoyancy-type liquid-level detector Any of several designs of level gage that depend for their operation on the buoyant force acting on a float or similar device located inside the tank or vessel.

burden 1. The amount of power consumed in the measuring circuit of an instrument, usually given as the volt-amperes consumed under normal operating conditions. 2. The property of a circuit connected to the secondary winding of an instrument transformer which determines active and reactive power at the transformer output terminals.

Bureau d'Orientation de la Normalisatin en Informatique The French national standards body for computer related standards.

burner 1. Any device for producing a flame using liquid or gaseous fuel. 2. A device in the firebox of a fossil-fuel-fired boiler that mixes and directs the flow of fuel and air to give rapid and complete combustion. 3. A worker who cuts metal using an oxyfuel-gas torch.

burner windbox A plenum chamber around a burner in which an air pressure is maintained to insure proper distribution and discharge of secondary air.

burner windbox pressure The air pressure maintained in the windbox or plenum chamber measured above atmospheric pressure.

burnish 1. To polish or make shiny. 2. Specifically, to produce a smooth, lustrous surface finish on metal parts by tumbling them with hardened metal balls or rubbing them with a hard metal pad.

burr 1. A thin, turned over edge or fin produced by a grinding wheel, cutting tool or punch. 2. A rotary tool having teeth similar to those on a hand file.

bursting In data processing, the act of separating continuous forms into single sheets.

burst pressure rating The pressure which may be applied to the sensing element or the case (as specified) of a transducer without rupture of either the sensing element or transducer case as specified. Note: (1) minimum number of applications and time duration of each application must be specified, (2) in the case of transducers intended to measure a property of a pressurized fluid, burst pressure is applied to the portion subjected to the fluid [S37.1].

bus 1. A group of wires or conductors, considered as a single entity, which interconnects parts of a system. 2. In a computer, signal paths such as the address bus, the data bus, etc. 3. A circuit over which data or power is transmitted; often one which acts as a common connection among a number of locations. Synonymous with trunk. 4. A communications path between two switching points. 5. A common connector circuit, usually multiwire, for transfer of power, data, timing, etc., between the several modules or units on the bus.

bus cycle The transfer of one word or byte between two devices.

bushing 1. A fixed member which supports and/or guides the closure member, valve stem and/or actuator stem. The bushing supports the nonaxial loads on these parts and is subject to relative motion of the parts [S75.05]. 2. A removable piece of soft metal or impregnated sintered-metal sleeve used as a bearing or guide. 3. A ring-shaped device made of ceramic or other nonconductive material used to support an electrical conductor while preventing it from becoming grounded to the support structure.

bus request The DEC PDP-11 priority system for determining which external device will obtain control of the UNIBUS to interrupt the CPU for service; there are seven bus requests and one bus grant.

butterfly valve 1. A valve with a circular body and a rotary motion disk closure member, pivotally supported by its stem [S75.05]. 2. A valve consisting of a disc inside a valve body which operates by rotating about an axis in the plane of the disc to shut off or regulate flow in a piping system; a similar device used in heating or ventilating ductwork is called a butterfly damper.

buttering Coating the faces of a weld joint prior to welding to preclude cross contamination of a weld metal and base metal.

butterworth The filter characteristic in which constant amplitude across the pass band is the objective; known also as constant amplitude (CA).

butt joint A joint between two members lying approximately in the same plane; in welded joints, the edges may be machined or otherwise prepared to create any of several types of grooves prior to welding.

buttstrap A narrow strip of boiler plate overlapping the joint of two butted plates, used for connecting by riveting.

butt weld A weld that joins the edges or ends of two pieces of metal having similar cross sections, without overlap or offset along the joint line.

by hand Denotes that an operation does not require the use of a tool, coin, or any other object that may serve as a tool [S82.01].

by-pass A passage for a fluid, permitting a portion or all of the fluid to flow around its normal pass flow channel.

bypass capacitor A capacitor connected in parallel with a circuit element to provide an alternative a-c current path of relatively low impedance.

by-product Incidental or secondary output of a chemical production or manufacturing process that is obtained in addition to the principal product with little or no additional investment or allocation of resources.

byte 1. A sequence of adjacent binary digits operated upon as a unit and usually shorter than a word [RP55.1]. 2. Generally accepted as an eight-bit segment of a computer word. 3. Eight contiguous bits starting on an addressable byte boundary; bits are numbered from the right, 0 through 7, with 0 the low-order bit. When interpreted arithmetically, a byte is a two's complement integer with significance increasing from bits 0 through 6; bit 7 is the sign bit. The value of the signed integer is in the range of –128 to 127 decimal. When interpreted as an unsigned integer, significance increases from bits 0 through 7 and the value of the unsigned integer is in the range 0 to 255 decimal. A byte can be used to store one ASCII character. 4. A collection of eight bits capable of representing an alphanumeric or special character.

C

CA See constant amplitude.

cable 1. A large, strong rope made of fiber or wire. 2. A rope or chain used to restrain a vessel at its mooring. 3. A composite electrical conductor consisting of one or more solid or stranded wires usually capable of carrying relatively large currents, covered with insulation and the entire assembly encased in a protective overwrap.

cache memory A small, high-speed memory placed between the slower main memory and the processor. A cache increases effective memory transfer rates and processor speed. It contains copies of data recently used by the processor and fetches several bytes of data from memory in anticipation that the processor will access the next sequential series of bytes.

cadmium plating An electroplated coating of cadmium on a steel surface which resists atmospheric corrosion. Applications include nuts, bolts, screws, and many hardware items in addition to enclosures.

cage 1. A part in a globe valve surrounding the closure member to provide alignment and facilitate assembly of other parts of the valve trim. The cage may also provide flow characterization and/or a seating surface for globe valves and flow characterization for some plug valves [S75.05]. 2. A circular frame for maintaining uniform separation between balls or rollers in a rolling-element bearing. Also known as separator.

cage guide A valve plug fitted to the inside diameter of the cage to align the plug with the seat [S75.05].

caking Producing a solid mass from a slurry or mass of loose particles by any of several methods involving filtration, evaporation, heating, pressure, or a combination of these.

calcine 1. To heat a material such as coke, limestone or clay without fusing it, for the purpose of decomposing compounds such as carbonates and driving off volatiles such as moisture, trapped gases and water of hydration. 2. To heat a material under oxidizing conditions. 3. The product of a calcining or roasting process.

calculating action A type of control system action in which one or more feedback signals are combined with one or more actuating signals to provide an output signal which is some function of the combination.

calculation A group of numbers and mathematical symbols that is executed according to a series of instructions.

calculus of variations The theory of maxima and minima of definite integrals whose integrand is a function of the dependent variables, the independent variables, and their derivatives.

calefaction 1. A warming process. 2. The resulting warmed condition.

calender 1. To pass a material such as rubber or paper between rollers or plates to make it into sheets or to make it smooth and glossy. 2. A machine for performing such an operation.

calibrate To ascertain outputs of a device corresponding to a series of values of the quantity which the device is to measure, receive, or transmit. Data so obtained are used to a) determine the locations at which scale graduations are to be placed; b) adjust the output, to bring it to the desired value, within a specified tolerance; and c) ascertain the error by comparing the device output reading against a standard [S51.1].

calibrated airspeed The airspeed of an aircraft as read from a differential-pressure airspeed indicator that has been corrected for instrument and installation errors; the reading equals true airspeed at standard sealevel temperature and pressure.

calibrating tank A liquid vessel of known capacity which is used to check the volumetric accuracy of positive-displacement meters. Also known as meter-proving tank.

calibration 1. A test during which known values of measurand are applied to the transducer and corresponding output reading are recorded under specified conditions [S37.1]. 2. The capability to adjust the instrument to "zero" and to set the desired "span" [S12.13]. 3. The procedure used to adjust the instrument for proper response (e.g., zero level, span, alarm and range) [S12.15]. 4. Determination of the experimental relationship between the quantity being measured and the output of the device which measures it; where the quantity measured is obtained through a recognized standard of measurement.

calibration curve 1. A graphical representation of the calibration report [S51.1] [S37.1]. 2. A graph of the performance of a turbine flowmeter, showing sensitivity as the ordinate and volume flow, flowmeter frequency, or frequency divided by kinematic viscosity as the abscissa, for a liquid of specified density, viscosity, and temperature [RP31.1]. 3. A plot of indicated value versus true value used to adjust instrument readings for inherent error; a calibration curve is usually determined for each calibrated instrument in a standard procedure and its validity confirmed or a new calibration curve determined by periodically repeating the procedure.

calibration cycle 1. The application of known values of the measured variable and the recording of corresponding values of output readings, over the range of the instrument, in ascending and descending directions [S51.1]. 2. The application of known values of measurand, and recording or corresponding output readings, over the full (or specified portion of the) range of a transducer in an ascending and descending direction [S37.1]. 3. The frequency that a device is due for calibration. This cycle could be dependent on calendar, cycles, or hours.

calibration gas The known concentration(s) of hydrogen sulfide gas used to set the instrument span or alarm level(s) [S12.15].

calibration record A record of the measured relationship of the transducer output to the applied measurand over the transducer range.

calibration report A table or graph of the measured relationship of an instrument as compared over its range against a standard [S51.1].

calibration simulation provisions Electrical connections or circuitry, contained within a transducer, designed to permit the calibration of the associated measuring system by causing output changes of known magnitude without varying the applied measurand [S37.1].

calibration system A complete system consisting of liquid storage, pumps, the filters; flow, pressure, and temperature controls; the quantity measuring apparatus and the associated electronic instruments used to calibrate turbine flowmeters [RP31.1].

calibration test A test using known weights and forces to load the scale in order to determine the performance of a belt-conveyor scale [RP74.01].

calibration traceability The relationship of the calibration of an instrument through a step-by-step process to an instrument or group of instruments calibrated and certified by the National Bureau of Standards. Note: The estimated error incurred in each step must be known [S51.1] [S37.1].

calibration uncertainty The maximum calculated error in the output values, shown in a calibration record, due to causes not attributable to the transducer [S37.1].

caliper A gaging device with at least one adjustable jaw used to measure linear dimensions such as lengths, diameters and thicknesses.

call 1. To transfer control to a specified closed subroutine. 2. In communications, the action performed by the calling party, or the operations necessary to making a call, or the effective use made of a connection between two stations.

calling sequence A specified arrangement of instructions and data necessary to set up and call a given subroutine.

calorie The mean calorie is 1/100 of the heat required to raise the temperature of 1 gram of water from 0 °C to 100 °C at a constant atmospheric pressure. It is about equal to the quantity of heat required to raise one gram of water 1 °C. A more recent definition is: a calorie is 3600/860 joules, a joule being the amount of heat produced by a watt in one second.

calorific value The number of heat units liberated per unit of quantity of a fuel burned in a calorimeter under prescribed conditions.

calorimeter 1. A device for determining the amount of heat liberated during a chemical reaction, change of state or dissolution process. 2. Apparatus for determining the calorific value of a fuel. 3. An instrument or detector which measures the amount of heat in a light beam—used to measure incident radiation if the percentage of absorbed radiation is known.

calorimetric detection A detector which operates by measuring the amount of heat absorbed—incident radiation must be absorbed as heat to be detected.

calorize To produce a protective coating of aluminum and aluminum-iron alloys on iron or steel (or, less commonly, on brass, copper or nickel); the calorized coating is protective at temperatures up to about 1800 °F.

cam A machine element that produces complex, repeating translational motion in a member known as a follower that slides or rolls along a shaped surface or in a groove that is an integral part of the cam; a cam is usually a rotating plate, eccentrically mounted on an axis perpendicular to the plate surface, with the follower resting against the contoured periphery of the plate; alternatively, it may be a rotating cylinder or reciprocating plate with a groove cut into its surface for the follower to rest in, or it may be some other shape.

camber 1. Deviation from a straight line, most often used to describe a convex, edgewise sweep or curve. 2. The angle of deviation from the vertical for the steerable wheels of an automobile or truck.

camera tube An electron-beam tube in which an optical image is converted to an electron-current or charge-density image, which is scanned in a predetermined pattern to provide an electrical output signal whose magnitude corresponds to the intensity of the scanned image.

cam follower The output link of a cam mechanism.

Campbell bridge A type of a-c bridge used to measure mutual inductance of coil or other inductor in terms of a mutual inductance standard.

camshaft The rotating member that drives a cam.

cam-type timer Any of several designs of timing device using a single contoured cam to continually adjust a process parameter, such as a setpoint, or employing several cams mounted on a single timer shaft to provide interlocked sequence control of a complex operation without using relays.

can A metal vessel or container, usually cylindrical, and usually having an open top or removable cover.

candela Metric unit for luminous intensity. The unit used to express the intensity of light visible to the human eye. It corresponds to the emission from 1/60th of a square centimeter of a black body operating at the solidification temperature of platinum, and emitting one lumen per steradian.

candlepower An obsolete unit of measure for luminous intensity.

canned 1. Describing a pump or motor enclosed within a watertight casing; in the case of a motor, it is usually enclosed within the same casing as the driven element (such as a pump) and designed so that its bearings are lubricated by the pumped liquid. 2. Describing a composite billet or slab consisting of a reactive metal core encased in metal that is relatively inert so that the reactive metal may be hot worked in air by rolling, forging or extrusion without excessive oxidation.

cannibalize To disassemble or remove parts from one assembly and use the parts to repair other, like assemblies.

cantilever A beam or other structural member fixed at one end and hanging free at the other end.

capacitance The ability of a condensor to store a charge before the terminals reach a potential difference of one volt. The greater the capacitance the greater the charge that can be stored.

capacitance meter An instrument for determining electrical capacitance of a circuit or circuit element. See also microfaradmeter.

capacitive Converting a change of measurand into a change of capacitance [S37.1].

capacitive instrument A measuring device whose output signal is developed by varying the capacitive reactance of a sensitive element.

capacitor A device used for storing an electrical charge.

capacity The rate of flow through a valve under stated test conditions [S75.05].

capacity factor The ratio of the average load carried to the maximum design capacity.

capacity lag In any process, the amount of time it takes to supply energy or material to a storage element at one point in the process from a storage point elsewhere in the process. Also known as transfer lag.

capillary 1. Having a very small internal diameter. 2. A tube with a very small diameter.

capillary action 1. Spontaneous elevation or

depression of a liquid level in a fine hair-like tube when it is dipped into a body of the liquid. 2. Capillary action is induced by differences in surface energy between the liquid and the tube material.

capillary drying Progressive removal of moisture from a porous solid by evaporation at an exposed surface followed by movement of liquid from the interior to the surface by capillary action until the surface and core reach the same stable moisture concentration.

capillary tube A tube sufficiently fine that capillary action is significant.

cap screw A threaded fastener similar to a bolt, but generally used without a nut by threading it into a tapped hole in one part of an assembly.

capstan A vertical-axis drum used for pulling or hauling; it may be power driven or it may be turned manually by means of a bar extending radially from a hole in the drum.

capsule A pressure sensing element consisting of two metallic diaphragms joined around their peripheries [S37.1].

carbide tool A cutting tool whose working edges and faces are made of tungsten, titanium or tantalum carbide particles, compacted and sintered into a hard, heat-resistant and wear-resistant solid by powder metallurgy; the heat-resistant properties of the material are derived in part from a matrix alloy, usually cobalt, which cements the carbide particles together.

carbon An element; the principal combustible constituent of all fuels.

carbon equivalent An empirical relationship that is used to estimate the ability to produce gray cast iron, or one that is used to rate weldability of alloy steels; for cast iron, the formula is $CE=TC+1/3(Si+P)$, where CE is the carbon equivalent, TC is the total carbon content, Si is the silicon content and P is the phosphorus content, all in wt%; for weldability, the formula is $CE=C+Mn/6+(Cr+Mo+V)/5+(Ni+Cr)/15$ where each symbol stands for the concentration of the indicated element in wt%.

carbonitriding A surface-hardening process in which a suitable ferrous material is heated at a temperature above the lower transformation temperature in an atmosphere that will cause simultaneous absorption of carbon and nitrogen at the surface and, by diffusion, create a concentration gradient; final properties are achieved by controlled cooling from temperature, and sometimes by subsequent tempering.

carbonization The process of converting coal to carbon by removing other ingredients.

carbon loss The loss representing the unliberated thermal energy occasioned by failure to oxidize some of the carbon in the fuel.

carbon-pile pressure transducer A resistive-type pressure transducer that depends for its operation on the change in resistance that occurs when irregular carbon granules or smooth carbon disks are pressed together; because of its low resistance, it can often provide sufficient output current to actuate electrical instruments without amplification.

carbon potential A measure of the ability of an environment to alter or maintain the surface-carbon content of ferrous alloys; the specific effect that occurs depends on temperature, time and steel composition as well as on carbon potential.

carbon steel An alloy of carbon and iron containing not more than 2% carbon, and which does not contain alloying elements other than a small amount of manganese.

carburetor A component of a spark-ignition internal combustion engine that mixes fuel with air, in proper proportions, and delivers a controlled quantity of the mixture to the cylinders.

carburizing A surface-hardening process in which a suitable ferrous material is heated at a temperature above the transformation range in the presence of carbon-rich environment, which may be produced from solid carbon, vaporized liquid hydrocarbons or gaseous hydrocarbons; following production of a carbon concentration gradient in the alloy, it is either quenched from the carburizing temperature and tempered, or reheated, quenched and tempered, to achieve desired properties in both the carbon-rich outer case and the carbon-lean inner core.

card A circuit board within a computer or other electronic instrument or system.

card hopper See hopper, card.

card reader (Hollerith cards) A hardware device for reading computer-standard punched cards for computer entry.

card stacker See stacker, card.

carriage 1. A mechanism that moves along a predetermined path in a machine to carry and position another component. 2. A mechanism designed to hold paper in the active portion of a printing or typing machine, and to advance the paper as necessary; sometimes, the mechanism also provides for automatically feeding new sheets of paper on demand.

carriage bolt   A threaded fastener with a plain (unslotted) head and a square shoulder below the head which keeps the bolt from turning as the nut is tightened; this type of bolt is designed primarily for bolting wood members, but can be used with metal members if the one next to the bolt head has a square bolt hole to accommodate the bolt's shoulder.

carriage return   The operation that causes printing to be returned to the left margin with or without line advance. Sometimes used to signify completion of manual data entry.

carriage return character (CR)   A format effector that causes the location of the printing or display position to be moved to the left margin with or without line advance.

carriage stop   A device attached to the outer way of a lathe bed which permits accurate and repeatable positioning of the tool carriage for cutting grooves, turning multiple diameters and lengths, and cutting off pieces of specific lengths.

carrier   A continuous frequency signal capable of being modulated to carry information.

carrier band   A single channel signaling technique in which the digital signal is modulated on a carrier and transmitted.

carrier frequency   The basic frequency or pulse repetition rate of a transmitted signal, bearing no intrinsic intelligence until it is modulated by another signal that does bear intelligence.

carrier sense multiple access with collision detect (CSMA/CD)   An access procedure where a device with data to transmit first listens to the medium. When the medium is not busy, the device starts transmitting. While the device is transmitting, it listens for collisions (simultaneous transmission by another station). If a collision occurs, the node stops transmitting, waits, and tries again.

carrier-to-noise ratio   Carrier amplitude divided by noise amplitude or carrier power divided by noise power.

carryover   The chemical solids and liquid entrained with the steam from a boiler.

cartridge   A small unit used for storing computer

programs or data values. The amount of information stored tends to be small and access times large compared to discs. However, they are widely used in hobbyist applications.

cartridge disk   A relatively low-capacity data- or program-storage medium; generally removable.

cartridge tape   Small magnetic tape for digital program storage; stores discrete records.

CASA/SME   See Computer and Automated Systems Association of the Society of Manufacturing Engineers.

cascade control   1. Control action in which the output of one controller is the set point for another controller [S77.42] 2. A control system composed of two loops where the setpoint of one loop (the inner loop) is the output of the controller of the other loop (the outer loop). 3. A control technique that incorporates a master and a slave loop. The master loop controls the primary control parameters and establishes the slave-loop set point. The purpose of the slave loop is to reduce the effect of disturbances on the primary control parameter and to improve the dynamic performance of the loop. See control, cascade.

cascade control action   Control action where the output of one controller is the setpoint for another controller.

cascaded   Describing a series of machines, devices, or machine elements, so arranged that the output from one feeds directly into the next.

cascade sequence   A welding sequence in which a continuous multiple-pass weld is built up by depositing weld beads in overlapping layers; usually, weld beads are laid in a backstep sequence, starting with the root bead that extends only part way along the joint length, then starting successive beads a short distance farther along the joint from the start of the previous bead. Compare with block sequence.

case   1. An enclosure designed to hold one or more components in a fixed position, usually by nestling into a conforming recess or resting on fixed supports; in some instances, components are attached directly to the enclosure; the entire unit may be kept in storage or taken to a jobsite, and the contents removed as needed; sometimes as with certain portable instruments, the contents can be used by merely opening the cover of the enclosure and making appropriate connections to the device inside. 2. A hardened outer layer on a ferrous alloy produced by suitable heat treatment, which sometimes involves altering the chemical composition of the outer layer before hardening.

CASE See Common Applications Service Elements.

case hardening Producing a hardened outer layer on a ferrous alloy by any of several surface-hardening processes, including carburizing, carbonitriding, nitriding, flame hardening and induction hardening. Also known as surface hardening.

case pressure See burst pressure rating, proof pressure, or reference pressure [S37.1].

casing A covering of sheets of metal or other material such as fire resistant composition board used to enclose all or a portion of a steam generating unit.

cassette 1. A light-tight container for holding photographic or radiographic film, or a photographic plate, and positioning it within a camera or other device for exposure. 2. A small, compact container holding magnetic tape along with supply and takeup reels so that it can be inserted and removed as a unit for quick loading and unloading of a tape recorder or playback machine; different sizes and styles of tape and cassette are used for audio, video and computer applications, depending on the hardware being used.

cassette tape Magnetic tape for digital data storage.

cast 1. To produce a solid shape from liquid or semisolid bulk material by allowing it to harden in a mold. 2. A tinge of a specific color; a slight overtint of a color different from the main color—for instance, white with a bluish cast.

castellated nut A hexagonal nut with a slotted cylindrical projection above one of the hexagonal sides; it is used in conjunction with a cotter pin or safety wire that passes through a lateral hole in the bolt or stud which is aligned with two of the slots in the nut; the cotter pin or safety wire keeps the nut from turning so the joint stays tight.

caster 1. The fore-and-aft angle of deviation from the vertical of the kingpin (or its equivalent) in an automobile or truck steering gear. 2. A wheel, usually small in diameter, which is mounted so it is free to swivel about a vertical axis; it is commonly used to support hand trucks, machinery or furniture.

casting 1. The process of making a solid shape by pouring molten metal into a cavity, or mold, and allowing it to cool and solidify. 2. A near-net-shape object produced by this process; a rough casting, cylindrical, square or rectangular in cross section and intended for subsequent hot working or remelting, is called an ingot.

casting alloy An alloy having suitable fluidity when molten and having suitable solidification characteristics to make it capable of producing shape castings; most casting alloys are not suitable for rolling or forging and can only be shaped by casting.

castings A designation for low-quality drill diamonds.

casting shrinkage 1. Total reduction in volume due to the three stages of shrinkage—during cooling from casting temperature to the liquidus, during solidification, and during cooling from the solidus to room temperature. 2. Reduction in volume at each stage in the solidification of a casting.

casting slip A slurry of clay and additives suitable for casting into molds to make unfired ceramic products.

casting wheel A large turntable with molds positioned around its periphery so that each can be moved, in turn, into position for receiving molten metal.

cast iron Any iron-carbon alloy containing at least 1.8% carbon and suitable for casting to shape.

catalog In data processing, the contents of a computer disk or tape.

cataphoresis Movement of suspended solid particles in a liquid medium due to the influence of electromotive force.

catastrophic failure 1. A sudden failure that occurs without prior warning, as opposed to a failure that occurs gradually by degradation. 2. Failure of a mechanism or component that renders an entire machine or system inoperable.

catenary The shape produced by holding a rope or cable at its ends and allowing the center section to sag under its own weight.

cathetometer An optical instrument for measuring small differences in height—for instance, the difference in height between two columns of mercury.

cathode 1. The metal plate or surface that acts as an electron acceptor in an electrochemical circuit; metal ions in an electrolytic solution plate on the cathode during electroplating, and hydrogen may be formed at the cathode during electroplating or electrochemical corrosion. 2. The positive electrode in a storage battery, or the negative electrode in an electrolytic cell. 3. The negative electrode in an x-ray tube or vacuum tube, where electrons enter the interelectrode space.

cathode corrosion 1. Corrosion of the cathode in an electrochemical circuit, usually involving the production of alkaline corrosion products. 2. Corrosion of the cathodic member of a galvanic couple.

cathode follower A type of electronic circuit in which the output load is connected in the cathode circuit of an electron tube or equivalent transistor and the input signal is impressed across a terminal pair where one is connected directly to the control grid and the other to the remote end of the output load.

cathode ray In an electron tube or similar device, a stream of electrons emitted by the cathode.

cathode-ray oscillograph An instrument that produces a record of a waveform by photographing its graph produced on a cathode-ray tube, or by otherwise recording such an image.

cathode-ray oscilloscope An instrument that indicates the shape of a waveform by producing its graph on the screen of a cathode-ray tube.

cathode ray tube (CRT) 1. An electronic vacuum tube containing a screen on which information may be stored for visible display by means of a multigrid modulated beam of electrons from the thermionic emitter, storage is effected by means of charged or uncharged spots. 2. A storage tube. 3. An oscilloscope tube. 4. A picture tube. 5. A computer terminal using a cathode ray tube as a display device.

cathodic coating A mechanical plate or electrodeposit on a base metal, with the coating being cathodic to the underlying base metal.

cathodic protection Preventing electrochemical corrosion of a metal object by making it the cathode of a cell using either a galvanic or impressed current.

cathodoluminescence Luminescence induced by exposure of a suitable material to cathode rays.

CATV Community Antenna Television. See broadband.

caulk 1. A heavy paste such as a mixture of a synthetic or rubber compound and a curing agent, or a natural product such as oakum, used to seal cracks or seams and make them airtight, steamtight or watertight. Also known as caulking compound; calk. 2. To seal a crack or seam with caulk.

caustic dip A strongly alkaline solution for immersing metal parts to etch them, to neutralize an acid residue, or to remove organic material such as grease or paint.

caustic embrittlement Intergranular cracking of carbon steel or Fe-Cr-Ni alloy exposed to an aqueous caustic solution at a temperature of at least 150 °F while stressed in tension; a form of stress-corrosion cracking. Also known as caustic cracking.

caustic soda The most important of the commercial caustic materials—it consists of sodium hydroxide that contains 76 to 78% sodium oxide.

Cavendish balance A torsional instrument for determining the gravitational constant by measuring the displacement of two spheres of known small mass, mounted on opposite ends of a thin rod suspended on a fine wire, when two spheres of known large mass are brought near the small spheres.

cavitation A two-stage phenomenon of liquid flow. The first stage is the formation of voids or cavities within the liquid system; the second stage is the collapse or implosion of these cavities back into an all-liquid state [S75.05].

cavitation erosion Progressive removal of surface material due to localized hydrodynamic impact forces associated with the formation and subsequent collapse of bubbles in a liquid in contact with the damaged surface. Also known as cavitation damage; liquid-erosion failure.

cavity resonator A space normally enclosed by an electrically conducting surface, which is used to store electromagnetic energy and whose resonant frequency is determined by the shape of the enclosure.

cavity-type wavemeter An instrument used to determine frequency in a waveguide system; typically, the position of a piston inside a cylindrical cavity is tuned to resonance, which is determined by a drop in transmitted power; the meter is then detuned for normal operation.

CBW See constant bandwidth.

CCR See control complexity ratio.

CD Constant delay. See bessel.

CD-ROM A compact disk used for computer data

storage. The letters stand for Compact Disk Read-Only Memory.

ceilometer A recording instrument for automatically determining cloud heights.

cell 1. One of a series of chambers in which a chemical or electrochemical reaction takes place—for example, the chambers of a storage battery or electrolytic refining bath. 2. One of the cavities in a honeycomb structure. 3. The storage of one unit of information, usually one character or one word. 4. A location specified by whole or part of the address and possessed of the faculty of store. Specific terms such as column, field, location, and block, are preferable when appropriate. See storage cell.

Celsius A scale for temperature measurement based on the definition of 0 °C and 100 °C as the freezing point and boiling point, respectively, of pure water at standard pressure.

cement 1. A dry, powdery mixture of silica, alumina, magnesia, lime and iron oxide which hardens into a solid mass when mixed with water; it is one of the ingredients in concrete and mortar. 2. An adhesive for bonding surfaces where intimate contact cannot be established and the adhesive must fill a gap over all or part of the faying surfaces.

cementation 1. High temperature impregnation of a metal surface with another material. 2. Conversion of wrought iron into steel by packing it in charcoal and heating it at about 1800 °F for 7 to 10 days.

cemented carbide A powder-metallurgy product consisting of granular tungsten, titanium or tantalum carbides in a temperature-resistant matrix, usually cobalt; used for high-performance cutting tools, punches and dies; the proportion of matrix material is small compared to the amount of carbide.

cent The interval between two sound frequencies, where the ratio of the two frequencies is the twelve-hundredth root of 2. Also equal to one-hundredth of a semi-tone.

center gage A gage used to check angles, such as the angle of a cutting-tool point or screw thread.

center of seismic mass The point within an acceleration transducer where acceleration forces are considered to be summed [S37.1].

center-to-end dimension The distance from the center line of a valve body to the extreme plane of a specific end connection. See face-to-face dimension and end-to-end dimension [S75.05].

centigrade A nonpreferred term formerly used to designate the scale now referred to as the Celsius scale.

centimetre-gram-second (CGS) A standard metric system of units used largely for scientific work prior to adoption of the international SI system currently preferred for both scientific and engineering work.

centralized To bring under one control.

centralized maintenance shops One maintenance shop that has responsibility to maintain all equipment in the facility. Usually several crafts work out of this one centralized maintenance shop.

central processing unit (CPU) 1. The brain of the computing machine, usually defined by the arithmetic and logic units (ALU) plus a control section, often called a processor. 2. The part of a computing system that contains the arithmetic and logical units, instruction control unit, timing generators, and memory and I/O interfaces. See also unit, central processing.

central station A power plant or steam heating plant generating power or steam for sale.

centrifugal fan Consists of a fan rotor or wheel within a housing discharging the air at right angle to the axis of the wheel.

centrifugal force A force acting in a direction along and outward on the radius of turn for a mass in motion.

centrifugal tachometer An instrument that measures the instantaneous angular speed of a rotating member such as a shaft by measuring the centrifugal force on a mass that rotates with it.

centrifuge A rotating device that separates suspended fine or colloidal particles from a liquid, or separates two liquids of different specific gravities, by means of centrifugal force.

ceramic 1. A heat-resistant natural or synthetic inorganic product made by firing a nonmetallic mineral. 2. A shape made by baking or firing a ceramic material, such as brick, tile or labware.

ceramic coating A protective coating made by thermal spraying a material such as aluminum or zirconium oxide, or by cementation of a material such as aluminum disilicide, on a metal substrate.

ceramic tool A cutting tool made from fused, sintered or cemented metallic oxides.

ceramic transducer See electrostriction transducer.

Cerenkov radiation Visible light produced when charged particles pass through a transparent medium at a speed exceeding the speed of light in the medium.

cermet A powder-metallurgy product consisting of ceramic particles bonded together with a metal matrix.

certification 1. The act of certifying. 2. The state of being certified. 3. Certification of instrumentation and control technicians in nuclear power plants. This criteria address qualifications based on education, experience, training, and job performance. 4. The attainment of certification is a means for individuals to indicate to the general public, co-workers, employers, and others that an impartial, nationally-recognized organization has determined that they are qualified to perform specific technical tasks by virtue of their technical knowledge and experience. 5. Certification bestows a sense of achievement upon the certificant, since it reflects professional advancement in a chosen field.

certify 1. To confirm formally as true, accurate, or genuine. 2. To guarantee as meeting a standard. 3. To issue a license or certificate.

CFR engine Cooperative Fuel Research engine—a standard test engine for determining the octane number of motor fuels.

CGS See centimetre-gram-second.

chad The piece of material removed when forming a hole or notch in a storage medium such as punched tape or punched cards [RP55.1].

chafing fatigue See fretting.

chain 1. A nonrigid series of metal links or rings that are interlinked with each other, or are pinned or otherwise held together, to make an elongated flexible member suitable for pulling, hauling, lifting, supporting or restraining objects, or for transmitting power. 2. A mesh of rods or plates used in place of a belt to convey objects or transmit power. 3. An organization in which records or other items of data are strung together by means of pointers.

chain-balanced density meter A submerged-float meter, using an iron-core float that moves up and down within a pickup coil; a slack chain attached to the bottom of the float applies more weight as the float rises and establishes a definite equilibrium position for any given fluid density within the range of the instrument.

chain block A lifting tackle, often suspended from an overhead track, which uses a chain instead of rope to lift heavy weights and is hand driven by pulling on an endless chain; some models are power driven. Also known as chain fall; chain hoist.

chain drive A device for transmitting power and motion without slipping which consists of an endless chain that meshes with driving and driven sprockets; chain drives are used on bicycles and motorcycles to provide the motive power, on conveyors to drive the belts and in hoisting mechanisms to provide the lifting power.

chain-float liquid-level gage A device for indicating liquid level in a tank which consists of a float connected to a counterweight by a chain running over a sprocket; as the float rises and falls with liquid level in the tank, the chain rotates the sprocket which in turn positions a pointer to indicate liquid level.

chalking A defect of coated metals caused by formation of a layer of powdery material at the metal-coating interface.

chamfer 1. A beveled edge that relieves an otherwise sharp corner. 2. A relieved angular cutting edge at a tooth corner on a milling cutter or similar tool.

channel 1. A path along which signals can be sent, e.g., data channel, output channel [RP55.1]. 2. The portion of a storage medium that is accessible to a given reading station [RP55.1]. 3. In communication, a means of one way transmission. Contrast with circuit [RP55.1]. 4. Sometimes called a point [RP55.1]. 5. A collection of instrument loops, including their sensing lines, that may be treated or routed as a group while being separated from instrument loops assigned to other redundant channels [S67.02][S67.10]. 6. An arrangement of components and modules as required to generate a single protective action signal when required by a generating station condition. A channel loses its identity where single-action signals are combined [S67.06]. 7. A path along which information, particularly a series of digits or characters, may flow. 8. One or more parallel tracks treated as a unit. 9. In a circulating storage, a channel is one recirculating path containing a fixed number of words stored serially by word. 10. A path for electrical communication. 11. A band of frequencies used for communication.

channel, input The analog data path between the field wiring connector or termination strip and the analog-to-digital converter or other quantizing device used in the subsystem. In typical subsystems, this path may include a filter, an analog signal multiplexer, and one or more amplifiers [RP55.1].

channel sampling rate The number of times a given data input is sampled during a specified time interval.

channel selector In an FM discriminator, the plug-in module which causes the device to select one of the channels and demodulate the subcarrier to recover data.

CHAR See character.

character 1. A term used to refer to a predefined group of pixels [S5.5]. 2. One symbol of a set of elementary symbols such as those corresponding to the keys on a typewriter. The symbols usually include the decimal digits 0 through 9, the letters A through Z, punctuation marks, operation symbols, and any other single symbols which a computer may read, store, or write [RP55.1]. 3. The electrical, magnetic, or mechanical profile used to represent a character in a computer, and its various storage and peripheral devices. A character may be represented by a group of other elementary marks, such as bits or pulses [RP55.1].

character codes The binary code patterns used to create characters in a computer.

character embossing A raising of the printing medium surface within the perimeter of a printer character caused by the impact of a type element against printing medium [RP55.1].

characteristic 1. The integral part of a common logarithm, i.e., in the logarithm 2.5, the characteristic is 2, the mantissa is 0.5. 2. Sometimes, that portion of a floating point number indicating the exponent. 3. A distinctive property of an individual, document, item, etc.

characteristic curve 1. A graph (curve) which shows the ideal values at steady-state, or an output variable of a system as a function of an input variable, the other input variables being maintained at specified constant values. When the other input variables are treated as parameters, a set of characteristic curves is obtained [S51.1]. 2. Of a photographic or radiographic film, the graph of relative transmittance of the emulsion versus exposure, or a graph of func-

tions of these two quantities. Also known as characteristic emulsion curve.

characteristic, equal percentage The inherent flow characteristic which, for equal increments of rated travel, will ideally give equal percentage changes of the existing flow coefficient (C_v) [S75.05].

characteristic, flow Indefinite term; see characteristic, inherent flow and characteristic, installed flow [S75.05].

characteristic, inherent flow The relationship between the flow rate through a valve and the travel of the closure member as the closure member is moved from the closed position to rated travel with constant pressure drop across the valve [S75.05].

characteristic, installed flow The relationship between the flow rate through a valve and the travel of the closure member as the closure member is moved from the closed position to rated travel when the pressure drop across the valve varies as influenced by the system in which the valve is installed [S75.05].

characteristic, linear flow An inherent flow characteristic which can be represented by a straight line on a rectangular plot of flow coefficient (C_v) versus percent rated travel. Therefore, equal increments of travel provide equal increments of flow coefficient (C_v) at constant pressure drop [S75.05].

characteristic, modified parabolic flow An inherent flow characteristic which provides fine throttling action at low valve plug travel and approximately a linear characteristic for upper portions of valve travel. It is approximately midway between linear and equal percentage [S75.05].

characteristic, quick opening flow An inherent flow characteristic in which there is a maximum flow with minimum travel [S75.05].

characterized cam A component in a valve positioner used to relate the closure component position to the control signal [S75.05].

characterized sleeve A part added to a plug valve to provide various flow characteristics [S75.05].

characterized trim Control valve trim that provides a predefined flow characteristic [S75.05].

charge 1. A defined quantity of an explosive. 2. The starting stock loaded into a batch process. 3. Material loaded into a furnace for melting or heat treating. 4. A measure of the accumulation or depletion of electrons at any given point. 5. The amount of substance loaded into a closed system, such as refrigerant into a refrigeration system. 6. The quantity of excess protons (positive charge) or excess electrons (negative charge) in a physical body, usually expressed in coulombs.

chart recorder A device for automatically plotting a dependent variable against an independent variable; the dependent variable is proportional to the input signal from a transducer; the independent variable may be proportional to a transducer signal, also, but is most often time or a time-dependent variable that can be produced by controlling the rate of advance of rolled chart paper.

chase 1. A vertical passage in a building that contains the pipes, wires and ducts which provide heat, ventilation, electricity, running water, drains and other building services. 2. The main body of a mold that contains one or more mold cavities. 3. To make a series of cuts, each following the path of a preceding cut, such as is done to produce a thread in lathe turning using a single-point tool. 4. To straighten and clean damaged or debris-filled threads on a screw or pipe end.

chassis 1. A frame or box-like sheet-metal support for mounting the components of an electronic device. 2. A frame for a wheeled vehicle that provides most of the stiffness and strength of the vehicle body, and supports the body, engine, and passenger or load compartment on the running gear.

CHEAPERNET An IEEE 802.3 standard for a low cost, 10 Mbit LAN, compatible with TOP.

check A process of partial or complete testing of the correctness of machine operations. The existence of certain prescribed conditions within the computer, or the correctness of the results produced by a program. A check of any of these conditions may be made automatically by the equipment or may be programmed. Related to marginal check.

check bit A binary check digit; often a parity bit. Related to parity check.

check digit In data transmission, one or more redundant digits appended to a machine word, and used in relation to the other digits in the word to detect errors in data transmission.

checker work An arrangement of alternately spaced brick in a furnace with openings through which air or gas flows.

checking A network of fine cracks in a coating or at the surface of a metal part; they may appear during processing but are more often associated with service, especially when it involves thermal cycling.

checkout 1. Determination of the working condition of a system. 2. A test or preliminary operation intended to determine whether a component or system is ready for service or ready for a new phase of operation.

check, parity A check that tests whether the number of ones (or zeros) in an array of binary digits is odd or even. Synonymous with odd-even check [RP55.1].

checkpoint A point in time in a machine run at which processing is temporarily halted, to make a record of the condition of all the variables of the machine run, such as the status of input and output devices and a copy of working storage. Check points are used in conjunction with a restart routine to minimize reprocessing time occasioned by functional failures. A checkpoint also may be a particular point in a program at which processing is halted for checking.

check problem A problem used to test the operation of a computer or to test a computer program; if the result given by the computer does not match the known result, it indicates an error in programming or operation.

checksum 1. A routine for checking the accuracy of data transmission by dividing the data into small segments, such as a disk sector, and computing a sum for each segment. 2. Entry at the end of a block of data corresponding to the binary sum of all information in the block. Used in error-checking procedures.

check, validity A check based upon known limits or upon given information or computer results, e.g., a calendar month will not be numbered greater than 12, and a week will not have more than 168 hours [RP55.1].

check valve A flow control device that permits flow in one direction and prevents flow in the opposite direction.

chemical affinity 1. The relative ease with which two elements or compounds react with each other to form one or more specific compounds. 2. The ability of two chemical elements to react to form a stable valence compound.

chemical analysis Determination of the principal chemical constituents.

chemical conversion coating A decorative or protective surface coating produced by inducing a chemical reaction between surface layers of a part and a specific chemical environment, such as in chromate treatment or phosphating.

chemical engineering A branch of engineering that deals with the design, operation and maintenance of plants and equipment for chemically converting raw materials into bulk chemicals, fuels and other similar products through the use of chemical reaction, often accompanied by a change in state or in physical form.

chemical feed pipe A pipe inside a boiler drum through which chemicals for treating the boiler water are introduced.

cherry picker Any of several types of small traveling cranes, especially one consisting of an open passenger compartment at the free end of a jointed boom.

chimney A brick, metal or concrete stack.

chimney core The inner cylindrical section of a double wall chimney, which is separated from the outer section by an air space.

chimney lining The material which forms the inner surface of the chimney.

chip Single large scale integrated circuit.

chip breaker An attachment or a relieving channel behind the cutting edge of a lathe tool to cause removed stock to break up into pieces rather than to come off as long, unbroken curls.

chipping 1. Using a manual or pneumatic chisel to remove seams, surface defects or excess metal from semifinished mill products. 2. Using a hand or pneumatic hammer with chisel-shaped or pointed faces to remove rust, scale or other deposits from metal surfaces.

choke A valve which increases suction to draw in an excess proportion of fuel and facilitate starting a cold internal combustion engine.

choke coil An inductor that allows direct current to pass but presents relatively large impedance to alternating current.

choked flow The condition that exists when, with the upstream conditions remaining constant, the flow through a valve cannot be further increased by lowering the downstream pressure.

chopper Any device for periodically interrupting a continuous current or flux.

Christiansen filter A device for admitting monochromatic radiation to a lens system; it consists of coarse powder of a transparent solid confined between parallel windows, with the spaces between particles being filled with a liquid whose refractive index is the same as that of the powder for a certain wavelength; only that wavelength is transmitted by the filter without deviation.

chromadizing Improving paint adhesion on aluminum and its alloys by treating the surface with chromic acid.

chromate treatment Applying a solution of hexavalent chromic acid to produce a protective conversion coating of trivalent and hexavalent chromium compounds.

chromatic aberration The focusing of light rays of different wavelengths at different distances from the lens. This is not a significant effect with a single wavelength laser source, but can be when working at different or multiple wavelengths.

chromaticity The color quality of light defined by the combination of its dominant wavelength and purity or by its chromaticity coordinates [S5.5].

chromaticity coordinate The ratio of any of the three tristimulus values of a color sample to the sum of the tristimulus values.

chromaticity diagram A graph of one of the chromaticity coordinates against another.

chromatography An instrumental procedure for separating components from a mixture of chemical substances which depends on selective retardation and physical absorption of substances by a porous bed of sorptive media as the substances are transported through the bed by a moving fluid; the sorptive bed (stationary phase) may be a solid or a liquid dispersed on a porous, inert solid; the moving fluid (moving phase) may be a liquid solution of the substances or a mixture of a carrier gas and the vaporized sample; a wide variety of detection techniques are used, some of which can be automated or microprocessor driven.

chrome See Munsell chroma.

chromium plating Electrodeposition of either a bright, reflective coating or a hard, less-reflective coating of chromium on a metal surface. Also known as chrome plating; chromium coating.

chromizing Producing an alloyed layer on the surface of a metal by deposition and subsequent diffusion of metallic chromium.

chromophore The group of atoms within a

molecule that contributes most heavily to its light-absorption qualities.

chronograph An instrument used to record the time at which an event occurs or the time interval between two events.

chronotron A device for measuring elapsed time between two events in which the time is determined by measuring the position of the superimposed loci of a pair of pulses initiated by the events.

Ci See curie.

CID See computer interface device.

cinder A particle of gas borne partially burned fuel larger than 100 microns in diameter.

cinder trap A dust collector having staggered elements in the gas passage which concentrates larger dust particles. A portion of the gas passes through the elements with the concentrated dust into a settling chamber, where change in direction and velocity drops out coarser particles.

Cipolletti weir An open-channel flow-measurement device similar to a rectangular weir but having sloping sides, which results in a simplified discharge equation.

circle of confusion A circular image in the focal plane of an optical system which is the image formed by that system of a distant point object.

circuit Any group of related electronic paths and components which electronic signals will pass to perform a specific function.

circuit analyzer A multipurpose assembly of several instruments or instrument circuits in one housing which are to be used in measuring two or more operating characteristics of an electronic circuit.

circuit breaker A device designed to open and close a circuit by non-automatic means and also to open the circuit automatically on a predetermined overload of current without injury to itself.

circuit diagram A line drawing of an electronic/electrical system which identifies components and diagrams how they are connected.

circuit-noise meter An instrument which uses frequency-weighting networks and other components to measure electronic noise in a circuit, giving approximately equal readings for noises that produce equal levels of interference.

circuit-to-ground voltage The rated value of voltage with respect to earth ground [S82.01].

circular-chart recorder A type of recording instrument where the input signal from a temperature, pressure, flow or other transducer moves a pivoted pen over a circular piece of chart paper that rotates about its center at a fixed rate with time.

circularity In data processing, a warning message that the commands for two separate but interdependent cells in a program cannot proceed until a value for one of the cells is determined.

circularly polarized light Light in which the polarization vector rotates periodically, but does not change magnitude, describing a circle. It can also be stated as the superposition of two plane-polarized (or linearly polarized) lightwaves, of equal magnitude, one 90° in phase behind the other.

circular mil A wire-gage measurement equal to the cross-sectional area of a wire one mil (0.001 in.) in diameter; actual area is 7.8540×10^{-7} in^2.

circular polarized wave An electromagnetic wave for which the electric field vector, magnetic field vector, or both, describe a circle.

circulating memory In an electronic memory device, a means of delaying information combined with a means for regenerating the information and reinserting it into the delaying means.

circulation The movement of water and steam within a steam generating unit.

circulation ratio The ratio of the water entering a circuit to the steam generated within that circuit in a unit of time.

circulator A pipe or tube to pass steam or water between upper boiler drums usually located where the best absorption is low. Also used to apply to tubes connecting headers of horizontal water tube boilers with drums.

cladding 1. Covering one piece of metal with a relatively thick layer of another metal and bonding them together; the bond may be produced by corolling or coextrusion at high temperature and pressure, or by explosive bonding. 2. The low refractive index material which surrounds the core of a fiber and protects against surface contaminant scattering.

cladding strippers Chemicals or devices which remove the cladding from an optical fiber to expose the light-carrying core. The term might sometimes be misapplied to chemicals or devices which remove the protective coating applied over cladding to protect the fiber from the environmental stress.

clamping circuit A circuit which maintains

either the maximum or minimum amplitude level of a waveform at a specific potential.

clamping plate A plate for attaching a mold to a plastics-molding or die-casting machine.

clamping pressure In die casting, injection molding and transfer molding, the force (or pressure) used to keep the mold closed while it is being filled.

clasp A nonthreaded fastener, usually hook-like, with a releasable catch.

class A amplifier An amplifier in which the grid bias and alternating grid voltages are such that plate current always flows in a specified tube.

class AB amplifier An amplifier in which the grid bias and alternating grid voltages are such that plate current in a specified tube flows considerably more than one-half but less than the entire electrical cycle.

class B amplifier An amplifier in which the grid bias is approximately equal to the cutoff value, therefore making the plate current in a specified tube approximately zero when the grid voltage is zero.

class C amplifier An amplifier in which the grid bias is considerably more negative than the zero plate current value.

Class I location A location in which flammable gases or vapors are or may be present in the air in quantities sufficient to produce ignitible mixtures.

Class II location A location that is hazardous because of the presence of combustible dust.

Class III location A location in which easily ignitible fibers or materials producing combustible flyings are handled, manufactured or used.

classification 1. The assignment of a hazard rating such as Division 1, Division 2 or non-hazardous [S12.11]. 2. Sorting particles or objects by specific criteria, such as size or function. 3. Separating a mixture into its constituents, such as by particle size or density. 4. Segregating units of product into various adjoining categories, often by measuring characteristics of the individual units, thus forming a spectrum of quality. Also termed grading.

classification of a location The assignment of a rating such as Division 1, Division 2, or non-hazardous.

clay atmometer A simple device for determining evaporation rate to the atmosphere, which consists of a porous porcelain dish connected to a calibrated reservoir filled with distilled water.

clean air 1. Air that is free of combustible gases and contaminating substances [S12.13]. 2. Air that is free of any substance that will adversely affect the operation or cause a response of the instrument [S12.15].

cleanout door A door placed so that accumulated refuse may be removed from a boiler setting.

cleanup 1. Removing small amounts of stock by an imprecise machining operation, primarily to improve surface smoothness, flatness or appearance. 2. The time required for an electronic leak-testing instrument to reduce its output signal to 37% of the initial signal transmitted when tracer gas is first detected. 3. The gradual disappearance of internal gases during operation of a discharge tube.

clear To erase the contents of a storage device by replacing the contents with blanks, or zeros. Clarified by erase.

clearance 1. The distance between two adjacent parts that do not touch. 2. Unobstructed space for insertion of tools or removal of parts during maintenance or repair.

clearance distance The shortest distance measured in air between conductive parts [S82.01].

clearance fit A type of mechanical fit in which the tolerance envelopes for mating parts always results in clearance when the parts are assembled.

clearance flow That flow below the minimum controllable flow with the closure member not seated [S75.05].

cleaver A device used to cut or break optical fibers in a precise way so the ends can be connected with low loss.

clevis A U-shaped metal fitting with holes at the open ends of the legs for insertion of a pin or bolt to make a closed link for attaching or suspending a load.

clinical thermometer A thermometer for accurately determining the temperature of the human body; most often, it is a mercury in glass maximum thermometer.

clinker A hard compact congealed mass of fused furnace refuse, usually slag.

clinometer A divided-circle instrument for determining the angle between mutually inclined surfaces.

clipboard In data processing, an area of information can be stored in order to use it later in a different application.

clipping circuit 1. A circuit that prevents the peak amplitude of a signal from exceeding some specific level. 2. A circuit that eliminates the tail of a signal pulse after some specific time. 3. A circuit element in a pulse amplifier that reduces the pulse amplitude at frequencies less than some specific value.

CLK See clock.

clock (CLK) 1. A master timing device used to provide the basic sequencing pulses for the operation of a synchronous computer. 2. A register which automatically records the progress of real time, or perhaps some approximation to it, records the number of operations performed, and whose contents are available to a computer program. 3. A timing pulse that coincides with or is phase-related to the occurrence of an event, such as bit rate or frame rate.

clock frequency The master frequency of periodic pulses which schedules the operation of the computer.

clock mode A system circuit that is synchronized with a clock pulse, that changes states only when the pulse occurs, and will change state no more than once for each clock pulse.

clock pulse A synchronization signal provided by a clock.

clock rate The time rate at which pulses are emitted from the clock. The clock rate determines the rate at which logical arithmetic gating is performed with a synchronous computer.

clock, real time A clock which indicates the passage of actual time, in contrast to a fictitious time set up by the computer program; such as, elapsed time in the flight of a missile, where in a 60-second trajectory is computed in 200 actual milliseconds, or a 0.1 second interval is integrated in 100 actual microseconds [RP55.1].

clock skew A phase shift between the clock inputs of devices in a single clock system; the result of variations in gate delays and stray capacitance in a circuit.

clone In data processing, an exact duplication of another computer device or software.

closed circuit 1. Any device or operation where all or part of the output is returned to the inlet for further processing. 2. A type of television system that does not involve broadcast transmis-sion, but rather involves transmission by cable, telephone lines or similar method.

closed die A forming or forging operation in which metal flow takes place only within the die cavity.

closed-fireroom system A forced draft system in which combustion air is supplied by elevating the air pressure in the fireroom.

closed loop 1. A combination of control units in which the process variable is measured and compared with the desired value (or set point). If the measured value differs from the desired value, a corrective signal is sent to the final control element to bring the controlled variable to the proper value. 2. An hydraulic or pneumatic system where flow is recirculated following the power cycle; the system contains a limited amount of fluid, which is continually reused. 3. Pertaining to a system with feedback type of control, such that the output is used to modify the input. 4. An operation by which the computer applies control action directly to the process without manual intervention. 5. A signal path which includes a forward path, a feed-back signal, and a summary point, and forms a closed circuit. See loop, closed.

closed-loop control See closed loop.

closed loop gain See gain, closed loop.

closed loop numerical control A type of numerical-control system in which position feedback, and often velocity feedback as well, is used to control the dynamic behavior and successive positions of machine slides or equivalent machine members.

closed-loop system A system with a feedback type of control, such that the output is used to modify the input.

closed pass A metal rolling arrangement in which a collar or flange on one roll fits into a groove on the opposing roll thus permitting production of a flash-free shape.

closed position A position that is zero percent open.

close-grained Consisting of fine, closely spaced particles or crystals.

close-tolerance forging Hot forging in which draft angles, forging tolerances and cleanup allowances are considerably smaller than those used for commercial-grade forgings.

closing plate A plate used to cover or close openings in non-pressure parts.

closing pressure In a safety relief valve, the static inlet pressure at the point where the disc has zero lift off the seat.

closure component The movable part of the valve that is positioned in the flow path to modify the rate of flow through the valve [S75.05]

closure component, characterized Closure component with contoured surface, such as the "vee plug," to provide various flow characteristics.

closure component, cylindrical A cylindrical closure component with a flow passage through it (or a partial cylinder).

closure component, eccentric Closure component face is not concentric with the shaft centerline and moves into seat when closing.

closure component, eccentric spherical disk Disk is spherical segment, not concentric with the disk shaft.

closure component, linear A closure component that moves in a line perpendicular to the seating plane.

closure component, rotary A closure component which is rotated into or away from a seat to modulate flow.

closure component, tapered Closure component is tapered and may be lifted from seating surface before rotating to close or open.

cloud chamber An enclosure filled with supersaturated vapor that can indicate the paths of energetic particles when vapor condenses along the trail of ionized molecules created as the particle passes through the enclosure.

clusec A unit of power used to express the pumping power of a vacuum pump; it equals about 1.333×10^{-6} watt, or the power associated with a leak rate of 10 ml/sec at a pressure of 1 millitorr.

clutch A machine element which allows a shaft in an equipment drive to be connected and disconnected from the power train, especially while the shaft is running.

CMOS See Complementary Metal Oxide Semiconductor.

CO_2 welding See gas metal-arc welding.

coal chemicals A group of chemicals used to make antiseptics, dyes, drugs and solvents that are obtained initially as by-products of the conversion of coal to metallurgical coke.

coalescence A term used to describe the bonding of materials into one continuous body, with or without melting along the bond line, as in welding or diffusion bonding.

coal gas Gas formed by the destructive distillation of coal.

Coanda effect A phenomenon of fluid attachment to one wall in the presence of two walls.

coarse aggregate Crushed stone or gravel, used in making concrete, which will not pass through a sieve with 1/4-in. (6 mm) holes.

coarse grained 1. Having a coarse texture. 2. Having a grain size, in metals, larger than about ASTM No. 5.

coarse vacuum An absolute pressure between about 1 and 760 torr.

coating A continuous film of some material on a surface.

coating (fiber optic) A layer of plastic or other material applied over the cladding of an optical fiber to prevent environmental degradation and to simplify handling.

coating (optics) A thin layer or layers applied to the surface of an optical component to enhance or suppress reflection of light, and/or to filter out certain wavelengths.

coaxial Having coincident axes; for example as in a cable where a central insulated conductor is surrounded by one or more metallic sheaths that act as ground leads or secondary conductors.

coaxial cable Cable with a center conductor surrounded by a dielectric sheath and an external conductor. Has controlled impedance characteristics that make it valuable for data transmission.

coaxial thermocouple element A coaxial thermocouple element consists of a thermoelement in wire form within a thermoelement in tube form and electrically insulated from the tube except at the measuring junction [S12.15].

COBOL See Common Business-Oriented Language.

cock A valve or other mechanism that starts, stops or regulates the flow of liquid, especially into or out of a tank or other large-volume container.

CODAB See configuration data block.

code 1. A system of symbols for meaningful communication. Related to instruction. 2. A system of symbols for representing data or instructions in a computer or a tabulating machine. 3. To translate the program for the solution of a problem on a given computer into a sequence of machine language, assembly language or pseudo instructions and addresses acceptable to that computer. Related to encode. 4. A machine language program.

code, Hollerith A widely used system of encoding alphanumeric information onto cards, hence hollerith cards are synonymous with punch cards [RP55.1].

codes In PCM telemetry, the manner in which ones and zeros in each binary number are denoted.

CODIL See Control Diagram Language.

coding The ordered list, in computer code or pseudo code, of the successive computer instructions representing successive computer operations for solving a specific problem.

coding sheet A fill-in form on which computer programming instructions are written.

coefferdam An earthwork or piling structure that prevents water from filling an excavation or keeps it from surrounding and undermining a pier or foundation. 2. A raised projection surrounding a hatch or trapdoor to keep water out of the opening.

coefficient, flow A constant (C_v), related to the geometry of a valve, for a given valve opening, that can be used to predict flow rate [S75.05].

coefficient of discharge The ratio of actual flow to theoretical flow. It includes the effects of jet contraction and turbulence.

coefficient, rated flow The flow coefficient (C_v) of the valve at rated travel [S75.05].

coefficient, relative flow The ratio of the flow coefficient (C_v) at a stated travel to the flow coefficient (C_v) at rated travel [S75.05].

coefficient, temperature/pressure See operating influence [S51.1].

coefficient, valve recovery See liquid pressure recovery factor [S75.05].

coercimeter An instrument for measuring the magnetic intensity of a magnet or electromagnet.

coextrusion 1. A process for bonding two metal or plastics materials by forcing them simultaneously through the same extrusion die. 2. The bimetallic or bonded plastics shape produced by such a process.

cog A tooth on the edge of a wheel.

cogwheel A wheel with radial teeth on its rim.

coherence A property of electromagnetic waves that are all the same wavelengths and precisely in phase with each other.

coherence length The distance over which light from a laser retains its coherence after it emerges from the laser.

coherent fiber bundle A bundle of optical fibers with input and output ends in the same spatial relationship to each other, allowing them to transmit an image.

coherent scattering Scattering of electromagnetic or particulate rays in which definite phase relationships exist between the incident and scattered waves; coherent waves scattered from two or more scattering centers are capable of interfering with each other.

coil breaks Creases or ridges in metal sheet or strip that appear as parallel lines across the direction of rolling, generally extending the full width of the material.

coil spring A flexible, elastic member in a helical or spiral shape which stores mechanical energy or provides a pulling or restraining force directly related to the amount of elastic deflection.

coincidence Existence of two phenomena or occurrence of two events simultaneously in time or space, or both.

coining Squeezing a metal blank between closed dies to form well-defined imprints on both front and back surfaces, or to compress a sintered powder-metal part to final shape; the process is usually done cold, and involves relatively small amounts of plastic deformation.

coke The solid residue remaining after most of the volatile constituents have been driven out by heating a carbonaceous material such as coal, pitch or petroleum residues; it consists chiefly of coherent, cellular carbon with some minerals and a small amount of undistilled volatiles.

coke oven gas Gas produced by destructive distillation of bituminous coal in closed chambers. Heating value 500-550 Btu/cu ft.

cold drawing Pulling rod, tubing or wire through one or more dies that reduce its cross section, without applying heat either before or during reduction.

cold extrusion Striking a cold metal slug in a punch-and-die operation so that metal is forced back around the die. Also known as cold forging; cold pressing; extrusion pressing; impact extrusion.

cold-finished Referring to a primary-mill metal product, such as strip, bar, tubing or wire, whose final shaping operation was performed cold; the material has more precise dimensions, and usually higher tensile and yield strength, than a comparable shape whose final shaping operation was performed hot.

cold forging See cold extrusion.

cold forming 1. Any operation to shape metal which is performed cold. 2. Shaping sheet metal, rod or wire by bending, drawing, stretching or other stamping operations without the application of heat.

cold galvanizing Painting a metal with a suspension of zinc particles in a solvent, so that a thin zinc coating remains after the organic solvent evaporates.

cold heading Cold working a metal by application of axial compressive forces that upset metal and increase the cross sectional area over at least a portion of the length of the starting stock. Also known as upsetting.

cold joint In soldering, making a soldered connection without adequate heating, so that the solder does not flow to fill the spaces, but merely makes a mechanical bond; a cold joint typically exhibits poor to nonexistent electrical conduction across the joint, is not leak tight, and may break loose under vibration or other mechanical forces.

cold junction See reference junction.

cold plate A mounting plate for electronic components which has tubing or internal passages through which liquid is circulated to remove heat generated by the electronic components during operation. Also known as liquid-cooled dissipator.

cold pressing See cold extrusion.

cold rolling Rolling metal at about room temperature; the process reduces thickness, increases tensile and yield strengths, improves fatigue resistance, and produces a smooth, lustrous or semilustrous finish.

cold trap A length of tubing between a vacuum system and a diffusion pump or instrument which is cooled by liquid nitrogen to help remove condensable vapors.

cold treatment Subzero treatment of a metal part—usually at -65 °F, -100 °F or liquid-nitrogen temperature—to induce metallurgical changes that either stabilize dimensions, complete a phase transformation or condition the metal and prepare it for further processing.

cold working Any plastic deformation of a metal carried out below its recrystallization temperature; the process always induces strain hard-

ening to a degree directly related to the percent reduction in cross section.

cold working pressure The maximum pressure rating of a valve or fitting coincident with ambient temperature, generally the range from -20 °F to +100 °F (-29 °C to +38 °C) [S75.05].

collar A rigid, ring-shaped machine element that is forced onto or clamped around a shaft or similar member to restrict axial motion, provide a locating surface or cover an opening.

collating sequence In data processing, the order of the ASCII numeric codes for the characters.

collator 1. A mechanical device at the output of a printing machine or copier which sorts multiple-page documents and arranges them into sets. 2. In data processing, a device for combining sets of data cards or other information-bearing elements into a desired sequence. 3. In data processing using electronic files, a program or routine used to merge two or more files into a single, ordered output file.

collector 1. Any of a class of instruments for determining electrical potential at a point in the atmosphere, and ultimately the atmospheric electric field; all collectors consist of a device for bringing a conductor rapidly to the potential of the surrounding air and an electrometer for measuring its potential with respect to the earth. 2. A device used for removing gas borne solids from flue gas. 3. One of the functional regions in a transistor.

collimate To make parallel.

collimation Producing a beam of light or other electromagnetic radiation whose rays are essentially parallel.

collimator An optical system which focuses a beam of light so all the rays form a parallel beam.

collision A close approach of two or more bodies (including energetic particles) that results in an interchange of energy, momentum or charge. See also elastic collision; inelastic collision.

colloid 1. A dispersion of particles of one phase in a second phase, where the particles are so small that surface phenomena play a dominant role in their chemical behavior; typical colloids include mists or aerosols (liquid dispersed phase in gaseous dispersion medium), smoke (solid in gas), foam (gas in liquid), emulsions (liquid in liquid), suspensions (solid in liquid), solid foam such as pumice (gas in solid), and solid solution such as colloidal gold in glass (solid in solid). 2. A finely divided organic substance which tends to inhibit the formation of dense scale and results in the deposition of sludge, or causes it to remain in suspension, so that it may be blown from the boiler.

Colmonoy A series of high nickel alloys (manufactured by Wall-Colmonoy Corp.) used for hard facing of surfaces subject to erosion.

color code 1. Any system of colors used to identify a specific type or class of objects from other, similar objects—for example, to differentiate steel bars of different grades in a warehouse. 2. A system of colors used to identify different piping systems from each other in a factory or other building—for example, red for fire protection, yellow for hazardous chemical, blue for potable water and green for compressed air.

color coding The use of different background and foreground colors to symbolically represent processes and process equipment attributes, such as status, quality, magnitude, identification, configuration, etc. [S5.5]

color filter A filter containing a colored dye, which absorbs some of the incident light and transmits the remainder.

colorimetry Any analytical process that uses absorption of selected bands of visible light, or sometimes ultraviolet radiation, to determine a chemical property such as the end point of a reaction or the concentration of a substance whose color is indicative of product purity or uniformity.

column A vertical structural member of substantial length designed to bear axial compressive loads.

column loading A factor that takes into consideration the quantity of liquid descending in the column and the quantity of vapor ascending in the column. If either the liquid or the vapor flow rate becomes too high, column flooding will occur.

coma A lens aberration in which light rays from an off-axis source which pass through the center of a lens arrive at the image plane at different distances from the axis than do rays from the same source which pass through the edges of the lens.

combination automatic controller A type of control system arrangement in which more than one closed control loop are coupled through pri-mary feedback or through any of the controller elements.

combination die A forging, forming or casting die with more than one cavity.

combination pliers A pliers whose jaws are designed for holding objects in one portion, and for cutting and bending wire in another portion.

combination scale An instrument scale consisting of two or more concentric or colinear scales, each graduated in equivalent values with two or more units of measure.

combination square A measuring and rough layout tool consisting of a special head and a short steel rule that, when used together, can check angles of both 90° and 45°.

combination wrench A fixed-size wrench having an open-end wrench at one end and box wrench at the other, usually both intended to fit the same size bolthead.

combustible The heat producing constituents of a fuel.

combustible dust classifications (a) Group E dusts are those having resistivities lower than 100 ohm-cm or which break down when subjected to 1000 volts/cm across a bulk sample when tested in accordance with ISA-S12.10 Appendix 8.7 Laboratory Resistivity Measurement of Dusts [S12.11]. (b) Group F dusts are those having resistivities between 100 ohm-cm and 100 meg-ohm-cm and which do not break down when subjected to 10000 volts/em across the bulk sample when tested in accordance with ISA S12.10 Appendix 8.7 Laboratory Resistivity Measurement of Dusts. This group includes the carbonaceous dusts, generally regarded as semiconductors [S12.11]. (c) Group G dusts are those having resistivities greater than 100 meg-ohm-cm and which do not break down when subjected to 10000 volts/cm across the bulk sample when tested in accordance with ISA S12.10 Appendix 8.7 Laboratory Resistivity Measurement of Dusts. This group includes the agricultural and plastic dusts generally regarded as insulators [S12.11].

combustible dust layer Any surface accumulation of combustible dust that is large enough to propagate flame or will degrade and ignite [S12.10].

combustible dusts Dusts which (when mixed with air in certain proportions) can be ignited and will propagate flame [S12.10].

combustible gas Any flammable or combustible gas or vapor that can, in sufficient concentration by volume in air, become the fuel for an explosion or fire. Materials which cannot produce sufficient gas or vapor to form a flammable mixture at ambient or operating temperatures and mists formed by the mechanical atomization of combustible liquids are NOT considered to be combustible gases [S12.13].

combustible loss The loss representing the unliberated thermal energy occasioned by failure to oxidize completely some of the combustible matter in the fuel.

combustion The rapid chemical combination of oxygen with the combustible elements of a fuel resulting in the production of heat.

combustion chamber Any chamber or enclosure designed to confine and control the generation of heat and power from burning fuels.

combustion engine An energy conversion machine that operates by converting to motion heat from the burning of a fuel.

combustion (flame) safeguard A system for sensing the presence or absence of flame and indicating, alarming or initiating control action.

combustion rate The quantity of fuel fired per unit of time, as pounds of coal per hour, or cubic feet of gas per minute.

combustion safety control—programming type A combustion safety control that provides for various operations at definite periods of time in predetermined sequences.

come-along 1. A lever-operated chain or wire-rope hoist for lifting or pulling at any angle, which has a reversible ratchet in the handle to permit using short strokes for tensioning or relaxing the fall. Also known as puller. 2. A device for gripping and applying tension to a length of cable, wire rope or chain by means of jaws that close when the user pulls on a ring.

COM file A computer file name ending in .COM which most often contains a machine code program. It is short for "command" file.

comfort curve A line on the graph of dry-bulb temperature versus wet-bulb temperature or relative humidity that represents optimum comfort for the average person who is not engaged in physical activity.

comfort zone The respective ranges of indoor temperature, relative humidity and ventilation rate (air-movement rate) that most persons consider acceptable for their normal degree of physical activity and mode of dress.

command 1. An electronic pulse, signal, or set of signals to start, stop, or continue some operation. It is incorrect to use command as a synonym for instruction. 2. The portion of an instruction word which specifies the operation to be performed [RP55.1]. 3. A signal that causes a computer to start, stop or to continue a specific operation.

command language A source language consisting primarily of procedural operations, each capable of invoking a function to be executed.

command resolution The maximum change in the value of a command signal which can be made without inducing a change in the controlled variable.

comment An expression which explains or identifies a particular step in a routine, but which has no effect on the operation of the computer in performing the instructions for the routine.

common A reference within a system having the same electrical potential throughout. Usually connected to ground at one point. Often different commons are used throughout a system such as power common, signal common, etc. depending on the accuracy to which the reference is held.

Common Applications Service Elements (CASE) One of the application protocols specified by MAP.

common area A section in memory that is set aside for common use by many separate programs or modules.

Common Business Oriented Language (COBOL) A specific language by which business data-processing procedures may be precisely described in a standard form. The language is intended not only as a means for directly presenting any business program to any suitable computer, for which a compiler exists, but also as a means of communicating such procedures among individuals.

common field A field that can be accessed by two or more independent routines.

common machine language In data processing, coded information which is in a form common to a related group of data-processing machines.

common mode In analog data, an interfering voltage from both sides of a differential input pair (in common) to ground.

common mode interference A form of interference which appears between the terminals of any measuring circuit and ground. See common mode voltage. See also interference, common mode.

common mode rejection The ability of circuit to discriminate against common mode voltage [RP55.1]. Note: It may be expressed as a dimensionless ratio, a scalar ratio, or in decibels as 20 times the $\log_{10}$ of that ratio [S51.1]. See voltage, common mode.

common-mode rejection ratio Abbreviated CMRR. A measure of the ability of a detector to damp out the effect of a common-mode-generated interference voltage; usually expressed in decibels.

common mode voltage (CMV) In-phase, equal-amplitude signals that are applied to both inputs of a differential amplifier, usually referred to as a guard shield or chassis ground. See voltage, common mode.

common port The port of a three-way valve that connects to the other two flow paths [S75.05].

communication Transmission of intelligence between points of origin and reception without alteration of sequence or structure of the information content.

communication link 1. Computer control is a device in which control and/or display actions are generated for use by other system devices. When used with other control devices on the communication link the computer normally performs or functions in a hierarchical relationship to the other control devices [S5.3]. 2. The physical means of connecting one location to another for the purpose of transmitting and receiving information. 3. The physical realization of a specified means by which stations communicate with each other. The specification normally cover the transmission medium, the protocol, mechanical and electrical interfaces and some aspects of functional capability. 4. A link may provide multiple channels for communications.

commutation Cyclic sequential sampling on a time-division basis of multiple data sources.

commutation duty cycle A channel dwell period, expressed as a percentage of a channel interval.

commutation frame period The time required for sequential sampling of all input signals; this would correspond to one revolution of a simple multicontact rotary switch.

commutation rate The number of commutator inputs sampled per specified time interval.

commutator 1. A device used to accomplish time-division multiplexing (TDM) by repetitive sequential switching. 2. A segmented ring, usually constructed of hard-drawn copper segments separated by an insulator such as mica, which is used to energize only the correct windings of a d-c generator or motor at any given instant.

compact 1. A powder-metallurgy part made by pressing metal powder, with or without a binder or other additives; prior to sintering it is known as a green compact, and after sintering as a sintered compact or simply a compact. 2. To consolidate earth or paving materials by weight, vibration, impact or kneading so that the consolidated material can sustain more load than prior to consolidation.

comparative tracking index (CTI) The numerical value of the maximum voltage in volts at which the material withstands 50 drops without tracking.

comparator A device for inspecting a part to determine any deviation from a specific dimension by electrical, optical, pneumatic or mechanical means.

compass 1. A drafting or layout tool for drawing circles, arcs or fillets which consists of a bow or radius bar connecting a pin center and a pan or drafting pencil. 2. An instrument for indicating relative direction on the earth's surface, usually measured with respect to a reference direction such as magnetic or true North.

compatibility The ability for two devices to communicate with each other in a manner that both understand.

compatibility interface A point at which hardware, logic, and signal levels are defined to allow the interconnection of independently designed and manufactured components.

compatible The state in which different kinds of computers or equipment can use the same programs or data.

compensated pendulum A pendulum made of two materials having different coefficients of linear expansion, and so constructed that the distance between the center of oscillation and the point of suspension remains the same over the normal range of ambient temperatures.

compensation In process instrumentation, provision of a special construction, a supplemental device, circuit, or special materials to counteract sources of error due to variations in specified operating conditions [S51.1] [S37.1].

compensation signals In telemetry, a set of reference signals recorded on tape along with the data, and used during playback to automatically compensate for any nonuniformity in tape speed.

compensator A device which converts a signal into some function of it which, either alone or in combination with other signals, directs the final controlling element to reduce deviations in the directly controlled variable [S51.1].

compensatory leads An arrangement of connecting elements between an instrument and a transducer or other observation device such that variations in the properties of any of the connecting elements—such as temperature effects that induce changes in resistance—are compensated so that they do not affect instrument accuracy.

compile 1. A computer function that translates symbolic language into machine language. 2. To prepare a machine-language program from a computer program written in another programming language by making use of the overall logic, structure of the program, or generating more than one machine instruction for each symbolic statement, or both, as well as performing the function of an assembler.

compiler 1. A program that translates a high-level source language (such as FORTRAN IV or BASIC) into a machine language suitable for a particular machine. 2. A computer program more powerful than an assembler. In addition to its translating function, which is generally the same process as that used in an assembler, it is able to replace certain items of input with a series of instructions, usually called subroutines. Thus, an assembler translates item for item, and produces as output the same number of instructions or constants which were put into it; a compiler will do more than this. The program which results from compiling is a translated and expanded version of the original. Synonymous with compiling routine and related to assembler.

compile time In general, the time during which a source program is translated into an object program.

compiling routine Same as compiler.

complement 1. A quantity expressed to the base n, which is derived from a given quantity by a particular rule. Frequently used to represent the negative of the given quantity. 2. A complement on n, obtained by subtracting each digit of the given quantity from n-1, adding unity to the least significant digit, and performing all resultant carries, e.g., the twos complement of binary 11010 is 00110. The tens complement of decimal 456 is 544. 3. A complement on n-1, obtained by subtracting each digit of the given quantity from n-1, e.g., the ones complement of binary 11010 is 00101. The nines complement of decimal 456 is 543.

Complementary Metal Oxide Semiconductor (CMOS) 1. One type of computer semiconductor memory. The main feature of CMOS memory is its low power consumption. 2. A type of semiconductor device not specifically memory.

complementary operator The logic operator which is the NOT of a given logic operator.

complementary wavelength The monochromatic wavelength of light which matches a standard reference light when combined with the sample color in suitable proportions as applied to colorimetry.

complete combustion The complete oxidation of all the combustible constituents of a fuel.

complete contraction A combination of both end and bottom contractions in a weir.

completion network In a strain gage signal conditioner, the one to three resistors which must be added to make a four-arm bridge (the transducer being the active arm or arms).

complex frequency A complex number used to characterize exponential or damped sinusoidal waves in the same way as an ordinary frequency is used to characterize a simple harmonic wave.

complex lens A lens system consisting of more than one optical element.

complex tone A sound wave produced by combining simple sinusoidal component waves of different frequencies.

compliance The reciprocal of stiffness [S51.1].

composite A material or structure made up of physically distinct components that are mechanically, adhesively or metallurgically bonded together; examples include filled plastics, laminates, filament-wound structures, cermets, and adhesive-bonded honeycomb-sandwich structures.

composite joint A connection between two parts that involves both mechanical joining and welding or brazing, and where both contribute to total joint strength.

composite subcarrier Two or more subcarriers that are combined in a frequency-division multiplexing (FDM) scheme.

composite wave filter A selective transducer made up of two or more filters—the filters being any combination of high-pass, low-pass, band-pass or band-elimination types.

compound angle The surface contour formed by two intersecting mitered angles.

compound die Any die so constructed that it performs more than one operation on a given part with a single stroke of the punch.

compound engine A multicylinder engine in which the working fluid—steam, air or hot gas—expands successively as it passes from one cylinder to another through the engine.

compound lever A device consisting of two or more levers, where force or motion is transferred from the arm of one lever to the next lever in the train.

compound screw A screw having threads of different pitches or opposite helixes on opposite ends of the shank.

compound semiconductor A semiconductor such as gallium arsenide which is made up of two or more materials, in contrast to simple single element materials such as silicon and germanium.

compressibility Volumetric strain per unit change in hydrostatic pressure.

compressibility factor (Z) A factor used to compensate for deviation from the laws of perfect gases. If the gas laws are used to compute the specific weight of a gas, the computed value must be adjusted by the compressibility factor Z to obtain the true specific weight.

compressible Capable of being compressed. Gas and vapor are compressible fluids.

compressible flow Fluid flow under conditions which cause significant changes in density.

compressional wave A wave in an elastic medium which causes an element of the medium to undergo changes in volume without rotating.

compression failure Buckling, collapse or fracture of a structural member that is loaded in compression.

compression member A beam, column or other structural component that is loaded in such a way as to be under predominantly compressive stress.

compression mold A type of plastics mold that is opened to introduce starting material, closed to shape the part, and reopened to remove the part and restart the cycle; pressure to shape the part is supplied by closing the mold.

compression ratio 1. In an internal combustion engine, the ratio of cylinder volume with the piston at bottom dead center to the volume with the piston at top dead center. 2. In powder metallurgy, the ratio of the volume of loose powder used to make a part to the volume of the pressed compact.

compression spring An elastic member, usually made by bending metal wire into a helical coil, that resists a force tending to compress it.

compression test A destructive test for determining fracture strength, yield strength, ductility and elastic modulus by progressively loading a short-column specimen in compression.

compressor 1. A device which the valve stem forces against the backside of the diaphragm to cause the diaphragm to move toward and seal against the internal flow passageway of the valve body [S75.05]. 2. A machine—usually a reciprocating-piston, centrifugal, or axial-flow design—which is used to increase pressure in a gas or vapor. Also known as compression machine. 3. A hardware or software process for removing redundant or otherwise uninteresting words from a stream, thereby "compressing" the data quantity.

Compton scattering A form of interaction between x-rays and loosely-bound electrons in which a collision between them results in deflection of the radiation from its previous path, accompanied by random phase shift and slight increase in wavelength.

computational process An instance of execution of a segment by a processor using a data area.

computational stability The degree to which a computational process remains valid.

computer 1. A data processor that can perform substantial computation, including numerous arithmetic or logic operations, without intervention by a human operator during the run. 2. A device capable of solving problems by accepting data, performing described operations on the data, and supplying the results of these operations. Various types of computers are calculators, digital computers, and analog computers. 3. See analog computer, digital computer, general purpose computer, hybrid computer, and stored program computer.

Computer and Automated Systems Association of the Society of Manufacturing Engineers (CASA/SME) A professional engineering association dedicated to the advancement of engineering technology. CASA/SME supports the administrative functions of the MAP/TOP users group.

computer code A machine code for a specific computer.

computer control Computer control is a device in which control and/or display actions are generated for use by other system devices. When used with other control devices on the communication link the computer normally performs or functions in a heirarchical relationship to the other control devices.

computer control system A system in which all control action takes place within the control computer. Single or redundant computers may be used [S5.5].

computer-dependent language A relative term for a programming language whose translation can be achieved only by a specific model (or models) of computer.

computer graphics Any display in pictorial form on a computer monitor that can be printed.

computer-independent language A language in which computer programs can be created without regard for the actual computers which will be used to process them. Related to transportability.

computer instruction A machine instruction for a specific computer.

Computer Integrated Manufacturing (CIM) A central computer gathers all types of data, provides information stored in the data base for decisions, and controls production input and output.

computer interface Serves as the interface device between the host computer and other devices on the data highway. It converts data from the protocol of the computer and that of the highway, and vice versa.

computer interface device (CID) Hardware that allows a general-purpose computer to share data with the rest of the distributed control system.

computer-limited Pertaining to a situation in which the time required for computation exceeds the time available.

computer network A complex consisting of two or more interconnected computing units.

computer networking Interconnection of two or more geographically separated computers so that information can be exchanged between them, usually under the direction of individual, autonomous control programs. See also distributed processing.

computer operator A person who performs standard system operations such as adjusting system operation parameters at the system console, loading a tape transport, placing cards in a card reader, and removing listings from the line printer.

computer part-programming In numerical control, the preparation of a part program to obtain a machine program using the computer and appropriate processor and post processor.

computer program A series of instructions or statements in a form acceptable to a computer prepared in order to achieve a certain result.

computer simulation A logical-mathematical representation of a simulation concept, system, or operation programmed for solution on an analog or digital computer.

computer word A sequence of bits or characters treated as a unit and capable of being stored in one computer location. Synonymous with machine word.

computing device A device or function that performs one or more calculations or logic operations, or both, and transmits one or more resultant output signals. A computing device is sometimes called a computing relay [S5.1].

computing instrument See instrument, computing.

conc See concentrated.

concatenate To combine several files into one file, or several strings of characters into one string, by appending one file or string after another.

concave A term describing a surface whose central region is depressed with respect to a flat plane approximately passing through its periphery.

concave curve A change in the angle of inclination of a belt conveyor where the center of the curve is above the conveyor [RP74.01].

concentrate 1. To separate metal-bearing minerals from the gangue in an ore. 2. The enriched product resulting from an ore-separation process. 3. An enriched substance that must be diluted, usually with water, before it is used.

concentration 1. The weight of solids contained

in a unit weight of boiler or feed water. 2. The number of times that the dissolved solids have increased from the original amount in the feedwater to that in the boiler water due to evaporation in generating steam.

concentricity The quality of two or more geometric shapes having the same center—usually, the term is applied to plane shapes or cross sections of solid shapes that are approximately circular.

concentric orifice plate A fluid-meter orifice plate having a circular opening whose center coincides with the axis of the center of the pipe it is installed in.

concrete A mixture of aggregate, water and a binder, usually Portland cement, that cures as it dries and becomes rock hard.

concurrent processing Two (or more) computer operations that appear to be processed simultaneously when in fact the CPU is rapidly switching between them.

cond See conductivity.

condensate 1. The liquid product of a condensing cycle. Also known as condensate liquid. 2. A light hydrocarbon mixture formed by expanding and cooling gas in a gas-recycling plant to produce a liquid output.

condensate pot A section of pipe (4 in. diameter) installed horizontally at the orifice flange union to provide a large-area surge surface for movement of the impulse line fluid with instrument element position change to reduce measurement error from hydrostatic head difference in the impulse lines.

condensate trap 1. A device to separate saturated water from steam in a pipe or piece of process equipment. 2. A device used to trap and retain condensate in a measurement impulse line to prevent hot vapors from reaching the instrument.

condensation-type hygrometer Any of several designs of dew-point instruments that operate by detecting the equilibrium temperature at which dew or frost forms on a thermoelectrically, mechanically or chemically cooled surface; surface condensation may be detected by optical, electrical or nuclear techniques.

condenser The heat exchanger, located at the top of the column, that condenses overhead vapors. For distillation, the common condenser cooling media are water, air, and refrigerants such as propane. The condenser may be partial or total. In a partial condenser only part of the vapors are condensed, with the remainder usually withdrawn as a vapor product.

condenser boiler A boiler in which steam is generated by the condensation of a vapor.

conditional branch See conditional transfer.

conditional jump See conditional transfer.

conditional stability 1. A linear system is conditionally stable if it is stable for a certain interval of values of the open-loop gain, and unstable for certain lower and higher values. 2. The property of a controlled process by which it can function in either a stable or unstable mode, depending on conditions imposed.

conditional transfer An instruction which, if a specified condition or set of conditions is satisfied, is interpreted as an unconditional transfer. If the conditions are not satisfied, the instruction causes the computer to proceed in its normal sequence of control. A conditional transfer also includes the testing of the condition. Synonymous with conditional lump and conditional branch and related to branch.

condition monitoring system A system designed to monitor the condition of a machine or process.

conducting polymer A plastics material having electrical conductivity approaching that of metals.

conduction 1. Flow of heat through or across a conductor. 2. The transmission of heat through and by means of matter unaccompanied by any obvious motion in the matter.

conduction band A partially filled or empty energy band in which electrons are free to move easily.

conduction error The error in a temperature transducer due to heat conduction between the sensing element and the mounting of the transducer [S37.1].

conduction pump A device for pumping a conductive liquid, such as a liquid metal, by passing an electric current across the stream of liquid and applying a magnetic field at right angles to the electrical current.

conductive dust A dust whose resistivity is less than 100 ohm-cm or which breaks down with 1000 volts per cm. applied across the bulk sample when tested in accordance with methods outlined in ISA-S12.10. Such dust is denoted Group E in the NEC [S12.11].

conductive elastomer An elastomeric material

that conducts electricity; usually made by mixing powdered metal into a silicone before it is cured.

conductively connected A part is conductively connected to another part if the current between the parts, with the equipment at reference test conditions, exceeds the limit for leakage current [S82.01].

conductivity 1. The amount of heat (Btu) transmitted in one hour through one square foot of a homogeneous material 1 in. thick for a difference in temperature of 1 °F between the two surfaces of the material. 2. The electrical conductance, at a specified temperature, between the opposite faces of a unit cube; usually expressed as ohm^{-1} cm^{-1}.

conductivity bridge A simple four-arm a-c bridge circuit in which a conductivity cell is the unknown circuit element; electrically, the cell is equivalent to a resistance and a capacitance in series; higher a-c frequencies lead to lower cell-polarization errors, but introduce greater errors due to capacitance impedance; the latter can be reduced by using a phase-sensitive detector.

conductivity-type moisture sensor An instrument for measuring moisture content of fibrous organic materials such as wood, paper, textiles and grain at moisture contents up to saturation.

conductometer An instrument that measures thermal conductivity, especially one that does so by comparing the rates at which different rods conduct heat.

conductor Any material through which electrical current can flow.

conduit 1. Any channel, duct, pipe or tube for transmitting fluid along a defined flow path. 2. A thin-wall pipe used to enclose wiring.

cone bearing A tapered sleeve bearing in the shape of a truncated cone that runs in a correspondingly tapered bearing block.

cone-plate viscometer An instrument for routinely determining the absolute viscosity of fluids in small sample volumes by sensing the resistance to rotation of a moving cone caused by the presence of the test fluid in a space between the cone and a stationary flat plate.

confidence level 1. The probability that the interval quoted will include the true value of the quantity being measured. 2. In acceptance sampling, the probability that accepted lots will be better than a specific value known as the rejectable quality level (RQL); a confidence level of

90% indicates that 90 out of every 100 lots accepted will have a quality better than the RQL. 3. In statistical work, the degree of assurance that a particular probability applies to a specific circumstance.

CONFIG.SYS A basic computer file that outlines how that particular device is designed to operate.

configurable 1. A system feature that permits selection through entry of keyboard commands of the basic structure and characteristics of a device or system, such as control algorithms, display formats, or input/output terminations [S5.3]. 2. A term applied to a device or system whose functional characteristics can be selected or rearranged through programming or other methods. The concept excludes rewiring as a means of altering the configuration [S5.1].

configuration 1. The arrangement of the parts or elements of something. 2. A low-level, fill-in-the-blank form of programming a process control device. 3. A particular selection of hardware devices or software routines and/or programs that function together. 4. A term applied to a device or system whose functional characteristics can be selected or rearranged through programming or other methods. 5. The hardware, firmware and/or software combinations make up a system.

configuration data block (CODAB) In TELEVENT, the data section that identifies the "personality" of a hardware-software combination.

configure The installation procedure that sets up software to operate on a particular computer and printer.

confined flow Flow of a continuous stream of fluid within a process vessel or conduit.

conformance A device conforms to the manufacturers specifications. See accuracy and error band.

conformity Of a curve, the closeness to which it approximates a specified curve (e.g., logarithmic, parabolic, cubic, etc.). Note: (a) It is usually measured in terms of nonconformity and expressed as conformity; e.g., the maximum deviation between an average curve and a specified curve. The average curve is determined after making two or more full range traverses in each direction. The value of conformity is referred to the output unless otherwise stated. (b) As a performance specification, conformity should be expressed as independent conformity, terminal-based conformity, or zero-based conformity. When expressed simply as conformity, it is assumed to be independent conformity. See linearity. [S51.1]

conformity, independent The maximum deviation of the calibration curve (average of upscale and downscale readings) from a specified characteristic curve so positioned as to minimize the maximum deviation [S51.1].

conformity, terminal-based The maximum deviation of the calibration curve (average of upscale and downscale readings) from a specified characteristic curve so positioned as to coincide with the actual characteristic curve at upper and lower range-values [S51.1].

conformity, zero-based The maximum deviation of the calibration curve (average of upscale and downscale readings) from a specified characteristic curve so positioned as to coincide with the actual characteristic curve at the lower range-value [S51.1].

conical orifice An orifice having a 45° bevel on the inlet edge to yield more constant and predictable discharge coefficient at low flow velocity (Reynolds number less than 10,000).

conical scan antenna An automatic-tracking antenna system in which the beam is driven in a circular path such that it forms a cone. The antenna is steered automatically so that the telemetry source is kept at the center of the cone.

coniscope See koniscope.

conjugate bridge An arrangement of electrical or electronic components in which the supply circuit and detector circuit are interchanged as compared with the normal arrangement for that type of bridge.

conjugate impedances An impedance pair having the magnitudes of resistance and reactive components of one equal to the corresponding values of the other, but whose reactive components are of opposite signs.

connect Establish linkage between an interrupt and a designated interrupt servicing program. See disconnect.

connecting rod Any straight link that transmits power or motion from one part of a mechanism to another, especially one that links a rotating member to a reciprocating member—for example, the link that attaches a piston to the crankshaft in a reciprocating internal-combustion engine.

connection head extension A connection head extension is a threaded fitting or an assembly of fittings extending between the thermowell or angle fitting and the connection head [ANSI-MC96.1].

connector 1. Any detachable device for providing electrical continuity between two conductors. 2. In fiber optics, a device which joins the ends of two optical fibers together temporarily.

consecutive access A method of data access that is characterized by the sequential nature of the I/O device involved; for example, a card reader is an example of a consecutive access device; each card must be read one after another, and no distinction is made between logical sets of data in or among the cards in the input hopper.

consistency A qualitative means of classifying substances, especially semisolids, according to their resistance to dynamic changes in shape.

console 1. A main control desk for an integrated assemblage of electronic equipment. Also known as control desk. 2. A grouping of control devices, instrument indicators, recorders and alarms, housed in a freestanding cabinet or enclosure, to create an operator's work station. 3. The cabinet or enclosure for a floor-model radio or television receiver, or similar electronic device.

constant 1. A value that remains the same throughout the distinct operation; opposite of variable. 2. A data item which takes as its value its name (hence, its value is fixed during program execution).

constant-amplitude (filter) A reference to the characteristic of a Butterworth filter. See Butterworth.

constant-bandwidth (CBW) The spacing of FM subcarriers equally with relation to each other; see proportional bandwidth.

constant-current potentiometer A type of null-balance instrument for determining an unknown d-c voltage, usually less than 10 V, under conditions that maintain constant current in the detector circuit; resolution up to one part in 10^3 can be achieved with a single potentiometer slidewire, and up to one part in 10^7 with a multidecade device.

constant-current transformer A type of transformer which automatically adjusts its output of its secondary circuit to maintain a constant current under varying load impedances when its primary windings are connected to a constant-voltage power supply.

constant-delay (filter) See Bessel.

constant-head meter A flow measurement device that maintains a constant pressure differential by varying the cross section of a flowpath through the meter, such as in a piston meter or rotameter.

constant-load balance A single-pan weighting device, having a constant load, in which the sample weight is determined by hanging precision weights from a counterpoised beam.

constant-resistance potentiometer A type of null-balance instrument for determining an unknown d-c voltage, usually less than 10 V, using a constant scaling resistor in parallel with the potentiometer circuit.

constant-volume gas thermometer A device for detecting and indicating temperature based on Charles' Law—the pressure of a confined gas varies directly with absolute temperature; in practical instruments, a bulb immersed in the thermal medium is connected to a Bourdon tube by means of a capillary; changes in temperature are indicated directly by movement of the Bourdon tube due to changes in bulb pressure.

constrained mechanism A mechanical device in which all members move only along predetermined paths.

constraint 1. The limit of normal operating range. 2. Anything which keeps a member under longitudinal tension from contracting laterally, which sets up a condition of biaxial tension in the member; the term is used most often in connection with welded joints that cannot shrink laterally as the weld solidifies and cools.

consumable electrode An arc-welding electrode that melts during welding to provide the filler metal.

consumable insert A piece of metal placed in the root of a weld prior to welding, and which melts during welding to supply part of the filler metal.

consumables Those materials or components that are depleted or require periodic replacement through normal use of the instrument [S12.15].

contact In hardware, a set of conductors that can be brought into contact by electromechanical action and thereby produce switching. In software, a symbolic set of points whose open or closed condition depends on the logic status assigned to them by internal or external conditions.

contact input See input, contact.

contact inspection In ultrasonic testing, a method of scanning a test piece which involves placing a search unit directly on a test piece surface covered with a thin film of couplant.

contactor A mechanical or electromechanical device for repeatedly making and breaking electrical continuity between two branches of a power circuit, thereby establishing or interrupting current flow.

contact output See output, contact.

contact rectifier A device for converting a-c electrical power to d-c power that is constructed of two different solids in contact with each other; rectification is accomplished because the selected combination of solids yields greater electrical conductivity in one direction across the interface between them than in the other direction.

contacts The electrically conducting parts in a contactor that repeatedly come in contact or separate to make or break electrical continuity.

contact sense module A device which monitors and converts program-specified groups of field-switch contacts into digital codes for input to the computer.

contact symbology Representation of logic schemes in contact or ladder diagram form.

contact thermography A method of measuring surface temperature in which the surface of an object is covered with a thin layer of luminescent material and then viewed under ultraviolet light in a darkened room; the brightness viewed indicates surface temperature.

contact tube In gas metal arc welding and flux cored arc welding, a metal part with a hole in it that provides electrical contact between the welding machine and the wire electrode that is fed continuously through the hole.

contact-type membrane switch A disk-shaped momentary-contact switch of multilayer construction; the active element consists of two conductive buttons separated by an insulating washer; finger pressure on one face of the disk brings the buttons into contact, completing the electrical circuit; when the pressure is released, the contacts separate, breaking the electrical circuit.

contact-wear allowance The thickness that may be lost due to wear from either of a pair of mating electrical contacts before they cease to adequately perform their intended function.

contaminant That which contaminates to make

impure or corrupt by contact or mixing [S71.04].

contaminate That which contaminates to make impure or corrupt by contact or mixing.

contamination Presence of an unwanted substance—usually, a substance that causes an undesired effect or interferes with a desired effect.

content-addressed storage See associative storage.

contention 1. A condition on a multidrop communication channel when two or more locations try to transmit at the same time. 2. Unregulated bidding for a line or other device by multiple users.

contiguous file A file consisting of physically adjacent blocks on a mass-storage device.

continuity tester A device for testing that a fiber optic communication system forms a continuous optical path between two points.

continuous blowdown The uninterrupted removal of concentrated boiler water from a boiler to control total solids concentration in the remaining water.

continuous dilution A technique of supplying a protective gas flow continuously to an enclosure, housing electrical circuitry, containing an internal potential source of flammable gas or vapor for the purpose of diluting any flammable gas or vapor which could be present to a level well below the lower-explosion-limit.

continuous-duty rating The maximum power or other operating characteristic that a specific device can sustain indefinitely without significant degradation of its functions.

continuous furnace A type of reheating furnace where the charge is loaded at one end, moves through the furnace to accomplish the intended treatment, and is discharged at the other end.

continuous mixer A type of mixer in which starting ingredients are fed continuously and the final mixture is withdrawn continuously, without stopping or interrupting the mixing process; generally, unmixed ingredients are fed at one end of the machine and blended progressively as they move towards the other end, where the mixture is discharged. Contrast with batch mixer.

continuous operation A process that operates on the basis of continuous flow, as opposed to batch, intermittent or sequenced operations.

continuous-path numerical control A type of numerical-control system involving not only specification of successive end positions of machine slides or equivalent machine members, but also automatic generation of the linear, circular or parabolic path to be followed in moving from one end position to the next. Also known as contouring numerical control.

continuous rating 1. The rating applicable to specified operation for a specified uninterrupted length of time [S37.1]. 2. A defined power input or set of operating variables that represent the maximum values for operating a device continuously for an indefinite time without reducing its normal service life.

continuous spectrum A distribution of wavelengths in a beam of electromagnetic radiation in which the intensity varies continuously with wavelength, exhibiting no characteristic structure such as a series of bands where the intensity does not abruptly change at discrete wavelengths. See also band spectrum.

continuous weld A welded joint where the fusion zone is continuous along the entire length of the joint.

contour control system A system of control in which two or more controlled motions move in relation to each other so that a desired angular path or contour is generated.

contouring numerical control See continuous-path numerical control.

contraction The narrowing of the stream of liquid passing through a notch of a weir.

contract maintenance 1. Maintenance not normally done by plant personnel. 2. A maintenance service organization which contracts to do specific maintenance.

contrast In a photographic or radiographic image, the ability to record small differences in light or x-ray intensity as discernible differences in photographic density.

contrast factor The slope of the central portion of a graph of photographic density versus exposure for a given photographic or radiographic emulsion.

control 1. Frequently, one or more of the components in any mechanism responsible for interpreting and carrying out manually initiated directions. 2. In some applications, a mathematical check. 3. Instructions which determine conditional jumps often are referred to as control instructions, and the time sequence of execution of instructions is called the flow of control. 4.

Any manual or automatic device for the regulation of a machine to keep it at normal operation. If automatic, the device is motivated by variations in temperature, pressure, water level, time, light, or other influences.

control accuracy The degree to which a controlled process variable corresponds to the desired value or setpoint.

control action Of a controller or a controlling system, the nature of the change of the output effected by the input. The output may be a signal or the value of a manipulated variable. The input may be an actuating error signal, the output of another controller, or the control loop feedback signal when the setpoint is constant [S51.1]. See proportional control action, integral control action, and derivative control action.

control action, derivative (rate) (D) In process instrumentation, control action in which the output is proportional to the rate of change of the input [S51.1].

control action, floating In process instrumentation, control action in which the rate of change of the output variable is a predetermined function of the input variable. The rate of change may have one absolute value, several absolute values, or any value between two predetermined values [S51.1].

control action, integral (reset) (I) Control action in which the output is proportional to the time integral of the input; i.e., the rate of change of output is proportional to the input [S51.1].

control action, proportional (P) Control action in which there is a continuous linear relation between the output and the input. This condition applies when both the output and input are within their normal operating ranges and when operation is at a frequency below a limiting value [S51.1].

control action, proportional plus derivative (rate) (PD) Control action in which the output is proportional to a linear combination of the input and the time rate of change of input [S51.1].

control action, proportional plus integral (reset) (PI) Control action in which the output is proportional to linear combination of the input and the time integral of the input [S51.1].

control action, proportional plus integral (reset) plus derivative (rate) (PID) Control action in which the output is proportional to a linear combination of the input, the time integral of input and the time rate of change of input [S51.1].

control, adaptive Control in which automatic means are used to change the type or influence (or both) of control parameters in such a way as to improve the performance of the control system [S51.1].

control agent The energy or material comprising the process element which is controlled by manipulating one or more of its attributes—the attribute(s) commonly termed the controlled variable(s).

control algorithm A mathematical representation of the control action to be performed.

Control and Instrumentation Engineer 1. Applies standard engineering standards and practices to the specification, sizing, and functional design of instrumentation hardware or control systems. Involves a clear understanding of the manufacturing or scientific process to be controlled. Serves as the key person on the instrumentation design and operation team, often supervising and reviewing the team's efforts. 2. Under supervision, participates in the design and planning of control and instrument systems as required by the project assignment, including: a) Collecting background information. b) Preparing drawings and calculations. c) Designing or modifying systems. d) Assisting in selection and procurement of equipment. e) Ensuring compliance with applicable standards and codes. f) Completing assigned tasks on schedule. g) Assisting technicians and designers as needed. h) Possible specializing in specific engineering discipline. 3. See Instrument Engineer.

control apparatus An assembly containing one or more control devices which acts to manipulate a controlled variable.

control block A storage area through which a particular type of information required for control of the operating system is communicated among its parts.

control board A panel that contains control devices, instrument indicators and sometimes recorders which display the status of a system or subsystem, and from which switches, dials and controllers can be manipulated to alter system operating variables. Also known as control panel; panel board.

control bus The data highway used for carrying control signals.

control calculations Installation-dependent calculations that determine output signals from the computer to operate the process plant. These may or may not use generalized equation forms such as PID forms.

control card A card which contains input data or parameters for a specific application of a general routine.

control, cascade Control in which the output of one controller is the set point for another controller [S51.1].

control center An equipment structure or group of structures, from which a process is measured, controlled and/or monitored [S51.1].

control character A character whose purpose is to control an action rather than to pass data to a program; ASCII control characters have an octal code between 0 and 37; normally typed by holding down the CTRL key on a terminal keyboard while striking a character key.

control chart A plot of some measured quantity, such as a dimension, versus sample number, time, or quantity of goods produced, which plot can be used to determine a quality trend or to make adjustments in process controls as necessary to keep the measured quantity within prescribed limits.

control circuit 1. A circuit in a control apparatus which carries the electrical signal used to determine the magnitude or duration of control action; it does not carry the main power used to energize instrumentation, controllers, motors or other control devices. 2. A circuit in a digital computer which performs any of the following functions—directs the sequencing of program commands, interprets program commands, or controls operation of the arithmetic element and other computer circuits in accordance with the interpretation.

control complexity ratio (CCR) A measure of the complexity of a particular control system's logic configuration.

control computer A process computer which directly controls all or part of the elements in the process. See process computer.

control counter A physical or logical device in a computer that records the storage locations of one or more instruction words which are to be used in sequence, unless a transfer or special instruction is encountered.

control device Any device—such as a heater, valve, electron tube, contactor, pump, or actuator—used to directly effect a change in some process attribute.

Control Diagram Language (CODIL) A process-oriented language and system offered by Leeds and Northrup Company.

control, differential gap Control in which the output of a controller remains at a maximum or minimum value until the controlled variable crosses a band or gap, causing the output to reverse. The controlled variable must then cross the gap in the opposite direction before the output is restored to its original condition [S51.1].

control, direct digital Control performed by a digital device which establishes the signal to the final controlling element [S51.1].

control, direct multiplex A control means using hardware, a computer program, or both, to directly interleave or simultaneously receive or transmit two or more signals on a single channel [RP55.1].

control drawing A drawing or other document provided by the manufacturer of the intrinsically safe or associated apparatus that details the allowed interconnections between the intrinsically safe and associated apparatus [RP12.6].

control electrode In an electron tube or similar device, an electrode whose potential can be varied to induce variations in the current flowing between two other electrodes.

control element A component of a control system that reacts to manipulate a process attribute when stimulated by an actuating signal.

control equipment Equipment which controls one or more output quantities to specific values, each value being determined by manual setting, local or remote programming, or by one or more input variables [S82.03].

control, feedback Control in which a measured variable is compared to its desired value to produce an actuating error signal which is acted upon in such a way as to reduce the magnitude of the error [S51.1].

control, feedforward Control in which information concerning one or more conditions that can disturb the controlled variable is converted, outside of any feedback loop, into corrective action to minimize deviations of the controlled variable. The use of feedforward control does not change system stability because it is not part of the feedback loop which determines the stability characteristics [S51.1].

control function   See control operation.

control grid   An element of an electron tube ordinarily positioned between the anode and cathode to act as a control electrode.

control, high limiting   Control in which the output signal is prevented from exceeding a predetermined high limiting value [S51.1].

control initiation   The signal introduced into a measurement sequence to regulate any subsequent control action as a function of the measured quantity.

control instruction   A computer instruction that directs the sequence of operations.

control key   A computer control key that when pressed with another key gives that key a different meaning.

controlled cooling   Cooling a part from elevated temperature in a specific medium to produce desired properties or microstructure, or to avoid cracking, distortion or high residual stress; the usual cooling mediums, in descending order of severity, are brine, water, soluble oil, fused salt, oil, fan-blown air and still air.

controlled medium   The process fluid or other substance containing the controlled variable.

controlled system   The body, machine or process which determines the relationship between an indirectly controlled variable and a corresponding directly controlled variable [S51.1]. See system, controlled.

controlled variable   1. The variable which the control system attempts to keep at the setpoint value. The setpoint may be constant or variable. 2. The part of a process which you want to control (flow, level, temperature, pressure, etc.). 3. A process variable which is to be controlled at some desired value by means of manipulating another process variable.

controller   1. A device or program which operates automatically to regulate a controlled variable. This term is adequate for the process industries where the word "controller" always means "automatic controller". In some industries, "automatic" may not be implied and the term "automatic controller" is preferred [S51.1]. 2. A device having an output that varies to regulate a controlled variable in a specified manner. A controller may be a self-contained analog or digital instrument, or it may be the equivalent of such an instrument in a shared-control system.

An automatic controller varies its output automatically in response to a direct or indirect input of a measured process variable. A manual controller is a manual loading station, and its output is not dependent on a measured process variable but can be varied only by manual adjustment. A controller may be integral with other functional elements of a control loop [S5.1]. 3. Any manual or automatic device or system of devices for the regulation of boiler systems to keep the boiler at normal operation. If automatic, the device or system is motivated by variations in temperature, pressure, water level, time, flow, or other influences [S77.42]. 4. A device for interfacing a peripheral unit or subsystem in a computer; for example, a tape controller or a disk controller. 5. Device which contains all the circuitry needed for receiving data from external devices, both analog and digital, processes the data according to pre-selected algorithms, then provides the results to external devices.

controller, derivative (D)   A controller which produces derivative control action only [S51.1].

controller, direct acting   A controller in which the value of the output signal increases as the value of the input (measured variable) increases [S51.1]. See controller, reverse acting.

controller, floating   A controller in which the rate of change of the output is a continuous (or at least a piecewise continuous) function of the actuating error signal. Note: The output of the controller may remain at any value in its operating range when the actuating error signal is zero and constant. Hence the output is said to float. When the controller has integral control action only, the mode of control has been called "proportional speed floating". The use of the term integral control action is recommended as a replacement for "proportional speed floating control" [S51.1].

controller, integral (reset) (I)   A controller which produces integral control action only. It may also be referred to as controller, proportional speed floating [S51.1].

controller, multiple-speed floating   A floating controller in which the output may change at two or more rates, each corresponding to a definite range of values of the actuating error signal [S51.1].

controller, multi-position   A controller having two or more discrete values of output [S51.1].

controller, on-off   A two-position controller in

which one of the two discrete values is zero [S51.1].

controller, program A controller which automatically holds or changes set point to follow a prescribed program for a process [S51.1].

controller, proportional (P) A controller which produces proportional control action only [S51.1].

controller, proportional plus derivative (rate) (PD) A controller which produces proportional plus derivative (rate) control action [S51.1].

controller, proportional plus integral (reset) (PI) A controller which produces proportional plus integral (reset) control action [S51.1].

controller, proportional plus integral (reset) plus derivative (rate) (PID) A controller which produces proportional plus integral (reset) plus derivative (rate) control action [S51.1].

controller, proportional speed floating See controller, integral (reset) (I) [S51.1].

controller, ratio A controller which maintains a predetermined ratio between two variables [S51.1].

controller, reverse acting A controller in which the value of the output signal decreases as the value of the input (measured variable) increases [S51.1]. See controller, direct acting.

controller, sampling A controller using intermittently observed values of a signal such as the set point signal and the actuating error signal, or the signal representing the controlled variable to effect control action [S51.1].

controller, self-operated (regulator) A controller in which all the energy to operate the final controlling element is derived from the controlled system [S51.1].

controller, single-speed floating A floating controller in which the output changes at a fixed rate increasing or decreasing depending on the sign of the actuating error signal. See controller, floating [S51.1].

controller, three-position A multi-position controller having three discrete values of output. This is commonly achieved by selectively energizing a multiplicity of circuits (outputs) to establish three discrete positions of the final controlling element [S51.1].

controller, time schedule A controller in which the set point or the reference input signal automatically adheres to a predetermined time schedule [S51.1].

controller, two-position A multi-position

controller having two discrete values of output [S51.1].

control limit An automatic safety control responsive to changes in liquid level, pressure or temperature or position for limiting the operation of the controlled equipment.

control limits In statistical quality control, the upper and lower values of a measured quantity that establish the range of acceptability; if any individual measurement falls outside this range, the part involved is rejected and if the sample average for the same measurement falls outside the range, the entire lot is rejected.

controlling extensions A controller which derives its input from the motion of the float can be installed within the extension housing.

controlling means The components of an automatic controller that are directly involved in producing an output control signal or other controlling action.

controlling system See system, controlling.

control logic The sequence of steps or events necessary to perform a particular function. Each step or event is defined to be either a single arithmetic or a single Boolean expression.

control loop 1. Two or more devices processing a single variable which may provide an input signal to a control system [S67.14]. 2. A combination of two or more instruments or control functions arranged so that signals pass from one to another for the purpose of measurement and/or control of a process variable. See closed loop and open loop.

control loop instability A regular oscillation of a feedback control system caused by excessive loop gain. It is independent of external disturbances [RP75.18].

control, low limiting Control in which output signal is prevented from decreasing beyond a predetermined low limiting value [S51.1].

control mode A specific type of control action such as proportional, integral, or derivative [S51.1].

control operation An action performed by a single device, such as the starting or stopping of a particular process. Conventionally, carriage return, fault change, rewind, end of transmission, etc., are control operations, whereas the actual reading and transmission of data are not.

control, optimizing Control that automatically seeks and maintains the most advantageous value of a specified variable, rather than maintaining it at one set value [S51.1].

control output module   A device which stores commands from the computer and translates them into signals which can be used for control purposes. It can generate digital outputs to control on-off devices or to pulse setpoint stations, or it can generate analog output (voltage or current) to operate valves and other process control devices.

control panel   1. A part of a computer console that contains manual controls. 2. See plugboard. 3. See console and automatic control panel.

control point   The setpoint or other reference value that an automatic controller acts to maintain as the measured value of a process variable under a given set of conditions.

control precision   The degree to which a given value of a controlled variable can be reproduced for several independent control initiations using the same control point and the same system operating conditions.

control program   1. A group of programs that provides such functions as the handling of input/output operations, error detection and recovery, program loading and communication between the program and the operator. IPL, supervisor, and job control make up the control program in the disk and tape operating systems. 2. Specific programs which control an industrial process.

control programming   Writing a user program for a computer which will control a process in the sense of reacting to random disturbances in time to prevent impairment of yield, or dangerous conditions.

control resolution   The smallest increment of change that can be induced in the controlled process variable as a result of control-system action.

control rod   A long piece of neutron-absorbing material that fulfills one or both of the functions of controlling the number of neutrons available to trigger nuclear fission or of absorbing sufficient neutrons to stop fission in case of an emergency.

control room area   See area, control room.

control, safety   Control (including relays, switches, and other auxiliary equipment used in conjunction therewith to form a safety control system) which are intended to prevent unsafe operation of the controlled equipment.

control, safety combustion   See combustion (flame) safeguard.

control, shared time   Control in which one controller divides its computation or control time among several control loops rather than by acting on all loops simultaneously [S51.1].

control signal override device   A device which overrides the effect of the control signal to the valve actuator to cause the closure member to remain stationary or assume a pre-selected position [S75.05].

control spring   A spring designed to produce a torque equal and opposite to the torque produced by an instrument's moving element for any position of the moving element within the limits of its operating range.

control station   A manual loading station that also provides switching between manual and automatic control modes of a control loop. It is also known as an auto-manual station. In addition, the operator interface of a distributed control system may be regarded as a control station [S5.1].

control, supervisory   Control in which the control loops operate independently subject to intermittent corrective action; e.g., set point changes from an external source [S51.1].

control system   A system in which deliberate guidance or manipulation is used to achieve a prescribed value of a variable [S51.1].

control system, automatic   A control system which operates without human intervention. See also control system [S51.1].

control system, multi-element (multi-variable)   A control system utilizing input signals derived from two or more process variables for the purpose of jointly affecting the action of the control system. Note: (a) Examples are input signals representing pressure and temperature, or speed and flow, etc. (b) A term used primarily in the power industry [S51.1].

control system, non-interacting   A control system with multiple inputs and outputs in which any given input-output pair is operating independently of any other input-output pair [S51.1].

Control Systems Engineer   1. Design, assemble, build, and operate instrumentation and control systems. 2. Applies standard engineering standards and practices to the specification, sizing, and functional design of instrumentation hardware and control systems. Involves a clear understanding of the manufacturing or scientific process to be controlled. Serves as the key person on the instrumentation design and operation team, often supervising and reviewing the team's efforts. 3. See Instrument Engineer.

control, time proportioning   Control in which the output signal consists of periodic pulses whose duration is varied to relate, in some prescribed manner, the time average of the output to the actuating error signal [S51.1].

control unit   1. That portion of a multipart gas detection instrument which is not directly responsive to the combustible gas but which responds to the electrical signal obtained from one or more detector heads to produce an indication, alarm, or other output function if gas is present at the detector head location [S12.13][S12.15]. 2. The portion of a computer which directs the sequence of operations, interprets the coded instructions, and initiates the proper commands to the computer circuits preparatory to execution. 3. A device designed to regulate the fuel, air, water, or electrical supply to the controlled equipment. It may be automatic, semi-automatic, or manual.

control valve   1. Any valve which controls pressure, rate of flow, or flow direction in a fluid- or gas-filled system. A final controlling element, through which a fluid or gas passes, which adjusts the size of flow passage as directed by a signal from a controller to modify the rate of flow of the fluid [S51.1]. 2. A device, other than a common, hand-actuated ON-OFF valve or self-actuated check valve, that directly manipulates the flow of one or more fluid process streams. It is expected that use of the designation "hand control valve" will be limited to hand-actuated valves that (a) are used for process throttling, or (b) require identification as an instrument [S5.1]. 3. A power operated device which modifies the fluid flow rate in a process control system. It consists of a valve connected to an actuator mechanism that is capable of changing the position of a flow controlling element in the valve in response to a signal from the controlling system [S75.05]. 4. A final controlling element, through which a fluid or gas passes, which adjusts the size of flow passage as directed by a signal from a controller to modify the rate of flow.

control valve gain   The change in the flow rate as a function of the change in valve travel. It is the slope of the installed or of the inherent valve flow characteristic curve and must be designated as installed or inherent [S75.05].

control variable   1. The variable which the control system attempts to keep at the setpoint value. 2. The part of a process which you want to control (flow, level, temperature, pressure, etc.) 3. A process variable which is to be controlled at some desired value by means of manipulating another process variable.

control, velocity limiting   Control in which the rate of change of a specified variable is prevented from exceeding a predetermined limit [S51.1].

convection   The transmission of heat by the circulation of a liquid or a gas such as air. Convection may be natural or forced.

convection cooling   Removing heat from a body by means of heat transfer using a moving fluid as the transfer medium, usually involving only the motion caused by differences in heat content between fluid near the hot surface and fluid at some distance from the surface.

convection-type superheater   See superheater.

convergence   The condition in which all the electron beams of a multibeam (color) cathode ray tube intersect at a specific point.

conversational mode   Communication between a terminal and a computer in which each entry from the terminal elicits a response from the computer and vice versa.

conversion coating   A protective surface layer on a metal that is created by chemical reaction between the metal and a chemical solution.

conversion time   The time required for a complete measurement by an analog-to-digital converter.

conversion (to engineering units)   Scaling signals from their raw input form to the form used internally, usually into floating point engineering units.

conversion transducer   Any transducer whose output-signal frequency is different from its input-signal frequency.

converter   1. A device that receives information in one form of an instrument signal and transmits an output signal of another form. An instrument which changes a sensor's output to a standard signal is properly designated as a transmitter, not a converter. Typically, a temperature element (TE) may connect to a transmitter (TT), not to a converter (TY). A converter is also referred to as a transducer; however, "transducer" is a completely general term, and its use specifically for signal conversion is not recommended [S5.1]. 2. A type of refining furnace where impurities are oxidized and removed by blowing air or oxygen through the molten metal. 3. A/D analog to digital; D/A digital to analog; I/P current to pneumatic pressure converter; P/I pneumatic pressure to current converter; P/V pneumatic pressure to voltage converter; V/P voltage to pneumatic pressure converter. 4. A converter is also referred to as a transducer; however, "transducer" is a completely general term, and its use specifically for signal conversion is not recommended.

converter, analog to digital (ADC) An instrument used to convert analog signals to digital coded values which are proportional to the analog input voltages [RP55.1].

converter, digital to analog (DAC) An instrument which converts digital information into analog signals which are proportional to the numerical value of the digital information [RP55.1].

convex A term describing a surface whose central region is raised with respect to a flat plane approximately passing through its periphery.

convex curve A change in the angle of inclination of a belt conveyor where the center of the curve is below the conveyor [RP74.01].

convex programming In operations research, a particular case of nonlinear programming in which the function to be maximized or minimized and the constrains are appropriately convex or concave functions of the controllable variables. Contrast with dynamic programming, integer programming, linear programming, mathematical programming, and quadratic programming.

conveyor A continuously moving materials-handling device for transferring large numbers of individual items or quantities of bulk solids from one location to another over a relatively short distance along a fixed path.

conveyor stringers Support members for the conveyor on which the idlers are mounted [RP74.01].

coolant 1. The fluid contained within the reactor coolant pressure boundary [S67.03]. 2. Any fluid used primarily to remove heat from an object and carry it away. 3. In a machining operation, any cutting fluid whose chief function is to keep the tool and workpiece cool.

Coolidge-type x-ray tube A high-vacuum tube in which electrons emitted from a high-voltage cathode impinge on a water-cooled metal target inclined with respect to the tube axis; x-rays emitted from the focal spot on the target are directed through a side window in the metal tube enclosure, where a material relatively transparent to x-rays—beryllium foil, mica, aluminum or special low-absorption glass—allows them to escape.

coprocessor A device added to a CPU that performs special functions more efficiently than the CPU alone.

copy In data processing, reproducing data from one storage device to another.

copy protection 1. The inability to copy a disk, particularly program disks, by the addition of codes on the disk. 2. The inability to copy data on a disk by the addition of an adhesive cover on a side slot of the disk.

corbinotron A device consisting of a corbino disc, made of high-mobility semiconductor material, and a coil that produces a magnetic field perpendicular to the plane of the disc.

cord-connected equipment Equipment that connects to a supply circuit receptacle by means of a permanently attached flexible power-supply cord and attachment plug or by means of a detachable power-supply cord [S82.01].

core 1. A strongly ferromagnetic material used to concentrate and direct lines of flux produced by an electromagnetic coil. 2. The inner layer in a composite material or structure. 3. The central portion of a case-hardened part, which supports the hard outer case and gives the part its toughness and shock resistance. 4. An insert placed in a casting mold to form a cavity, recess or hole in the finished part. 5. A rod or closed tube inserted in a tube to reduce the flow area. See magnetic core. 6. Magnetic memory elements; typically the main memory in a computer system.

cored electrode A tubular welding electrode containing flux or some other material in the central cavity.

cored solder Wire solder having a flux-filled central cavity.

core dump See storage dump.

core (fiber optic) The inner portion of an optical fiber which carries light along the length of the fiber. Light is confined to the core by a difference in refractive index between core and cladding, with the latter having a lower index.

core iron A grade of soft steel suitable for making cores used in electromagnetic devices such as chokes, relays and transformers.

core memory The most common form of main memory storage used by a central processing unit, in which binary data are represented by switching the polarity of magnetic cores.

core resident A term pertaining to programs or data permanently stored in core memory for fast access.

core storage See magnetic core.

core wire Copper wire having a steel core, often used to make antennas.

coring A metallurgical condition where individual grains or dendrites vary in composition from center to grain boundary due to nonequilibrium cooling during solidification in an alloy that solidifies over a range of temperatures.

Coriolis effect An accelerating force acting on any body moving freely above the earth's surface due to the fact that the earth is rotating with respect to a given axis through its center; it is the Coriolis effect that causes, for instance, a level bubble carried in an airplane to be deflected perpendicular to the direction of flight, and a river in the Northern Hemisphere to scour its right bank more than its left bank whereas a river in the Southern Hemisphere scours its left bank more than its right. Coriolis effect is the basis for mass-flow meters.

Coriolis force Results from Coriolis acceleration acting on a mass moving with a velocity radially outward in a rotating plane.

Coriolis-type mass flowmeter An instrument for measuring mass flowrate by determining the torque from radial acceleration of the fluid.

Corliss valve A type of valve used to admit steam to, or exhaust it from, a reciprocating engine cylinder.

corner-cube prism A prism in which three flat surfaces meet at right angles, as they would if they were the corner of the cube. Incident light through a planar face is reflected back to the source.

corner frequency In the asymptotic form of Bode diagram, that frequency indicated by a break point, i.e., the junction of two confluent straight lines asymptotic to the log gain curve [S51.1].

corner taps The differential pressure signal location in an orifice flange union defined by the corner formed between the orifice plate and the internal diameter of the flange.

corona voltmeter A type of voltmeter that uses the inception of corona to determine the crest value of voltage in an a-c electric current.

Corporation for Open Systems (COS) An organization formed in 1985 to coordinate member company efforts in the selection of standards and protocols, conformance testing, and the establishment of certification.

correction In process instrumentation, the algebraic difference between the ideal value and the indication of the measured signal. It is the quantity which added algebraically to the indication gives the ideal value. A positive correction denotes that the indication of the instrument is less than the ideal value. Correction=Ideal value-indication [S51.1]. See error.

correction time See time, settling.

corrective action The change produced in a controlled variable in response to a control signal.

corrective maintenance 1. An activity that is not normal in the operation of the equipment and which requires access to the interior. Such activities are expected to be performed by qualified personnel who are aware of the hazards involved. Such activities typically include locating causes of faulty performance, replacement of defective components, adjustment of service controls, or the like [S12.12]. 2. Maintenance specifically intended to eliminate an existing fault. Synonymous with emergency maintenance. Contrast with preventive maintenance.

corrective network An electronic network incorporated into a circuit to improve its transmission or impedance properties, or both.

correlation Measurement of the degree of similarity of two images as a function of detail and relative position of the images. It is obtained by multiplying the Fourier transforms of the two images, then taking the the Fourier transform of the product.

correlation check A procedure whereby the performance and accuracy of a calibration system is checked against another calibration system using "master flowmeters" as the standards [RP31.1].

correlator A logic device which compares a series of bits in a data stream with a known bit sequence and puts out a signal when correlation is achieved. One use of the correlator is as a PCM frame synchronizer.

corresponding states A principle that states that two substances should have similar properties at corresponding conditions with reference to some basic properties, e.g., critical pressure and critical temperature.

Corrodekote test An accelerated corrosion test for electrodeposits in which a specimen is coated with a slurry of clay in a salt solution, and then is exposed for a specified time in a high-humidity environment.

corrosion 1. Deterioration of a substance (usually a metal) because of a reaction with its environment [S71.04]. 2. The wasting away of metals due to chemical action in a boiler; usually caused by the presence of O_2, CO_2 or an acid.

corrosion fatigue A synergistic interaction of the failure mechanisms corrosion and fatigue such that cracking occurs much more rapidly than would be predicted by simply adding their separate effects; failure by corrosion fatigue requires the simultaneous presence of a cyclic stress and a corrosive environment.

corrosion protection Preventing corrosion or reducing the rate of corrosive attack by any of several means including coating a metal surface with a paint, electroplate, rust-preventive oil, anodized coating or conversion coating; adding a corrosion-inhibiting chemical to the environment; using a sacrificial anode; or using an impressed electric current.

corrosive Any substance or environment that causes corrosion.

corrosive flux A soldering flux that removes oxides from the base metal when the joint is heated to apply solder; the flux is usually composed of inorganic salts and acids which are corrosive and must be removed before placing the soldered components in service to ensure maximum service life.

corrosiveness The degree to which a substance causes corrosion.

corrugated fastener A thin, corrugated strip of steel used to fasten two pieces of wood together by hammering it into the wood approximately at right angles to the joint line.

corrugating Forming sheet metal into a series of alternating parallel ridges and grooves; forming may be done by rolling the metal between matched grooved rolls or by forming it in a press brake equipped with a special-shaped punch and die.

corrupt In data processing, the inclusion of errors in programs or data.

COS See Corporation for Open Systems.

cosmic rays Penetrating ionizing radiation whose ultimate origin is outside the earth's atmosphere; some of the constituents of cosmic rays can penetrate many feet of material such as rock.

cotter A tapered part similar to a wedge or key that can be driven into a tapered hole to hold an assembly together.

cottered joint A joint in which power is transferred across the joint via shear force transverse to the longitudinal axis of a bar (usually tapered along one side to ensure a tight fit) known as a cotter, which holds the joint together.

cotter pin A split pin, usually formed by folding a length of half-round wire back on itself; the pin is inserted into a hole and then is bent to keep a castle nut from turning on a bolt, to hold a cotter securely in place, to hold hinge plates together, or to pin various other machine parts together. Also known as cotter key.

Cottrell precipitator A device for removing dust or mist from a gas by passing the gas through a vertical, electrically grounded pipe where the particulates become ionized by corona discharge from an axial wire maintained at a high negative voltage; the ionized particles migrate to the pipe's inner wall where they collect for later removal by mechanical means.

coulomb Metric unit for quantity of electricity.

coulombmeter An instrument for measuring the quantity of electricity (in coulombs) by integrating a stored charge in a circuit that has a high impedance.

coulometer An electrolytic cell constructed and operated to measure a quantity of electricity in terms of the electrochemical action it produces.

coulometric titration A method of wet chemical analysis in which the amount of an unknown substance taking part in a chemical reaction is determined by measuring the number of coulombs required to reach the end point in electrolysis.

count In computer programming, the total number of times a given instruction is performed.

counter 1. A device or register in a digital processor for determining and displaying the total number of occurrences of a specific event. 2. In the opposite direction. 3. Device or PC program element that can total binary events and perform ON/OFF actions based on the value of the total. 4. A device, register, or location in storage for storing numbers or number representations in a manner which permits these numbers to be increased or decreased by the value of another number, or to be changed or reset to zero or to an arbitrary value.

counterbore A drilled or bored flat-bottomed hole, often concentric with another, smaller hole.

countercurrent flow Flow of two fluids in opposite directions within the same device, such as a tube-in-shell heat exchanger. Contrast with counterflow.

counterflow Flow of a single fluid in opposite directions in adjacent portions of the same device, such as a U-bend tube. Contrast with countercurrent flow.

counter, input The storage and buffer device between an external pulse source and the computer; e.g., a real time clock or some other totalizing unit [RP55.1].

countershaft A secondary shaft, driven by the main shaft of a machine, and used to supply power to one or more machine parts.

countersink A chamfer around the edge of a circular hole, which removes burrs, provides a seat for a flat-head screw or other fastener, or provides a tapered surface for a machine center to rest in.

counterweight 1. A mass which counterbalances the weight of the lifting device or load platform of an elevator or hoist so that the engine must only work against the payload, friction, and any remaining unbalanced machine loads. 2. Any mass incorporated into a mechanism to compensate for an out-of-balance condition and maintain static equilibrium. Also known as counterbalance; counterpoise.

counting rate The average number of ionizing events that occur per unit time, as determined by a counting tube or similar device.

counting-rate meter An instrument whose indicated output is related to the average rate of occurrence of ionizing events.

counting scale Any of several designs of weighing device where the total weight of a large number of identical parts is compared with the weight of one part or the weight of a small, easily counted number of parts, and the number of parts in the unknown quantity determined by automatic indication, readout or calculation.

counts 1. An alternate form of representing raw data corresponding to the numerical represen-

tation of a signal received from or applied to external hardware. 2. The accumulated total of a series of discrete inputs to a counter. 3. The discrete inputs to an accumulating counter. See digitized signal.

couplant A substance used to transmit sound waves from an ultrasonic search unit to the surface of a test piece, thus reducing losses and improving test accuracy; usual couplants include water, oil, grease, paste or other liquid or semi-solid substances.

coupled control-element action A type of control system action in which two or more actuating signals or control element actions are used in concert to operate one control device.

coupled reference input See cascade action.

coupler 1. In data processing, a device that joins similar items. 2. In fiber optics, a device which joins together three or more fiber ends—splitting the signal from one fiber so it can be transmitted to two or more other fibers. Directional, star, and tee couplers are the most common.

coupling 1. Any device that connects the ends of adjacent parts; the connection may be rigid, allowing little or no relative movement, or it may be flexible, accommodating misalignment and other sources of relative movement. 2. A mechanical fastening between two shafts that provides for the transmission of power and motion. Also known as shaft coupling. 3. Interdependence in a computer system.

covering power The ability of an electroplating solution to give a satisfactory plate at low current densities, such as occur in recesses, but not necessarily to build up a uniform coating. Contrast with throwing power.

cover plate 1. Any flat metal or glass plate used to cover an opening. 2. Specifically, a piece of glass used to protect the tinted glass in a welder's helmet or goggles from being damaged by weld spatter.

cowling A metal cover, usually one that provides a streamlined enclosure for an engine.

CP/M An operating system for microcomputers.

C power supply An electrical power supply connected between the cathode and grid of a vacuum tube to provide a grid-bias voltage.

CPU See central processing unit; see also unit, central processing.

CPU-bound A state of program execution in which all operations are dependent on the activity of the central processor, for example, when a large number of calculations are being performed; compare to I/O-bound.

crack 1. A fissure in a part where it has been broken but not completely severed into two pieces. 2. The fissure or chink between adjacent components of a mechanical assembly. 3. To incompletely sever a solid material, usually by overstressing it. 4. To open a valve, hatch, door or other similar device a very slight amount.

cracked flow A nonstandard term. See clearance flow [S75.05].

cracked residue The fuel residue obtained by cracking crude oils.

cracking The thermal decomposition of complex hydrocarbons into simpler compounds or elements.

cracking process A method of manufacturing gasoline and other hydrocarbon products by heating crude petroleum distillation fractions or residues in the presence of a catalyst so that they are broken down into lighter hydrocarbon products, some of which can be distilled off.

crane A hoisting machine with a power-driven horizontal or inclined boom and lifting tackle.

crane hoist A mobile hoisting machine used principally for lifting loads by means of cables; it consists of a mobile undercarriage and support structure, a power unit and winch enclosed in a cab or house (often one that swivels on the undercarriage), a movable boom and various lifting, boom positioning and support cables.

crane scale A type of lifting device integral with or attached to a crane hook and having an internal load cell that automatically weighs a load as it is lifted; where a strain-gage load cell is used, weight can be indicated or recorded remotely.

crank A mechanical link that can revolve about a center of rotation.

crankpin A cylindrical projection on a crank for attaching a connecting rod.

crankshaft 1. A straight shaft to which one or more cranks are attached. 2. A cast, forged or machined shaft with integral cranks, such as is used in a reciprocating automobile engine.

crank throw 1. The web or arm of a crank. 2. The radial displacement of the crankpin from the crankshaft axis.

crank web The portion of a crank that connects a crankpin to the crankshaft or to another adjacent crankpin. Also known as crank throw.

crash A computer hardware or software malfunction that causes the system to be reset or restarted.

crater 1. A spot on the face of a cutting tool where it has been worn by contact with chips. 2. A depression at the finishing end of a weld bead.

crazing 1. A network of fine, shallow cracks at the surface of a coating, solid metal or plastics material. 2. Development of such a network.

CRC See cyclic redundancy check.

create To open, write data to, and close a file for the first time.

creep 1. A change in output occurring over a specific time period while the measurand and all environmental conditions are held constant [S37.1]. 2. Time-dependent plastic strain occurring in a metal or other material under stress, usually at elevated temperature.

creepage distance The shortest distance measured over the surface of insulation between conductive parts. Air gaps shorter than 1.0 millimeter (0.04 inch) are not considered to interrupt the surface path [S82.01].

creep at load The change in output occurring with time under rated load and with all environmental conditions and other variables remaining constant [S37.8].

creep recovery The change in zero-measurand output occurring with time after removal of rated load, which had been applied for an identical period of time as employed in evaluating creep at load [S37.8].

crest 1. The top of a screw thread. 2. The bottom edge of a weir notch, sometimes referred to as the sill.

crest voltmeter An instrument whose indicated value is the average positive peak amplitude of a sinusoidal a-c electric voltage.

crest width The distance along the crest between the sides.

crevice corrosion A type of concentration-cell corrosion associated with the stagnant conditions in crevices, fissures, pockets and recesses away from the flow of a principal fluid stream, where concentration or depletion of dissolved salts, ions or gases such as oxygen leads to deep pitting.

crimping 1. Forming small corrugations in order to set down and lock a seam, create an arc in a metal strip or reduce the radius of an existing arc or circle. 2. Causing something to become wavy, crinkled or warped. 3. Pinching or pressing together to seal or unite, especially the longitudinal seam of a tube or cylinder.

critical cooling rate The minimum cooling rate that will suppress undesired transformations during a hardening heat treatment.

critical damping See damping.

critical dimension 1. Generally, any physical measurement whose value or accuracy is considered vital to the function of the involved component or assembly. 2. In a waveguide, the cross-sectional dimension which determines the waveguide's critical frequency.

critical flow This is a somewhat ambiguous term that signifies a point at which the characteristics of flow suffer a finite change. In the case of a liquid, critical flow could mean the point at which the flow regime changes from laminar to transitional. It more often is used to mean choked flow. In the case of a gas, critical flow may mean the point at which the velocity at the vena contracta attains the velocity of sound, or it may mean the point at which the flow is fully choked.

critical frequency The frequency below which a traveling wave of a given mode cannot be maintained in a given waveguide.

critical pressure The equilibrium pressure of a fluid that is at its critical temperature.

critical-pressure ratio The ratio of downstream pressure to upstream pressure which corresponds to the onset of turbulent flow in a moving stream of fluid.

critical speed The speed of angular rotation at which a shaft becomes dynamically unstable due to lateral resonant vibration.

critical strain The amount of prior plastic strain that is just sufficient to trigger recrystallization when a deformed metal is heated.

critical temperature The temperature of a fluid above which the fluid cannot be liquefied by pressure alone.

critical velocity For a given fluid, the average linear velocity marking the upper limit of streamline flow and the lower limit of turbulent flow at a given temperature and pressure in a given confined flowpath.

cross-assembler An assembler program run on a larger host computer and used for producing machine code to be executed on another usually smaller, computer.

cross-axis acceleration See transverse acceleration.

crossbar micrometer An instrument for determining differences in right ascension and declination of celestial objects; it consists of two bars mounted perpendicular to each other in the focal plane of a telescope and inclined at 45° to the east-west path of the stars.

cross-compiler A computer program run on a larger host computer and used for translating a high level language program into the machine code to be executed on another computer.

cross drum boiler A section header or box boiler in which the axis of the horizontal drum is at right angles to the center lines of the tubes in the main bank.

crosshair An inscribed line or a thin hair, wire or thread used in the optical path of a telescope, microscope or other optical device to obtain accurate sightings or measurements; sometimes, a pair of hairs at right angles are used, which is the original source of the term.

crosshead 1. A sliding block that moves back and forth between guides and that contains a wrist pin for converting reciprocating motion to rotary motion. 2. A device designed to extrude material at an angle which is used most extensively at the discharge end of an extruder in a wire-coating operation.

cross-modulation Carrier and signal harmonics of one or more channels appearing in other channels of a system; in the case of a large number of cross-modulation products, the resultant cross-talk noise approaches the characteristics of fluctuation noise (AM).

crossover frequency 1. The frequency at which a dividing network delivers equal power to upper-band and lower-band channels. 2. The frequency at which the asymptotes to the constant-amplitude and constant-velocity portions of the frequency-response curve of an acoustic recording system intersect. Also known as transition frequency; turnover frequency.

crossover network A selective network that divides the audio-frequency output of an amplifier into two or more bands of frequencies to supply two or more loudspeakers. Also known as dividing network; loudspeaker dividing network.

cross section 1. For a given confined flowpath or a given elongated structural member, the dimensions, shape or area determined by its intersection with a plane perpendicular to its longitudinal axis. 2. In characterizing interactions between moving atomic particles, the probability per unit flux and per unit time that a given interaction will occur.

cross sensitivity, cross-axis sensitivity See transverse sensitivity.

crosstalk 1. The unwanted energy transferred from one circuit, called the "disturbing" circuit, to another circuit, called the "disturbed" circuit [RP55.1]. 2. The unwanted signals in a channel that originate from one or more channels in the same communication system. 3. Signals electrically coupled from another circuit, usually undesirably, but sometimes for useful purposes.

cross-wire weld A resistance weld made by passing a controlled electric current through the junction of a pair of crossed wires or bars; used extensively to make mesh or screening.

crown 1. The part of a drill bit that is inset with diamonds. 2. The vertex of a structural arch or arched surface. 3. The domed top of a furnace or kiln. 4. The central portion of sheet material that is slightly thicker than at the edges. 5. Any raised central portion of a nominally flat surface.

crown glass An optical glass of alkali-lime-silica composition with index refraction usually 1.5 to 1.6.

crown sheet In a firebox boiler, the plate forming the top of the furnace.

CRT See Cathode Ray Tube.

CRT display 1. Cathode ray tube (video screen). 2. Hardware display on a cathode-ray-tube alphanumerically and/or graphically.

crucible A pot or vessel made of a high-melting-point material, such as a ceramic or refractory metal, used for melting metals and other materials.

crude oil Unrefined petroleum.

crush 1. A casting defect caused by displacement of sand as the mold is closed. 2. Buckling or breaking of a section of a casting mold caused by incorrect register as the mold is closed.

cryogenic Any process carried out at very low temperature, usually considered to be –60 °F (–50 °C) or lower.

cryogenic fluid A liquid which boils below –123 °Kelvin (–238 °F, –150 C) at one atmosphere absolute pressure.

cryometer A thermometer for measuring very low temperatures.

cryoscope A device for determining the freezing point of a liquid.

cryostat An apparatus for establishing the very low-temperature environment needed for carrying out a cryogenic operation.

crystalline fracture A type of fracture surface appearance characterized by numerous brightly reflecting facets resulting from cleavage fracture of a polycrystalline material.

crystal oscillator A device for generating an a-c signal whose frequency is determined by the properties of a piezoelectric crystal.

crystal spectrometer An instrument that uses diffraction from a crystal to determine the component wavelengths in a beam of x-rays or gamma rays.

CSMA/CD See carrier sense multiple access with collision detect.

CSU (cumulative sum) algorithm See compressor.

CTI See comparative tracking index.

cubicle 1. Any small room or enclosure. 2. An enclosure, usually free standing, that houses high-voltage electrical equipment.

cumulative dose The total amount of penetrating radiation absorbed by the whole body, or by a specific region of the body, during repeated exposures.

cup fracture A mixed mode fracture in ductile metals, usually observed in round tensile specimens, in which part of the fracture occurs under plane-strain conditions and the remainder under plane-stress conditions, such that in a round tensile bar one of the mating fracture surfaces looks like a miniature cup and the other like a truncated cone. Also known as cup-and-cone fracture.

cupping 1. The first step in deep drawing. 2. The fracture of severely worked rod or wire where one of the fracture surfaces is roughly conical and the other cup-shaped.

Curie Abbreviated Ci. The standard unit of measure for radioactivity of a substance; it is defined as the quantity of a radioactive nuclide that is disintegrating at the rate of 3.7×10^{10} disintegrations per second.

curing 1. Allowing a substance such as a polymeric adhesive or poured concrete to rest under controlled conditions, which may include clamping, heating or providing residual moisture, until it undergoes a slow chemical reaction to reach final bond strength or hardness. 2. In thermoplastics molding, stopping all movement for an interval prior to releasing mold pressure so that the molded part has sufficient time to stabilize.

current The rate of flow of an electrical charge in an electric circuit analogous to the rate of flow of water in a pipe.

current amplification For a given amplifier, the ratio of current delivered to the output circuit to the corresponding current supplied to the input circuit.

current loop (20mA) A serial transmission standard widely used for v.d.u.'s and teletypes. 0 and 1 are represented by the absence or presence of a current (20mA).

current meter Any of a wide variety of devices for measuring a-c or d-c electric current—including moving-coil, moving iron, electronic and electrodynamic instruments. See velocity-type flowmeter.

current-to-pressure transducer (I/P) A device which receives an analog electrical signal and converts it to a corresponding air pressure.

current transformer An instrument transformer designed to have its primary winding connected in series with a circuit carrying the current being measured or controlled.

current word address (CWA) The memory address of a word that is currently being operated on.

cursor A symbol used in the operation of keyboard-video displays to indicate on the display screen the physical location of the next character to be entered [RP55.1].

curvature of field A defect in an optical lens or system which causes the focused image of a plane field to lie along a curved surface rather than a flat plane.

curve-fit The process of determining the coefficients in a curve by mathematically fitting a given set of data to that curve class; for example, linear curve-fit, or n^{th}-order polynomial curve-fit.

custody transfer The act of transferring ownership of a fluid for money or the equivalent.

custom LSI A large scale integrated circuit designed for a specific purpose and which hence has a dedicated function.

cutoff 1. The parting line on a compression-molded plastics part. Also known as flash groove; pinch-off. 2. The point in the stroke of an engine where admission of the working fluid to the cylinder is shut off. 3. The time required to shut off the flow of working fluid into a cylinder.

cutoff tool A lathe tool with a narrow cutting edge used to sever a finished piece from remaining bar stock. Also known as parting tool.

cutoff valve A quick-acting valve used to stop the flow of working fluid into an engine cylinder.

cutoff wheel A thin abrasive wheel used to cut stock or to make slots in a part.

cutter A cutting tool, especially a rotary, toothed cutting wheel.

cutter bar A supporting member for the cutting tool in a lathe or other machine tool.

cutting angle The angle between the face of a cutting tool and the uncut stock surface.

cutting edge 1. In a diamond or ceramic tool, the point or edge of the insert material that actually cuts the work. 2. Generally, the sharpened edge of any cutting tool that contacts the work during machining.

cutting fluid In a metal-cutting operation, any liquid that is introduced into the area where the tool contacts the work, especially a liquid used to provide lubrication at the cutting edge, to carry away the heat generated during machining, and to flush out chips or other machining debris. Some cutting fluids have chemical compounds which react with the tool and material being cut to enhance cutting action.

cutting speed The relative velocity between cutting tool and workpiece along the main direction of cutting. Also known as peripheral speed.

cutting tool A sharp-edged single-point or toothed tool that comes in contact with the workpiece and removes stock in a machining operation. Also known as cutter.

cutting torch A device for producing a controlled flame which has an additional supply line for introducing a jet of oxygen into the flame; it cuts metal and other materials by first heating a small area, then rapidly oxidizing and melting the material along a thin line when the jet of oxygen is turned on. Usually a special plasma torch is needed for stainless steel because of its oxidation resistance.

CVT (current value table) See multiplex processor.

CWA See current word address.

cyaniding A surface-hardening process similar to carbonitriding that produces a carbon- and nitrogen-rich surface layer on steel by immersing parts in a bath of molten cyanide salts; can also be done in the gas phase.

cybernetics The branch of learning which brings

together theories and studies on communication and control in living organisms and machines.

cycle 1. An interval of space or time in which one set of events or phenomena is completed. 2. Any set of operations that is repeated regularly in the same sequence. The operations may be subject to variations in each repetition. 3. In any repetitive variable process, variation of a given variable through one complete range of values. 4. To run a machine through a complete set of operating steps. 5. The fundamental time interval for operations inside the computer. 6. A condition in a sequential circuit; from an initial, unstable state the circuit passes through more unstable states before reaching a stable state.

cycle-index The number of times a cycle has been executed or the difference, or the negative of the difference, between the number that has been executed and the number of repetitions desired.

cycle life The specified minimum number of cycles over which a device will operate as specified without changing its performance beyond specified tolerance [S75.05].

cycle redundancy check (CRC) An error detection scheme, usually hardware implemented, in which a check character is generated by taking the remainder after dividing all the serialized bits in a block of data by a predetermined binary number. This remainder is then appended to the transmitted date and recalculated and compared at the receiving point to verify data accuracy.

cycle stealing 1. A control feature which delays execution of a program to allow an I/O device to communicate with main storage without changing the logical condition of the CPU [RP55.1]. 2. Data transferred over the data bus during a direct memory access while little disruption occurs to the normal operation of the microprocessor.

cycle time 1. The time required by a computer to read from or write into the system memory. If system memory is core, the read cycle time includes a write-after-read (restore) subcycle. 2. Cycle time is often used as a measure of computer performance, since this is a measure of the time required to fetch an instruction.

cyclic A condition of either steady-state or transient oscillation of a signal about the nominal value.

cyclic code A form of gray code, used for expressing numbers in which, when coded values are arranged in the numeric order of real values, each digit of the coded value assumes its entire range of values alternately in ascending and descending order.

cyclic redundancy check (CRC) An error-checking technique in which a checking number is generated by taking the remainder after dividing all the bits in a block (in serial form) by a predetermined binary number. Can easily be achieved by shift operations.

cyclic redundancy check character (CRC) A character used in a modified cyclic code for error detection and correction.

cyclic shift A shift in which the data moved out of one end of the storing register are re-entered into the other end, as in a closed loop.

cycling Periodic repeated variation in a controlled variable or process action.

cycling life The specified minimum number of full scale excursions or specified partial range excursions over which a device will operate as specified without changing its performance beyond specified tolerances [S51.1] [S75.05].

cyclograph A device for electromagnetically sorting or testing metal parts by means of the pattern produced on a cathode-ray tube when a sample part is placed in an electromagnetic sensing coil; the CRT pattern is different in shape for different values of carbon content, case depth, core hardness or other metallurgical properties.

cyclotron A device that utilizes an alternating electric field between electrodes positioned in a constant magnetic field to accelerate ions or charged subatomic particles to high energies.

cylinder 1. A domed, closed storage tank for hot water. Also known as storage calorifier. 2. A strong, thick-walled container for storing and transporting compressed gases. 3. A round, straight-walled cavity, closed at one or both ends, that a piston rides in to convert the potential energy in pressurized gas to linear mechanical motion and power, or to utilize mechanical power to compress a gas.

cylinder block A massive piece of metal, usually made by casting, that contains the piston chambers of a multicylinder engine or compressor. Also known as block; engine block.

cylinder bore The inside diameter of a piston chamber.

cylinder, disk All like-numbered tracks on a disk pack; a portion of the disk which can be recorded or reproduced without moving the heads.

cylinder head The cap, which usually has a specially shaped recess, used to close the end of a piston chamber in a reciprocating engine, pump or compressor; usually, it provides valve openings, spark-plug taps and other penetrations necessary for machine operation.

cylinder liner A separate cylindrical sleeve that is inserted into a piston chamber to provide a cylinder wall with properties different from those of the cylinder block. Normally used to furnish a better wearing material for piston rings than the block, i.e. a cast iron liner in an aluminum block.

cylindrical cam A mechanism consisting of a cylinder which rotates on its longitudinal axis and causes linear motion parallel to that axis in a cam follower which rolls in a groove cut in the cylindrical surface.

cylindrical lens A lens which is cylindrical in cross section, so it is curved in one direction but not in the perpendicular direction, used to expand a laser beam into a plane of light.

D

D/A See digital-to-analog.

daisy chain 1. A serial interconnection of devices. Signals are passed from one device to another, generally in the order of high priority to low priority. 2. A method of propagating signals along a bus, often used in applications in which devices are connected in series.

daisywheel printer Using a rotating wheel to type, a printer providing slow but good quality print output.

Dalton's law A scientific principle that the total pressure exerted by a mixture of gases equals the sum of the partial pressures that would be exerted if each of the individual gases present were to occupy the same volume by itself.

damped frequency See frequency, damped.

damped wave A wave in which the source amplitude diminishes with each succeeding cycle.

dampener A device for progressively reducing the amplitude of spring oscillations after abrupt application or removal of a load.

damper A device for introducing a variable resistance for regulating the volumetric flow of gas or air: a) butterfly type damper—A single blade damper pivoted about its center; b) curtain type damper—A damper, composed of flexible material, moving in a vertical plane as it is rolled; c) flat type damper—A damper consisting of one or more blades each pivoted about one edge.; d) louvre type damper—A damper consisting of several blades each pivoted about its center and linked together for simultaneous operation; e) slide type damper—A damper consisting of a single blade which moves substantially normal to the flow.

damper loss The reduction in the static pressure of a gas flowing across a damper.

damping 1. (noun) The progressive reduction or suppression of oscillation in a device or system [S51.1]. It is built into electrical circuits and mechanical systems to prevent rapid or excessive corrections which may lead to instability or oscillatory conditions, e.g., connecting a resistor on the terminals of a pulse transformer to remove natural oscillations or placing a moving element in oil or sluggish grease to prevent mechanical overshoot of the moving parts [S51.1]. 2. (adj) Pertaining to or productive of damping. Note 1: The response to an abrupt stimulus is commonly said to be "critically damped" when the time response is as fast as possible without overshoot; "underdamped" when overshoot occurs, or "overdamped" when response is slower than critical [S51.1]. 3. The energy dissipating characteristic which, together with natural frequency, determines the limit of frequency and the response-time characteristics of a transducer. Note: (a) In response to a step change of measurand, an underdamped (periodic) system oscillates about its final steady value before coming to rest at that value; and overdamped (aperiodic) system comes to rest without overshoot; and a critically damped system is at the point of change between the underdamped and overdamped conditions. (b) viscous damping uses the viscosity of fluids (liquids or gases) to effect damping. (c) magnetic damping uses the current induced in electrical conductors by changes in magnetic flux to effect damping [S37.1]. 4. Reducing or eliminating vibrations, especially reducing noise or reverberations by using sound-absorbing materials.

damping factor 1. For the free oscillation of a second order linear system, a measure of damping, expressed (without sign) as the quotient of the greater by the lesser of a pair of consecutive swings of the output (in opposite directions) about an ultimate steady-state value [S51.1]. 2.

97

In any damped oscillation, the ratio of the amplitude of any given half-cycle to the amplitude of the succeeding half-cycle.

damping fluid   A fluid used to damp the single-degree-of-freedom spring/mass system, usually surrounding the reference side (transduction element side) of the sensing element [S37.6].

damping integrity   The ability of the accelerometer to produce a predicted output, with no transients, during or after changes in the attitude of the transducer, due to bubbles, contamination, etc. [S37.5]

damping magnet   A permanent magnet used in conjunction with a moving conductor to produce an opposing torque when there is relative motion between the magnet and the conductor; a secondary function is to dissipate kinetic energy resulting from eddy currents that may be induced in the moving conductor.

damping ratio   1. The ratio of the actual damping to the damping required for critical damping [S37.1]. 2. The ratio of the deviations of the indicator following an abrupt change in the measurand in two consecutive swings from the position of equilibrium, the greater deviation being divided by the lesser. The deviations are expressed in angular measure.

damping, relative   For an underdamped system, a number expressing the quotient of the actual damping of a second-order linear system or element by its critical damping [S51.1].

dark current   The current that flows in photosensitive detectors when there is no incident radiant flux (total darkness).

d'Arsonval galvanometer   A galvanometer made by suspending a light coil of wire on thin gold or copper ribbons in the field of a permanent magnet; when current is carried to the coil via the suspending ribbons, the coil rotates, and the amount of rotation is indicated by reflecting a beam of light from a small mirror carried on the coil onto a fixed linear scale. Also known as light-beam galvanometer.

d'Arsonval movement   The mechanism of a permanent-magnet moving-coil instrument such as a d'Arsonval galvanometer.

DAS   See data acquisition system.

dashpot   1. A mechanical damping device consisting of a cylinder and piston apparatus arranged so as to dampen the movement of a valve stem. A less preferred term. See snubber. [S75.05].

2. A fluid-filled cylinder containing a loose-fitting piston that is used to damp vibratory motion or to change the effect of a sharp change in load from an instantaneous change in position to a more gradual change.

dashpot relay   A timing device relying upon the restrictive action of an orifice upon a fluid to provide the delay. When the relay coil is energized, the armature piston moves against a reservoir of fluid, forcing it through a restriction. This slows down the action. Timing is achieved by variations in orifice size.

data   1. Information of any type. 2. A common term used to indicate the basic elements that can be processed or produced by a computer.

data acquisition   The function of obtaining data from sources external to a microprocess or a computer system, converting it to binary form, and processing it.

data acquisition system   A system used for acquiring data from sensors via amplifiers and multiplexers and any necessary analog to digital converters.

data averaging   An optional mode of operation for an automatic data logger which allows readings from two or more data acquisition channels to be averaged in each scan or, alternatively, readings from each of several channels to be averaged over a preselected number of successive scans.

data bank   A comprehensive collection of data, for example, several automated files, a library, or a set of loaded disks. Synonymous with database.

database   1. Any body of information. 2. A specific set of information available to a computer. 3. A collection of interrelated data stored together with controlled redundancy to serve one or more applications; the data are stored so that they are independent of programs that use the data; a common and controlled approach is used in adding new data and in modifying and retrieving existing data within a data base. A system is said to contain a collection of databases if they are disjointed in structure.

database management   A system that provides meaningful information from the data included in a database.

data block   In TELEVENT, a short section of memory in which data relating to events or operating programs are stored.

data bus The highway connecting the various microcomputer components carrying the data signals.

data capture (logging) The systematic collection of data to use in a particular data processing routine, such as monitoring and recording temperature changes over a period of time.

data channel A bidirectional data path between I/O devices and the main memory of a digital computer. Data channels permit one or more I/O operations to proceed concurrently with computation thereby enhancing computer performance.

data code A structured set of characters used to represent the data items of a data element, for example, the data codes 1,2,...,7 may be used to represent the data items Sunday, Monday,..., Saturday.

data collection The act of bringing data from one or more points to a central point.

data communication The transmission of data from one point to another.

data compression The elimination of redundant data without loss of information; a few standard telemetry data compression algorithms are ZFN, ZVP, ZVA, FFN, FFP, FFA, FVP, and FVA.

data converter Any of numerous devices for transforming analog signals to digital signals, or vice versa.

data directory A listing of data stored in a database.

data display module A device which stores computer output and translates this output into signals which are distributed to a program-determined group of lights, annunciators, numerical indicators, and cathode ray tubes in operator consoles and remote stations.

data distributor A manually or automatically controlled unit that is used to distribute specific data channels to quick-look devices.

data element A scalar, array, or structure.

data error A deviation from correctness in data, usually an error, which occurred prior to processing the data.

data file In a computer, a portion of memory allocated to a specific set of organized data, including codes that identify the file name and sometimes the file type. Also referred to as data set.

data gathering See data collection.

data handling See data processing.

data highway A communication link between separate stations tied with a multidrop cable and/or optical connections. It eliminates a need for separate, independently wired data links. Each station on a highway can function independently.

data input/output unit (DI/OU) A device that interfaces to the process for the sole purpose of acquiring or sending data.

data link 1. Equipment which permits the transmission of information in data format. 2. Facility for transmission of information. Also used to refer to layer 2 of the Open Systems Interconnection definitions. 3. A fiber optic signal transmission system which carries information in digital or analog form. The term usually refers to short distance communications, spanning distances less than a kilometer.

data logger 1. A system or subsystem with a primary function of acquisition and storage of data in a form that is suitable for later reduction and analysis, such as computer-language tape. 2. A computer system designed to obtain data from process sensors and to provide a log of the data. Many data loggers can carry out some filtering and linearising of the data.

data logging Recording of data about events that occur in time sequence. See also data collection.

data management A general term that collectively describes those functions of the control program that provide access to data sets, enforce data storage conventions, and regulate the use of input/output devices.

data plate A plate bearing the name of the manufacturer and other information related to the product as may be required by various regulations or codes. See also nameplate [S75.05].

data processing The execution of a systematic sequence of operations performed upon data. Synonymous with information processing.

data processing system A network of machine components capable of accepting information, processing it according to plan, and producing the desired results.

data processor A device capable of performing data processing, for example, a desk calculator, punched-card machine, or computer.

data protection Any method to preserve computer data from destruction or misuse. Backing up computer files is one example.

data reduction The process of transforming masses of raw test or experimentally obtained data, usually gathered by automatic recording equipment, into useful, condensed, or simplified intelligence.

data set (DS) 1. A collection of data in one of several prescribed arrangements to which the system has access. 2. A device which performs the modulation/demodulation and control functions necessary to provide compatibility between data processing equipment and communications facilities. See also subset.

data signaling rate In communications, the data transmission capacity of a set of parallel channels. The data signaling rate is expressed in bits per second.

data sink In communications, a device capable of accepting data signals from a transmission device. It also may check these signals and originate error control signals. Contrast with data source.

data source In communications, a device capable of originating data signals for a transmission device. It also may accept error control signals. Contrast with data sink.

data stream The movement of a group of measurements in one multiplexer.

data structures The storage of related data in computer memory by use of arrays, records or data lists.

data terminal equipment Either a data source or a data sink, or both.

data transmission The sending of data from one part of a system to another part.

data type Any one of several different types of data, such as integer, real, double precision, complex, logical, and Hollerith. Each has a different mathematical significance and may have different internal representation.

data unpacking The process of recovering individual items of data from packed information [RP55.1].

datum 1. A point, direction or level used as a convenient reference for measuring angles, distances, heights, speeds or similar attributes. 2. Any value that serves as a reference for measuring other values of the same quantity.

datum plane A permanently established reference level, usually average sea level, used for determining the value of a specific altitude, depth sounding, ground elevation or water-surface elevation. Also known as chart datum; datum level; reference level; reference plane.

daughter A nuclide formed as a result of nuclear fission or radioactive decay.

day A unit of time whose exact value depends on the system of time measurement being used—apparent solar time, mean solar time, apparent sidereal time, universal time, ephemeris time, or atomic time; except for atomic time, the basis of the definition is the period during which the earth makes one rotation on its axis; for general purposes, one day equals 24 h or 86,400 s.

dB See decibel.

d controller See controller, derivative (D).

dc output With integral demodulator, rectifier or frequency integrator [S37.1].

DDC See Direct Digital Control.

DDCMP See Digital Data Communications Message Protocol.

dead band In process instrumentation, the range through which an input signal may be varied, upon reversal of direction, without initiating an observable change in output signal [S75.05] [S51.1]. Note 1: There are separate and distinct input-output relationships for increasing and decreasing signals. 2: Dead band produces phase lag between input and output. 3: Dead band is usually expressed in percent of span [S51.1].

dead center 1. Either of two positions of a crank where the turning force between the crank and its connecting rod are zero; it occurs when the centerline of the crank and the centerline of the connecting rod lie in the same plane. 2. A nonrotating center for holding a rotating workpiece.

dead-end shutoff A nonstandard term used to refer to control valve leakage. Refer to ANSI/FCI 70-2 for specifications of leakage classifications.

dead-end tube A tube with a closed end—for example, a tube in a porcupine boiler.

dead-front switchboard A switching panel constructed so that all of the live terminations are made on the rear of the panel.

deadman's brake A safety device that automatically stops a vehicle when the driver does not have his foot on the pedal; it is also used on other operator-controlled mechanisms such as cranes and lift trucks.

deadman's handle A hand grip or handle that an operator must squeeze or press on continuously to keep a machine running.

dead man timer (DMT) Circuit monitors operation of the processor cards and signals if a failure occurs.

dead reckoning Determining a navigational position by constructing distance vectors on a map or similar representation, starting from a known position and calculating the distance vectors from a log of headings and speeds versus time.

dead room See anechoic chamber.

dead time 1. The interval of time between initiation of an input change or stimulus and the start of the resulting response. 2. Any definite delay deliberately placed between two related actions in order to avoid overlap that might cause confusion or to permit a particular different event, such as a control decision, switching event or similar action to take place. See time, dead.

dead-time correction A correction applied to an instrument reading to account for events or stimuli actually occurring during the instrument's dead time.

dead volume The total volume of the pressure port cavity of a transducer with room barometric pressure applied [S37.1].

deadweight gage A device used to generate accurate pressures for the purpose of calibrating pressure gages; freely balanced weights (dead weights) are loaded on a calibrated piston to give a static hydraulic pressure output.

dead zone Also called dead band. A range of values around the set point. When the controlled variable is within this range, no control action takes place. See zone, dead.

deaeration Removing a gas—air, oxygen or carbon dioxide, for example—from a liquid or semisolid substance, such as boiler feedwater or food.

debug 1. To locate and correct any errors in a computer program. 2. To detect and correct malfunctions in the computer itself. Related to diagnostic routine. 3. To submit a newly designed process, mechanism or computer program to simulated or actual operating conditions for the purpose of detecting and eliminating flaws or inefficiencies.

debuggers System programs that enable computer programs to be debugged.

debugging The process of detecting, diagnosing and then correcting program faults.

debugging aid routine A routine to aid programmers in the debugging of their routines. Some typical aid routines are storage printout, tape print-out, and drum print-out routines.

debugging on-line See on-line debugging.

deburr To remove burrs, fins, sharp edges and the like from corners and edges of parts or from around holes, by any of several methods, often involving the use of abrasives.

decade A group or assembly of ten units, e.g., a counter which counts to ten in one column or a resistor box which inserts resistance quantities in multiples of powers of 10.

decade scaler A scaling device that produces one output pulse for each ten input pulses.

decalescence Darkening of a metal surface upon undergoing a phase transformation on heating; the phenomenon is caused by isothermal absorption of the latent heat of transformation.

decanting Boiling or pouring off liquid near the top of a vessel that contains two immiscible liquids or a liquid-solid mixture which has separated by sedimentation, without disturbing the heavier liquid or settled solid.

decarburizing Removing carbon from the surface layer of a steel or other ferrous alloy by heating it in an atmosphere that reacts selectively with carbon; atmospheres that are relatively rich in water vapor or carbon dioxide are typical deoxidizing atmospheres.

decay The spontaneous transformation of a nuclide into one or more other nuclides either by emitting one or more subatomic particles or gamma rays from its nucleus or by nuclear fission; radioactive decay of a specific nuclide is characterized by its half life—the time it takes for one-half of the original mass to spontaneously transform.

decay time The time in which a voltage or current pulse will decrease to one tenth of its maximum value. Decay time is proportional to the time constant of the circuit.

decelerating electrode An intermediate electrode in an electron tube which is maintained at a potential that induces decelerating forces on a beam of electrons.

decelerometer An instrument for measuring the rate at which speed decreases.

decentralized 1. To distribute the functions among several authorities. 2. Decentralized maintenance distributes maintenance functions among areas of responsibilities or areas of the physical plant.

decibel (dB) 1. A unit of level, where: Level in dB $= 10 \log 10\, P_1/P_{ref}$. $P_1 = $ a power, or, quantity directly proportional to power. P_{ref} = a reference power, or, a corresponding reference quantity proportional to power[S51.1]. 2. A measure of magnitude ratio; magnitude ratio is dB = $20 \log_{10}$ (magnitude ratio) [S26]. 3. A unit for measuring relative strength of a signal parameter, such as power, voltage, etc. The number of decibels is twenty (ten for power ratio), times the logarithm (base 10) of the ratio of the measured quantity to the reference level. The reference level must always be indicated, such as 1 milliamp for current ratio. See also power level.

decibel meter An instrument calibrated in logarithmic steps and used for measuring power levels, in decibel units, of audio or communication circuits.

decimal 1. Pertaining to a characteristic or property involving a selection, choice, or condition in which there are ten possibilities. 2. Pertaining to the numeration system with a radix of ten. 3. See binary code decimal.

decimal balance A type of balance having one arm ten times as long as the other, so that heavy objects can be balanced with light weights.

decimal coded digit A digit or character defined by a set of decimal digits, such as a pair of decimal digits specifying a letter or special character in a system of notation.

decimal digit In decimal notation, one of the characters 0 through 9.

decimal notation A fixed radix notation, where the radix is ten; for example, in decimal notation, the numeral 576.2 represents the number 5x10 squared plus 7x10 to the first power, plus 6x10 to the zero power, plus 2x10 to the minus 1 power.

decimal number A number, usually of more than one figure, representing a sum, in which the quantity represented by each figure is based on the radix of ten. The figures used are 0, 1, 2, 3, 4, 5, 6, 7, 8, and 9.

decimal numbering system A system of reckoning by 10 or the powers of 10 using the digits 0-9 to express numerical quantities.

decimal numeral A decimal representation of a number.

decimal point The radix point in decimal representation.

decimal-to-binary conversion The process of converting a number written to the base ten, or decimal, into the equivalent number written to the base two, or binary.

decision instruction An instruction that effects the selection of a branch in a program, for example, a conditional jump instruction.

decision table A table of all contingencies that are to be considered in the description of a problem, together with the actions to be taken. Decision tables are sometimes used in place of flow charts for problem description and documentation.

DECK See digital to analog converter, an electronic device that converts a digital signal, often from a computer, into a proportional analog voltage or current.

deck A collection of cards, commonly a complete set of cards which have been punched for a definite service or purpose.

deck scale A low-profile weighing device used for moderate to heavy loads—up to 20,000 lb; because the load platform is 2 to 10 in. above floor level, loads must be lifted onto the scale or ramps must be provided to enable wheeled vehicles to move onto the platform and off again; the frame of a deck scale rests directly on the existing floor, rather than in a pit, and most models can be moved to different locations as needed.

declaration As used in many programming languages, a statement that is not to be executed, but usually is used for descriptive purposes.

declinometer An instrument similar to a surveyor's compass used for determining the variation of magnetic directions from true directions; the horizontal circle is constructed so that the line of sight can be aligned with the magnetic needle or with any other desired setting.

decode 1. To apply a code so as to reverse some previous encoding. 2. To determine the meaning of individual characters or groups of characters in a message. 3. To determine the meaning of instructions from the status of bits which describes the instruction, command, or operation to be performed.

decoder 1. A device which determines the meaning of a set of signals and initiates a computer operation based thereon. 2. A matrix of switching elements which selects one or more output channels according to the combination of input signals present. Contrasted with encoder and clarified by matrix. 3. A device used to change computer data from one coded format to another.

decollate The separation of multi-part computer forms.

decommutation A reversal of the commutation

process; separation of information in a commutated data stream into as many independent information channels as were originally commutated.

decommutator Equipment for the separation or demultiplexing of commutated signals.

decompression Any method for relieving pressure.

decontamination Removing or neutralizing an unwanted chemical, biological or radiological substance.

decoupling The technique of reducing process interaction through coordination of control loops.

decoupling control A technique in which interacting control loops are automatically compensated when any one control loop takes a control action.

decrement 1. The quantity by which a variable is decreased. 2. A specific part of an instruction word in some binary computers, thus a set of digits.

decremeter An instrument for measuring the damping of a train of waves by determining its logarithmic decrement.

decryption Translating computer data from an unreadable format to a readable format.

dedicated In data processing, a device that performs only one function.

deep drawing A press operation for forming cup-shaped or deeply recessed parts from sheet metal by forcing the metal to undergo plastic deformation between dies without substantial thinning.

default 1. The value of an argument, operand, or field assumed by a program if a specific assignment is not supplied by the user. 2. The alternative assumed when an identifier has not been declared to have one of two or more alternative attributes.

default directory In MS-DOS, the directory in which the computer looks for files if no directory is specified.

default drive In MS-DOS, the disk drive the computer will use to search for files if no disk drive is specified.

defect A departure of any quality characteristic from its specified or intended value that is severe enough to constitute cause for rejecting the object or service.

definition 1. The resolution and sharpness of an image, or the extent to which an image is brought into sharp relief. 2. The degree with which a communication system reproduces sound images

or messages.

deflashing Removing fins or protrusions from the parting line of a die casting or molded plastics part.

deflecting electrode An intermediate electrode in an electron tube whose surrounding electric field induces constant or variable deflecting forces on an electron beam.

deflecting force In a direct-acting recording instrument, the force produced at the marking device, for any position of the scale, by its positioning mechanism acting in response to the electrical quantity being measured.

deflecting yoke An assembly of one or more coils that induce a magnetic field to deflect an electron beam in a manner related to the oscillating frequency and magnitude of the current flowing through the coils.

deflection 1. Movement of a pointer away from its zero or null position. 2. Elastic movement of a structural member under load. 3. Shape change or change in diameter of a tubular member without fracturing the material.

deflection factor The reciprocal of the instrument sensitivity.

deflection polarity In an oscilloscope, the relationship between direction of electron-beam displacement and polarity of applied signal voltage.

deflectometer An instrument for determining minute elastic movements that occur when a structure is loaded.

deflector A device for changing direction of a stream of air or of a mixture of pulverized fuel and air.

defocus To cause a beam of electrons, light, x-rays or other type of radiation to depart from accurate focus at a specific point in space, ordinarily the surface of a workpiece or test object.

defrost To remove ice from a surface, usually by melting or sublimation.

deg or ° See degree.

degas To remove dissolved, entrained or adsorbed gas from a solid or liquid.

degasification Removal of gases from samples of steam taken for purity tests. Removal of CO_2 from water as in the ion exchange method of softening.

degasifier 1. An element or compound added to molten metal to remove dissolved gases. 2. A process or type of vessel that removes dissolved gases from molten metal.

degenerate waveguide modes A set of waveguide modes having the same propagation constant for all frequencies of interest.

degeneration Negative feedback.

degradation failure Gradual shift of an attribute or operating characteristic to a point where the device no longer can fulfill its intended purpose.

degreasing An industrial process for removing grease, oil or other fatty substances from the surfaces of metal parts, usually by exposing the parts to condensing vapors of a polyhalogenated hydrocarbon solvent.

dehumidification Reducing the moisture content of air, which increases its cooling power.

deicing Using heat, chemicals or mechanical rupture to remove ice deposits, especially those that form on motor vehicles and aircraft at low temperatures or high altitudes.

deionization time The time it takes for the grid in a gas tube to regain control of tube output after the anode current has been interrupted.

delamination Separation of a material into layers, especially a material such as a bonded laminate.

delay The interval of time between a changing signal and its repetition for some specified duration at a downstream point of the signal path; the value L in the transform factor exp (-Ls). See time, dead [S51.1].

delay distortion A form of distortion in a transmitted radio wave that occurs when the rate of change of phase shift with frequency is not constant over the transmission-frequency range.

delayed combustion A continuation of combustion beyond the furnace. See also secondary combustion.

delay-interval timer A timing device which is electrically reset to delay energization or deenergization of a circuit for an interval of time up to 10 min following a specific event such as restoration of power after a power failure or turning a manual switch off.

delay line A transmission medium which delays a signal passing through it by a known amount of time; typically used in timing events.

delay-line memory A type of circulating memory having a delay circuit as the chief element in the path of circulation.

delay-line register An acoustic or electric delay line, one or more words long, combined with appropriate input, output and circulation circuits.

delay modulation A method of data encoding for serial data transmission or recording; a logic ONE (or ZERO) is represented by a signal transition at midbit time and a logic ZERO (or ONE) followed by a logic ZERO (or ONE) is represented by a transition at the end of the first ZERO (or ONE) bit.

delay-on-make timer A timing device that holds its main contacts open for a preset period of time after it receives an initiating signal, then closes the contacts and allows current to flow in the main circuit; when the timer receives a stopping signal, the contacts open and after a short interval the timer automatically resets so it can repeat the cycle.

deletion punch A record elimination feature used on paper tape I/O devices to cause all tape channels to be punched [RP55.1].

delimiter A character that separates, terminates, or organizes elements of a character string, statement, or program.

delta network A set of three circuit branches connected in series, end-to-end, to form a mesh having three nodes.

demand meter Any of several types of instruments used to determine the amount of electricity used over a fixed period of time, usually for the purpose of establishing a customer's bill.

demodulation The process of retrieving intelligence (data) from a modulated carrier wave. The reverse of modulation.

demodulator A device which recovers information from a carrier or subcarrier. A telemetry receiver has a demodulator; an FM discriminator is a demodulator.

demultiplexer 1. The device which enables the telemetry operator to observe individual measurements from within a multiplexer. The opposite of a "multiplexer." 2. A device which separates two or more signals which have been multiplexed together for transmission through a single optical fiber. 3. A reverse multiplexer which allows the transfer of data from one microprocessor port to a number of output devices such as actuators.

densimeter An instrument for determining the density of a substance in absolute units, or for determining its specific gravity—that is, its relative density with respect to that of pure water. Also known as density gage; density indicator; gravitometer.

densitometer An instrument for determining

optical density of photographic or radiographic film by measuring the intensity of transmitted or reflected light.

density 1. The mass of a unit volume of a liquid at a specified temperature. The units shall be stated, such as kilograms per meter^3. The form of expression shall be: Density _____ kg/m^3 at _____ Kelvin [RP31.1]. 2. A physical property of materials measured as mass per unit volume. 3. The weight of a substance for a specified volume at a definite temperature, for example, grams per cubic centimeter at 20 °C. 4. Closeness of texture or consistency. 5. Degree of opacity, often referred to as optical density.

density bottle See specific gravity bottle.

density correction Any correction made to an instrument reading to compensate for the deviation of density from a fixed reference value; it may be applied because the fluid being measured is not at standard temperature and pressure, because ambient temperature affects density of the fluid in a fluid-filled instrument, or because of other similar effects.

density transmitter An instrument used to determine liquid density by measuring the buoyant force on an air-filled float immersed in a flowing liquid stream.

depolarizers Optical components which scramble the polarization of light passing through them, effectively turning a polarized beam into an unpolarized beam.

deposit 1. Any substance intentionally laid down on a surface by chemical, electrical, electro-chemical, mechanical, vacuum or vapor transfer methods. 2. Solid or semisolid material accumulated by corrosion or sedimentation on the interior of a tube or pipe.

deposited metal In a weldment, filler metal added to the joint during welding.

deposit gage Any instrument used for assessing atmospheric quality by measuring the amount of particulate matter that settles out on a specific area during a defined period of time.

deposition rate 1. The amount of filler metal deposited per unit time by a specific welding procedure, usually expressed in pounds per hour. 2. The rate at which a coating material is deposited on a surface, usually expressed as weight per unit area per unit time, or as thickness per unit time.

deposition sequence The order in which increments of a weld deposit are laid down.

depth gage An instrument or micrometer device capable of measuring distance below a reference surface to the nearest 0.001 in.; it is most often used to measure the depth of a blind hole, slot or recess below the normal part surface surrounding it, or to measure the height of a shoulder or projection above the adjacent part surface.

depth of engagement The radial contact distance between mating threads.

depth of fusion The distance from the original surface that the molten zone extends into the base metal during welding.

depth of thread The radial distance from crest to root of a screw thread.

derandomizer The circuit which removes the effect of data randomizing, thereby recovering data which had been randomized for tape storage.

derivative 1. Mathematically it is the reciprocal of rate. 2. This control action will cause the output signal to change according to the rate at which input signal variations occur during a certain time interval.

derivative action A type of control-system action in which a predetermined relation exists between the position of the final control element and the derivative of the controlled variable with respect to time.

derivative action gain See gain, derivative action (rate gain) [S51.1].

derivative action time See time derivative action [S51.1].

derivative action time constant See time constant, derivative action [S51.1].

derivative control Change in the output that is proportional to the rate of change of the input. Also called rate control. See control action, derivative (D).

derivative control action (rate action) Control action in which the output is proportional to the rate of change of the input. See control action; see also control action, derivative.

derivative controller See controller, derivative.

derivative control mode A controller mode in which controller output is directly proportional to the rate of change of controlled variable error.

derivative time The time interval by which rate action advances the effect of proportional action on the final control element.

descaling Removing adherent deposits from a metal surface, such as thick oxide from hot rolled or forged steel, or inorganic compounds from the interior of boiler tubes; it may be done by chemical attack, mechanical action, electrolytic dissolution or other means, alone or in combination.

describing function   For a nonlinear element in sinusoidal steady state, the frequency response obtained by taking only the fundamental component of the output signal. The describing function depends on the frequency and on the amplitude of the input signal, or only on the amplitude of the input signal.

design capacity   The maximum weight load that the scale is designed to weigh in one hour within the designated class accuracy. It is customarily 125 percent of normal capacity and is also known as "scale capacity" [RP74.01].

design load   The load for which a steam generating unit is designed, considered the maximum load to be carried.

design pressure   The maximum allowable working pressure permitted under the rules of the ASME Construction Code. See pressure, design.

design steam temperature   The temperature of steam for which a boiler is designed.

design stress   The maximum permissible load per unit area a given structure can withstand in service, including all allowances for such things as unexpected or impact loads, corrosion, dimensional variations during fabrication and possible underestimation of service loading.

design thickness   The sum of thickness required to support service loads. This method of specifying material thickness is used particularly when designing boilers, chemical process equipment, and metal structures that will be exposed to atmospheric environments, soils or seawater.

desired value   See value, desired.

desk top publishing   The computer merging of text and graphics to produce manuals and leaflets.

desorption   Removing adsorbed material.

destructive testing   Any method of determining a material property, functional attribute or operational characteristic which renders the test object unsuitable for further use or severely impairs its intended service life.

detectability   The quality of a measured variable in a specific environment that is determined by relative freedom from interfering energy or other characteristics of the same general nature as the measured variable.

detector   A device which detects light, generating an electrical signal which can be measured or otherwise processed. See transducer.

detector-amplifier   A device in which an optical detector is packaged together with electronic amplification circuitry.

detector head   That gas-responsive portion of a multipart gas detection instrument that is located in the area where sensing the presence of combustible gases is desired. Its location may be integral with or remote from its control unit [S12.13] [S12.15]. Note: The detector head may incorporate additional circuitry such as signal-processing or amplifying components or circuits in addition to the gas-sensing element in the same housing [S12.13].

detent   A catch or lever that initiates or prevents movement in a mechanism, especially an escapement.

detergent   A natural material or synthetic substance having the soaplike quality of being able to emulsify oil and remove soil from a surface.

deterioration   Decline in the quality of a device, mechanism or structure over time due to environmental effects, corrosion, wear or gradual changes in material properties; if allowed to continue unchecked, deterioration often leads to degradation failure.

deutron detector   A type of specialized radiation detector used in some nuclear reactors to detect the concentration of deuterium nuclei present.

developed boiler horsepower   The boiler horsepower generated by a steam generating unit.

development system   A system used to develop both the hardware and software for a microcomputer system. The development system may contain an editor, assembler and/or high level language, compiler, debugging and in-circuit emulation facilities.

deviation   1. Any departure from a desired value or expected value or pattern [S51.1]. 2. The difference between control or set point, and a value of process variable. 3. In quality control, any departure of a quality characteristic from its specified value. 4. A statistical quantity that gives a measure of the random error which can be expected in numerous independent measurements of the same value under the same conditions of measurement.

deviation alarm   1. Alarm that is set whenever the deviation exceeds the preset limits. 2. An alarm caused by a variable departing from its desired value by a specified amount.

deviation controller   A type of automatic control device which acts in response to any difference between the value of a process variable and the instrument setpoint, independent of their actual values.

deviation ratio The ratio given by $M = f/f_{max}$, where f is the maximum frequency difference between the modulated carrier and the unmodulated carrier, and f_{max} is the maximum modulation frequency.

deviation, steady-state The system deviation after transients have expired. See offset [S51.1].

deviation, system In process instrumentation, the instantaneous value of the directly controlled variable minus the set point. See also signal, actuating error [S51.1].

deviation, transient In process instrumentation, the instantaneous value of the directly controlled variable minus its steady-state value [S51.1].

device 1. An apparatus for performing a prescribed function [S51.1]. 2. A component or assembly designed to perform a specific function by harnessing mechanical, electrical, magnetic, thermal or chemical energy. 3. Any piece of machinery or computer hardware that can perform a specific task. 4. A component in a control system, such as; primary element, transmitter, controller, recorder, or final control element.

device control character One of a class of control characters intended for the control of peripheral devices associated with a data processing or telecommunication system, usually for switching devices "on" or "off".

device controller A hardware unit that electronically supervises one or more of the same types of devices; acts as the link between the CPU and I/O devices.

device driver A program/routine that controls the physical hardware activities on a peripheral device; a device driver is generally the device-dependent software interface between a device and the common, device-independent I/O code in an operating system.

device flags One-bit registers which record the current status of a device.

device handler A program/routine that drives or services an I/O device; a device handler is similar to a device driver but provides more control and interfacing functions than a device driver.

device independence The ability to request input/output operations without regard to the characteristics of the input/output devices.

dewars Insulated thermos-like containers for cryogenic liquids, which can be designed to house detectors or lasers requiring cooling.

dewatering 1. Removing water from solid or semisolid material—for instance, by centrifuging, filtering, settling or evaporation. 2. Removing water from a riverbed, pond, caisson or other enclosure by pumping or evaporation.

dew cell An instrument consisting of two bare electrical wires wound spirally around an electrical insulator and covered by wicking wetted with an aqueous solution containing an excess of LiCl; dew point of the surrounding atmosphere is determined by passing an electric current between the two wires, which raises the temperature of the LiCl solution until its vapor pressure is the same as that of the ambient atmosphere.

dewetting 1. Generally, loss of surface attraction between a solid and a liquid. 2. Specifically, flow of solder away from a soldered joint upon reheating.

dew point 1. The temperature, referred to a specific pressure, at which water vapors condense [S7.3]. 2. The temperature at which condensation starts.

dew point (at line pressure) The dew point value of the air at line pressure of the compressed air system (usually measured at the outlet of the dryer system, or at any instrument air supply source, prior to pressure reduction). When presenting or referencing dew point, the value shall be given in terms of the line pressure; e.g., $-40\,°C$ ($-40\,°F$) dew point at 100 psig [S7.3].

dew-point recorder An instrument that determines dew-point temperature by alternately heating and cooling a metal plate and using a photocell to automatically detect and record the temperature at which condensed moisture appears and disappears on the target. Also known as mechanized dew-point meter.

dew-point temperature That temperature at which condensation of moisture from the vapor phase begins.

DFT See diagnostic function test.

diagnostic 1. Pertaining to the detection and isolation of a malfunction or mistake. 2. Program or other system feature designed to help identify malfunctions in the system. An aid to debugging.

diagnostic alarm Alarm that is set whenever the diagnostic program reports a malfunction.

diagnostic function test (DFT) A program to test overall system reliability.

diagnostic message An error message in a programming routine to help the programmer identify the error.

diagnostic programs 1. A troubleshooting aid for locating hardware malfunctions in a system. 2. A program to aid in locating coding errors in newly developed programs. 3. Computer programs that isolate equipment malfunctions or programming errors.

diagnostic routine A routine used to locate a malfunction in a computer, or to aid in locating mistakes in a computer program. Thus, in general, any routine specifically designed to aid in debugging or trouble shooting. Synonymous with malfunction routine and related to debug.

diagnostics Information concerning known failure modes and their characteristics that can be used in troubleshooting and failure analysis to help pinpoint the cause of a failure and aid in defining suitable corrective measures.

diagnostics, or diagnostic software The program by which a computer or other programmable device or system can literally "check itself," to diagnose any defects that may be present.

diagnostic test The running of a machine program or routine for the purpose of discovering a failure or a potential failure of a machine element, and to determine its location or its potential location.

diagonal stay A brace used in fire-tube boilers between a flat head or tube sheet and the shell.

dial 1. Generally, any circular scale. 2. The graduated scale adjacent to a control knob that is used to indicate the value or relative position of the control setting.

dial indicator 1. Any meter or gage with a graduated circular face and a pivoted pointer to indicate the reading. 2. A type of measuring gage used to determine fine linear measurements, such as radial or lateral runout of a rotating member, by resting a feeler against a surface and noting the change in position of a pivoted pointer relative to the calibrated gage face as the part is rotated; the gage also can be adapted to other setups where precise relative position is to be determined.

diamagnetic material A substance whose specific permeability is less than 1.00 and is therefore weakly repelled by a magnetic field.

diamond-pyramid hardness A material hardness determined by indenting a specimen with a diamond-pyramid indenter having a 136° angle between opposite faces then calculating a hardness number by dividing the indenting load by the pyramidal area of the impression. Also known as Vickers hardness.

diamond-turned mirror A mirror in which the surface has been formed by machining away material with a diamond tool.

diamond wheel A grinding wheel for cutting very hard materials which uses synthetic diamond dust as the bonded abrasive material.

diaphragm 1. A sensing element consisting of a thin, usually circular, plate which is deformed by pressure differential applied across the plate [S37.1]. 2. A thin, flexible disc that is supported around the edges and whose center is allowed to move in a direction perpendicular to the plane of the disc; it is used for a wide variety of purposes, such as detecting or reproducing sound waves, keeping two fluids separate while transmitting pressure or motion between them, or producing a mechanical or electrical signal proportional to the deflection produced by differential pressure across the diaphragm. 3. A partition of metal or other material placed in a header, duct or pipe to separate portions thereof.

diaphragm motor A diaphragm mechanism used to position a pneumatically operated control element in response to the action of a pneumatic controller or pneumatic positioning relay.

diaphragm seal A thin flexible sheet of material clamped between two body halves to form a physical barrier between the instrument and process fluid.

diaphragm type valve A fluid powered device in which the fluid acts upon a flexible member, the diaphragm, to provide linear motion to the actuator stem [S75.05].

diaphragm valve A valve with a flexible linear motion closure member that is forced into the internal flow passageway of the body by the actuator [S75.05].

dichroic filter A filter which selectively transmits some wavelengths of light and reflects others. Typically such filters are based on multilayer interference coatings.

dichromate treatment A technique for producing a corrosion-resistant conversion coating on magnesium parts by boiling them in a sodium dichromate solution.

dictionary A list of code names used in a computer routine or system and their intended meaning in that routine or system.

die　A tool, usually containing at least one cavity, that imparts shape to solid, molten or powdered metal, or to elastomers or plastics, primarily because of the shape of the tool itself; a die is used together with a punch or a matching die in such operations as stamping, forging, forming, blanking, die casting, plastics molding and coining; in certain operations—die casting, powder metallurgy and plastics forming, for instance—dies are sometimes referred to as molds.

die block　A heavy block, usually of tool steel, into which the desired impressions are sunk, formed or machined, and which is bolted to the bed of a press.

die body　The stationary part of a powder pressing or extrusion die.

die casting　1. A casting process in which molten metal is forced under pressure into the cavity of a metal mold. 2. A part made by this process.

die chaser　One of the cutting parts of a threading die.

die clearance　The amount of lateral clearance between mated die parts when the dies are closed; commonly expressed as clearance per side.

die cushion　A press accessory located beneath or within a bolster or die block, and actuated by air, oil, rubber or springs, to provide additional motion or pressure during stamping.

die forging　1. The process of forming shaped metal parts by pressure or impact between two dies. 2. A part formed in this way.

die holder　A plate or block mounted between the die block and press bed.

die insert　A removable part of a die or punch.

dielectric　An insulating material, or a material that can sustain an electric field with very little dissipation of power.

dielectric coating　An optical coating made up of one or more layers of dielectric (nonconductive) materials. The layer structure determines what fractions of incident light at various wavelengths are transmitted and reflected.

dielectric constant　A material characteristic expressed as the capacitance between two plates when the intervening space is filled with a given insulating material divided by the capacitance of the same plate arrangement when the space is filled with air or is evacuated.

dielectric strength　See breakdown voltage rating and insulation resistance.

die scalping　Improving the surface quality of bar stock, rod tubing or wire by drawing it through a sharp-edged die to remove a thin surface layer containing minor defects.

die set　A tool or tool holder consisting of a die base and punch plate for attaching matched upper and lower dies, and that can be inserted into a press and removed from it as a single unit.

diesinking　Making a shaped recess in a working face of a die, usually by mechanical, electrochemical or spark discharge machining.

die slide　A device that slides into and out of the bed of a power press, carrying the lower die and providing for easy access and improved safety in feeding stock or removing stamped parts.

die welding　Forge welding using shaped dies.

difference　The output equals the algebraic difference between the two inputs.

difference limen　The increment in a stimulus which is barely noticed in a specified fraction of independent observations where the same increment is imposed.

differential　Any arrangement of epicyclic gears that allows two driven shafts to revolve at different speeds, with the speed of the main driving shaft being the algebraic mean of the speeds of the driven shafts. Also known as differential gear.

differential amplifier　A device which compares two input signals and amplifies the difference between them.

differential analyzer　A computer (usually analog) designed and used primarily for solving many types of differential equations.

differential delay　The difference between the maximum and the minimum frequency delays occurring across a band.

differential gap　The smallest increment of change in a controlled variable required to cause the final control element in a two-position control system to move from one position to its alternative position.

differential gap control　See control, differential gap.

differential input　The difference between the instantaneous values of two voltages both being biased by a common mode voltage.

differential input (to a signal conditioner)　1. An input in which both sides are isolated from the chassis and power supply ground. The signal is applied as a differential voltage across the two sides. 2. Allows an analog-to-digital converter to measure the difference between two input signals.

differential instrument Any instrument that has an output signal or indication proportional to the algebraic difference between two input signals.

differential mode interference See interference, normal mode.

differential motion A mechanism in which the net motion of a single driven element is the difference between motions that would be imparted by each of two driving elements acting alone.

differential (of a control) The difference between cut in and cut out points.

differential pressure 1. The difference in pressure between two points of measurement. 2. The static pressure difference generated by the primary device when there is no difference in elevation between the upstream and downstream pressure taps.

differential-pressure gage Any of several instruments designed to measure the difference in pressure between two enclosed spaces, independent of their absolute pressures.

differential-pressure transmitter Any of several transducers designed to measure the pressure difference between two points in a process and transmit a signal proportional to this difference, without regard to the absolute pressure at either point.

differential-pressure-type liquid-level meter Any of several devices designed to measure the head of liquid in a tank above some minimum level and produce an indication proportional to this value; alternatively, the head below some maximum level can be measured and similarly displayed.

differential quantum efficiency Used in describing quantum efficiency in devices having nonlinear output/input characteristics, the slope of the characteristic curve is the differential quantum efficiency.

differential screw A type of compound screw which produces a motion equal to the difference in motion between the two components of the compound screw.

differential windlass A windlass that has a barrel with two sections of different diameter; the pulling rope passes around one section, then through a pulley and around the other section; the pulley is attached to the load.

differentiator A device whose output function is proportional to the derivative, i.e., the rate of change, of its input function with respect to one or more variables (usually with respect to time).

diffracted beam In x-ray crystallography, a beam of radiation composed of a large number of scattered rays mutually reinforcing one another.

diffracted wave The wave component existing in the primary propagation medium after an interaction between the wave and a discontinuity or a second medium; the diffracted wave coexists in the primary medium with incident waves and with waves reflected from suitable plane boundaries.

diffraction 1. Deviation of light from the paths and foci prescribed by rectilinear propagation; phenomenon responsible for bright and dark bands found within a geometrical shadow. 2. A phenomenon associated with the scattering of waves when they encounter obstacles whose size is about the same order of magnitude as the wavelength; in effect, each scattering point produces a secondary wave superimposed on the unscattered portion of the incident wave, the intensity of the scattered wave varying with direction from the scattering point; diffraction effects form the basis for x-ray crystallography, and they also tend to produce aberrations that must be dealt with in the design and construction of high-quality acoustical and optical systems.

diffraction grating An array of fine, parallel, equally spaced reflecting or transmitting lines which diffract light into a direction characteristic of the spacing of the lines and the wavelength of the diffracted light.

diffraction-limited beam A beam with a far-field spot size dependent only on the theoretical diffraction limit, which is the function of output wavelength divided by output aperture diameter.

diffraction x-ray machine An apparatus consisting of an x-ray tube, power supply, controls and auxiliary equipment used in the study of crystals, semiconductors and polymeric materials.

diffused-semiconductor strain gage A component used in manufacturing transducers, principally diaphragm-type pressure transducers, that consists of a slice of silicon about 2.5 to 22 mm in diameter into which an impurity element such as boron has been diffused; modern photolithographic-masking techniques make it possible to simultaneously produce hundreds of full four-arm Wheatstone bridge patterns, complete with leadwire soldering pads, on a single slice of silicon about 50 to 75 mm (2 to 3 in.) in diameter.

diffuse-field response A frequency response of a piezoelectric sound-pressure transducer with the sound incident from random directions [S37.10].

diffuser 1. A duct, chamber or enclosure in which low-pressure, high-velocity flow of a fluid, usually air, is converted to high-pressure, low-velocity flow. 2. As applied to oil or gas burners, a metal plate with openings so placed as to protect the fuel spray from high velocity air while admitting sufficient air to promote the ignition and combustion of fuel. Sometimes termed impeller.

diffusion 1. A method by which the atmosphere being monitored gains access to the gas-sensing element by natural molecular movement [S12.13] [S12.15]. 2. Conversion of gas-flow velocity into static pressure, as in the diffuser casing of a centrifugal fan. 3. The movement of ions from a point of high concentration to low concentration. 4. Migration of atoms, molecules or ions spontaneously, under the driving force of compositional differences, and using only the energy of thermal excitation to cause atom movements.

diffusion pump A vacuum pump in which a stream of heavy particles such as oil or mercury vapors carries gas molecules out of the vacuum chamber.

digit 1. A character used to represent one of the nonnegative integers smaller than the radix, for example, in decimal notation, a digit to one of the characters from 0 to 9 [RP55.1]. Synonymous with numeric character. 2. See binary digit, equivalent binary digits, sign position and significant digits.

digital 1. A term applied to a signal or device that uses binary digits to represent continuous values or discrete states [S12.11]. 2. Pertaining to data in the form of digits. Contrast with analog [RP55.1]. 3. A method of measurement using precise quantities to represent variables. 4. Binary. 5. A reference to the representation of data by discrete pulses, as in the presence or absence of a signal level to indicate the 1's and 0's of binary data. 6. A type of readout in which the data is displayed as discrete, fully-informed alphanumeric characters.

digital back-up An alternate method of digital process control initiated by use of special purpose digital logic in the event of a failure in the computer system.

digital computer 1. A computing device that uses numerical digits to represent discretely all variables. 2. A computer in which discrete representation of data is mainly used. 3. A computer that operates on discrete data by performing arithmetic and logic processes on these data. Contrast with analog computer.

digital controller A control device consisting of a microprocessor plus associated A/D input converters and D/A output converters; it receives one or more analog inputs related to current process variables, uses the digitized information to compute an output signal using a predetermined control algorithm, and converts the result to an analog signal which operates the final control element; the device also may be adapted to furnish additional outputs such as alarms, totalizer signals and displays.

digital data Data represented in discrete discontinuous form, as contrasted with analog data represented in continuous form. Digital data is usually represented by means of coded characters, for example, numbers, signs, symbols, etc.

Digital Data Communications Message Protocol (DDCMP) A character-oriented communications protocol standard.

digital delay generator An electronic instrument which can be programmed digitally to delay a signal by a specific interval—time delay generator.

digital differential analyzer 1. An incremental computer in which the principal type of computing unit is a digital integrator whose operation is similar to the operation of an integrating mechanism. 2. A differential analyzer that uses digital representation for the analog quantities.

Digital Equipment Corporation (DEC) Manufacturer of the PDP-11 series computer systems and peripheral devices.

digital filter An algorithm which reduces undesirable frequencies in the signal.

digital indicator A device that displays the value of a measured variable in digitized form; in most instances, the measurement range is not displayed simultaneously, which is considered an inherent disadvantage.

digital input A number value input. See input, digital.

digital logic A signal level is represented as a number value with a most significant and least significant bit. Binary digital logic uses numbers consisting of strings of 1's and 0's.

digital manometer   A manometer equipped with a sonar device which measures column height and produces a digitized display.

digital motor   See stepping motor.

digital multiplexer   A data selection device that permits sharing a common information path between multiple groups of digital devices, such as from a computer CPU to any of several groups of digital output devices.

digital output   Transducer output that represents the magnitude of the measurand in the form of a series of discrete quantities coded in a system of notation. Note: distinguished from analog output [S37.1]. See output, digital.

digital readout   An electrically powered device which interprets a continuously variable signal and displays its amplitude, or another signal attribute, as a series of numerals or other characters that correspond to the measured value and can be read directly; the accuracy of measurement is limited by the decimal position of the rightmost character in the display rather than by characteristics of the measurement circuit alone.

digital resolution   The value of the least significant digit in a digitally coded representation.

digital signal   A discrete or discontinuous signal, one whose various states are discrete intervals apart. See signal, digital.

digital speed transducer   See digital tachometer.

digital subset   See data set.

digital tachometer   Any of several instruments designed to determine rotational speed and display the indication in digital form.

digital-to-analog converter (D/A or DAC)   1. A device, or sub-system that converts binary (digital) data into continuous analog data, as, for example, to drive actuators of various types, motor-sped controllers, etc. 2. An electronic device that converts a binary-coded word to an analog voltage proportional to the binary value of that word. See converter, digital to analog.

digital valve   A single valve casing containing multiple solenoid valves whose flow capacities vary in binary sequence $(1, 2, 4, 8, 16, ...)$; to regulate flow, the control device sends operating signals to various combinations of the solenoids; applications are limited to very clean fluids at moderate temperatures and pressures, but within these limitations precise flow control and rapid

response are possible—an eight-element valve, for example, yields flow resolution of 0.39% (1 part in 256).

digitize   To convert an analog measurement of a physical variable into a numerical value, thereby expressing the quantity in digital form. See analog-to-digital converter.

digitized signal   Representation of information by a set of discrete values, in accordance with a prescribed law. Every discrete value represents a definite range of the original undigitized signal. See analog-to-digital converter.

digitizer   A device which converts an analog measurement into digital form.

dilatant substance   A material which flows under low shear stress but whose rate of flow decreases with increasing shear stress.

dilatometer   An apparatus for accurately measuring thermal expansion of materials.

dilution   1. Adding solvent to a solution to lower its concentration. 2. Melting low-alloy base metal or previously deposited weld metal into high-alloy filler metal to produce a weld deposit of intermediate composition.

dimensional stability   The ability of a material to retain its size and shape over an extended period of time under a defined set of environmental conditions, especially temperature.

dimetcote   An inorganic zinc coating composed of two materials, (1) a reactive liquid and (2) a finely divided powder which are mixed together. The mixture reacts in place with a steel surface to form an insoluble coating.

diminished radix complement   A number obtained by subtracting each digit of the given number from one less than the radix; typical examples are the nine's-complement in decimal notation and one's-complement in binary notation.

DIN   Abbreviation for the standards institution of the Federal Republic of Germany (West Germany).

diode   A two-electrode electronic component containing merely an anode and a cathode.

diode laser   A laser in which stimulated emission is produced at a p-n junction in a semiconductor material. Only certain materials are suited for diode-laser operation, among them gallium arsenide, indium phosphide, and certain lead salts.

diode laser array   A device in which the output of several diode lasers is brought together in one beam. The lasers may be integrated on the same substrate, or discrete devices may be coupled optically and electronically.

diopter A measurement of refractive power of a lens equal to the reciprocal of the focal length in meters. A lens with 20-centimeter focal length has power of five diopters, while one with a 2-meter focal length has a power of 0.5 diopter.

DI/OU See data input/output unit, a device that interfaces to the process for the sole purpose of acquiring or sending data.

DIP See Dual In-line Package.

dip brazing Producing a brazed joint by immersing the assembly in a bath of hot molten chemicals or hot metal; a chemical bath may provide the brazing flux; molten metal, the brazing alloy.

dip coating Covering the surface of a part by immersing it in a bath containing the coating material.

dip needle A device for indicating the angle, in a vertical plane, between a magnetic field and the horizontal plane.

dipole antenna A center-fed antenna which is approximately half as long as the wavelength of the radio waves it is primarily intended to transmit or receive.

dip soldering A process similar to dip brazing, but using a lower-melting filler metal.

dip tube See bubble tube.

dir In MS-DOS, the command that will cause file directories to be displayed.

direct access The retrieval or storage of data by a reference to its location on a volume, rather than relative to the previously retrieved or stored data.

direct access device See random access device.

direct access storage device (DASD) A data storage unit on which data can be accessed directly at random without having to progress through a serial file such as tape; a disk unit is a direct-access storage device.

direct acting controller A controller in which the value of the output signal increases as the value of the input (measured variable or controlled variable) increases. See controller, direct acting.

direct acting recorder A recorder in which the pen or other writing device is directly connected to, or directly operated by, the primary sensor.

direct action A controller in which the value of the output signal increases as the value of the input (measured variable or controlled variable)

increases.

direct address An address which indicates the location where the referenced operand is to be found or stored with no reference to an index register. Synonymous with first-level address.

direct addressing An addressing mode in which the instruction operand specifies the location of the data to be used.

direct code A code which specifies the use of actual computer command and address configurations.

direct-connected An arrangement whereby a meter or other driving mechanism is connected to a driven mechanism without intervening gears, pulleys or other speed-changing devices.

direct coupling The association of two circuits which is accomplished by capacitance, resistance or self-inductance common to both circuits.

direct-current amplifier An amplifier designed to amplify signals of infinitesimally small frequency.

direct digital control (DDC) 1. A computer control technique that sets the final control-elements position directly by the computer output. 2. A control system in which the computer carries out the functions normally performed by conventional controllers, for example, three term control. 3. A term used to imply that a digital controller is connected directly to a final control element or actuator in a manufacturing process, e.g., a valve in a process stream, an electric drive motor mechanically operating on a process. Used to distinguish from analog control. 4. A method of control in which all control outputs are generated by the computer directly, with no other intelligence between the central computer and the process being controlled. See control, direct digital.

direct drive Any powered mechanism where the driven portion is on the same shaft as the driving portion, or is coupled directly to the driving portion.

direct entry In data processing, the input of data directly to computer memory and disk, in contrast to earlier methods of keying to punched cards which were then read into a computer.

direct extensions A device that provides flow rate indication by means of viewing the position of the extension of the metering float within a glass extension tube.

directional control valve A valve whose chief function is to control the direction of flow within a fluid system.

directional coupler 1. A device for separately sampling either the forward or backward oscillations in a transmission line. 2. A fiber optic coupler is directional if it preferentially transmits light in one direction.

directional gyroscope A navigational instrument for indicating direction; it contains a free gyroscope which holds its position in azimuth, thus allowing the instrument scale to indicate deviation from the reference direction.

directional property Any mechanical or physical property of a material whose value varies with orientation of the test axis within the test specimen.

direction of polarization The direction of the electric field vector of an electromagnetic wave.

direction of propagation The direction of average energy flow with respect to time at any point in a homogeneous, isotropic medium.

directive An operator command that is recognized by computer software.

directivity The solid angle, or the angle in a specified plane, over which sound or radiant energy incident on a transducer is measured within specified tolerances in a specified band of measurand frequencies [S37.1].

directivity characteristic (directional response pattern) A plot of the sensitivity level of piezoelectric sound-pressure transducer vs. the angle of sound incidence on its sensing element relative to the sensitivity level in a specified direction, and at a specified frequency [S37.10].

directly controlled system See system, directly controlled.

directly controlled variable See variable, directly controlled.

direct memory access (DMA) 1. A method of fast data transfer between the peripherals and the computer memory. The transfer does not involve the CPU. 2. Pertains to hardware that enables data to be entered into computer memory without involving the CPU; this is the method used by most telemetry/computer systems.

direct multiplex control See control, direct multiplex.

direct numerical control (DNC) A distributed numerical control system in which the supervisory computer controls several CNC or NC machines.

directory 1. A file with the layout for each field of the record which it describes. 2. The layout of

a record within a file. 3. A table that contains the names of, and pointers to, files on a mass-storage device.

directory device A mass-storage retrieval device, such as disk or DECtape, that contains a directory of the files stored on the device.

directory service The network management function that provides all addressing information required to access an application process. See PSAP address.

direct power generation Any method of producing electric power directly from thermal or chemical energy without first converting it to mechanical energy; examples include thermopiles, primary batteries, and fuel cells.

direct process Any method for producing a commercial metal directly from metal ore, without an intervening step such as roasting or smelting that produces semirefined metal or another intermediate product.

direct process piping That piping between the process and the control center which contains process fluid [RP60.9].

direct-reading gage Any instrument that indicates a measured value directly rather than by inference—for instance, indicating liquid level by means of a sight glass partly filled with liquid from the tank or by means of a pointer directly connected to a float in the tank.

direct record In instrumentation tape, the mode in which tape magnetization is directly related to data voltage level.

direct storage access (DSA) See access, direct memory.

direct storage access channel (DSAC) A channel for direct access to storage. See access, direct memory; see also channel [RP55.1].

direct wave A wave that is propagated through space without relying on the properties of any gas or other substance occupying the space.

direct-writing recorder A pen-and-ink recorder in which the position of the pen on the chart is controlled directly by a mechanical link to the coil of a galvanometer, or indirectly by a motor controlled by the galvanometer.

DIS See Draft International Standard.

disable 1. To remove or inhibit a computer hardware or software feature. 2. Disallow the processing of an established interrupt until interrupts are enabled. Contrasted with enable. See disarm.

disarm Cause an interrupt to be completely ignored. Contrasted with arm. See disable.

disassemble To reduce an assembly to its component parts by loosening or removing threaded fasteners, pins, clips, snap rings or other mechanical devices—in most instances, for some purpose such as cleaning, inspection, maintenance or repair followed by reassembly.

discharge head The pressure at which a pump discharges freely to the atmosphere, usually measured as feet of water above the intake level.

disconnect 1. To disengage the apparatus used in a connection and to restore it to its ready condition when not in use. 2. Disengaging the linkage between an interrupt and a designated interrupt servicing program. See connect.

disconnect switch An electrical switch for interrupting power supplied to a machine; it is usually separate from the machine controls (often mounted nearby on the wall) and serves mainly to deenergize the equipment for safety during setup or maintenance.

discontinuity Any feature within a bulk solid that acts as a free surface; it may be a crack, lap, seam, pore or other physical defect, or it may be a sharp boundary between the normal structure and an inclusion or other second phase; a discontinuity may or may not impair the usefulness of a part.

discrete 1. Pertaining to distinct elements or to representation by means of distinct elements, such as characters. 2. In data processing, data organized in specific parts. 3. An individual bit from a selected word. 4. Discrete manufacturing refers to the manufacture of distinct products or parts.

discrete component circuit A circuit implemented by uses of individual transistors, resistors, diodes, capacitors, etc. Contrasted with integrated circuit.

discrete increment Providing an output which represents the magnitude of the measurand in the form of discrete or quantized values [S37.1].

discrete instrument Pertaining to distinct elements or to representation by means of distinct elements.

discrete programming See integer programming.

discriminator A hardware device used to demodulate a frequency-modulated carrier or subcarrier to produce analog data.

disdrometer An apparatus capable of measuring and recording the size distribution of raindrops in the atmosphere.

disengage To intentionally pull apart two normally meshing or interlocking parts, such as gears or splines, especially for the purpose of interrupting the transmission of mechanical power.

disengaging surface The surface of the boiler water from which steam is released.

dish antenna An antenna in which a parabola-shaped "dish" serves as the reflector to increase antenna gain.

dishing A metalforming operation that forms a shallow concave surface.

disinfectant A chemical agent that destroys microorganisms, bacteria, and viruses or renders them inactive.

disk 1. An essentially flat, circular shaped part which modifies the flow rate with either linear or rotary motion [S75.05]. 2. A high-speed rotating magnetic platter for storing computer data.

disk brake A mechanical brake in which the friction elements, normally called pads, press against opposite sides of a spinning disk attached to the rotating element to slow or stop its motion.

disk cam A flat cam with a contoured edge that rotates about an axis perpendicular to the plane of the cam, communicating radial linear motion to a follower that rides on the edge of the cam.

disk clutch A device for engaging or disengaging a connection between two shafts where the chief clutch element is a pair of disks, one coupled to each shaft, and which transmit power when engaged by means of disk-face linings made of friction materials.

disk coupling A flexible coupling in which power and motion is transmitted by means of a disk made of elastomeric or other flexible material.

disk directory Table for storing the location of files held on the disk.

disk drive 1. The mechanism which moves the disk in a disk storage unit, usually including the spindle, drive motor, read-record heads, and head actuating mechanism. The term is sometimes used to include the logic control unit and other electronic circuits included in the drive unit. 2. A device that reads and writes computer data on disks.

diskette A round, flat, flexible platter coated with magnetic material and used for storage of software or data.

disk formatting See format.

disk map The organization of information stored on disks.

disk meter A flow-measurement device that contains a nutating disk mounted in such a way that each time the disk nutates, a known volume of fluid passes through the meter.

disk operating system (DOS) A collection of system programs for operating the microcomputer system.

disk pack A large disk with very high storage capacity.

disk spring A mechanical spring consisting of a dished circular plate and washer supported in such a way that one opposing force is distributed uniformly around the periphery and the second acts at the center. Washer-type disk springs are sometimes known as Belleville washers.

disk valve A valve with a closure member that consists of a disk which moves with a rotary motion against a stationary disk, each disk having flow passages through it [S75.05].

dispatching priority A number assigned to tasks, and used to determine precedence for the use of the central processing unit in a multitask situation.

dispersing prism A prism designed to spread out the wavelengths of light to form a spectrum.

dispersion 1. Any process that breaks up an inhomogeneous, lumpy mixture and converts it to a smooth paste or suspension where particles of the solid component are more uniform and small in size. 2. Breaking up globs of oil and mixing them into water to make an emulsion. 3. Intentionally breaking up concentrations of objects of substances and scattering them over a wide area. 4. The process by which an electromagnetic signal is distorted because the various frequency components of that signal have different propagation characteristics. 5. The relationship between refractive index and frequency (or wavelength). 6. In wave mechanics, linear dispersion is the rate of change of distance along a spectrum with frequency, whereas reciprocal linear dispersion is the rate of change of frequency with distance along a spectrum.

dispersion limited operation Denotes operation when the dispersion of the pulse, rather than its amplitude, limits the distance between repeaters. In this regime of operation, waveguide and material dispersion preclude an intelligent decision on the presence or absence of a pulse.

displacement 1. The change in position of a body or point with respect to a reference point. Note: Position is the spatial location of a body or point with respect to a reference point [S37.1]. 2. The volume swept out by a piston as it moves inside a cylinder from one extreme of its stroke to the other extreme. 3. For a reciprocating engine, pump or compressor, the volume swept out by one piston as it moves from top dead center to bottom dead center, multiplied by the number of cylinders. 4. Forcing a fluid or granular substance to move out of a cavity or tube by forcing more of the substance in, or by means of a piston or inflatable bladder that moves or expands into the space.

displacement antiresonance A condition of antiresonance where the external sinusoidal excitation is a force and the specified response is displacement at the point where the force is applied.

displacement meter A meter that measures the amount of a material flowing through a system by recording the number of times a vessel or cavity of known volume is filled and emptied.

displacement resonance A condition of resonance where the external sinusoidal excitation is a force and the specified response is displacement at the point where the force is applied.

displacement-type density meter A device that measures liquid density by means of a float and balance beam used in conjunction with a pneumatic sensing system; the float is confined within a small chamber through which the test liquid continually flows, so that density variations with time can be determined.

displacer-type liquid-level detector A device for determining liquid level by means of force measurements on a cylindrical element partly submerged in the liquid in a vessel; as the level in the vessel rises and falls, the displacement (buoyant) force on the cylinder varies and is measured by the lever system, torque tube or other force measurement device.

displacer-type meter An apparatus for detecting liquid level or determining gas density by measuring the effect of the fluid on the buoyancy of a displacer unit immersed in it.

display 1. A visual presentation of data. 2. In data processing, the visible representation of data on a screen.

display tube A cathode ray tube used to display information.

display unit A device which provides a temporary visual representation of data. Compare hard copy. See cathode ray tube.

dissector tube A camera tube which produces an output signal by moving the electron-optical image formed by photoelectric emission on a continuous photocathode surface past an aperture.

dissociation The process by which a chemical compound breaks down into simpler constituents, as the CO_2 and H_2O at high temperature.

dissolved gases Gases which are "in solution" in water.

dissolved solids Those solids in water which are in solution.

dissymmetrical transducer A transducer in which interchanging at least one pair of specified terminals will change the output signal delivered when the input signal remains the same.

distance/velocity lag A delay attributable to the transport of material or to the finite rate of propagation of a signal [S51.1].

distillate 1. The distilled product from a fractionating column. 2. The overhead product from a distillation column. When a partial condenser is used, there may be both a liquid and a vapor distillate stream. 3. In the oil and gas industry the term distillate refers to a specific product withdrawn from the column, usually near the bottom.

distillate fuel Any of the fuel hydrocarbons obtained during the distillation of petroleum which have boiling points higher than that of gasoline.

distillation 1. A unit operation used to separate a mixture into its individual chemical components. 2. Vaporization of a substance with subsequent recovery of the vapor by condensation. 3. Often used in less precise sense to refer to vaporization of volatile constituents of a fuel without subsequence condensation.

distilled water Water produced by vaporization and condensation with a resulting higher purity.

distortion 1. Deformation of signal shape by device or system to which it is applied [S26]. 2. An undesired change in the waveform of a given signal. 3. A lens defect that causes the images of straight lines to appear geometrically other than straight lines. See harmonic content.

distortion meter An instrument that visually indicates the harmonic content of an audio-frequency signal.

distributed In a control system, refers to control achieved by intelligence that is distributed about the process to be controlled, rather than by a centrally located single unit.

distributed control system 1. That class of instrumentation (input/output devices, control devices and operator interface devices) which in addition to executing the stated control functions also permits transmission of control, measurement, and operating information to and from a single or a plurality of user specifiable locations, connected by a communication link [S5.3]. 2. A system which, while being functionally integrated, consists of subsystems which may be physically separate and remotely located from one another [S5.1]. 3. Comprised of operator consoles, a communication system, and remote or local processor units performing control, logic, calculations and measurement functions. 4. Two meanings of distributed shall apply: a) Processors and consoles distributed physically in different areas of the plant or building, b) Data processing distributed such as several processors running in parallel, (concurrent) each with a different function. 5. A system of dividing plant or process control into several areas of responsibility, each managed by its own controller (processor), with the whole interconnected to form a single entity usually by communication buses of various kinds.

distributed database Relational computer data that can be stored in more than one networked computer, but accessed entirely by one computer.

distributed digital control systems (DDCS) See distributed control system.

distributed processing Interconnection of two or more computers so that they can work together on the same problem, not necessarily under the direction of a single control program. See also computer networking.

distributed system An arrangement whereby the computer processing power is distributed instead of being centralized.

distributor 1. Any device for apportioning current or flow among various output paths. 2. In an automotive engine, a device for sending an ignition spark to the individual cylinder in a fixed order at a rate determined by engine speed.

disturbance An undesired change in a variable applied to a system which tends to affect adversely the value of a controlled variable [S51.1].

disturbance resolution The minimum change caused by a disturbance in a measured variable which will induce a net change of the ultimately

117

controlled variable.

disturbance variable A measured variable that is uncontrolled and that affects the operations of the process.

dither A useful oscillation of small magnitude, introduced to overcome the effect of friction, hysteresis, or recorder pen clogging. See also hunting [S51.1].

dithering The application of intermittent or oscillatory forces just sufficient to minimize static friction within the transducer [S37.1].

divergence The spreading out of a laser beam with distance, measured as an angle.

divergence loss The portion of energy in a radiated beam which is lost due to nonparallel transmission, or spreading.

diversion valve A type of fluidic control device that uses the Coanda effect to either switch flow from one outlet port to another or proportion flow between two divergent outlet ports.

diversity combiner The device that accepts two radio signals from a single source that have been received with polarization, frequency, or space diversity, and combines them to yield an output that is better than either original signal.

diversity reception The use of two or more radio receivers, each being connected to different antennas, to improve the signal level. The antennas have diversity in space, phasing, and polarity.

divider A layout tool resembling a draftsman's compass which is used in toolmaking or sheet-metal work to draw circles or arcs, or to scribe hole spacings or other linear dimensions.

dividing network See crossover network.

division 1 The classification assigned to a location where either there is a high probability of a dust hazardous atmosphere occurring frequently, or regularly, or where the dust is electrically conductive [S12.11].

division 2 The classification assigned to a location where there is a low probability of a dust hazardous atmosphere occurring and/or a high probability of the presence of a hazardous dust layer [S12.11].

DMA See direct memory access.

document 1. A medium and the data recorded on it for human use, for example, a report sheet, a book. 2. By extension, any record that has permanence and that can be read by man or machine.

documentation 1. The creating, collecting, organizing, storing, citing, and disseminating of documents, or the information recorded in documents. 2. A collection of documents or information on a given subject. 3. Often used in specific reference to computer program explanation.

Dodge-Romig tables A set of standard tables with known statistical characteristics that are used in lot-tolerance and AOQL acceptance sampling.

dog Any of several simple devices for fastening, gripping or holding.

DO loop A FORTRAN statement which directs the computer to perform that sequence to which it is keyed.

dominant wavelength The wavelength of monochromatic light which matches a given color when combined in suitable proportions with a standard reference light.

dope A cellulose ester lacquer used as an adhesive or coating.

doped germanium A type of detector in which impurities are added to germanium to make the material respond to infrared radiation at wavelengths much longer than those detectable by pure germanium.

doping 1. Adding a small amount of a substance to a material or mixture to achieve a special effect. 2. Coating a mold or mandrel to prevent a molded part from sticking to it.

Doppler-effect flowmeter A device that uses ultrasonic techniques to determine flow rate; a continuous ultrasonic beam is projected across fluid flowing through the pipe, and the difference between incident-beam and transmitted-beam frequencies is a measure of fluid flow rate.

Doppler shift 1. A phenomenon that causes electromagnetic or compression waves emanating from an object to have a longer wavelength if the object moves away from an observer than would be the case if the object were stationary with respect to the observer, and to have a shorter wavelength if the object moves toward the observer; it is the physical phenomenon that forms the basis for analyzing certain sonar data and certain astronomical observations. 2. A change in the wavelength of light caused by the motion of an object emitting (or reflecting) the light. Motion toward the observer causes a shift toward shorter wavelengths, while motion away causes a shift toward longer wavelengths.

DOS See disk operating system.

dose The amount of radiation received at a specific location per unit area or unit volume, or the amount received by the whole body.

dose meter Any of several instruments for directly indicating radiation dose.

dose rate Radiation dose per unit time.

dose-rate meter Any of several instruments for directly indicating radiation dose rate.

dot matrix Characters formed by a matrix of dots.

dot-matrix printer A printer that produces letters, numbers and symbols from a two-dimensional group of dot patterns.

double acting 1. An actuator in which the power supply acts both to extend and retract the actuator stem [S75.05]. 2. Acting in two directions—for example, as in a reciprocating compressor where each piston has a working chamber at both ends of the cylinder, in a pawl that drives in both directions, or in a forging hammer that is raised and driven down by air or steam pressure.

double acting positioner A positioner is double acting if it has two outputs, one with "direct" action and the other with "reversed" action [S75.05].

double-action forming A metalforming process in which one stroke of the press performs two die operations.

double amplitude The peak-to-peak value.

double-buffered I/O An input or output operation that uses two buffers to transfer data; while one buffer is being used by the program, the other buffer is being read from or written to by an I/O device.

double-density A type of computer diskette that has twice the storage capacity of a single-density diskette.

double groove weld A weldment in which the joint is beveled or grooved from both sides to prepare the joint for welding.

double pole A type of device such as a switch, relay or circuit breaker that is capable of either closing or opening two electrical paths.

double precision 1. Pertaining to the use of two computer words to represent a number. 2. In floating-point arithmetic, the use of additional bytes or words representing the number, in order to double the number of bits in the mantissa.

double sampling A type of sampling inspection in which the lot can be accepted or rejected based on results from a single sample, or the decision can be deferred until the results from a second sample are known.

double sided A computer diskette that stores data on both sides.

doublet lens A lens with two components of different refractive index—generally designed to be achromatic.

double-welded joint A weldment in which the joint is welded from both sides.

double window fibers Optical fibers which are designed for transmission at two wavelength regions, 0.8 to 0.9 micrometer and around 1.3 micrometers.

dowel 1. A headless, cylindrical pin used to locate parts in an assembly or to hold them together. 2. A round wood stick or metal rod used to make dowel pins.

dowel screw A dowel that is threaded at both ends.

down 1. Any machinery or equipment that is not operating. 2. In data processing, computer hardware that is not running.

downhand welding See flat-position welding.

download Data or program transfer, usually from a larger computer to a PC.

downstream seating Seating is accomplished by pressure differential thrust across the ball in the closed position, moving the ball slightly downstream into tighter contact with the seat ring seal which is supported by the body [S75.05].

downtime The time when a piece of equipment is not available due to various causes, such as maintenance, set up, power failure or equipment malfunction.

dowtherm A constant boiling mixture of phenyl oxide and diphenyl oxide used in high-temperature heat transfer systems (boiling point 494 °F, 257 °C).

DP See draft proposal.

DP cell A pressure transducer that responds to the difference in pressure between two sources. Most often used to measure flow by the pressure difference across a restriction in the flow line.

draft Also spelled draught. 1. The side taper on molds and dies that makes it easier to remove finished parts from the cavity. 2. The depth to which a boat or other vessel is submerged in a body of water; the value varies with vessel weight and water density. 3. Drawing a product in a die. 4. The small, positive pressure that propels exhaust gas out of a furnace and up the stack. 5. The difference between atmospheric pressure and some lower pressure existing in the furnace or gas passages of a steam generating unit. 6. A

119

preliminary document.

draft differential The difference in static pressure between two points in a system.

draft gage 1. A type of manometer used to measure small gas heads, such as the draft pressure in a furnace. 2. A hydrostatic indicator used to determine a ship's depth of submergence.

Draft International Standard (DIS) The second stage of the ISO standard process.

draft loss A decrease in the static pressure in a boiler or furnace due to flow resistance.

draft proposal (DP) The first stage of the ISO standard process.

drag 1. The bottom part of the flask for a casting mold. See also cope. 2. Resistance of a vehicle body to motion through the air due to total force acting parallel to and opposite to the direction of motion. 3. Generally, any resistance to the motion of a solid shape through a body of fluid. 4. In data processing, the movement of an object on a screen by using a mouse.

drag-body flowmeter A device that measures the net force on a submerged solid body in a direction parallel to the direction of flow, and converts this value to an indication of flow or flow rate.

drain 1. A pipe that carries away waste solutions or effluent. 2. To empty a tank or vessel by means of gravity flow into a waste system or auxiliary holding vessel. 3. A valved connection at the lowest point for the removal of all water from the pressure parts.

draw 1. To pull a load. 2. To form cup-shaped parts from sheet metal. 3. To reduce the size of wire or bar stock by pulling it through a die. 4. To remove a pattern from a sand-mold cavity. 5. A fissure or pocket in a casting caused by inadequate feeding of molten metal during solidification. 6. A shop term referring to temper.

draw bead 1. A bead or offset used for controlling metal flow during sheet-metal forming. 2. A contoured rib or projection on a draw ring or holddown to control metal flow in deep drawing.

drawbench The stand that holds a die and draw head used for reducing the size of wire, rod, bar stock or tubing.

drawdown The curvature of the liquid surface upstream of the weir plate.

drawdown ratio The ratio of die opening to product thickness in a deep drawing operation.

drawhead 1. The die holder on a drawbench. 2. A group of rollers through which strip, tubing or

solid stock is pulled to form angle stock.

drawing back 1. A shop term for tempering. 2. Reheating hardened steel below the critical temperature to reduce its hardness.

drawing compound A lubricating substance such as soap or oil applied to prevent draw marks, scoring or other defects caused by metal-to-metal contact during a stamping, wiredrawing or similar metalforming operation.

drawing tower Equipment for making optical fibers, in which optical fibers are drawn from heated glass preforms.

draw mark Any surface flaw or blemish that occurs during drawing, including scoring, galling, pickup or die lines.

draw radius The curvature at the edge of the cavity in a deep-drawing die.

draw ring A ring-shaped die part which the punch pulls the draw blank over during a drawing operation.

dress 1. To shape a tool such as a grinding wheel. 2. To restore a tool to its original contour and sharpness.

drift 1. An undesired change in output over a period of time, which change is unrelated to the input, environment, or load [S51.1] [S67.04] [S75.05]. 2. An undesired change in output over a period of time, which change is not a function of the measurand [S37.1]. Drift is usually expressed as the change in output over a specified time with fixed input and operation conditions. Drift is usually used in connection with analog transducers, analyzers, etc.

driftpin A round, tapered metal rod that is driven into matching holes in mating parts to stretch them and bring them into alignment, such as for riveting or bolting.

drift plug A tapered rod that can be driven into a pipe to straighten it or flare its end.

drift, point The change in output over a specified period of time for a constant input under specified reference operating conditions. Point drift is frequently determined at more than one input, as for example: at 0%, 50% and 100% of range. Thus, any drift of zero or span may be calculated. Typical expression: The drift at midscale for ambient temperature (70 ± 2 °F) for a period of 48 hours was within 0.1% of output span [S51.1].

drill A cylindrical tool with one or more cutting edges on one end, which is used to make or enlarge holes in solid material by rotating it about its longitudinal axis and applying axial force.

drill drift A flat, tapered piece of steel used to remove taper shank drills and other tools from their tool holders.

drill gage A flat, thin steel plate with numerous holes of accurate sizes that can be used to check the size of drills.

drill jig A tool constructed to guide a drill during repeated drilling of the same size holes, either at many locations in a given piece or at the same location in many identical pieces, especially where exceptional straightness or accuracy of location is desired.

drill press A vertical drilling machine so constructed as to hold the work stationary and apply vertical force to press a rotating drill into the work.

drill sleeve A hollow, tapered cylinder used as an adapter to fit the shank of a taper-shank drill or other tool into the spindle of a drill press or similar machine tool.

drip tight A nonstandard term used to refer to control valve leakage. Refer to ANSI/FCI 70-2 for specification of leakage classifications [S75.05].

drive In data processing, a device that manipulates a diskette, disk, or magnetic tape so the computer can read or write data to them. See disk drive and tape drive.

drive fit A type of interference fit requiring light to moderate force to assemble.

driven gear The member(s) of a gear train that receive power and motion from another gear.

driver 1. A software element that converts operator instructions into suitable language to drive a hardware device (unit or stream drivers, for example). 2. A small program or routine that handles the control of an external peripheral device or executes other programs.

drive shaft A shaft which transmits power and motion from a motor or engine to the other elements of a machine.

driving pinion The gear in a gear train that receives power and motion by means of a shaft connected to the source of power and transmits the power and motion via its teeth to the next gear in the train.

driving-point impedance The complex ratio of applied sinusoidal voltage, force or pressure at the driving point of a transducer to resulting current, velocity or volume velocity, respectively, at the same point, all inputs and outputs being terminated in some specified manner.

driving-point reactance The imaginary component of driving-point impedance.

driving-point resistance The real component of driving-point impedance.

drone A remotely controlled, self-powered aircraft or missile.

droop See offset.

droop rate The rate at which the voltage output of a storage device decays [RP55.1].

drop leg The section of measurement piping below the process tap location to the instrument.

dropout Any discrete variation in signal level during the reproduction of recorded data that results in a data-reduction error.

drop tight A nonstandard term used to refer to control valve leakage. Refer to ANSI/FCI 70-2 for specification of leakage classifications [S75.05].

drosometer An instrument for measuring the amount of dew that condenses on a given surface.

drum 1. Any machine element consisting essentially of a thin-walled, hollow cylinder. 2. A thin-walled, cylindrical container, especially a flat-ended shipping container holding liquids or bulk solids and having a capacity of 12 to 110 gallons (50 to 400 litres). 3. The cylindrical member around which a hoisting rope is wound. 4. A high capacity computer storage device.

drum baffle A plate or series of plates or screens placed within a drum to divert or change the direction of the flow of water or water and steam.

drum brake A mechanical brake in which the friction elements, normally called shoes, press against the inside surface of a cylindrical member (the drum) attached to the rotating element to slow or stop its motion.

drum course A cylindrical section of a drum.

drum head A plate closing the end of a boiler drum or shell.

drum internals All apparatus within a drum.

drum operating pressure The pressure of the steam maintained in the steam drum or steam-and-water drum of a boiler in operation.

drum (steam) A closed vessel designed to withstand internal pressure. A device for collecting and separating the steam/water mixture circulated through the boiler [S77.42].

dry air Air with which no water vapor is mixed. This term is used comparatively, since in nature there is always some water vapor included in air, and such water vapor, being a gas is dry.

dry ash Refuse in the solid state, usually in granular or dust form.

dry assay Determining the amount of a metal or compound in an alloy, ore or metallurgical residue by means that do not involve the use of liquid to separate or analyze for constituents.

dry back The baffle provided in a firetube boiler joining the furnace to the second pass to direct the products of combustion, that is so constructed to be separate from the pressure vessel and constructed of heat resistant material, (generally refractory and insulating material).

dry-back boiler The baffle provided in a firetube boiler joining the furnace to the second-pass to direct the products of combustion, that is so constructed to be separate from the pressure vessel and constructed of heat resistance material. (Generally refractory and insulating material).

dry basis A method of expressing moisture content where the amount of moisture present is calculated as a percentage of the weight of bone-dry material; used extensively in the textile industry.

dry blast cleaning Using a dry abrasive medium such as grit, sand or shot to clean metal surfaces by driving it against the surface with a blast of air or by centrifugal force.

dry-bulb temperature The temperature of the air indicated by thermometer not affected by the water vapor content of the air.

dry corrosion Atmospheric corrosion taking place at temperatures above the dew point.

dry gas Gas containing no water vapor.

dry-gas loss The loss representing the difference between the heat content of the dry exhaust gases and their heat content at the temperature of ambient air.

drying oven A closed chamber for driving moisture from surfaces or bulk materials by heating them at relatively low temperatures.

dry pipe A perforated pipe in the steam space above the water level in a boiler which helps keep entrained liquid from entering steam outlet lines.

dry steam Steam containing no moisture. Commercially dry steam containing not more than one half of one percent moisture.

dry steam drum A pressure chamber, usually serving as the steam offtake drum, located above and in communication with the steam space of a boiler steam-and-water drum.

dry test meter A type of meter used extensively to determine gas flow rates for billing purposes and to calibrate other flow-measuring instruments; it has two chambers separated by a flexible diaphragm which is connected to a dial by means of a gear train; in operation, the chambers are filled alternately, with a flow control valve switching from one chamber to the other as the first becomes completely filled, while flow rate is indicated indirectly from movement of the diaphragm.

DSL (delta slope) algorithm See compressor.

dual-axis tracking antenna A tracking antenna which is steered automatically in both azimuth and elevation.

dual-beam analyzer A type of radiation-absorption analyzer that compares the intensity of a transmitted beam with the intensity of a reference beam of the same wavelength.

Dual In-line Package (DIP) A common way of packaging semiconductor components.

dual-output Providing two separate and noninteracting outputs which are functions of the applied measurand [S37.1].

dual ramp ADC Technique for converting analog data into digital format. The unknown voltage is input to a ramp generator and integrated for a specified time. At the end of this time a counter is started and a reference voltage applied to cause a controlled ramp down. The counter is stopped when the voltage becomes zero. The count gives the digital number output.

dual sealing valve A valve which uses a resilient seating material for the primary seal and a metal-to-metal seat for a secondary seal [S75.05].

dual-slope converter An integrating analog-to-digital converter in which the unknown signal is converted to a proportional time interval.

dual system Special configurations that use two computers to receive identical input and execute the same routines, with the results of such parallel processing subject to comparison. Exceptional high-reliability requirements usually are involved.

duct An enclosed fluid-flow passage, which may be any size up to several feet in cross section, usually constructed of galvanized sheet metal and not intended to sustain internal pressures of more than a few psi; the term is most often applied to passages for ventilating air, and to intakes and exhausts for engines, boilers and furnaces.

ductile iron The term preferred in the United States for cast iron containing spheroidal nodules of graphite in the as-cast condition. Also known as nodular cast iron; nodular iron; spherulitic-graphite cast iron.

ductility The property of a metal that indicates its relative ability to deform without fracturing; it is usually measured as percent elongation or reduction of area in a uniaxial tensile test.

Dumet wire Wire made of Fe-42Ni, covered with a layer of copper, which is used to replace expensive platinum as the seal-in wire in incandescent lamps and vacuum tubes; the copper coating prevents gassing at the seal.

dummy 1. A device constructed physically to resemble another device, but without the operating characteristics. 2. A cathode, usually corrugated to give varying current densities, which is plated at low current densities to preferentially remove impurities from an electroplating solution. 3. A substitute cathode used during adjustment of the operating conditions in electroplating. 4. An artificial address, instruction, or record of computer information inserted solely to fulfill prescribed conditions, such as to achieve a fixed word length or block length, but without itself affecting machine operations except to permit the machine to perform desired operations.

dummy argument A variable such as the one which appears in the argument list of a function definition but which is replaced by the actual argument when the function is used.

dummy block A thick plate the same shape as the extrusion billet which is placed between billet and ram to prevent the latter from overheating.

dummy instruction An artificial instruction or address inserted in a list to serve a purpose other than for execution as an instruction.

dump 1. A printout of computer memory or a file in hexadecimal and character form. 2. The transfer of data without regard for its significance. Same as storage dump.

dump valve A large valve in the bottom of a tank or container that can quickly empty the tank in an emergency.

duodecimal number A number, consisting of successive characters, representing a sum, in which the individual quantity represented by each character is based on a radix of twelve. The characters used are 0, 1, 2, 3, 4, 5, 6, 7, 8, 9, T (for ten) and E (for eleven). Related to number systems.

duplex 1. Pertaining to a twin, pair, or a two-in-one situation, e.g., a channel provided simultaneous transmission in both directions or a second set of equipment to be used in event of the failure of the primary device. 2. Referring to any item or process consisting of two parts working in connection with each other.

duplex cable A cable which contains two optical fibers in a single cable structure. Light is not coupled between the two fibers; typically one is used to transmit signals in one direction and the other to transmit in the opposite direction.

duplex connector A connector which simultaneously makes two connections, joining one pair of optical fibers with another.

duplex control A control in which two independent control elements share a common input signal for the operation of separate final control elements both of which influence the value of the controlled condition.

duplexed system A system with two distinct and separate sets of facilities, each of which is capable of assuming the system function while the other assumes a standby status. Usually, both sets are identical in nature.

duplex, full Method of operation of a communication circuit where each end can simultaneously transmit and receive [RP55.1].

duplex, half Permits one direction, electrical communication between stations. Technical arrangements may permit operation in either direction but not simultaneously [RP55.1].

duplex mode The communication link which allows simultaneous transmission and receipt of data.

duplex operation Operating an associated transmitter and receiver which are designed for concurrent transmission and reception.

duplex process Any integrated process in which a manufacturing operation is carried out by two procedures in series—for example, refining steel by the Bessemer process followed by producing ingots or continuously cast slabs by the basic oxygen or electric furnace process.

duplex pump A reciprocating or diaphragm pump having two parallel flow paths through the same housing, with a common inlet and a common outlet.

dust 1. Any finely divided solid material 420 μm or smaller in diameter (material passing a U.S. No. 40 Standard Sieve)[S12.10]. 2. Particles of gas borne solid matter larger than one micron in diameter.

dust counter A photoelectric instrument that measures the number and size of dust particles in a known volume of air. Also known as Kern counter.

dust-ignition-proof enclosure One which excludes ignitable amounts of dusts or amounts which might affect performance or rating and which, when installation and protection are in conformance with the NEC will not permit arcs, sparks or heat otherwise generated or liberated inside of the enclosure, to cause ignition of exterior accumulations or atmospheric suspensions of a specified dust on or in the vicinity of the enclosure [S12.11].

dust loading The amount of dust in a gas, usually expressed in grains per cu ft or lb per thousand lb of gas.

dust tight enclosure An enclosure of substantial mechanical construction provided with gaskets or otherwise designed to exclude dust. It has no open through holes and no knock outs. Conduit entrance is by tapped threads with a minimum of 3-1/2 threads engaged or by a gasket, bonded conduit hub. It has a substantial door or cover made dust tight to the enclosure by a securely fastened gasket or by width and closeness of fit of the mating flanges. Door or cover fasteners are of substantial construction and are permanently captive. The door or cover is permanently captive to the enclosure. Threaded-hub conduit connections are made dust tight by welding or gasketing. Threaded hub conduit connections are solidly bonded to the enclosure by welding or bonding through proper fittings. Such enclosures are NEMA 3, 3X, 4, 4X, 6, 12 or 13 enclosures [S12.11].

dutch oven A furnace that extends forward of the wall of a boiler setting. It usually is of all refractory construction, although in some cases it is water cooled.

duty The specification of service conditions which defines the type, duration and constancy of applied load or driving power.

duty cycle 1. For a device that operates repeatedly, but not continuously, the time intervals involved in starting, running and stopping plus any idling or warm-up time. 2. For a device that operates intermittently, the ratio of working time to total time, usually expressed as a percent. Also known as duty factor. 3. The percent of total operating time that current flows in an electric resistance welding machine.

duty cyclometer A meter for directly indicating duty cycle.

dwell 1. A contour on a cam that causes the follower to remain at maximum lift for an extended portion of the cycle. 2. In a hydraulic or pneumatic operating cycle, a pause during which pressure is neither increased nor decreased.

dwell period The time spent by a commutator at a given channel position.

dwell time In any variable cycle, the portion of the cycle when all controlled variables are held constant—for example, to allow a parameter such as temperature or pressure to stabilize, or to allow a chemical reaction to go to completion.

dye penetrant A low-viscosity liquid containing a dye used in nondestructive examination to detect surface discontinuities such as cracks and laps in both magnetic and nonmagnetic materials.

dynamically relocatable coding Coding for a computer which has special hardware to perform the derelativization. With an appropriately designed computer system, coding can be loaded into various sections of core, appropriate addresses changed, and the program executed.

dynamic calibration A calibration procedure in which the quantity of liquid is measured while liquid is flowing into or out of the measuring vessel.

dynamic characteristics Those characteristics of a transducer which relate to its response to variations of the measurand with time [S37.1].

dynamic compensation A technique used in control to compensate for dynamic response differences to different input streams to a process. A combination of lead and lag algorithms will handle most situations.

dynamic gain See gain, dynamic.

dynamic load That portion of a service load which varies with time, and cannot be characterized as a series of different, unvarying (static) loads applied and removed successively.

dynamic memory Same as dynamic storage.

dynamic model A model in which the variables are functions of time. Contrast with static model and steady-state model.

dynamic optimization A type of control, frequently multivariable and adaptive in nature, which optimizes some criterion function in bringing the system to the setpoints of the controlled variables. The sum of the weighted, time-absolute errors is an example of a typical criterion function to be minimized. Contrast with steady-state optimization.

dynamic pressure The increase in pressure above the static pressure that results from complete transformation of the kinetic energy of the fluid into potential energy.

dynamic programming In operations research, a procedure for optimization of a multi-stage problem wherein a number of decisions are available at each stage of the process. Contrast with convex programming, integer programming, linear programming, mathematical programming, nonlinear programming, and quadratic programming.

dynamic RAM Random access memory that needs to be refreshed at regular time intervals. It involves the extra complexity of refresh circuits but higher densities can be achieved.

dynamic range The range of signals which is accepted by a device without manual adjustment.

dynamic response 1. The behavior of the output of a device as a function of the input, both with respect to time. See response, dynamic [S67.04]. 2. The behavior of an output in response to a changing input.

dynamic sensitivity In leak testing, the minimum leak rate that a particular device is capable of detecting.

dynamic stability The property which permits the response of a positively damped physical system to asymptotically approach a constant value when the level of excitation is constant. Compare with static stability.

dynamic stop A loop stop consisting of a single jump instruction.

dynamic storage The storage of data on a device or in a manner that permits the data to move or vary with time, and thus the data is not always available instantly for recovery, e.g., acoustic delay line, magnetic drum, or circulating or recirculating of information in a medium. Synonymous with dynamic memory.

dynamic storage allocation A storage allocation technique in which the location of programs and data is determined by criteria applied at the moment of need.

dynamic subroutine A subroutine which involves parameters, such as decimal point position or item size, from which a relatively coded subroutine is derived. The computer itself is expected to adjust or generate the subroutine according to the parametric values chosen. Contrasted with static subroutine.

dynamic test A test of a device or mechanism conducted under variable loading or stimulation.

dynamic unbalance A condition in rotating equipment where the axle of rotation does not exactly coincide with one of the principal axes of inertia for the mechanism; it produces additional forces and vibrations which, if severe, can lead to failure or malfunction.

dynamic variable Process variables that can change from moment to moment due to unspecified or unknown sources.

dynamometer 1. An electrical instrument in which current, voltage or power is measured by determining the force between a fixed coil and a moving coil. 2. A special type of rotating machine used to measure the output or driving torque of rotating equipment.

E

e The base of natural logarithms.

E See modulus of elasticity.

earing Forming a scalloped edge around a deep-drawn sheet-metal part due to directional properties in the blank material.

EAROM See electrically alterable read-only memory.

earphone An electrically driven acoustic transducer intended to operate in the audio frequency range and to be held against the ear while in use.

EBCDIC See Extended Binary Coded Decimal Interchange Code.

ebullition The act of boiling or bubbling.

eccentric Describing any rotating mechanism whose center of rotation does not coincide with the geometric center of the rotating member.

eccentric orifice An orifice whose center does not coincide with the centerline of the pipe or tube; usually, the eccentricity is toward the bottom of a pipe carrying flowing gas and toward the top of a pipe carrying liquid, which tends to promote the passage of entrained water or air rather than allowing entrained water or gas to build up in front of the orifice.

echo 1. A reflected wave returned with sufficient amplitude and phase shift delay to be detected as a wave distinct from the wave originally transmitted. 2. In MS-DOS, the echo command prints the text of the command on the screen as they are executed.

echo check A check of accuracy of transmission in which the information which was transmitted to an output device is returned to the information source and compared with the original information to insure accuracy of output.

echo ranging A form of active sonar, in which the sonar equipment generates pulses of sound, then determines the distance to underwater objects by precisely measuring the time it takes for a pulse to reach an object, be reflected and return to a known location, usually one adjacent to the transmitter; by using a narrow, focused sound beam, direction to the object also can be found.

echosonogram A graphic display, such as an echocardiogram, determined with ultrasonic pulse-echo techniques.

echo sounding Determining the depth of water below a vessel or platform by sonic or ultrasonic pulse-echo techniques.

ECM See electrochemical machining.

economic lot size The number of items to be manufactured in each setup, or the number to be purchased on each order, that will keep the costs of manufacturing, purchasing, setup, inspection and warehousing to a minimum over a certain time, usually a year or longer.

economizers Heat exchangers used to recover excess thermal energy from process streams. Economizers are used for feed preheat and as column reboilers. In some systems the reboiler for one column is the condenser for another.

eddy A whirlpool of fluid.

eddy current An electric current set up in the near-surface region of a metal part by induction resulting from the electromagnetic field of an external coil carrying an alternating current; eddy currents are used to generate heat or electromagnetic fields for use in such applications as induction heating, electromagnetic sorting and testing of materials, vibration damping in spacecraft, and various types of instrumentation.

eddy-current tachometer A device for measuring rotational speed which has been used extensively in automotive speedometers; it consists of a permanent magnet revolving in close proximity to an aluminum disk which is pivoted to turn against a spring; as a magnet revolves, it induces eddy currents in the disk, setting up torque that acts against the spring; the amount of disk deflection is indicated by a moving pointer directly coupled to the disk.

edge filter An interference filter which abruptly shifts from transmitting to reflecting over a narrow range of wavelengths.

edit 1. To rearrange data or information. Editing may involve the deletion of unwanted data, the selection of pertinent data, the application of format techniques, the insertion of symbols such as page numbers and typewriter characters, the application of standard processes, such as zero suppression, and the testing of data for reasonableness and proper range. Editing sometimes may be distinguished between input edit (rearrangement of source data) and output edit (preparation of table formats) [RP55.1]. 2. To organize data for subsequent processing on a computer.

editor 1. A routine which performs editing operations. 2. A system program for amending the source code programs in high level language or assembly languages.

EDM See electrical discharge machining.

EDP See electronic data processing.

eductor 1. A device that withdraws a fluid by aspiration and mixes it with another fluid. 2. Using water, steam or air to induce the flow of other fluids from a vessel. See injector.

EEPROM See electrically erasable and programmable read-only memory.

effective address 1. A modified address. 2. The address actually considered to be used in a particular execution of a computer instruction.

effective bandwidth An operating characteristic of a specific transmission system equal to the bandwidth of an ideal system whose uniform pass-band transmission equals maximum transmission of the real system and whose transmitted power is the same as the real system for equal input signals having a uniform distribution of energy at all frequencies.

effective cutoff frequency A transducer characteristic expressed as the frequency where the insertion loss between two terminating impedances exceeds the loss at some reference frequency in the transmission band by a specified value.

effective value The root-mean-square value of a cyclically varying quantity; it is determined by finding the average of the squares of the values throughout one cycle and taking the square root of the average.

efficiency 1. The ratio of output to the input. The efficiency of a steam generating unit is the ratio of the heat absorbed by water and steam to the heat in the fuel fired. 2. In manufacturing, the average output of a process or production line expressed as a percent of its expected output under ideal conditions. 3. The ratio of useful energy supplied by a dynamic system to the energy supplied to it over a given period of time.

effluent Liquid waste discharged from an industrial processing facility or waste treatment plant.

effluvium Waste by-products of food or chemical processing.

E format In FORTRAN, an exponential type of data conversion denoted by Ew.d where w is the number of characters to be converted as a floating point number with d spaces reserved for the digits to the right of the decimal point, e.g., E11.4 yields 000.5432E03 as input, 543.2 internally and 0.5432E+03 as output.

E.I.A. Electronics Industry Association who provide standards for such things as interchangeability between manufacturers.

EIA interface Serial word transfer in ASCII characters with RS-232C logic levels, as between the computer and a manual operator.

ejection 1. Physical removal of an object from a specific site—such as removal of a cast or molded product from a die cavity by hand, compressed air or mechanical means. 2. Emergency expulsion of a passenger compartment from an aircraft or spacecraft. 3. Withdrawal of fluid from a chamber by the action of a jet pump or eductor.

ejector A device which utilizes the kinetic energy in a jet of water or other fluid to remove a fluid or fluent material from tanks or hoppers.

ejector condenser A direct-contact condenser in which vacuum is maintained by a jet of high-velocity injection water, which simultaneously condenses the steam and discharges water, condensate and noncondensible gases to the atmosphere or to the next stage ejector.

ejector half The movable portion of a diecasting or plastics-forming mold.

ejector pin A pin or rod that is driven into a hole in the rear of a mold cavity to remove the finished piece. Also known as knockout pin.

ejector plate The plate in a die-casting or plastics molding machine that backs up the ejector pins and holds the ejector assembly together.

ejector rod A rod or rodlike member that automatically operates the ejector assembly when a mold is opened.

elastic chamber The portion of a pressure-measuring system that is filled with the medium whose pressure is being measured, and that expands and collapses elastically with changes in pressure; examples include Bourdon tube, bellows, flat or corrugated diaphragm, spring-loaded piston, or a combination of two or more single elements, which may be the same or different types.

elastic collision A collision between two or more bodies in which the internal energy of the participating bodies remains constant, and in which the kinetic energy of translation for the combination of bodies is conserved.

elastic scattering A collision between two particles, or between a particle and a photon, in which total kinetic energy and momentum are conserved.

elastomer A material that can be stretched to approximately twice its original length with relatively low stress at room temperature, and which returns forcibly to about its original size and shape when the stretching force is released.

elastomeric energized liner A resilient elastomeric ring under the main liner in a butterfly valve body is compressed by the disk acting through the main liner, thus generating a resilient sealing action between the disk and the main liner [S75.05].

elbow 1. A fitting that connects two pipes at an angle, usually 90° but may be any other angle less than 100°. 2. Any sharp bend in a pipe.

elbow meter A pipe elbow that is used as a flow measurement device by placing a pressure tap at both the inner and outer radius and measuring the pressure differential caused by differences in flow velocity between the two flow paths.

electrical apparatus 1. All items applied as a whole or in part for the utilization of electrical energy. These include, among others, items for generation, transmission, distribution storage measurement, control and consumption of electrical energy and items for telecommunications. 2. Intrinsically safe electrical apparatus and intrinsically safe parts of associated electrical apparatus shall be placed in one of two categories, 'ia' or 'ib'. The requirements of this standard apply to both categories, unless otherwise specified.

electrical apparatus category 'ia' An electrical apparatus that is incapable of causing ignition in normal operation, with a single fault and with any combination of two faults applied, with the following safety factors: 1.5—in normal operation and with one fault; 1.0—with two faults.

electrical apparatus category 'ib' An electrical apparatus that is incapable of causing ignition in normal operation and with a single fault applied, with the following safety factors: 1.5—in normal operation and with one fault; 1.0—with one fault, if the apparatus contains no unprotected switch contacts in parts likely to be exposed to a potentially explosive atmosphere and the fault is self-revealing.

electrical circuits—low-voltage circuit A circuit involving a potential of not more than 30 volts and supplied by a primary battery or by a standard Class 2 transformer or other suitable transforming device, or by a suitable combination of transformer and fixed impedance having output characteristics in compliance with what is required for a Class 2 transformer. A circuit derived from a source of supply classified as a high-voltage circuit, by connecting resistance in series with the supply circuit as a means of limiting the voltage and current, is not considered to be a low-voltage circuit.

electrical conductivity A material characteristic indicative of the relative ease with which electrons flow through the material—usual units are %IACS, which relates the conductivity to that of annealed pure copper; it is the reciprocal of electrical resistivity.

electrical discharge machining A machining method in which stock is removed by melting and vaporization under the action of rapid, repetitive spark discharges between a shaped electrode and the workpiece through a dielectric fluid flowing in the intervening space, often referred to by its abbreviation, EDM; process variations include electrical discharge grinding and electrical discharge drilling. Also known as electro-erosive machining; electron discharge machining; electrospark machining.

electrical engineering A branch of engineering that deals with practical applications of electricity, especially the generation transmission and utilization of electric power by means of current flow in conductors.

electrical insulation 1. Basic: Insulation applied to live parts to provide basic protection against electric shock. 2. Supplementary: Independent insulation applied in addition to basic insulation in order to provide protection against electric shock in the event of a failure of basic insulation. 3. Protective (double): Insulation comprising both basic insulation and supplementary insulation. 4. Reinforced: A single insulation system applied to live parts, which provides protection against electric shock equivalent to double insulation [S82.01].

electrically alterable read-only memory (EAROM) A type of computer memory that is normally unchangeable, its contents can be changed only under special conditions.

electrically erasable and programmable read-only memory (EEPROM) A later version of EAROM that is simpler to use.

electrically operated extensions Usually a highly sensitive induction type device for signaling high or low flows or deviations from any set flow. The device consists of a sensing coil positioned around the extension tube of the rotameter. Movement of the metering float into the field of the coil causes a low level signal change which is usually amplified to a level suitable for performing annunciator or control functions.

electrical pumping Deposition of energy into a laser medium by passing an electrical current or discharge, or a beam of electrons through the material.

electrical resistivity A material characteristic indicative of its relative resistance to the flow of electrons—usual units are ohm-m (SI) or ohms per circular-mil foot (U.S. customary); it is the reciprocal of electrical conductivity.

electrical spacing 1. Basic: Physical separation between conductive parts which provides basic protection against electrical shock. 2. Supplementary: Physical separation used in addition to basic insulation in order to provide protection against electric shock in the event of a failure of the basic insulation. 3. Protective: Increased physical separation between conductive parts which provides protection against electric shock equivalent to double insulation [S82.01].

electrical steel Low carbon steel that contains 0.5 to 5% Si or other material; produced specifically to have enhanced electromagnetic properties suitable for making the cores of transformers, alternators, motors and other iron-core electric machines. Contrast with electric steel.

electrical terminals 1. Field wiring terminal: Any terminal to which a supply circuit wire is intended to be connected by an installer in the field. 2. Measuring or testing terminal: An external terminal or connector of the equipment to which connection is made to serve the equipment's function. 3. Measuring or testing grounded terminal: An external terminal of a grounded measuring or testing circuit which is internally connected to the equipment's protective grounding system and which is intended to be connected to the grounded side of the external circuit to which it is attached. 4. Protective-grounding terminal: A terminal connected to the equipment protective-grounding system and intended to be externally connected to earth ground [S82.01].

electrical tests 1. Routine test: A test that is performed on each piece of equipment during the production process. 2. Type test: A test that is performed on one or more pieces of equipment, representative of a type, to determine whether the design, construction, and manufacturing methods comply with the requirements according to standard [S82.01].

electric boiler A boiler in which electric heating means serve as the source of heat.

electric chart drive A clocklike mechanism driven by an electric motor which advances a circular or strip chart at a preset rate.

electric contact Either of two opposing, electrically conductive buttons or other shapes which allow current to flow in a circuit when they touch each other; they are usually attached to a spring-loaded mechanism that is mechanically or electromagnetically operated to control whether or not the contacts touch.

electric controller An assembly of devices and circuits which turns electric current to an electrically driven system off and on in response to a stimulus; in most instances, the assembly also monitors and regulates one or more characteristics of the electric supply—voltage or amperage, for example.

electric/electronic extensions A system that converts float position to a proportional electric signal (either a-c or d-c), or to a proportional shift or unbalance in impedance which is balanced by a corresponding shift in impedance in the receiving instrument.

electric field A condition within a medium or evacuated space which imposes forces on stationary or moving electrified bodies in direct relation to their electric charges.

electric field strength The magnitude of an electric field vector.

electric heating Any method for converting

electric energy into heat, but especially those methods involving resistance to the passage of electric current.

electric hygrometer An instrument that uses an electrically powered sensing means to determine ambient atmospheric humidity.

electric instrument An indicating device for measuring electrical attributes of a system or circuit. Contrast with electric meter.

electricity meter A device for indicating the time integral of an electrical quantity.

electric meter A recording or totalizing instrument that measures the amount of electric power generated or used as a function of time. Contrast with electric instrument.

electric-resistance-type liquid-level detector A device for detecting the presence of liquid at a given point; it consists of an electric probe which is insulated from the side of a vessel and positioned so that the end of the probe is at the desired liquid level; in operation, a small electric voltage is impressed between the probe and the vessel; if liquid exists at the probe level, current flows in the circuit, but does not flow if the liquid is below probe level; simple electrical systems are used when the solution has a resistivity less than 20,000 ohm-cm, but electronic systems can extend the range to 20-million ohm-cm; two probes, widely separated, can be used to control level between high and low limits or to provide high-level and low-level alarms.

electric steel Any steel melted in an electric furnace, which allows close control of composition. Also known as electric-furnace steel. Contrast with electrical steel.

electric stroboscope A device that uses an electric oscillator or similar element to produce precisely timed pulses of light; oscillator frequency can be controlled over a wide range so that the device can be used to determine the frequency of a mechanical oscillation—rpm of a rotating shaft or frequency of a mechanical vibration, for instance—by determining light-pulse frequency at which the object appears motionless.

electric tachometer An electrically powered instrument for determining rotational speed, usually in rpm.

electric telemeter An apparatus for remotely detecting and measuring a quantity—including the detector intermediate means, transmitter, receiver and indicating device—in which the transmitted signal is conducted electrically to the remote indicating or recording station.

electric thermometer An instrument that uses electrical means to measure and indicate temperature.

electric transducer A type of transducer in which all input, output and intermediate signals are electric waves.

electroacoustic transducer A type of transducer in which the input signal is an electric wave and the output a sound wave, or vice versa.

electrochemical cleaning Removing soil by the chemical action induced by passing an electric current through an electrolyte. Also known as electrolytic cleaning.

electrochemical coating A coating formed on the surface of a part due to chemical action induced by passing an electric current through an electrolyte.

electrochemical corrosion 1. Corrosion of metal caused by current flowing through an electrolyte between anode and cathode areas [S71.04]. 2. Corrosion of a metal due to chemical action induced by electric current flowing in an electrolyte. Also known as electrolytic corrosion.

electrochemical machining A machining method in which stock is removed by electrolytic dissolution under the action of a flow of electric current between a tool cathode and the workpiece through an electrolyte flowing in the intervening space. Abbreviated ECM. Also known as electrochemical milling; electrolytic machining.

electrochemical recording A type of recording system in which a signal-controlled electric current passes through a sensitized recording medium, usually in sheet form, inducing a chemical reaction to occur in the medium.

electrochemical transducer A device that uses a chemical change to measure an input parameter, and that produces an output electrical signal proportional to the input parameter.

electrode An electrically conductive member that emits or collects electrons or ions, or that controls movement of electrons or ions in the interelectrode space by means of an electric field.

electrode characteristic The relation between electrode voltage and electrode amperage for a given electrode in a system, with the voltages of all other electrodes in the system held constant; the electrode characteristic is usually shown as a graph.

electrode force The force that tends to compress

the electrodes against the workpiece in electric-resistance spot, seam or projection welding. Also known as welding force.

electrodeposition Any electrolytic process that results in deposition of a metal from a solution of its ions; it includes processes such as electroplating and electroforming. Also known as electrolytic deposition.

electrode voltage The electric potential difference between a given electrode and the system cathode or a specific point on the cathode—the latter is especially applicable when the cathode is a long wire or filament.

electrodynamic instrument An electrical instrument having both a fixed and a moving coil, both of which carry all or part of the current to be measured; if the coils are connected in series, interaction of the fields induced by the coils produces a torque proportional to the square of the current (as in an a-c voltmeter), and if connected in parallel, proportional to the product of the two coil currents (as in an a-c ammeter); in both cases, the indication is an effective (rms) value.

electroemissive machining See electrical discharge machining.

electroforming Shaping a component by electrodeposition of a thick metal plate on a conductive pattern; the part may be used as formed, or it may be sprayed on the back with molten metal or other material to increase its strength.

electrogalvanizing Coating a metal with electrodeposited zinc.

electrograph 1. A tracing produced on prepared sensitized paper or other material by passing an electric current or electric spark through the paper. 2. A plot or graph produced by an electrically controlled stylus or pen.

electro hydraulic type valve A self contained device which responds to an electrical signal, positioning an electrically operated hydraulic pilot valve to allow pressurized hydraulic fluid to move an actuating piston, bellows, diaphragm or fluid motor to position a valve stem [S75.05].

electroless plating Deposition of a metal from a solution of its ions by chemical reduction induced when the basis metal is immersed in the solution, without the use of impressed electric current.

electrolytic cleaning See electrochemical cleaning.

electrolytic corrosion See electrochemical corrosion.

electrolytic deposition See electrodeposition.

electrolytic etching Engraving a pattern on a metal surface by electrolytic dissolution.

electrolytic grinding A combined grinding and electrochemical machining operation in which an electrically conductive grinding wheel is made the cathode and the workpiece the cathode, and an electric current is impressed between them in the presence of a chemical electrolyte.

electrolytic hygrometer An apparatus for determining water-vapor content of a gas by directing it at known flow rate through a teflon or glass tube coated on the inside with a thin film of P_2O_5 (phosphorus pentoxide), which absorbs water from the flowing gas; the water is dissociated by a d-c voltage impressed on a winding embedded in the hygroscopic film that dissociates the water, and the resulting current represents the number of molecules dissociated; a calculation based on flow rate, current and temperature yields water concentration in ppm.

electrolytic machining See electrochemical machining.

electrolytic pickling Removal of scale and surface deposits by electrolytic action in a chemically active solution.

electrolytic powder Metal powder that is produced directly or indirectly by electrodeposition.

electromagnet Any magnet assembly whose magnetic field strength is determined by the magnitude of an electric current passing through some portion of the assembly.

electromagnetic Converting a change of measurand into an output induced in a conductor by a change in magnetic flux, in the absence of excitation [S37.10].

electromagnetic field sensitivity The maximum output of a transducer in response to a specified amplitude and frequency of magnetic field, usually expressed in gauss equivalent to a stated fraction of 1 g [RP37.2].

electromagnetic instrument Any instrument in which the indicating means or recording means is positioned by mechanical motion controlled by the strength of an induced electromagnetic field.

electromagnetic interference See interference, electromagnetic.

electromagnetic pulse (EMP) A type of disturbance that leads to noise in radio-frequency electric or electronic circuits.

electromagnetic radiation An all-inclusive

term for any wave having both an electric and a magnetic component; the spectrum of electromagnetic waves includes—in order of increasing photon energy, increasing frequency and decreasing wavelength—radio waves, infrared, visible light, ultraviolet, x-rays, gamma rays and cosmic rays.

electromagnetic wave A wave in which both the electric and magnetic fields vary periodically, usually at the same frequency.

electromechanical transducer A type of transducer in which the input signal is an electric wave and the output mechanical oscillation, or vice versa.

electro mechanical type valve A device which uses an electrically operated motor driven gear train or screw to position the actuator stem. Such actuators may operate in response to either analog or digital electrical signals. The electromechanical actuator is also referred to as a motor gear train actuator [S75.05].

electromechanics The technology associated with mechanical devices and systems that are electromagnetically or electrostatically actuated or controlled.

electrometallurgy The technology associated with recovery and processing of metals using electrolytic and electrical methods.

electrometer An instrument for measuring electric charge, usually by means of the forces exerted on one or more charged electrodes in an electric field.

electrometer tube A high-vacuum tube that can measure extremely small d-c voltages or amperages because of its exceptionally low control-electrode conductance.

electron An elementary subatomic particle having a rest mass of 9.107×10^{-28} g and a negative charge of 4.802×10^{-10} statcoulomb. Also known as negatron. A subatomic particle of identical weight and positive charge is termed a positron.

electron beam A narrow, focused ray of electrons which streams from a cathode or emitter and which can be used to cut, machine, melt, heat treat or weld metals.

electron-beam instrument Any instrument whose operation depends on using variable electric or magnetic fields, or both, to deflect a beam of electrons; the electron beam may be of constant intensity or it may vary in intensity according to a control signal.

electron-beam tube Any of several types of electron tube whose performance depends on formation and control of one or more electron beams.

electron device Any device whose operation depends on conduction by the flow of electrons through a vacuum, gas-filled space or semiconductor material.

electron emission Ejection of free electrons from the surface of an electrode into the adjacent space.

electron gun An electron-tube subassembly that generates a beam of electrons, and may additionally accelerate control, focus or deflect the beam.

electronic data processing (EDP) Data processing performed largely by electronic equipment.

electronic engineering A branch of engineering that deals chiefly with the design, fabrication and operation of electron-tube or transistorized equipment, which is used to generate, transmit, analyze and control radio-frequency electromagnetic waves or similar electrical signals.

electronic heating Producing heat by the use of radio-frequency current generated and controlled by an electron-tube oscillator or similar power source. Also known as high-frequency heating; radio-frequency heating.

electronic mail The use of a large centralized computer to store messages for users of the electronic mail network.

electronic measuring equipment Equipment which, by means of incorporating electronic devices, serves to measure or to observe quantities or to supply electrical quantities for measuring purposes. Electronic devices are parts or assemblies of parts which use electron or hole conduction in semiconductors, gases, or a vacuum [S82.03].

electronic photometer See photoelectric photometer.

electronic switch A circuit element causing a start and stop action or a switching action electronically, usually at high speeds.

electronic transition A transition in which an electron in an atom or molecule moves from one energy level to another.

electron metallography Using an electron microscope to study the structure of metals and alloys.

electronmicroprobe analysis A technique for

determining concentration and distribution of chemical elements over a microscopic area of a specimen by bombarding the specimen with high-energy electrons in an evacuated chamber and performing x-ray fluorescent analysis of secondary x-radiation emitted by the specimen.

electron microscope Any of several designs of apparatus that use diffracted electron beams to make enlarged images of tiny objects.

electron multiplier tube A type of electron tube that uses cascaded secondary emission to amplify small amperages.

electron tube Any device whose operation depends on conduction by the flow of electrons through a vacuum or gas-filled space within a gastight envelope.

electron volt Abbreviated eV (preferred) or EV. A unit of energy equal to the work done in accelerating one electron through an electric potential difference of one volt.

electro-optic effect A change in the refractive index of a material under the influence of an electric field. Kerr and Pockels effects are respectively quadratic and linear in electric field strength.

electropainting Electrodeposition of a thin layer of paint on metal parts which are made anodic. Also known as electrophoretic painting.

electrophonic effect A hearing sensation that results when a-c electric current of suitable amplitude and frequency passes through an animal's tissues.

electroplating Electrodeposition of a thin layer of metal on a surface of a part that is in contact with a solution, or electrolyte, containing ions of the deposited metal; in most electroplating processes, the part to be plated is the cathode, and the concentration of metal ions in the solution is maintained by placing a sacrificial anode of the deposited metal in the electrolyte.

electropneumatic controller An electrically powered controller in which some or all of its basic functions are performed by pneumatic devices.

electropolishing Smoothing and polishing a metal surface by closely controlled electrochemical action similar to electrochemical machining or electrolytic pickling.

electroscope An instrument for detecting an electric charge by observing the effects of mechanical force exerted between two or more electrically charged bodies.

electrospark machining See electrical discharge machining.

electrostatic field interference A form of interference induced in the circuits of a device due to the presence of an electrostatic field. It may appear as common-mode or normal-mode interference in the measuring circuits [S51.1].

electrostatic instrument Any instrument whose operation depends on forces of electrostatic attraction or repulsion between charged bodies.

electrostatic lens A set of electrodes arranged so that their composite electric field acts to focus a beam of electrons or other charged particles.

electrostatic memory A memory device which retains information by means of electrostatic charge, usually involving a special type of cathode-ray tube and its associated circuits.

electrostatic microphone An electroacoustic transducer for converting sound into an electrical signal by means of variation in electrostatic capacitance of the active transducer element.

electrostatic painting A spray painting process in which the paint particles are charged by spraying them through a grid of wires which is held at a d-c potential of about 100 kV; the parts being painted are connected to the opposite terminal of the high-voltage circuit so that they attract the charged paint particles; the process yields more uniform coverage than conventional spray painting, especially at corners, edges, recesses and oblique surfaces.

electrostatic precipitator A device for removing dust and other finely divided matter from a flowing gas stream by electrostatically charging them and then passing the gas stream over charged collector plates which attract and hold the particles.

electrostatic voltmeter An instrument for measuring electrical potential by means of electrostatic forces between elements in the instrument.

electrostriction transducer A device which consists of a crystalline material that produces elastic strain when subjected to an electric field, or that produces an electric field when strained elastically. Also known as piezoelectric transducer.

electrothermal process Any process that produces heat by means of an electric current—using an electric arc, induction or resistance method—especially when temperatures higher than those obtained by burning a fuel are required.

electrothermal recording A form of electro-chemical recording in which chemical change is induced by thermal effects associated with the passage of electrical current.

electrothermic instrument Any instrument whose operation depends on the heating effect associated with the passage of electric current.

element 1. A component of a device or system [S51.1]. 2. In data processing, one of the items in an array. 3. A substance which cannot be de-composed by chemical means into simpler sub-stances.

element, final controlling The forward con-trolling element which directly changes the value of the manipulated variable [S51.1].

element, primary The system element that quantitatively converts the measured variable energy into a form suitable for measurement. Note: For transmitters not used with external primary elements, the sensing portion is the primary element [S51.1].

element, reference-input The portion of the controlling system which changes the reference-input signal in response to the set point [S51.1].

element, sensing The element directly respon-sive to the value of the measured variable. It may include the case protecting the sensitive portion [S51.1].

elements, feedback Those elements in the con-trolling system which act to change the feed-back signal in response to the directly controlled variable [S51.1].

elements, forward controlling Those ele-ments in the controlling system which act to change a variable in response to the actuating error signal [S51.1].

elements of industrial process measurement and control systems Functional units or in-tegrated combinations thereof which ensure the transducing, transmitting or processing of mea-sured values, control quantities or variable, and reference variables. A valve actuator in combi-nation with a current to pressure transducer, valve positioner, or a booster relay is considered an element which receives the standard pneu-matic transmission signal or standard electric current transmission signal.

elements of process control systems Ele-ments which ensure the transducing, transmit-ting and processing of measured values, control quantities, controlled variables and reference variables. (transmitters, indicators, controllers, recorders, computers, actuators, signal condi-tioners [S50.1].)

elevated range See range, suppressed-zero.

elevated span See range, suppressed-zero.

elevated-zero range See range, elevated-zero.

elevation Vertical distance above a reference level, or datum, such as sea level. See range, suppressed-zero.

elevation error A type of error in temperature-measuring or pressure-measuring systems that incorporate capillary tubes partly filled with liquid; the error is introduced when the liquid-filled portion of the system is at a different level than the instrument case, the amount of error varying with distance of elevation or depression.

elinvar An iron-nickel-chromium alloy that also contains varying amounts of manganese and tungsten and that has low thermal expansion and almost invariable modulus of elasticity; its chief uses are for chronometer balances, watch balance springs, instrument springs and other gage parts.

ellipsometer An optical instrument which mea-sures the constants of elliptically polarized light. It is most often used in thin-film measurements.

elliptically polarized light Light in which the polarization vector rotates periodically, chang-ing in magnitude with a period of 360° so it describes an ellipse. The result of two plane polarized beams of light (each approximately a sine wave) perpendicular to each other and hav-ing a constant phase difference; the resultant planes polarized wave in the direction of the common beams well describe an ellipse. A spe-cial case called circular polarization occurs when the amplitudes of the two planes' polarized waves are equal and the phase difference is an odd multiple of $\pi/2$.

elliptically polarized wave Any electro-magnetic wave whose electric or magnetic field vector, or both, at a given point describes an ellipse.

elongation Axial plastic strain, usually ex-pressed as a percent of the original gage length in a uniaxial tension test to fracture.

elutriation Separation of fine, light particles from coarser, heavier particles by passing a slow stream of fluid upward through a mixture so that the finer particles are carried along with it.

embrittlement cracking A form of metal fail-ure that occurs in steam boilers at riveted joints and at tube ends, the cracking being predominantly intercrystalline.

emergency maintenance An urgent need for repair or upkeep that was unpredicted or not previously planned work. See corrective maintenance.

emery An abrasive material composed of pulverized, impure corundum—used in various forms including cloth or paper with an adhesive-bonded layer of emery grains, and compacted emery-binder mixtures shaped into cakes, sticks, stones, grinding wheels and other implements.

EMI Electromagnetic interference. See interference, electromagnetic.

emission characteristic The relation between rate of electron emission and some controlling factor—temperature, voltage or current of a filament or heater, for instance—for a specific element of a system, all other factors being held constant.

emissivity 1. A material characteristic determined as the ratio of radiant-energy emission rate due solely to temperature for an opaque, polished surface of a material divided by the emission rate for an equal area of a blackbody at the same temperature. 2. The rate at which electrons are emitted from a solid or liquid surface when additional energy is imparted to the system by radiant energy such as heat or light or by energetic particles such as a beam of electrons.

emittance An alternative term for emissivity.

EMR The earlier name of the Data Systems Division within Fairchild-Weston, a Schlumberger corporation, supplier of telemetry/computer systems.

emulate To imitate one system with another such that the imitating system accepts the same data, executes the same programs, and achieves the same results as the imitated system.

emulator 1. A device or program that emulates, usually done by microprogramming the imitating system. Contrasted to simulate. 2. A computer that behaves very much like another computer by means of suitable hardware and software.

emulsifier A substance that can be mixed with two immiscible liquids to form an emulsion. Also known as disperser; dispersing agent.

emulsion characteristic curve A graph of relative transmittance of a developed photographic or radiographic emulsion versus exposure; alternatively, a graph of a function of transmit-

tance versus a function of exposure.

enable 1. To restore a computer system to ordinary operating conditions. 2. To "arm" a software or hardware element to receive and respond to a stimulus. 3. Allow the processing of an established interrupt. 4. Remove a blocking device, i.e. switch, to permit operation. Contrasted with disable. See arm.

enamel 1. A type of oil paint that contains a finely ground resin and that dries to a harder, smoother, glossier finish than other types of paint. 2. Any relatively glossy coating, but especially a vitreous coating on metal or ceramic obtained by covering it with a slurry of glass frit and firing the object in a kiln to fuse the coating. Also known as glaze; porcelain enamel.

encapsulated body liner In a butterfly valve body, all surfaces of the body are covered by a continuous surface layer of a different material, usually an elastomeric or plastic material. A soft elastomer behind a harder encapsulating material may be used to provide interference for disk and stem sealing areas [S75.05].

encipher See encode.

encode 1. To apply a code, frequently one consisting of binary numbers, to represent individual characters or groups of characters in a message. Synonymous with encipher. 2. To represent computer data in digital form. 3. To substitute letters, numbers, or characters for other numbers, letters, or characters, usually to intentionally hide the meaning of the message except to certain individuals who know the enciphering scheme.

encoder 1. A device capable of translating from one method of expression to another method of expression, e.g., translating a message, "add the contents of A to the contents of B," into a series of binary digits. Contrasted with decoder. 2. A device that transforms a linear or rotary displacement into a proportional digital code. 3. A hardware device that converts analog data into digital representations.

encrustation The buildup of slag, corrosion products, biological organisms such as barnacles, or other solids on a structure or exposed surface.

encryption Converting data into codes that cannot be read without a key or password.

end In computer programming, a word indicating the completion of a program structure.

end around carry A carry from the most significant digit place to the least significant digit place.

end connection The configuration provided to make a pressure tight joint to the pipe carrying the fluid to be controlled [S75.05].

end connections, flanged End connections incorporating flanges that mate with corresponding flanges on the piping.

end connections, split clamp End connections of various proprietary designs using split clamps to apply gasket or mating surface loading.

end connections, threaded End connections incorporating threads, either male or female.

end connections, welded End connections that have been prepared for welding to the line pipe or other fittings. May be butt weld (BW), or socket weld (SW).

end device The last device in a chain of devices that performs a measurement function, the last device being the one performing final conversion of a measured value into an indication, record or control-system input signal.

end device, end instrument See transducer.

end-of-file (EOF) A magnetic marker on tape that signifies where a data file ends.

end of tape (EOT) A unique reflective marker near the end of a reel of magnetic tape to warn the computer that the end is approaching.

endothermic reaction A reaction which occurs with the absorption of heat.

end play Axial movement in a shaft-bearing assembly due to clearances within the assembly.

end point In titration, an experimentally determined point close to the equivalence point, which is used as the signal to terminate titration; it is used instead of equivalence point in most calculations, and corrections for the error between end point and equivalence point usually are not applied.

endpoint control The exact balancing of process inputs required to satisfy its stoichiometric demands.

end-point line The straight line between the end points [S37.1].

endpoint linearity The linearity of the object taken between the end points of calibration.

end points The outputs at the specified upper and lower limits of the range. Unless otherwise specified, end points are averaged during any one calibration [S37.1].

end scale value The value of an actuating electrical quantity that corresponds to the high end of the indicating or recording scale on a given instrument.

end-to-end dimension See face-to-face dimension and center-to-end dimension [S75.05].

endurance limit The maximum stress below which a material can presumably withstand an infinite number of stress cycles; if the stress is not completely reversed, the minimum stress also should be given. See also fatigue strength.

energy The capacity of a body for doing work or its equivalent—it may be classified as potential or kinetic, depending on whether it is associated with bodies at rest or bodies in motion; or it may be classified as chemical, electrical, electromagnetic, electrochemical, mechanical, radiant, thermal or vibrational, or any other type, depending on its source or nature.

energy balance The balance relating the energy in and energy out of a column. In control applications the energy balance manipulative variables are reflux and boilup.

energy beam An intense ray of electromagnetic radiation, such as a laser beam, or of nuclear particles, such as electrons, that can be used to test materials or to process them by cutting, drilling, forming, welding or heat treating.

energy density Light energy per unit area, expressed in joules per square meter—equivalent to the radiometric term "irradiance".

energy exchanger A generic term for any of several devices whose primary function is to transfer energy from one medium to another—examples include heat exchangers, boilers, and electrical transformers.

engine A machine whose chief purpose to to convert various forms of energy, such as heat or chemical energy, into mechanical power, and to perform work by imparting mechanical force and motion to other mechanisms.

engine block See cylinder block.

engineering plastics Plastics materials that are suitable for making into structural members and machine elements.

engineering time The total machine downtime necessary for routine testing, good or bad, for machine servicing due to breakdowns, or for preventive servicing measures. This includes all test time, good or bad, following breakdown and subsequent repair of preventive servicing. Synonymous with servicing time.

engineering units Terms of data measurement, as degrees Celsius, pounds, grams, and so on.

engine lathe A manually operated lathe whose headstock is driven by a gear train, by a stepped pulley mechanism, or by a combination of gears and pulleys.

Engler viscosity A standard time-based viscosity scale used primarily in Europe.

Enhanced Performance Architecture (EPA) EPA is an extension to MAP that provides for low delay communication between nodes on a single segment. See MAP/EPA and MINI-MAP.

enhancement, serial data A method whereby a continuous string of logical ONEs or ZEROs is modified to introduce bit transitions to enable bit synchronization for recording purposes; also preserves bandwidth. For example, in an incoming serial data stream, a number of words are all logic ZEROES, and therefore a DC level that the bit synchronizer cannot synchronize on; the data are enhanced by making the LSB of the words a logic ONE.

enter key The key on a computer terminal that is pressed to enter data into a computer.

entity An active element within an OSI layer (e.g. Token Bus MAC is an entity in the Layer 2).

entrainment The conveying of particles of water or solids from the boiler water by the steam.

entry Any item of computer data to be stored and processed.

entry conditions The initial data and control conditions to be satisfied for successful execution of a given routine.

entry data The initial data required for successful execution of a given routine. See entry conditions.

entry name The alphanumeric name given to an entry point. See entry point.

entry point In a routine, any place to which control can be passed.

envelope 1. Generally, the boundaries of an enclosed system or mechanism. 2. Specifically, the glass or metal housing of an electron tube, or the glass enclosure of an incandescent lamp. Also known as bulb.

environment 1. The ambient natural and artificial conditions that surround a piece of operating equipment. 2. Ambient conditions (including temperature, pressure, humidity, radioactivity, and corrosiveness of the atmosphere) surrounding the valve and actuator. Also, the mechanical and seismic vibration transmitted through the piping or heat radiated toward the actuator from a valve body [S75.05].

environmental area See area, environmental.

environmental conditions Specified external conditions (shock, vibration, temperature, etc.) to which a transducer may be exposed during shipping, storage, handling, and operation [S37.1].

environmental conditions, operating Environmental conditions during exposure to which a transducer must perform in some specified manner [S37.1].

environmental engineering A branch of engineering that deals with the technology related to control of the surroundings which humans live in, especially the control or mitigation of contamination or degradation of natural resources such as air quality and water purity.

environmental influence See operating influence.

environmental test Any laboratory test conducted under conditions that simulate the expected operating environment in order to determine the effect of the environment on component operation or service life.

EOF marker In data processing, a code written after the last record of a file to indicate the end of that file.

EPA 1. See Enhanced Performance Architecture.

EP lubricant See extreme-pressure lubricant—an oil or grease containing additives that enhance the ability of the lubricant to adhere to a surface and reduce friction under high bearing loads.

epoxy adhesive An adhesive made of epoxy resin.

Eppley pyrheliometer A thermoelectric device for measuring direct and diffuse solar radiation; radiation is directed onto two concentric silver rings, the outer covered with MgO and the inner covered with lampblack, and a thermopile is used to determine the difference in temperature between the two rings.

equalizer 1. A device that connects parts of a boiler to equalize pressures. 2. The electronic circuit in a tape reproducer whose gain across the spectrum of interest compensates for the unequal gain characteristic of the record/reproduce heads, thereby providing "equalized" gain across the band.

equal percent characteristic See characteristic, equal percentage.

equilibrium state Any set of conditions that results in perfect stability—mechanical forces that completely balance each other and do not produce acceleration, or a reversible chemical reaction in which there is no net increase or decrease in the concentration of reactants or reaction products, for instance.

equipment 1. An assembly of electrical or electronic components or circuits intended to perform a complete function apart from being a substructure of a system [S82.01]. 2. A generic term for any apparatus, assembly, mechanism or machine, or for a group of units constructed similarly, or for a group of units performing similar functions.

equipment compatibility The characteristic of computers by which one computer may accept and process data prepared by another computer without conversion or code modification.

equipment failure A fault in the equipment, excluding all external factors, which prevents continued performance.

equivalence point Point on the titration curve where the acid ion concentration equals the base ion concentration.

equivalent binary digits The number of binary digits required to express a number in another base to the same precision, e.g., approximately 3 1/3 binary digits are required to express in binary form each digit of a decimal number. For the case of binary coded decimal notation, the number of binary digits required is usually four times the number of decimal digits.

equivalent evaporation Evaporation expressed in pounds of water evaporated from a temperature of 212 °F to dry saturated steam at 212 °F.

equivalent network A network that can perform the functions of another network under certain conditions; the two networks may be of different forms—one mechanical and one electrical, for instance.

equivalent volume The volume of a gas enclosed in a rigid cavity which would give the same acoustical input impedance as that of the piezo-electrical sound-pressure transducer [S37.10].

erasable storage 1. A storage device whose data can be altered during the course of a computation, e.g., magnetic tape, drum and cores. 2. An area of storage used for temporary storage.

erase To obliterate information from a storage medium, for example, to delete or to overwrite.

erg The unit of energy in the CGS system; it is the amount of energy consumed (work) when a force of one dyne is applied through a distance of one centimeter.

ergonomics The science of designing machines and work environments to suit the needs of people.

Erichsen test A cupping test for determining the suitability of metal sheet for use in a deep drawing operation; it is expressed as the depth in millimeters of a cup-shaped impression in a sheet of metal, supported on a ring, which is deformed at the center by a spherical tool until it breaks or tears.

erosion 1. Deterioration by the abrasive action of fluids, usually accelerated by the presence of solid particles in suspension [S71.04]. 2. The wearing away of refractory or of metal parts by gas borne dust particles. 3. Progressive destruction of a structural member by the abrasive action of a moving fluid, often one that contains solid particles in suspension; if the fluid is a gas, erosion may be caused by liquid droplets carried in the moving gas stream.

erosion-corrosion Progressive destruction of a structural member by the combined effects of corrosion and erosion acting simultaneously.

erosion resistant trim Valve trim that has been designed with special surface materials or geometry to resist the erosive effects of the controlled fluid flow [S75.05].

err See error.

error 1. In process instrumentation, the algebraic difference between the indication and the ideal value of the measured signal. It is the quantity which algebraically subtracted from the indication gives the ideal value. A positive error denotes that the indication of the instrument is greater than the ideal value. See correction [S51.1]. 2. The algebraic difference between the indicated value and the true value of the measurand. Note: (a) It is usually expressed in percent of the full scale output, sometimes expressed in percent of the output reading of the transducer. (b) A theoretical value may be specified as true value [S37.1]. 3. The general term referring to any deviation of a computed or a measured quantity from the theoretically correct or true value. 4. The part of the error due to a particular identifiable cause, e.g., a truncation error, or a rounding error. In a restricted sense, that deviation due to unavoidable random disturbances, or to the use of finite approximations to what is defined by an infinite series. 5. The amount by which the computed or measured quantity differs from the theoretically correct or true value. 5. In a single automatic control loop, the setpoint minus the controlled variable measurement.

error band The band of maximum deviations of output values from a specified reference line or curve due to those causes attributable to the transducer. Note: (a) The band of allowable deviations is usually expressed as "± _____ percent of full scale output", whereas in test and calibration reports the band of maximum actual deviations is expressed as "+ _____ percent, – _____ percent of full scale output." (b) The error band should be specified as applicable over at least two calibration cycles, so as to include repeatability, and verified accordingly [S37.1].

error burst In data transmission, a sequence of signals containing one or more errors but counted as only one unit in accordance with some specific criterion or measure. An example of a criterion is that if three consecutive correct bits follow an erroneous bit, an error burst is terminated.

error checking Data quality assurance usually attempted by calculating some property of the data block before transmission. The resulting property or check character is also sent to the receiver, where it may be inspected and compared with a recalculated value based on the received data.

error correcting code A code in which each acceptable expression conforms to specific rules of construction that also define one or more equivalent nonacceptable expressions, so that if certain errors occur in an acceptable expression, the result will be one of its equivalents, and thus the error can be corrected.

error curve A graphical representation of errors obtained from a specified number of calibration cycles [S37.1]. See calibration curve.

error detecting code A code in which each expression conforms to specific rules of construction, so that if certain errors occur in an expression, the resulting expression will not conform to the rules of construction, and thus the presence of the errors is detected. Synonymous with self-checking code.

error detection routine A routine used to detect whether or not an error has occurred, usually without special provision to find or indicate its location.

error, environmental Error caused by a change in a specified operating condition from reference operating condition. See operating influence [S51.1].

error, frictional Error of a device due to the resistance to motion presented by contacting surfaces [S51.1].

error, hysteresis See hysteresis.

error, hysteretic See hysteresis.

error, inclination The change in output caused solely by an inclination of the device from its normal operating position [S51.1].

error indication Ideal value.

error maximum (data processing) The maximum error EMAX of the analog subsystem is defined as the deviation between the true value of the input signal and the particular output reading within the distribution of output readings furthest displaced from the true value [RP55.1].

error message An audible or visual indication of a software or hardware malfunction, or a nonacceptable data entry attempt.

error, mounting strain Error resulting from mechanical deformation of an instrument caused by mounting the instrument and making all connections. See also error, inclination [S51.1].

error position The change in output resulting from mounting or setting an instrument in a position different from that at which it was calibrated. See also error, inclination [S51.1].

error range 1. The range of all possible values of the error of a particular quantity. 2. The difference between the highest and the lowest of these values.

error ratio The ratio of the number of data units in error to the total number of data units.

error signal The output of a comparing element. See signal, error.

error, span The difference between the actual span and the ideal span. It is usually expressed as a percent of ideal span [S51.1].

error, systematic An error which, in the course of a number of measurements made under the same conditions of the same value of a given quantity, either remains constant in absolute value and sign or varies according to a definite law when the conditions change [S51.1].

error, zero In process instrumentation, error of a device operating under specified conditions of use, when the input is at the lower range-value. It is usually expressed as percent of ideal span [S51.1].

escape key A key on a computer keyboard that

returns the operator to the prior step in a command sequence.

escapement A ratchet device that permits motion only in one direction, such as the device that controls motion in the works of a mechanical watch or clock.

escutcheon A decorative shield, flange, or border around a panel-mounted part such as a dial or control knob. Also known as escutcheon plate.

Etalon A type of Fabry-Perot interferometer in which the distance between two highly reflecting mirrors is fixed. It is used to separate light in different wavelengths when the wavelengths are closely spaced.

etch cleaning Removing soil by electrolytic or chemical action that also removes some of the underlying metal.

etch cracks Shallow cracks in the surface of hardened steel due to hydrogen embrittlement that sometimes occurs when the metal comes in contact with an acidic environment.

etching 1. Controlled corrosion of a metal surface to reveal its metallurgical structure. 2. Controlled corrosion of a metal part to create a design; the design may consist of alternating raised and depressed areas, or it may consist of alternating polished and roughened areas, depending on the conditions and corrodent used.

European Workshop on Industrial Computer Systems (EWICS) The European industrial computer control standards group.

eutectic 1. A process by which a liquid solution undergoes isothermal decomposition to form two homogeneous solids—one richer in solute than the original liquid, and one leaner. 2. The composition of the liquid which undergoes eutectic decomposition and possesses the lowest coherent melting point of any composition in the range where the liquid remains single-phase. 3. The solid resulting from eutectic decomposition, which consists of an intimate mixture of two phases.

eutectoid A decomposition process having the same general characteristics as a eutectic, but taking place entirely within the solid state.

evaluation kit A small microcomputer system used for learning the instruction set of a given microcomputer. It usually includes light emitting diodes, a keyboard, a monitor/debugger in ROM, a small amount of RAM and some input-output ports.

evaporated make-up Distilled water used to supplement returned condensate for boiler feed water.

evaporation The change of state from a liquid to a vapor.

evaporation gage See atmometer.

evaporation rate The number of pounds of water evaporated in a unit of time.

evaporative cooling 1. Lowering the temperature of a mass of liquid by evaporating part of it, using the latent heat of vaporization to dissipate a significant amount of heat. 2. Cooling ambient air by evaporating water into it. 3. See vaporization cooling.

evaporator Any of several devices where liquid undergoes a change of state from liquid to gas under relatively low temperature and low pressure.

evaporimeter See atmometer.

event 1. The occurrence of some programmed action within a process which can affect another process. 2. In TELEVENT, an occurrence recognized by telemetry hardware such as frame synchronization, buffer complete, start, halt, and the like.

event oriented Pertaining to a physical occurrence.

event recorder An instrument that detects and records the occurrence of specific events, often by recording on-off information against time to show when an event starts and stops, and how often it recurs.

EWICS See European Workshop on Industrial Computer Systems.

exception reporting An information system which reports on situations only when actual results differ from planned results. When results occur within a normal range they are not reported.

excess air Air supplied for combustion in excess of that theoretically required for complete oxidation.

excimer laser A laser in which the active medium is an "excimer" molecule—a diatomic molecule which can exist only in its excited state. The internal physics are conducive to high powers in short pulses, with wavelengths in the ultraviolet.

excitation 1. The external supply applied to a device for its proper operation. See also excitation, maximum [S51.1]. 2. The external electrical voltage and/or current applied to a transducer for its proper operation. Note: (a) In the sense of a physical quantity to be measured by a transducer, use measurand. (b) Usually expressed as range(s) of voltage and/or current values. (c) Also see maximum excitation [S37.1]. 3. Voltage supplied by a signal conditioner to certain types of physical measurement transducers (bridges, for example).

excitation, maximum The maximum value of excitation parameter that can be applied to a device at rated operating conditions without causing damage or performance degradation beyond specified tolerances [S51.1].

excitation voltage A precision voltage applied to transducers; when pressure, strain, or the like are sensed by the transducer, a small portion of this voltage appears on the signal lines; the value of this signal voltage is proportional to the stimulus applied.

exclusions See exclusive OR.

exclusive OR A logical operator which has the property that if P and Q are two statements, then the statement P*Q, where the * is the exclusive OR operator, is true if either P or Q, but not both are true, and false if P and Q are both false or both true.

executable statement Constituent of a program specifying the action of the program. Contrasted with a nonexecutable statement which describes the use of the program, the characteristics of the operands, editing information, statement functions, or data arrangement.

execute 1. To interpret machine instructions or higher-level statements and perform the indicated operations on the operands specified. 2. In computer terminology, to run a program.

execution The act of performing programmed actions. Execution time is the interval required for a specified action to be performed.

execution of an instruction The set of elementary steps carried out by the computer to produce the result specified by the operation code of the instruction.

execution time 1. Time required to execute a program. 2. The period during which a program is being executed. 3. The time at which execution of a program is initiated. 4. The period of time required for a particular machine instruction. See also instruction time.

executive 1. Short for executive routine. See routine, executive [RP16.4]. 2. The controlling program or set of routines in an operating system; the executive coordinates all activities in the system including I/O supervision, resource al-

location, program execution, and operator communication. See also monitor.

executive commands In TELEVENT, the several commands that establish modes of operation, such as SETUP, END, CONNECT, and the like.

executive mode A central processor mode characterized by the lack of memory protection and relocation by the normal execution of all defined instruction codes.

executive program A program which controls the execution of all other programs in the computer based on established hardware and software priorities and real time or demand requirements.

executive software That portion of the operational software which controls on-line, response-critical events, responding to urgent situations as specified by the application program. This software is also known as the real-time executive.

executive system An integrated collection of service routines for supervising the sequencing of programs by a computer.

EXE file In MS-DOS, the designation .EXE follows most application program file names.

exfoliation corrosion A type of corrosion that proceeds parallel to the surface of a material, causing thin outer layers to be undermined and lifted by corrosion products.

exhaust 1. Discharge of working fluid from an engine cylinder or from turbine vanes after it has expanded to perform work on the piston or rotor. 2. The fluid discharged. 3. A duct for conducting waste gases, fumes or odors from an enclosed space, especially the discharge duct from a steam turbine, gas turbine, internal combustion engine or similar prime mover; gas movement may be assisted by fans.

exhaust-gas analyzer An instrument that measures the concentrations of various combustion products in waste gases to determine the effectiveness of combustion.

exhaust steam Steam discharged from a prime mover.

exhaust stroke The portion of the cycle in an engine, pump or compressor that expels working fluid from the cylinder.

exhaust valve A valve in the headspace of a cylinder that opens during the exhaust stroke to allow working fluid to pass out of the cylinder.

exit The time or place at which the control sequence ends or transfers out of a particular computer program or sub-routine.

exothermic reaction A reaction which occurs with the evolution of heat.

exotic fuels High-energy fuels, especially the hydroborons, which have higher calorific values than the corresponding hydrocarbons, and which at one time were proposed for use in high-performance aircraft and missiles.

expanded joint The pressure tight joint formed by enlarging a tube end in a tube seat.

expanded memory The ability to add usable memory to a computer.

expanded metal A form of coarse screening made by lancing sheet metal in alternating rows of short slits, each offset from the adjacent rows, then stretching the sheet in a direction transverse to the rows of slits so that each slit expands to give a roughly diamond-shaped opening.

expanded plastic A light, spongy plastics material made by introducing air or gas into solidifying plastic to make it foamy. Also known as foamed plastic; plastic foam.

expander The tool used to expand tubes.

expanding Increasing the diameter of a ring-shaped or cylindrical part, usually by placing it over a circular segmented die and forcing the segments to move radially in a controlled manner to stretch the part circumferentially.

expansion 1. Increasing the volume of a working fluid, with a corresponding decrease in pressure, and usually with an accompanying decrease in temperature, as in an engine, turbine or other prime mover. 2. Generally, any increase in volume or dimension which causes a body to occupy more physical space.

expansion factor (Y) Correction for the change in density between two pressure-measurement stations in a constricted flow.

expansion joint The joint to permit movement due to expansion without undue stress.

experience Applicable work in design, construction, preoperational and startup testing activities, operation, maintenance, onsite activities, or technical services. Observation of others performing work in the above areas is not experience. This experience can be obtained during startup or operations in a nuclear power plant, in fossil power plant, in other industries, or in the military.

expert system A computer program that uses stored data to reach conclusions, unlike a data base which presents data unchanged.

explosion Combustion which proceeds so rapidly that a high pressure is generated suddenly.

explosion door A door in a furnace or boiler setting designed to be opened by a pre-determined gas pressure.

explosion proof apparatus Apparatus enclosed in a case that is capable of withstanding an explosion of a specified gas or vapor which may occur within it and of preventing the ignition of a specified gas or vapor surrounding the enclosure by sparks, flashes, or explosion of the gas or vapor within, and which operates at such an external temperature that a surrounding flammable atmosphere will not be ignited thereby.

explosion welding A solid-state process for creating a metallurgical bond by driving one piece of metal rapidly against another with the force of a controlled explosive detonation.

explosive cladding Producing a bimetallic material by explosion welding a thin layer of one metal on a substrate; it is used most advantageously to yield a material with one surface having a unique property, such as resistance to corrosion by certain strong chemicals, while the bulk of the material possesses good fabrication and structural properties.

explosive forming Shaping parts in dies through the use of explosives to generate the forming pressure; most often, a sheet metal part is placed over an open die and covered with a sheet of explosive, which is then detonated to drive the metal into the die.

exponent In floating-point representation, one of a pair of numerals representing a number that indicates the power to which the base is raised. Synonymous with characteristic.

exponential notation A way to express very large or small numbers in data processing.

exponentiation A mathematical operation that denotes increases in the base number by a previously selected factor.

exposure 1. For a photographic or radiographic emulsion, the product of incident radiation intensity and interval of time it is allowed to impinge on the emulsion. 2. A term loosely used to indicate time of exposure in photography.

exposure time The elapsed time during which radiant energy is allowed to impinge on photographic or radiographic emulsion.

expression 1. A combination of operands and operators that can be evaluated to a distinct result by a computing system. 2. Any symbol representing a variable or a group of symbols representing a group of variables possibly combined by symbols representing operators in accordance with a set of definitions and rules. 3. In computer programming, a set of symbols that can have a specific value.

Extended Binary Coded Decimal Interchange Code (EBCDIC) An 8-bit code that represents an extension of a 6-bit BCD code, which has been widely used in computers of the first and second generations. EBCDIC can represent up to 256 distinct characters and is the principal code used in many of the current computers.

extended instruction set (EIS) The software in the DEC system that provides hardware with fixed-point arithmetic and direct implementation of multiply, divide, and multiple shifting.

extension 1. A device for translating float motion into a useful secondary function, for either indicating, alarming, transmitting or other secondary functions. An extension usually consists of an extension tube, an extension housing, and the necessary adaptor to the primary rotameter, but may be any auxiliary device fixed to the rotameter which performs specific functions [RP16.4]. 2. A multiple character set that follows a computer filename that further clarifies the filename.

extension bonnet A bonnet with a packing box that is extended above the bonnet joint of the valve body so as to maintain the temperature of the packing above or below the temperature of the process fluid. The length of the extension bonnet is dependent upon the difference between the fluid temperature and the packing design temperature limit as well as upon the valve body design [S75.05].

extension furnace See dutch oven.

extensions, controlling A controller which derives its input from the motion of the float can be installed within the extension housing [RP16.4].

extensions, integrating An integrator which derives its input from the motion of the float can be installed within the extension housing [RP16.4].

extension spring A tightly coiled helical spring designed to resist a tensile force.

extensions, recording The recorder is attached directly to the meter body with the recorder pen positioned by the metering float through a magnetic coupling [RP16.4].

extensometer 1. An apparatus for studying seismic displacements by measuring the change in distance between two reference points that are separated by 20 to 30 metres or more. 2. An instrument for measuring minute elastic and plastic strains in small objects under stress, especially the strains prior to fracture in standard tensile-test specimens.

externally fired boiler A boiler in which the furnace is essentially surrounded by refractory or water-cooled tubes.

externally quenched counter tube A radiation counter tube equipped with an external circuit that inhibits reignition of the counting cycle by internal ionizing events.

external memory See external storage.

external-mix oil burner A burner having an atomizer in which the liquid fuel is struck, after it has left an orifice, by a jet of high velocity steam or air.

external multiplexors Scanivalves, switching temperature indicators, and other devices that permit input of several signals on one computer input channel.

external party line (XPL) A logic level from telemetry equipment that causes the buffered data channel to switch input ports (to merge time, for example).

external start (XST) The hardware-generated pulse that causes the system to start receiving data (concurrent with word 1 of a frame, and often frame 1 of a subframe).

external storage 1. The storage of data on a device which is not an integral part of a computer but in a form prescribed for use by the computer. 2. A facility or device, not an integral part of a computer, on which data usable by a computer is stored, such as off-line magnetic tape units, or punch card devices. Synonymous with external memory and contrasted with internal storage.

external treatment Treatment of boiler feed water prior to its introduction into the boiler.

extract instruction An instruction that requests the formation of a new expression from selected parts of given expressions.

extractive distillation A distillation technique (employing the addition of a solvent) used when the boiling points of the components being separated are very close [within 3 °C (5 °F)] or the components are constant boiling mixtures. In extractive distillation, which is a combination of fractionation and solvent extraction, the solvent is generally added to the top of the column and recovered from the bottom product by means of subsequent distillation. The chemical added is a solvent only to the less-volatile components. See azeotrope.

extra hard temper A level of hardness and strength in nonferrous alloys and some ferrous alloys corresponding approximately to a cold worked state one-third of the way from full hard to extra spring temper.

extra spring temper A level of hardness and strength for nonferrous alloys and some ferrous alloys corresponding to a cold worked state above full hard beyond which hardness and strength cannot be measurably increased by further cold work.

extrusion 1. A process for forming elongated metal or plastic shapes of simple to moderately complex cross section by forcing ductile, semi-soft solid material through a die orifice. 2. A length of product made by this process.

extrusion billet A slug of metal, usually heated into the forging temperature range, which is forced through a die by a ram in an extrusion process.

extrusion pressing See cold extrusion.

eyebar A metal bar having a hole through an enlarged section at each end of the bar.

eyebolt A bolt with a loop formed at one end in place of a head.

eyelet A small ring or barrel-shaped piece of metal used to reinforce a hole, especially in fabric.

eyeleting Forming a lip around the rim of a hole in sheet metal.

F

F See farad and Fahrenheit.

fabrication 1. A general term for parts manufacture, especially structural or mechanical parts. 2. Assembly of components into a completed structure.

fabry-perot A pair of highly reflecting mirrors, whose separation can be adjusted to select light of particular wavelengths. When used as a laser resonator, this type of cavity can narrow the range of wavelengths emitted by the laser.

face 1. An exposed structural surface. 2. In a weldment, the exposed surface of the fusion zone.

faceplate 1. A circular plate attached to the spindle of a lathe with the plane of the plate perpendicular to the spindle axis; it is used to attach and align certain types of workpieces. 2. A protective cover for holes in an equipment enclosure. 3. A glass or plastic window in personal protective gear such as welding helmets, respirator masks or diving masks. 4. A two-dimensional array of separate optical fibers, fused together, serving to strongly direct light forward.

facet The plane surface of a crystal or fracture surface.

face-to-face dimension The dimensions from the face of the inlet opening to the face of the outlet opening of a valve or fitting. See end-to-end dimension [S75.05].

facing 1. Machining a flat planar surface in lathe turning by positioning a single-point tool against the workpiece at the axis of rotation and moving the tool radially outward so that it cuts a spiral path in a plane perpendicular to the axis of rotation. 2. Fine molding sand applied to the surface of the mold cavity.

facing, flange The finish on the end connection gasket surfaces of flanged or flangeless valves [S75.05].

facsimile A system for utilizing telephone transmission apparatus to send written or pictorial information to a remote location; it consists of a transmitter, which scans the hard-copy record and converts its image into an electrical signal wave, and a receiver, which converts the electrical wave into its final pictorial form and registers it on a record sheet.

fading A drop in signal intensity, or a slow undulation, caused by changes in the properties of the transmission medium.

Fahrenheit A temperature scale where the freezing point of pure water occurs at 32 °F and the span between freezing point and boiling point of pure water at standard pressure is defined to be 180 scale divisions (180 degrees).

fail-close A condition wherein the valve closure member moves to a closed position when the actuating energy source fails. See normally closed [S75.05].

fail closed A condition wherein the valve closure component moves to a closed position when the actuating energy source fails.

fail-in place A condition wherein the valve closure component stays in its last position when the actuating energy source fails [S75.05].

fail-open A condition wherein the valve closure member moves to an open position when the actuating energy source fails. See normally open [S75.05].

fail-safe 1. A characteristic of a particular valve and its actuator, which upon loss of actuating energy supply, will cause a valve closure member to fully close, fully open or remain in fixed position. Fail-safe action may involve the use of auxiliary controls connected to the actuator [S75.05]. 2. Any protection against effects of failure of the equipment, such as, fuel shut-off in the event of loss of flame in a furnace.

fail-safe device A component, system or control device so designed that it places the controlled parameter in a safe condition in case of a power interruption, controller malfunction or failure of a load-carrying member.

failure mode The position to which the valve closure component moves when the actuating energy source fails [S75.05].

failure rate A measure of component reliability usually expressed as the probability of failure after a specified length of time in service.

fall The chain, rope or wire rope used in lifting tackle.

fall block In lifting tackle, a pulley block attached to the load and rising or descending with it.

fall time The time required for the output voltage of a digital circuit to change from a logical high level (1) to a logical low level (0).

false add To form a partial sum, that is, to add without carries.

false brinelling Fretting between the rolling elements and races of ball or roller bearings.

false set Rapid hardening of freshly mixed cement, mortar or concrete with a minimum evolution of heat; plasticity can be restored by mixing without adding more water.

fan A rotating mechanism, usually consisting of a paddle wheel or screw, with or without a casing, used to induce movement (currents) in air or other gas, such as in a circulation, ventilation or exhaust system where large volumes must be delivered.

fang bolt A bolt having a triangular head with sharp projections at the corners which is used primarily to attach metal parts to wood.

fan inlet area The inside area of the fan outlet.

fan performance A measure of fan operation in terms of volume, total pressures, static pressures, speed, power input, mechanical and static efficiency, at a stated air density.

fan performance curves The graphical presentation of total pressure, static pressure, power input, mechanical and static efficiency as ordinates and the range of volumes as abscissa, all at constant speed and air density.

fan requirements—recommended or specified The fan requirements recommended by the manufacturer of the steam generating equipment which will include necessary tolerances to overcome unfavorable operating conditions.

farad Metric unit of electrical capacitance.

Faraday rotation A rotation of the plane of polarization of light caused by the application of a magnetic field to the material transmitting the light.

Faraday rotator A device which relies on the Faraday effect to rotate the plane of polarization of a beam of light passing through it. Faraday rotator glass is a type of glass with composition designed to display the Faraday effect.

far field Distant from the source of light. This qualification is often used in measuring beam quality, to indicate that the measurement is made far enough away from the laser that local aberrations in the vicinity of the laser have been averaged out.

far-infrared laser Generically, this term could be taken to mean any laser emitting in the far infrared, a vaguely defined region of wavelengths from around 10 micrometers to 1 millimeter. This family of lasers require optical pumping by an external laser—usually carbon dioxide.

fast break In magnetic particle testing of ferromagnetic materials, interrupting the current in the magnetizing coil to induce eddy currents and strong magnetization as the magnetizing field collapses.

fastener 1. Any of several types of devices used to hold parts firmly together in an assembly; some fasteners hold parts firmly in position, but allow free or limited relative rotation. 2. A device for holding a door, gate or similar structural member closed.

Fast-Fourier Transform A type of frequency analysis on data that can be done by computer using special software, or by an array processor or by a special-purpose, hardware device.

fatigue Progressive fracture of a material by formation and growth of minute cracks under repeated or fluctuating stresses whose maximum value is less than the material's tensile strength, and is often wholly within the elastic-stress range.

fatigue life The number of stress cycles that a material can sustain prior to fracture for a given set of fatigue conditions.

fatigue notch factor The ratio of the fatigue strength of an unnotched specimen to the fatigue strength of a notched specimen of the same material and condition; the notch used is of a specified size and contour, and the strengths are compared at the same number of stress cycles.

fatigue notch sensitivity An estimate of the effect of a notch or hole on the fatigue properties of a material; it is expressed as $q = (K_f - 1)/(K_t - 1)$, where q is the fatigue notch sensitivity, K_f is the fatigue notch factor, and K_t is the stress concentration factor for a specimen of the material containing a notch of a specific size and shape.

fatigue strength The maximum stress that ordinarily leads to fatigue fracture in a specified number of stress cycles; if the stress is not completely reversed during each stress cycle, the minimum stress also should be given. See also endurance limit.

fault 1. A physical condition that causes a device, a component, or an element to fail to perform in a required manner, for example, a short circuit, a broken wire, an intermittent connection. 2. A defect of any component upon which the intrinsic safety of a circuit depends. 3. The failure of any part of a computer system.

fault tolerance That property of a system which permits it to carry out its assigned function even in the presence of one or more faults in the hardware or software components.

FAX Sending digitized copies of documents by telephone.

faying surface Either of two surfaces in contact with each other in a welded, fastened or bonded joint, or in one about to be welded, fastened or bonded.

FCC See frame code complement.

F-center laser A solid state laser in which optical pumping by light from a visible-wavelength laser produces tunable near-infrared emission from defects—called "color centers" or "F centers"—in certain crystals.

FDDI See fiber distributed data interface.

FDM See Frequency-Division Multiplex.

feasibility study Any evaluation of the worth of a proposed project based on specific criteria.

feed 1. The act of supplying material to a process or to a specific processing unit. 2. The material supplied. Also known as feedstock. 3. Forward motion tending to advance a tool or cutter into the stock in a machining operation.

feedback 1. Process signal used in control as a measure of response to control action. 2. The part of a closed-loop system which automatically brings back information about the condition under control. 3. Part of a closed loop system which provides information about a given condition for comparison with the desired condition.

feedback control An error driven control system in which the control signal to the actuators is proportional to the difference between a command signal and a feedback signal from the process variable being controlled. See control, feedback.

feedback control signal The output signal which is returned to the input in order to achieve a desired effect, such as fast response.

feedback elements See elements, feedback.

feedback loop The components and processes involved in correcting or controlling a system by using part of the output as input. See loop, closed (feedback loop).

feedback oscillator An amplifier circuit in which an oscillating output signal is coupled in phase with the input signal, the oscillation being maintained at a frequency determined by frequency-selective parameters of the amplifier and its feedback circuits.

feedback ratio In a control system, the ratio of the feedback signal to a corresponding reference input.

feedback signal A signal derived from some attribute of the controlled variable, or from a control-system output, which is combined with one or more input or reference signals to produce a composite actuating signal. See signal, feedback.

feeder 1. A conveyor adapted to control the rate of delivery of bulk materials, packages or objects to a specific point or operation. 2. A device for the controlled delivery of materials to a processing unit. 3. In metalcasting, a runner or riser so placed that it can deliver molten metal to the contracting mass of metal as it cools and solidifies, thus preventing voids, porosity or shrinkage cavities.

feed-forward 1. An industry standard process control program, in which mathematically predicted errors are corrected before they occur; usually mainly for process loops with long lags or response times. 2. Open loop control.

feedforward control A method of control that compensates for a disturbance before its effect is felt in the output. It is based on a model that relates the output to the input where the disturbance occurs. In distillation the disturbances are usually feed rate and feed compositions. Steady-state feedforward models are usually combined with dynamic compensation functions to set the manipulative variables and combined with feedback adjustment (trim) to correct for control model-accuracy constraints. See control, feedforward.

feedforward control action Control action in which information concerning one or more external conditions that can disturb the controlled variable is converted into corrective action to minimize deviations of the controlled variable. Feedforward control is usually combined with other types of control to anticipate and minimize deviations of the controlled variable.

feedhead A reservoir of molten metal that extends above a casting to supply additional molten metal and compensate for solidification shrinkage. Also known as riser; sinkhead.

feed pipe A pipe through which water is conducted into a boiler.

feed rate The relative velocity between tool holder and workpiece along the main direction of cutting in a machining operation.

feedscrew An externally threaded rod used to control the advance of a tool or tool slide on a lathe, diamond drilling rig, percussion drill or other equipment.

feedstock Material delivered to a process or processing unit, especially raw material delivered to a chemical process or reaction vessel.

feedthrough error A signal caused by coupling from reference input to output when the digital-to-analog converter logic inputs are all low. Expressed in mV or dB relative to V REF.

feed trough A trough or pan from which feed water overflows in the drum.

feedwater Process water supplied to a vessel such as a boiler or still, as opposed to circulating water or cooling water.

feedwater control system A control system using input signals derived from the process for the purpose of regulating feedwater flow to the boiler to maintain adequate drum level according to the manufacturer's recommendations [S77.42].

feed-water treatment The treatment of boiler feed water by the addition of chemicals to prevent the formation of scale or eliminate other objectionable characteristics.

female fitting An element of a connection in pipe, tubing, electrical conductors or mechanical assemblies that surrounds or receives the mating (male) element; for example, the internally threaded end of a pipe fitting is termed female.

fermat principle Also called the principle of least time; a ray of light traveling from one point to another, including reflections and refractions which may be suffered, follows that path which requires the least time. Stated another way, the optical path is an extreme path, in the terminol-

ogy of the calculus of variations.

ferric percentage Actual ferric iron in slag, expressed as percentage of the total iron calculated as ferric iron.

ferroalloy An alloy, usually a binary alloy, of iron and another chemical element which contains enough of the second element for the alloy to be suitable for introducing it into molten steel to produce alloy steel, or in the case of ferrosilicon or ferroaluminum to produce controlled deoxidation.

ferrodynamic instrument An electrodynamic instrument in which the presence of ferromagnetic material (such as an iron core for an electromagnetic coil) enhances the forces ordinarily developed in the instrument.

ferrography Wear analysis conducted by withdrawing lubricating oil from an oil reservoir and using a ferrograph analyzer to determine the size distribution of wear particles picked up as the oil circulates between moving mechanical parts; the technique also may be used to assess deterioration of human joints or joint-replacement prostheses by analyzing for the presence of bone, cartilage and prosthetic-material fragments in human synovial fluid.

ferromagnetic material Any material that exhibits the phenomena of magnetic hysteresis and saturation, and whose permeability depends on the magnetizing force, all of which are exhibited by the chemical element iron.

ferrometer An instrument for measuring magnetic permeability and hysteresis in iron, steel and other ferromagnetic materials.

ferrous alloy Any alloy containing at least 50% of the element iron by weight.

ferrule 1. A metal ring or cap that is fitted onto the end of a tool handle, post or other similar member to strengthen and protect it. 2. A bushing inserted in the end of a boiler flue to spread and tighten it. 3. A tapered bushing used in compression-type tubing fittings to provide the wedging action that creates a mechanical seal. 4. An element of a fiber optic connector, typically used to house or align fibers.

fetch The process of obtaining the data from an address memory location.

F format In FORTRAN Fw.d indicates that w characters are to be converted to a floating point mixed number with d spaces reserved for digits to the right of the decimal point, e.g., F6.3 yields 24683 as input, 24.683 internally and 24.683 as output.

FFT See Fast-Fourier Transform.

fiber 1. The characteristic of wrought metal that indicates directionality, and can be revealed by etching or fractography. 2. The pattern of preferred orientation in a polycrystalline metal after directional plastic deformation such as by rolling or wiredrawing. 3. A filament or filamentary fragment of natural or synthetic materials used to make thread, rope, matting or fabric. 4. In stress analysis, a theoretical element representing a filamentary section of solid material aligned with the direction of stress; usually used to characterize nonuniform stress distributions, as in a beam subjected to a bending load.

fiber metal A material composed of metal fibers which have been pressed or sintered together, and which may also have been impregnated with resin, molten metal or other material that subsequently hardened.

fiber optic gyroscope A device in which changes in the wavelength of light going in different directions through a long length of optical fiber wound many times around a ring is used to measure rotation speed.

fiber optics A medium that uses light conducted through glass or plastic fibers for data transmission, optical measurements, or optical observations.

fiber optic system 1. A relatively new method of data transmission. Light transmitting fibers are used for connecting sensors to the computer. Fiber optic systems have very good immunity from noise. 2. Any system employing fiber optics to provide analytical observation of measurements employing the transmission properties of glass, plastic, polycrystalline, or crystals materials.

fiber sensor A sensing device in which the active element is an optical fiber or an element attached directly to an optical fiber. The quantity being measured changes the optical properties of the fiber in a way that can be detected and measured.

fibrous composite A material consisting of natural, synthetic or metallic fibers embedded in a matrix, usually a matrix of molded plastics material or hardenable resin.

fibrous fracture A type of fracture surface appearance characterized by a smooth, dull gray surface.

fibrous structure 1. In fractography, a ropy fracture-surface appearance, which is generally synonymous with silky or ductile fracture. 2. In forgings, a characteristic macrostructure indicative of metal flow during the forging process, which is revealed as a ropy appearance on a fracture surface or as a laminar appearance on a macroetched section; a ropy appearance on the fracture surface of a forging does not carry the same implication as a ropy fracture of other wrought metals, and should not be considered the same as a silky or ductile fracture. 3. In wrought iron, a microscopic structure consisting of elongated slag fibers embedded in a matrix of ferrite.

fidelity The degree to which a system, subsystem or component accurately reproduces the essential characteristics of an input signal in its output signal.

field The part of a computer record containing a specific portion of information.

field bus A standard under development in ISA SP50 for a bus to interconnect process control sensors, actuators, and control devices.

field coil A stationary or rotating electromagnetic coil.

field contact follower See auxiliary output.

field contact, normally open (NO) A field contact that is open for a normal process condition and closed when the process condition is abnormal [S18.1].

field contact, normally closed (NC) A field contact that is closed for a normal process condition and open when the process condition is abnormal [S18.1].

field contact (trouble or signal contact) The electrical contact of the device sensing the process condition. The contact is either open or closed. Annunciator field contacts are identified in relation to process conditions and annunciator operation, not the disconnected position of the devices [S18.1].

field contact voltage (trouble or signal contact voltage) The voltage applied to field contacts [S18.1].

field curvature Formation of an image that lies on a curved surface rather than a flat plane. For single and double element lenses, curvature is always inward, but for other types the curvature can be in either direction.

field emission Induced electron emission from an unheated metal surface resulting from application of a strong electric field.

field excitation Controlling the speed of a series-wound electric motor or a diesel-electric locomotive engine by changing the relationship between armature current and field strength, either through the use of shunts to reduce field current or through the use of field taps.

field-free emission current The electron current flowing from a cathode when the electric gradient at the cathode surface is zero.

field installable Nominally, a fiber optic splice or cable is field installable if it can be mounted by technicians working in the field, without a lab full of equipment at hand.

field of view The solid angle, or the angle in a specified plane, over which radiant energy incident on a transducer is measured within specified tolerances [S37.1].

field piping That piping connecting the control center to items external to the control center [RP60.9].

field-replaceable unit Computer hardware modules that are easily replaced.

field weld A weld made at a construction or installation site as opposed to a weld made in a fabrication shop.

FIFO See first in, first out.

filament A very fine single strand of metal wire, extruded plastic or other material.

filament winding Fabricating a composite structure by winding a continuous fiber reinforcement on a rotating core under tension; the reinforcement usually consists of glass, boron or silicon carbide thread, either previously impregnated with resin or impregnated during winding.

filament-wound structure A composite structure made by fabricating one or more structural elements by filament winding, then curing them and assembling them, or assembling them first and curing the entire structure.

filar micrometer An attachment for a microscope or telescope consisting of two parallel fine wires or knife edges in an eyepiece, one of them in a fixed position and the other capable of being moved in a direction perpendicular to its length by means of a very accurate micrometer screw; the device is used to make accurate measurements of linear distances in the optical field of view; actual distances are determined by dividing the micrometer reading by the magnification of a microscope, although in some cases the micrometer scale is calibrated for direct reading

at a specific magnification.

file 1. A collection of related records treated as a unit [S61.2]. 2. An organized structure consisting of an arbitrary number of records, for storing information on a bulk storage device, e.g., disc, drum, core, or tape.

file gap An interval of space or time associated with a file to indicate or signal the end of the file; related to gap.

file handling The manipulation of data files by various methods. It generally involves read, write, and compare.

file name extension An addition to a computer file name that indicates the file type, such as .BAT, .COM, etc.

file specification A name that uniquely identifies a file maintained by an operating system. A file specification generally consists of at least three components; a device name identifying the volume on which the file is stored, a file name, and a file name extension. In addition, depending on the system, a file specification can include a user file directory name, or UIC, and a version number.

file structured device A device on which data are organized into files; the device usually contains a directory of the files stored on the device.

file transfer access and management (FTAM) One of the application protocols specified by MAP and TOP.

filled composite A plastics material made of short-strand fibers or a granular solid mixed into thermoplastic or thermosetting resin prior to molding.

filled-system thermometer Any of several devices consisting of a temperature-sensitive element (bulb), an element sensitive to changes in pressure or volume (Bourdon tube, bellows or diaphragm), capillary tubing and an indicating or recording device; the bulb, capillary tube and pressure- or volume-sensitive element are partly or completely filled with a fluid that changes its volume or pressure in a predictable manner with changes in temperature.

filler 1. An inert material added to paper, resin, elastomers and other materials to modify their properties or improve quality in end products. 2. A material used to fill holes, cracks, pores and other surface defects before applying a decorative coating such as paint. 3. A metal or alloy deposited in a joint during welding, brazing or soldering; usually referred to as filler metal.

fillet 1. A concave transition surface between two surfaces that meet at an angle. 2. A molding or corner piece placed at the junction of two perpendicular surfaces to lessen the likelihood of cracking.

fillet weld A roughly triangular weld that joins two members along the intersection of two surfaces that are approximately perpendicular to each other.

fill-in-the-blank programming language A nonprocedural programming language in which programs are developed by filling out data sheets for an existing program. Examples: BICEPS, PROSPRO, CODIL.

film 1. A flat, continuous sheet of thermoplastic resin or similar material that is extremely thin in relation to its width and length. 2. A very thin coating, deposit or reaction product that completely covers the surface of a solid.

filmogen The material or binder in paint that imparts continuity to the coating.

film strength 1. Generally, the resistance of a film to disruption. 2. In lubricants, a measure of the ability to maintain an unbroken film over surfaces under varying conditions of load and speed.

filter 1. In electronic, acoustic and optical equipment, a device that allows signals of certain frequencies to pass, while rejecting signals having frequencies in another range. 2. A device used in a frequency transmission circuit to exclude unwanted frequencies and to keep the channels separate. 3. A device to suppress interference which would appear as noise. 4. A machine word that specifies which parts of another machine word are to be operated upon, thus the criterion for an external command. Synonymous with mask. 5. A porous material or structural element designed to allow fluids to pass through it while collecting and retaining solids of a certain particle size or larger.

filter aid An inert powdery or granular material such as diatomaceous earth, fly ash or sand which is added to a liquid that is about to be filtered in order to form a porous bed on the filter surface, thereby increasing the rate and effectiveness of the filtering process.

filter, bandpass A circuit which is tuned to pass all frequencies between certain points in the spectrum.

filter cake The solid or semisolid material retained on the surface of a filter after a liquid containing suspended solids has passed through.

filter capacitor A capacitor used as an element of an electronic filter circuit.

filter inductor An inductor used as an element of an electronic filter circuit.

filter, low pass A circuit which is tuned to pass all frequencies lower than a specified cutoff point.

filter medium The portion of a filter or filtration system that actually performs the function of separating out the solid material; it may consist of metal or nonmetal screening, closely woven fabric, paper, matted fibers, a granular bed, a porous ceramic cup or plate, or other porous component.

fin 1. A thin, flat or curved projecting plate, typically used to stabilize a structure surrounded by flowing fluid or to provide an extended surface to improve convective or radiative heat transfer. 2. A defect consisting of a very thin projection of excess material at a corner, edge or hole in a cast, forged molded or upset part, which must be removed before the part can be used.

final control element 1. The device that directly controls the value of the manipulated variable of a control loop. Often the final control element is a control valve [S5.1]. 2. An instrument that takes action to adjust the manipulated variable in a process. This action moves the value of the controlled variable back towards the set point. 3. The last system element that responds quantitatively to a control signal and performs the actual control action. Examples include valves, solenoids and servometers. 4. The device that exerts a direct influence on the process. 5. Unit of a control loop (such as a valve) which manipulates the control agent.

final controlling element The element in a control system that directly changes the value of the manipulated variable. See element, final controlling.

fine grinding 1. Mechanical reduction of a powdery material to a final size of at least -100 mesh, usually in a ball mill or similar grinding apparatus. 2. In metallography or abrasive finishing, producing a surface finish of fine scratches by use of an abrasive having a particle size of 320 grit or smaller.

fineness Purity of gold or silver expressed in parts per thousand; for instance, gold having a fineness of 999.8 has only 0.02%, or 200 parts per million, of impurities by weight.

fines 1. In a granular substance having mixed

particle sizes, those particles smaller than the average particle size. 2. Fine granular material which passes through a standard screen on which the coarser particles in the mixture are retained. 3. In a powdered metal, the portion consisting of particles smaller than a specified particle size.

finger plate   A plate used to restrict the upward motion of the diaphragm and prevent diaphragm extrusion into the bonnet cavity in the full open position [S75.05].

finish   1. A chemical or other substance applied to the surface of virtually any solid material to protect it, alter its appearance, or modify its physical properties. 2. The degree of reflectivity of a lustrous material, especially metal; it is usually described by one of the following imprecise terms, listed in order of increasing luster and freedom from scratches—machined, ground brushed, matte, dull lustrous, bright, polished and mirror. 3. Generally, the surface quality, condition or appearance of a metal or plastic part.

finish grinding   The final step in a grinding operation, which imparts the desired surface appearance, contour and dimensions.

finishing temperature   In a rolling or forging operation, the metal temperature during the last reduction and sizing step, or the temperature at which hot working is completed.

fin tube   A tube in a boiler having water on the outside and carrying the products of combustion on the inside.

fire assay   Determining the metal content of an ore or other substance through the use of techniques involving high temperatures.

fire box   The equivalent of a furnace. A term usually used for the furnaces of locomotive and similar types of boilers.

fire crack   A crack starting on the heated side of a tube, shell, or header resulting from excessive temperature stresses.

fired pressure vessel   A vessel containing a fluid under pressure exposed to heat from the combustion of fuel.

fire point   The lowest temperature at which, under specified conditions, fuel oil gives off enough vapor to burn continuously when ignited.

fireproof   Resistant to combustion or to damage by fire under all but the most severe conditions.

fire-resistant   Resistant to combustion and to heat of standard intensity for a specified time without catching fire or failing structurally.

fire retardant   1. Treated by coating or impregnation so that a combustible material—wood, paper or textile, for instance—catches fire less readily and burns more slowly than untreated material. 2. The substance used to coat or impregnate a combustible material to reduce its tendency to burn.

fire tube   A tube in a boiler having water on the outside and carrying the products of combustion on the inside.

fire tube boiler   A boiler with straight tubes, which are surrounded by water and steam and through which the products of combustion pass.

firing rate control   A pressure, temperature or flow controller which controls the firing rate of a burner according to the deviation from pressure or temperature set point. The system may be arranged to operate the burner on-off, high-low or in proportion to load demand.

firmware   Programs or instructions that are permanently stored in hardware memory devices (usually read-only memories) that control hardware at a primitive level.

first alert   See first out.

first-in, first-out (FIFO)   An ordered queue. A discipline wherein the first transaction to enter a queue is also the first to leave it. Contrast with last-in, first-out.

first-level address   See direct address.

first-order system   A system definable by a first-order differential equation.

first out (first alert)   A sequence feature that indicates which of a group of alarm points operated first [S18.1].

first out reset   See reset.

first word address (FWA)   A program/routine.

Fisher loop test   One of several Wheatstone bridge test arrangements commonly used to determine the distance to a fault (grounded or crossed wires) in a communications cable.

fisheye   1. An area on a fracture surface having a characteristic white crystalline appearance, usually caused by internal hydrogen cracking. 2. A small globular mass in a blended material such as plastic or glass which is not completely homogeneous with the surrounding material; it is particularly noticeable in transparent or translucent materials.

fishing tool   An elongated or telescoping tool with a magnet, hook or grapple at one end which is used to retrieve objects from inaccessible places.

fish plate Either of the two plates bolted or riveted to the webs of abutting rails or beams on opposite sides to secure a mechanical joint.

fishtail Excess metal at the trailing end of an extrusion or a rolled billet or bar, which is generally cropped and either discarded or recycled into a melting operation.

fissure A small, cracklike surface discontinuity, often one whose sides are slightly opened or displaced with respect to each other.

fit The closeness of mating parts in an assembly, as determined by their respective dimensions and tolerances; fits may be classified as running (sliding), locational, transition or force (shrink) fits, depending on the size and direction (positive for running or negative for force fits) of the dimensional allowance; fits may also be termed clearance or interference depending on whether there is always a gap between mating parts or always interference, as long as the parts are within specified tolerances.

fitting An auxiliary part of standard size and configuration that can be used to facilitate assembly; in constructing a system of pipe or tubing, for example, connections are more easily made if standard elbows, tees, unions and couplings are used to connect straight lengths of pipe, rather than bending the pipe or making special preparations before welding lengths of pipe together.

fixed carbon The carbonaceous residue less the ash remaining in the test container after the volatile matter has been driven off in making the proximate analysis of a solid fuel.

fixed equipment Equipment which is designed to be fastened or otherwise physically secured to a supporting device [S82.01].

fixed length record A record where the number of characters is constant.

fixed pitch A typeface in which all letters are the same width.

fixed point 1. A reproducible standard value, usually derived from a physical property of a pure substance, which can be used to standardize a measurement or check an instrument calibration. 2. Pertaining to a numeration system in which the position of the radix point is fixed with respect to one end of the numerals, according to some convention. See fixed-point arithmetic.

fixed-point arithmetic 1. A method of calculation in which operations take place in an invariant manner, and in which the computer does not consider the location of the radix point. This is illustrated by desk calculators or slide rules with which the operator must keep track of the decimal point, and similarly with many automatic computers, in which the location of the radix point is the programmer's responsibility. Contrasted with floating-point arithmetic. 2. A type of arithmetic in which the operands and results of all arithmetic operations must be properly scaled so as to have a magnitude between certain fixed values.

fixed-point data In data processing, the representation of information by means of the set of positive and negative integers. It is faster than floating point data and requires fewer circuits to implement.

fixed point notation In data processing, numbers that are expressed by a set of digits with the decimal point in the correct position.

fixed-point part In a floating-point representation, the numeral of a pair of numerals representing a number, that is the fixed-point factor by which the power is multiplied. Synonymous with mantissa.

fixed-program computer A computer in which the sequence of instructions are permanently stored or wired in, and perform automatically and are not subject to change either by the computer or the programmer except by rewiring or changing the storage input. Related to wired program computer.

fixed restrictor A fixed physical restriction to fluid flow.

fixed storage A storage device that stores data not alterable by computer instructions, for example, magnetic core storage with a lockout feature, or photographic disk.

fixed word length Having the property that a machine word always contains the same number of characters or bits.

fixture 1. An auxiliary component or operator aid attached to a structure or machine enclosure—a light or tool shelf, for instance. 2. A special holder that positions the work in a machining operation but does not guide the tool.

flag 1. A bit of information attached to a character or word to indicate the boundary of a field. 2. An indicator frequently used to tell some later part of a program that some condition occurred earlier. 3. An indicator used to identify the members of several sets which are intermixed. 4. A storage bit whose location is usually reserved to indicate the occurrence or nonoccurrence of some condition, e.g., a Halt/Run flag would be 1 when the processor was halted and 0 when in the RUN condition.

flag register   This is an 8-bit register in which each bit acts as a flag.

flake   1. Dry, unplasticized cellulosic plastics base material. 2. Plastics material in chip form used as feed in a molding operation. 3. An internal hydrogen crack such as may be formed in steel during cooling from high temperature. Also known as fisheye; shattercrack; snowflake. 4. Metal powder in the form of fish-scale particles. Also known as flaked powder.

flame   A luminous body of burning gas or vapor.

flame cutting   Using an oxyfuel-gas flame and an auxiliary oxygen jet to sever thick metal sections or blanks.

flame detector   A device which indicates if fuel, such as liquid, gaseous, or pulverized, is burning, or if ignition has been lost. The indication may be transmitted to a signal or to a control system.

flame hardening   A form of surface hardening that uses the inherent hardenability of a steel or other hardenable alloy to produce a hardened surface layer by spot-heating the metal with a fuel-gas flame to a shallow depth and then rapidly cooling the heated metal.

flame photometer   An instrument for determining compositions of solutions by spectral analysis of the light emitted when the solution is sprayed into a flame.

flame plate   A baffle of metal or other material for directing gases of combustion.

flame propagation rate   Speed of travel of ignition through a combustible mixture.

flame spraying   1. Applying a plastic coating on a surface by projecting finely powdered plastic material mixed with suitable fluxes through a cone of flame toward the target surface. 2. Thermal spraying by feeding an alloy or ceramic coating material into an oxyfuel-gas flame; compressed gas may or may not be used to atomize the molten material and propel it onto the target surface.

flame-spray strain gage   A fine-wire strain gage element attached to a substrate by flame spraying a ceramic encapsulation over the element, which attaches it without damaging either the gage or the substrate; this attachment technique produces a bond suitable for operating over the temperature range –270 to 820 °C (–450 to 1500 °F).

flame treating   Making inert thermoplastics parts receptive to inks, lacquers, paints or adhesives by bathing them in open flames to promote surface oxidation.

flammability   Susceptibility to combustion.

flammable liquid   A liquid, usually a liquid hydrocarbon, that gives off combustible vapors.

flammable range   The range of flammable vapor concentrations or gas-air mixtures in which propagation of flame will occur on contact with a source of ignition [S12.13].

flanged body   Valve body with full flanged end connections [S75.05].

flanged ends   Valve end connections incorporating flanges which allow pressure seals by mating with corresponding flanges on the piping [S75.05].

flangeless control valve   A valve without integral line flanges, which is installed by bolting between companion flanges, with a set of bolts, or studs, generally extending through the companion flanges [S75.05].

flange retained liner   A liner retained in the body of a butterfly valve by the pipe flanges or by a continuous or segmented ring. The segmented ring provides a means of adjusting the liner to disk interference to achieve improved sealing. The bore of the pipe flanges is smaller in diameter than the body bore, therefore the flanges retain the liner in the body [S75.05].

flange taps   See orifice flange taps.

flank   1. On a cutting tool, the end surface adjacent to the cutting edge. 2. On a screw thread, the side of the thread.

flap valve   A valve with a hinged flap or disc that swings in only one direction.

flareback   A burst of flame from a furnace in a direction opposed to the normal flow, usually caused by the ignition of an accumulation of combustible gases.

flared tube-end   The projecting end of a rolled tube which is expanded or rolled to a conical shape.

flaring   Increasing the diameter at the end of a pipe or tube to form a conical section.

flash   1. In plastics molding, elastomer molding or metal die casting, a portion of the molded material that overflows the cavity at the mold parting line. 2. A fin of material attached to a molded, cast or die forged part along the parting line between die halves, or attached to a resistance flash welded, upset welded or friction welded part along the weld line.

flash converter A converter in which all bit choices are made at the same time.

flasher A device that causes visual displays to turn on and off repeatedly. Types of flashing include fast flashing, flashing, slow flashing, and intermittent flashing [S18.1].

flashing Steam produced by discharging water at saturation temperature into a region of lower pressure.

flashlamp A gas-filled lamp which is excited by an electrical pulse passing through it to emit a short bright flash of light. A broad range of wavelengths are produced, with their precise nature depending on the gas or gases used.

flash line A raised line on the surface of a molded or die cast part that corresponds to the parting line between mold faces.

flash plating Electrodeposition of a very thin film of metal, usually just barely enough to completely cover the surface.

flash point The lowest temperature at which, under specified conditions, a liquid gives off enough vapor to flash into momentary flame when ignited.

flash welding A resistance welding process commonly applied to wide, thin members, irregularly shaped parts and tube-to-tube joints, in which the faying surfaces are brought into close proximity, electric current is passed between them to partly melt the surfaces by combined arcing and resistance heating, and the surfaces are then upset forged together to complete the bond.

flask A wood or metal frame for holding a sand mold in foundry work; it is open ended, and usually consists of two halves—the cope (upper half) and the drag (lower half), although three or more flask sections are occasionally used.

flat die forging Shaping metal by compressing or hammering it between simple flat or regularly contoured dies.

flat-position welding Welding from above the work, with the face of the weld in the horizontal plane. Also known as downhand welding.

flattening Straightening metal sheet by passing it through a set of staggered and opposing rollers that bend the sheet slightly to flatten it without reducing its thickness.

flattening test A test that evaluates the ductility, formability and weld quality of metal tubing by flattening it between parallel plates to a specified height.

flatting agent A chemical additive that promotes a nonglossy, matte finish in paints and varnishes.

flaw A discontinuity or other physical attribute in a material that exceeds acceptable limits; the term flaw is nonspecific, and more specific terms such as defect, discontinuity or imperfection are often preferred.

flexible lip seal A seal ring retained in the body bore with raised flexible lip which contacts an offset disk in the closed position yet is clear of the disk in other positions [S75.05].

flexible manufacturing systems (FMS) A manufacturing system under computer control with automatic material handling. The system is primarily designed for batch manufacturing.

flexivity Temperature rate of flexure for a bimetal strip of given dimensions and material composition.

flight monitor package A special software system that enables several operators to monitor data from an aircraft or other source in real time.

flinching In quality control inspection, failure of an inspector to call a borderline defect a defect.

flint glass An optical glass which contains lead or other elements which raise its refractive index between 1.6 and 1.9; higher than other types of optical glass.

flip-flop 1. A bistable device, i.e., a device capable of assuming two stable states. 2. A bistable device which may assume a given stable state depending upon the history of pulses of one or more input points and having one or more output points. The device is capable of storing a bit of information. 3. A control device for opening or closing gates, i.e., a toggle.

float 1. Any component having positive buoyancy—for example, a hollow watertight body that rests on the surface of a liquid, partly or completely supported by buoyant forces. 2. See plummet.

float chamber A vessel in which a float regulates the liquid level.

float control A type of control apparatus in which the control signal is regulated by a float riding up and down with liquid level. Contrast with floating control.

157

float gage Any of several types of devices that use pulleys, levers or other mechanisms to transmit the position of a float to a scale that indicates liquid level in a tank or vessel.

floating A condition of a line in a logic circuit that is not grounded or tied to any established potential.

floating action A type of control-system action in which a fixed relationship exists between a measured deviation and the rate of motion of the final control element.

floating ball A full ball positioned within the valve that contacts either of two seat rings and is free to move toward the seat ring opposite the pressure source when in the closed position to effect tight shutoff [S75.05].

floating control A control device in which the output control signal is proportional to the difference between an indicator signal and the controller's setpoint; this difference is often referred to as an error signal; in operation, floating control reduces the tendency to overshoot the setpoint because it reduces power input to the system as the controlled variable approaches the setpoint value. Contrast with float control.

floating control action See control action, floating.

floating controller See controller, floating.

floating control mode A controller mode in which an error in the controlled variable causes the output of the controller to change at a constant rate. The error must exceed preset limits before controller change starts.

floating plug A short nosed mandrel attached to a rod which is inserted into pipe or tubing during reduction by drawing. Also known as plug die.

floating point 1. An arithmetic notation in which the decimal point can be manipulated; values are sign, magnitude, and exponent $(+.833 \times 10^2)$. 2. A form of number representation in which quantities are represented by a bounded number (mantissa) and a scale factor (characteristic or exponent) consisting of a power of the number base, e.g., $127.6 = 0.1276*10**3$ where the bounds are 0 and 1 for the mantissa and the base is ten.

floating point arithmetic A method of calculation which automatically accounts for the location of the radix point. This usually is accomplished by handling the number as a signed mantissa times the radix raised to an integral exponent, e.g., the decimal number +88.3 might be written as $+.883*10**2$. The binary number $-.0011$ as $-.11*2**-2$. Contrasted with fixed-point arithmetic.

floating point base In floating-point representation, the fixed positive integer that is the understood base of the power. Synonymous with floating-point radix.

floating point notation In data processing, numbers that are expressed as a fraction coupled with an interger exponent of the base.

floating-point routine A set of subroutines which causes a computer to execute floating-point arithmetic. These routines may be used to simulate floating-point operations on a computer with no built in floating-point hardware.

floating rate The rate of motion of the final control element in a proportional-speed floating-control system which corresponds to a given deviation.

floating speed The rate of motion of the final control element in a single-speed or multispeed floating-control system.

floatless level control Any device for measuring or controlling liquid level in a tank or vessel without the use of a float—methods include manometers, electrical probes, capacitance devices, radiation instruments, and sonic or ultrasonic instruments.

float switch An on-off switch activated by the position of a float.

float valve An on-off-type valve whose action is triggered by the rise or fall of a float.

flooding The consequence of excessive column liquid loading where, in effect, the liquid on trays becomes too deep for the vapor to pass through or where the vapor flow rate is too high, creating an excessive differential pressure or a decrease in the differential temperature across the column.

floppy disk Any flexible platter with a magnetic coating that can accept computer data. Also called a diskette.

flospinning Forming cylindrical, conical or curvilinear parts from light plate by power spinning the metal over a rotating mandrel.

flotation A process for separating particulate matter in which differences in surface chemical properties are used to make one group of particles float on water while other particles do not; it is used primarily to separate minerals from gangue but is also used in some chemical and biological processes. In mining engineering, also known as froth flotation.

flow 1. The rate of flow of a liquid expressed in volume units per unit of time. Examples are: meter^3/second (m^3/s) [RP31.1]. 2. The order of events in the computer solution to a problem. 3. The movement of material in any direction.

flowability A general term describing the ability of a slurry, plasticized material or semisolid to behave like a fluid.

flow brazing A brazing process in which the joint is heated by pouring hot molten nonferrous filler metal over the assembled parts until brazing temperature is attained.

flow chart 1. A system-analysis tool that provides a graphical presentation of a procedure. Includes block diagrams, routine sequence diagrams, general flow symbols, and so forth. 2. A chart to represent, for a problem, the flow of data, procedures, growth, equipment, methods, documents, machine instructions, etc. 3. A graphical representation of a sequence of operations by using symbols to represent the operations such as COMPUTE, SUBSTITUTE, COMPARE, JUMP, COPY, READ, WRITE, etc. 4. The graphical representation of the processing steps performed. 5. The sequence of logic operations implemented in hardware, software, firmware, or manual procedures.

flow coat To apply a coating by pouring liquid over an object and allowing the excess to drain off.

flow coefficient A constant (Cv), related to the geometry of a valve, for a given valve opening, that can be used to predict flow rate [S75.11].

flow compensation Using secondary signals to correct flow values for changes in density or viscosity.

flow control Any method for controlling the flow of a material through piping, ductwork or channels.

flow control orifice The part of the flow passageway that, with the closure member, modifies the rate of flow through the valve. The orifice may be provided with a seating surface, to be contacted by or closely fitted to the closure member, to provide tight shutoff or limited leakage [S75.05].

flow diagram See flow chart.

flow line 1. The connecting line or arrow between symbols on a flow chart. 2. A mark on a molded plastic part where two flow fronts met during molding. Also known as weld mark. 3. In me-

chanical metallurgy, a path followed by minute volumes of metal during forming. 4. Texture line, often revealed by etching a surface or section, which indicates the direction of metal flow during hot or cold working.

flow marks Wavy surface marks on a molded thermoplastic part resulting from improper flow of resin during molding.

flowmeter 1. A device that measures the rate of flow or quantity of a moving fluid in an open or closed conduit. It usually consists of both a primary and a secondary device. It is acceptable in practice to further identify the flowmeter by its applied theory; as differential pressure, velocity, area, force, etc. or by its applied technology as orifice, turbine, vortex, ultrasonic, etc. Examples include turbine flowmeter, magnetic flowmeter, fluidic pressure flowmeter, etc. [S51.1] 2. An instrument used to measure linear, non-linear or volumetric flow rate or discharge rate of a fluid flowing in a pipe. Also known as fluid meter.

flowmeter primary device The device mounted internally or externally to the fluid conduit which produces a signal with a defined relationship to the fluid flow in accordance with known physical laws relating the interaction of the fluid to the presence of the primary device. The secondary device may consist of one or more elements as needed to translate the primary device signal into standardized or nonstandarized display or transmitted units [S51.1].

flowmeter secondary device The device that responds to the signal from the primary device and converts it to a display or to an output signal that can be translated relative to flow rate or quantity.

flow mixer A device for mixing two solids, liquids, or gases together in which the mixing action occurs as the materials pass through the device. Also known as line mixer.

flow nozzle A type of differential pressure producing element having a contoured entrance. Characterized by its ability to be mounted between flanges and have a lower permanent pressure loss than an orifice plate.

flow rate 1. The time rate of motion of a fluid, usually contained in a pipe or duct, expressed as fluid quantity per unit time [S37.1]. 2. The quantity of fluid which moves through a pipe or channel within a given period of time.

flow rate instability (bistable flow) An abrupt change in the control valve flow rate that occurs independent of changes in valve position. It may be caused by variable wall attachment of the fluid stream at the valve orifice, by flashing, or by cavitation [RP75.18].

flow-rate range   Range of flow rates bounded by the minimum and maximum flow rates.

flow soldering   See wave soldering.

flow straightener   A supplementary length of straight pipe or tube, containing straightening vanes or the equivalent, which is installed directly upstream of the turbine flowmeter for the purpose of eliminating swirl from the fluid entering the flowmeter [RP31.1].

flow transmitter   A device that senses the flow of liquids in a pipe and converts the sensor output into electric signals proportional to flow rate which can be transmitted to a remote indicator or controller.

flue   A conduit or duct for conveying combustion products from a furnace chamber or firebox to the point of discharge to the atmosphere.

flue dust   The particles of gas-borne solid matter carried in the products of combustion.

flue gas   The gaseous products of combustion in the flue to the stack.

flue-gas analyzer   An instrument that monitors the composition of flue gas as it passes out of a boiler or heating unit; the readout is used to guide adjustment of combustion controls to achieve maximum combustion efficiency or heat output.

fluid   A gas or liquid, both of which have the property of undergoing continuous deformation when subjected to any finite shear stress as long as the shear stress is maintained.

fluid coupling   A device for transmitting rotational motion and power between shafts by means of the acceleration or deceleration of oil or another suitable liquid. Also known as hydraulic coupling.

fluidics   The technology associated with the application of fluid dynamics to perform sensing, control, data acquisition, information processing and device actuation functions without relying on moving mechanical parts.

fluidity   The degree to which a substance flows freely.

fluidized bed   A dynamic mixture of a gas and/or vapor and minute solid particles of such a size that the mixture resembles a fluid in motion.

fluid meter   See flowmeter.

fluid motor type valve   A fluid powered device which uses a rotary motor to position the actuator stem [S75.05].

flume   An adaptation of the venturi concept of flow constriction applied to open-channel flow measurement.

fluorescence   1. Emission of electromagnetic radiation from a surface upon absorption of energy from other electromagnetic or particulate radiation, the emission being sustained only so long as the stimulating radiation impinges on the material. 2. The electromagnetic radiation produced by the above process. 3. Characteristic x-rays produced due to absorption of higher-energy x-rays.

fluorescence spectroscopy   The study of materials by light which they emit when irradiated by other light. Many materials emit visible light after they have been illuminated by ultraviolet light. The intensity and wavelengths of the emitted light can be used to identify the material and its concentration.

fluorometer   An instrument for measuring the fluorescent radiation emitted by a material when excited by monochromatic incident radiation—usually filtered radiation from a mercury-arc lamp or from a tungsten or molybdenum x-ray tube. Also spelled fluorimeter.

fluoroplastics   A family of plastics resins based on fluorine substitution of hydrogen atoms in certain hydrocarbon molecules.

fluoroscopy   X-ray examination similar to radiography, but in which the image is produced on a fluorescent screen instead of on radiographic film.

flush   The injecting of a fluid into the sample line where the flow of sample fluid can be directed to a portable container. It may be referred to as "sample point." [S67.10]

flushing   Removing debris, deposits, wear particles, or used lubricating oil from a piping system, chamber or mechanism by circulating a liquid such as a solvent oil or water, then draining the system to carry off unwanted substances.

flushing connection   A connection on the instrument, manifold or piping to permit periodic back-flow of an external fluid for clearing purposes.

flush left   In a printout, the alignment of type so that the left start-point of each line is the same.

flush right   In a printout, the alignment of type so that the end-point of each line is the same.

Most books are typeset with lines flushed both left and right.

flute 1. In a drill, reamer or tap, a channel or groove in the body of the tool which exposes the cutting edge and provides a passage for cutting fluid and chips. 2. In a milling cutter or hob, the chip space between the back of one cutting tooth and the face of the following tooth.

fluted-rotor flowmeter A type of flow-measurement device in which fluid is trapped between two fluted rotors which are dynamically balanced but hydraulically unbalanced so that they turn at a rate proportional to the volume rate of fluid flow.

flutter 1. In tape recorders, the higher-frequency variations in record and/or reproduce speed which cause time base errors in the record/reproduce process. 2. Irregular alternating motion of a control surface, often due to turbulence in a fluid flowing past it. 3. Repeated speed variation in computer processing.

flux 1. In metal refining, a substance added to the melt to remove undesirable substances such as sand, dirt or ash, and sometimes to absorb undesirable elements or compounds such as sulfur in steelmaking or iron oxide in copper refining. 2. In welding, brazing and soldering, a substance preplaced in the joint or fed into the molten zone to prevent formation of oxides or other undesirable compounds, or to dissolve them and make it easy to remove them. 3. In magnetic or electromagnetic applications, the integral of magnetic field strength over the cross sectional area of the field.

flux-cored arc welding Abbreviated FCAW. A form of electric arc welding in which the electrode is a continuous tubular wire of filler metal whose central cavity contains welding flux; welding may be performed with or without a shielding gas such as CO_2 or argon.

flux gate A detector that produces an electric signal whose magnitude and phase are proportional to the magnitude and direction of an external magnetic field aligned with the detector's axis.

flux guide A shaped piece of metal used in magnetic or electromagnetic applications to direct magnetic flux along preferred paths or to prevent it from spreading beyond specific boundaries.

fluxmeter An instrument for measuring the intensity of magnetic flux.

fly A fan with two or more blades that is used in timepieces or light machinery to control rotational speed by means of air resistance.

fly cutting Machining using a rotating single-point tool or a milling cutter having only one tooth.

flywheel A balanced rotating element attached to a shaft which utilizes inertial forces to maintain uniform rotational speed and damp out small variations in power generated by the driving elements.

FM discriminator A device that converts frequency variations to proportional variations in voltage or current.

FM/FM Frequency modulation of a carrier by subcarriers that are frequency modulated by information.

FM (frequency modulation) The process (or the result of the process) in which the frequency deviates from the unmodulated carrier in proportion to the instantaneous value of the modulating signal.

FMS See flexible manufacturing system.

FM (tape record/reproduce) The tape record/reproduce process whereby data modulate an FM oscillator for recording, and are demodulated by an FM discriminator.

F number The ratio of the focal length of a lens to its diameter.

foaming 1. Any of various methods of introducing air or gas into a liquid or solid material to produce a foam. 2. The continuous formation of bubbles which have sufficiently high surface tension to remain as bubbles beyond the disengaging surface.

foam-in-place A method widely used to apply foamed insulation to homes and industrial equipment, in which two or more reactive substances are deposited onto a surface to be covered where the foaming reaction takes place.

focal length The distance from the focal point of a lens or lens system to a reference plane at the lens location, measured along the principle axis of the lens system.

focal point 1. The location on the opposite side of a lens or lens system where rays of light from a distant object meet at a point. Also known as focus. 2. The point in space where a beam of electromagnetic energy (such as light, x-rays or laser energy) or of particles (such as electrons) has its greatest concentration of energy; it corresponds to the point where a converging beam of energy undergoes a transition to become a diverging beam.

focal spot The area of the target in an x-ray tube where the stream of electrons from the cathode strikes the target.

focus 1. To adjust the position of a lens with respect to an imaging surface so that sharp features of the object appear sharp in the image. 2. A focal point.

focusing coil An assembly containing one or more electromagnetic coils which is used to focus an electron beam.

focusing electrode An electrode configured so that its electric field acts to control the cross-sectional area of an electron beam.

focusing magnet An assembly containing one or more permanent magnets or electromagnets which is used to focus an electron beam.

fog A defect in developed radiographic, photographic or spectrographic emulsions consisting of uniform blackening due to unintentional exposure to low-intensity light or penetrating radiation.

fogged metal A metal surface whose luster has been greatly reduced by the creation of a film of oxide or other reaction products.

fog quenching Rapidly cooling an item by subjecting it to a fine mist, usually of water.

foil Very thin metal sheet, usually less than 0.006 in. (0.15 mm) thick.

foil strain gage A type of metallic strain gage usually made in the form of a back-and-forth grid by photoetching a precise pattern on foil made of a special alloy having high resistivity and low temperature coefficient of resistivity.

folding error An error in sampling an electronic signal arising from failure to sample at a high enough rate (sampling rate should be at least double the maximum signal frequency), so that the sampling device perceives high-frequency components of the signal as low-frequency components. Also known as aliasing error.

foldover A device characteristic exhibited when a further change in the input produces an output signal which reverses its direction from the specified input-output relationship [S67.04]. See aliasing.

font Typesetting characters of a particular style.

foot A fundamental unit of length in the British and U.S. Customary systems of measurement equal to 12 inches.

foot-pound A force of one pound applied to a lever one foot long.

force-balance transmitter A transmitter design technique utilizing feedback of the output signal to balance the primary input signal from the measuring element. The balanced output signal is proportional to the measured variable.

forced circulation Using a pump or fan to move fluid through a conduit or process vessel—for instance, air or gases through a furnace or combustion chamber (often referred to as forced draft), ambient or conditioned air through ductwork (often referred to as forced ventilation), or a mixture of water and steam through tubes in a boiler.

forced draft fan A fan supplying air under pressure to the fuel burning equipment.

forced oscillation Oscillation of a system attribute where the period of oscillation is determined by an external periodic force.

force factor 1. The complex ratio of the force required to block the mechanical system of an electromechanical transducer to the corresponding current in the electrical system. 2. The complex ratio of open-circuit voltage in the electrical system of an electromechanical transducer to corresponding velocity in the mechanical system.

force fit A class of interference fit involving relatively large amounts of negative allowance, which requires large amounts of force to assemble and results in relatively large induced stresses in the assembled parts.

forcing Applying control signals greater than those warranted by a given deviation in the controlled variable in order to induce a more rapid rate of adjustment in the controlled variable.

Ford cup viscometer A time-to-discharge apparatus used primarily for determining the viscosity of paints and varnishes.

foreground 1. The information element on a background field [S5.5]. 2. The area in memory designated for use by high-priority programs; the program, set of programs, or functions that gain the use of machine facilities immediately upon request.

foreground/background A control system that uses two computers, one performing the control functions and the other used for data logging, off-line evaluation of performance, financial operations, and so on. Either computer is able to perform the control functions.

foreground/background processing A computer system organized so that primary tasks dominate computer processing time when required, and secondary tasks fill the remaining time.

foreground program A time-dependent program initiated via request, whose urgency preempts operation of a background program. Contrast with background program.

forehand welding Welding in which the palm of the welder's torch or electrode hand faces the direction of weld travel; it has special significance in oxyfuel-gas welding, where the welding flame is directed ahead of the weld puddle and provides preheating. Contrast with backhand welding.

fore pump A vacuum pump operated in series with another vacuum pump to produce vacuum at the discharge of the latter, where the second pump is not capable of discharging gases at atmospheric pressure.

fore vacuum A space on the exhaust side of a vapor jet or pump where the ambient static pressure is below atmospheric.

forging 1. Using compressive force to plastically deform and shape metal; it is usually done hot, in dies or between rolls. 2. A shaped part made by impact, compression or rolling; if by rolling, the part is usually referred to as a roll forging.

forging range In hot forging, the optimum temperature range for shaping the metal.

forging stock A piece of semifinished metal used to make a forging. Also known as forging billet.

format 1. To prepare a diskette for acceptance of computer data. 2. A specific arrangement of computer data. 3. The arrangement of a tape record, buffer, or the like for compatibility with processing or storage standards. 4. The arrangement of programming elements comprising any field, record, file, or volume. 5. The basic parameters of a telemetry/data-acquisition sample plan; for example: number of words-per-frame, frames-per-subframe in a sample plan.

formatted ASCII A mode in which data are transferred; a file containing formatted ASCII data is generally transferred as strings of seven-bit ASCII characters (bit eight is zero) terminated by a line-feed, formfeed, or vertical tab. Special characters, such as null, RUBOUT, and tab may be interpreted specially.

formatted binary A mode in which data are transferred; formatted binary is used to transfer check-summed binary data (eight-bit characters) in blocks. Formatting characters are start-of-block indicators, byte count and check-sum values.

formatter A hardware or software process of arranging data on tape or disk, or in a buffer.

form grinding Producing a contoured surface on a part by grinding it with an abrasive wheel whose face has been shaped to the reverse of the desired contour.

forming Applying pressure to shape a material by plastic deformation without intentionally altering its thickness.

forming die A die for producing a contoured shape in sheet metal or other material.

form tool A single-edge, nonrotating cutting tool that produces its inverse or reverse form on a workpiece.

formula A set of parameters that distinguish the products defined by procedures. It may include types and quantities of ingredients, along with information such as the magnitude of process variables. It may effect procedures.

formula translating system (FORTRAN) A procedure-oriented language for solution of arithmetic and logical programs.

FORTRAN The computing language defined as full FORTRAN ISO R1539-1972 which is the same as American National Standard FORTRAN ANSI X3.9-1966 [S61.2]. See formula translating system.

FORTRAN compiler A processor program for FORTRAN.

forward controlling elements See elements, forward controlling.

fossil fuel Coal or petroleum hydrocarbon fuel as distinguished from nuclear fuel.

fouling 1. Growth of adherent plant or animal life on submerged structures, which often leads to biological corrosion or degradation of performance, such as reduction of heat transfer or increase of fluid friction. 2. The accumulation of refuse in gas passages or on heat absorbing surfaces which results in undesirable restrictions to the flow of gas or heat.

foundry A commercial enterprise, plant or portion of a factory where metal or glass is melted and cast.

four-ball tester An apparatus for determining lubrication efficiency by driving one ball against three stationary balls clamped together in a cup filled with the test lubricant; effectiveness of lubrication is expressed relatively in terms of wear-scar diameters on the stationary balls.

Fourier optics 1. Optical components used in making Fourier transforms and other types of optical processing operations. 2. A prism or grating monochromator which essentially performs a Fourier transform on the light incident upon the entrance slit.

fraction 1. In classification of powdered or granular solids, the proportion of the sample (by weight) that lies between two stated particle sizes. 2. In chemical distillation, the proportion of a solution of two liquids consisting of a specific chemical substance.

fractional distillation A thermal process whereby a mixture of liquids that boil at different temperatures is heated at a series of increasing temperatures, and the distillates boiled off at each temperature are collected separately.

fractionating column An apparatus for fractional distillation in which rising vapor and falling liquid are brought into intimate contact.

fraction defective In quality control, the average number of units of product containing one or more defects for each 100 units of product in a given lot.

fractography The study of fracture surfaces, especially for the purpose of determining the causes of failure and relating these causes to macrostructural and microstructural characteristics of parts and materials.

fracture test A method for determining composition, grain size, case depth or material soundness by breaking a test specimen and examining the fracture surface for certain characteristic features.

fragmentation In data processing, the effect of often used files that are growing in size to become non-contiguous when stored on a soft or hard disk.

frame 1. The image in a computer display terminal. 2. In time-division multiplexing, one complete commutator revolution that includes a single synchronizing signal or code.

frame code complement (FCC) The subframe synchronization method whereby the frame synchronization code is complemented to signal the beginning of each subframe.

frame rate (FRATE) The rate, or the pulses that clock that rate, of rotation of a data multiplexer "wheel."

frame synchronization pattern A unique code, coded pulse, or interval to mark the start of a commutation frame period.

frame synchronizer (FSY) Telemetry hardware that recognizes the unique signal that indicates the beginning of a frame of data; a typical frame synchronizer "searches" for the code, "checks" the recurrence of the code in the same position for several frame periods, and then "locks" on the code.

framework The load carrying members of an assembled structure.

framing error An error resulting from transmitting or receiving data at the wrong speed. The character of data will appear to have an incorrect number of bits.

free ash Ash which is not included in the fixed ash.

free-electron laser A laser in which stimulated emission is produced by a beam of free electrons passing through a magnetic field which periodically (in space rather than time) alters its polarity.

free field Ideally, a wave field or potential-energy field in a homogeneous, unbounded medium, but practically, a field where boundary effects are negligible over the useful portion of the medium.

free-field frequency response The free-field frequency response of a piezoelectric sound-pressure transducer is the ratio, as a function of frequency, of the transducer's output in a sound field to the free-field sound pressure existing at the transducer location in the absence of the transducer [S37.10].

free-field grazing incidence response A free-field frequency response of a piezoelectric sound-pressure transducer with the sound incident parallel to a specified sensing surface of the microphone [S27.10].

free field (sound) A free sound field is one existing in a homogeneous, isotropic medium free of any acoustically-reflecting boundaries [S37.10].

free fit A type of clearance fit having a relatively large allowance; it is used when accuracy of assembly is not essential, or when large temperature variations may occur, or both. Also known as free-running fit.

free flow A condition in which the liquid surface downstream of the weir plate is far enough below the crest so that air has free access beneath the nappe.

free gyroscope A gyro wheel mounted in two or more gimbal rings so that its spin axis can maintain a fixed position in space.

free impedance A transducer characteristic

equal to the input impedance when the load impedance is zero.

free-machining A material description that indicates some alteration of chemical composition to substantially improve machinability—such as by the addition of sulfur, phosphorus or lead to steel, or lead to nonferrous metals. Also termed free-cutting.

free oscillation Periodic variation of some system variable when externally applied forces consist of either those that do no work or those that are derived from an invariant potential.

free water The amount of water released when a wet solid is dried to its equilibrium moisture content.

freeze To hold the contents of a register (time, for example) until they have been transferred to another device.

freeze-up 1. Abnormal operation of a refrigeration unit because ice has formed on the heat absorbing elements. 2. Stoppage of rotational motion due to radial expansion or adhesive welding between a bearing and its journal. Also called seizure.

freezing point The temperature at which equilibrium is attained between liquid and solid phases of a pure substance; the term also is applied to compounds and alloys that undergo isothermal liquid-solid phase transformation.

french coupling A coupling with a right and left hand thread.

frequencies The harmonics of a periodic variable.

frequency 1. The number of cycles a periodic variable passes through per unit time. 2. Rate of signal oscillation in Hertz.

frequency band The continuum between two specified limiting frequencies.

frequency, damped The apparent frequency of a damped oscillatory time response of a system resulting from a non-oscillatory stimulus [S51.1].

frequency departure The amount that a carrier frequency or center frequency varies from its assigned value.

frequency deviation The peak difference between the instantaneous frequency of a modulated wave and the frequency of the unmodulated carrier wave.

frequency distortion A form of distortion in which the relative magnitudes of the components of a complex wave are changed during transmission.

frequency divider An electronic circuit or device whose output-signal frequency is a proper fraction of its input-signal frequency.

frequency-divisional multiplexing The combination of two or more signals at different frequencies so they can be transmitted as one signal. This can be done electronically, or it can be done optically by using two or more light sources of different wavelengths.

frequency-division multiplex (FDM) A system for the transmission of information about two or more quantities (measurands) over a common channel, by dividing the available frequency bands; amplitude, frequency, or phase modulation of the subcarriers may be employed.

frequency, gain crossover 1. On a Bode diagram of the transfer function of an element or system, the frequency at which the gain becomes unity (and its decibel value zero). 2. Of integral control action, the frequency at which the gain becomes unity [S51.1].

frequency meter An instrument for determining the frequency of a cyclic signal, such as an alternating current or radio wave.

frequency-modulated output An output in the form of frequency deviations from a center frequency, where the deviation is a function of the applied measurand (e.g., angular speed and flow rate) [S37.1].

frequency modulation 1. A type of electronic circuit which produces an output signal whose frequency has been modified by one or more input signals. See also modulated wave. 2. In telemetry, modulation of the frequency of an oscillator to indicate data magnitude.

frequency monitor An instrument that determines the amount that a frequency deviates from its assigned value.

frequency multiplication The generation of harmonics of the frequency of a lightwave by nonlinear interactions of the lightwave with certain materials. Frequency doubling is equivalent to dividing the wavelength in half. High power beams are needed for the nonlinear interaction to occur.

frequency multiplier An electronic circuit or device whose output-signal frequency is an exact multiple of its input-signal frequency.

frequency, natural The frequency of free (not forced) oscillations of the sensing element of a fully assembled transducer. Note: (a) It is also defined as the frequency of a sinusoidally ap-

165

plied measurand at which the transducer output lags the measurand by 90 degrees. (b) Applicable at room temperature unless otherwise specified. (c) Also see frequency, resonant and frequency, ringing which are considered of more practical value than natural frequency [S37.1].

frequency, phase crossover Of a loop transfer function, the frequency at which the phase angle reaches ±180° [S51.1].

frequency, resonant The measurand frequency at which a transducer responds with maximum output amplitude. Note (a) When major amplitude peaks occur at more than one frequency, the lowest of these frequencies is the resonant frequency. (b) A peak is considered major when it has an amplitude at least 1.3 times the amplitude of the frequency to which specified frequency response is referred. (c) For subsidiary resonance peaks see resonances [S37.1].

frequency response 1. The change with frequency of the sensitivity with respect to the reference sensitivity, for a sinusoidally varying acceleration applied to a transducer within a stated rate of frequencies, usually specified as "within ± _____ percent of the reference sensitivity from _____ to _____ cps." [RP37.2]. 2. The change with frequency of the output/measurand amplitude ratio (and of the phase difference between output and measurand), for a sinusoidally varying measurand applied to a transducer within a stated range of measurand frequencies. Note: (a) (S) It is usually specified as "within ± _____ percent (or ± _____ db) from _____ to _____ hertz." (b) (S) Frequency response should be referred to a frequency within the specified measurand frequency range and to a specific measurand value [S37.1]. 3. A measure of the effectiveness with which a circuit or device transmits signals of different frequencies, usually expressed as a graph of magnitude or phase of an output signal as a function of frequency. Also known as amplitude-frequency response; sine-wave response. 4. For a linear system in sinusoidal steady-state, the ratio of the Fourier-transform of the output signal to the Fourier-transform of the corresponding input signal. 5. The response of a component, instrument, or control system to input signals at varying frequencies.

frequency response, calculated The frequency response of a transducer calculated from its transient response, its mechanical properties, or its geometry, and so identified [S37.1].

frequency response characteristic The frequency-dependency relation, in both amplitude and phase, between steady-state sinusoidal inputs and the resulting fundamental sinusoidal outputs. Frequency response is commonly plotted on a Bode diagram [S51.1].

frequency response method A method of tuning a process control loop for optimum operation by proper selection of controller settings. This method is based on a study of the frequency response of the open process control loop.

frequency, ringing The frequency of the oscillatory transient occurring in the transducer output as a result of a step change in measurand [S37.1].

frequency shift keying (FSK) Modulation accomplished by switching from one discrete frequency to another discrete frequency.

frequency stability 1. A measurement of how well the output frequency (or equivalently emitted wavelength) of a laser stays constant. In some types the emitted wavelength tends to drift because of factors such as changing temperature of the laser itself. 2. Statement of deviation with time, temperature or supply voltages of an electronic oscillator when compared to a standard.

frequency swing A characteristic of a frequency-modulation system equal to the difference between the maximum and minimum design values of instantaneous frequency in the modulated wave.

frequency telemetering A system for transmitting measurements where the information values are represented by frequencies within a specific band, the specific frequency being determined by the percent of full scale equivalent to the current value of the measured variable.

frequency, undamped (frequency, natural) 1. Of a second-order linear system without damping, the frequency of free oscillation in radians or cycles per unit of time [S51.1]. 2. Of any system whose transfer function contains the quadratic factor $s^2+2z\omega_n s+\omega_n^2$ the value ω_n, where s=complex variable; z=constant; ω_n=natural frequency in radians per second. 3. Of a closed-loop control system or controlled system, a frequency at which continuous oscillation (hunting) can occur without periodic stimuli [S51.1]. Note: In linear systems, the undamped frequency is the phase crossover frequency. With proportional control action only, the undamped frequency of a linear system may be obtained in most cases by raising the proportional gain until continuous oscillation occurs [S51.1].

Fresnel lens A lens in which the surface is composed of a number of concentric lens sections with the same focal length desired for the larger lens. Typically for high quality optical applications, the smaller lenses are concentric circles. This technique is used to compress a short focal length optical component into a thickness much less than a plane-convex lens of the same material and focal length.

fretting A form of wear that occurs between closely fitting surfaces subjected to cyclic relative motion of very small amplitude; it is usually accompanied by corrosion, especially of the very fine wear debris. Also known as chafing fatigue; fretting corrosion; friction oxidation; molecular attrition; wear oxidation; and in rolling-element bearings, false brinelling.

friable Capable of being easily crumbled, pulverized or otherwise reduced to powder.

friction See friction error.

frictional error See error, frictional.

friction error The maximum change in output, at any measurand value within the specified range, before and after minimizing friction within the transducer by dithering [S37.1].

friction feed printer A printer using the pressure of a platen to advance the paper.

friction-free error band The error band applicable at room conditions and with frictions within the transducer minimized by dithering [S37.1].

friction oxidation See fretting.

friction saw A toothless circular saw that cuts materials largely by fusion due to frictional heat along the line of contact with the piece being cut.

friction tape A type of cotton tape impregnated with a sticky, moisture-resistant compound, which is used to cover and insulate exposed electrical connections or terminations; it has been largely replaced by electrical tape made of polyvinyl chloride resin backed with a sticky adhesive.

friction-tube viscometer A device for measuring viscosity by determining the pressure drop across a friction tube as the fluid is pumped through it.

frigorimeter A thermometer for measuring low temperatures.

frit Fusible ceramic mixture used to make ceramic glazes and porcelain enamels.

frit seal A hermetical seal for enclosing integrated circuits and other electronic components, which is made by fusing a mixture of metallic powder and glass binder.

from-to tester A type of electronic test equipment for checking continuity between two points in a circuit.

front-end processor 1. A device that receives computer data from other input devices, organizes such data as specified, and then transmits this data to another computer for processing. 2. The computer equipment used to receive plant signals, including analog-to-digital converters and the associated controls.

frost plug A device for determining liquid level when the contents of a tank are at a temperature below 0 °C; a side tube resembling a sight glass but having a series of closed tubes (plugs) at different levels instead of the glass; the tubes below liquid level are cooled so that moisture from the atmosphere forms frost on them, while the tubes above liquid level remain frost free.

frothing Production of a layer of relatively stable bubbles at an air-liquid interface; it can be accomplished by any of several methods, including aeration, agitation or chemical reaction; in many instances it is an undesired side effect of an operation, but sometimes it is an essential element of the operation, as in froth flotation to separate a mineral from its ore.

FTAM See file transfer access and management.

fuel Any material that will burn or otherwise react to release heat energy—common fuels include coal, charcoal, wood and petroleum products (fossil fuels), which burn, and uranium, which undergoes nuclear fission.

fuel-air mixture Mixture of fuel and air.

fuel-air ratio The ratio of the weight, or volume, or fuel to air.

fuel gas A combustible gaseous substance that is used as a fuel.

fuel oil Any oily hydrocarbon liquid having a flash point of at least 100 °F (38 °C) which can be burned to generate heat.

fulchronograph An instrument for recording lightning strikes electromagnetically.

fulgurator An atomizer used in flame analysis to spray the salt solution to be analyzed into the flame.

full adder A computer logic device that accepts two addends and a carry input, and produces a sum and a carry output.

full annealing An imprecise term that implies heating to a suitable temperature followed by controlled cooling to produce a condition of minimum strength and hardness.

full ball A closure member that is a complete spherical surface with a flow passage through it. The flow passage may be round, contoured or otherwise modified to yield a desired flow characteristic [S75.05].

full duplex 1. Communications that appear to have information transfer in both directions (transmit and receive) at the same time. 2. The electronic transmission of data simultaneously in two directions. See duplex, full.

Fuller's earth A highly absorbent, claylike material formerly used to remove grease from woolen cloth, but now used principally as a filter medium.

full face gasket A flat gasket which contacts the entire flat contact surface of two mating flanges, extending past the bolt holes. This term applies to flat face flanges only [S75.05].

full hard temper A level of hardness and strength for nonferrous alloys and some ferrous alloys corresponding to a cold worked state beyond which the material can no longer be formed by bending.

full-height drive A 5 1/4-inch disk drive that is 3 1/4-inches or 1 1/2-inches wide when installed.

full range (F.R.) The algebraic difference between the minimum and maximum values for which a device is specified [RP55.1].

full scale error The difference between the actual and ideal ADC or DAC output for full scale input. Expressed in millivolts, percent of FSR or LSBs. Also called gain error.

full scale (F.S.) The maximum absolute value for which a device is specified [RP55.1]. See range.

full-scale gas concentration 100 percent of the actual marked full-scale concentration value [S12.13].

full-scale output The algebraic difference between the end points. Sometimes expressed as ± (half the algebraic difference) e.g., ±2.5 volts [S37.1].

full-scale value 1. The largest value of a measured quantity that can be indicated on an instrument scale. 2. For an instrument whose zero is between the ends of the scale, the sum of the absolute values of the measured quantity cor-

responding to the two ends in the scale.

full-wave rectifier An electronic circuit that converts an a-c input signal to a d-c output signal, with current flowing in the output circuit during both halves of each cycle in the input signal.

function 1. The purpose of, or an action performed by, a device [S5.1]. 2. A specific purpose of an entity or its characteristic action. 3. In communications, a machine action such as a carriage return or line feed. 4. A closed subroutine which returns a value to the calling routine upon conclusion. 5. The operation called for in a computer software instruction.

functional design The specification of the working relations between the parts of a system in terms of their characteristic actions.

functional diagram A diagram that represents the functional relationships among the parts of a system.

functional program A routine or group of routines which, when considered as a whole, completes some task with a minimum of interaction of other functional programs other than to obtain data and signal completion of its task. For example, a group of routines which take data from an analog scanner and store it on a bulk storage device might be considered to be a functional program.

functional requirements A specification of required functional behavior, operation, performance, or purpose.

functional specification A document that tells exactly what the system should do, what will be supplied to the system, and what is expected to come out of it.

functional test See test.

function keys Special keys on a computer keyboard that instruct the computer to perform a specific operation.

function subprogram The function subprogram is an independently written program and is treated as such by the compiler. It may consist of any number of statements which are executed when it is called. See also subroutine and subprogram.

function switch A circuit having a fixed number of inputs and outputs designed such that the output information is a function of the input information, each expressed in a certain code, signal configuration, or pattern.

function table 1. The two or more sets of information so arranged that an entry in one set selects one or more entries in the remaining sets. 2. A dictionary. 3. A device constructed of hardware, or a subroutine, which can either decode multiple inputs into a single output or encode a single output into multiple outputs. 4. A tabulation of the values of a function for a set of values of the variable.

fundamental frequency The frequency of a sinusoidal function having the same period as a complex periodic quantity.

fundamental mode 1. The mode of a waveguide having the lowest critical frequency. 2. A type of sequential circuit in which there is only one input change at a time and no further change occurs until all states are stabilized.

fundamental natural frequency The lowest frequency in a set of natural frequencies.

furnace An apparatus for liberating heat and using it to produce a physical or chemical change in a solid or liquid mass; most often, the heat is produced by burning a fossil fuel, passing electric current through a heavy-duty resistance element, generating and sustaining an electric arc, or electromagnetically inducing large eddy currents in the charge.

furnace draft The draft in a furnace, measured at a point immediately in front of the highest point at which the combustion cases leave the furnace.

furnace volume The cubical contents of the furnace or combustion chamber.

fuse 1. Any of several devices for detonating an explosive—for example, by elapsed time, command, impact, proximity or thermal effects. 2. A link in a series with a source of power which will open a circuit if a fault of predetermined magnitude occurs in the powered device.

fused fiber optics A number of separate fibers which are melted together to form a rigid fused bundle to transmit light. Fused fiber optics may be used for transmitting images or simply illumination; they are not necessarily coherent bundles of fibers.

fused silica The term usually applied to synthetic fused silica, formed by the chemical combination of silicon and oxygen to produce a high purity silica. Optical glass is made by the melting of high purity sands, while fused quartz is made by crushing and melting natural quartz. See silica glass.

fused slag Slag which has coalesced into a homogeneous solid mass by fusing.

fuse-protected shunt diode barrier A network identical to a fuse-protected shunt diode barrier, except that the fuse is replaced by a resistor [RP12.6].

fuse pull-out A removable fuseholder which can be removed to replace fuses or to open an electrical circuit.

fusibility Property of slag to fuse and coalesce into a homogeneous mass.

fusible alloy An alloy with a very low melting point, in some instances approaching 150 °F (65 °C), usually based on Bi, Cd, Sn or Pb; the fusible alloys have varied uses, the most widely known being solders and fusible links for automatic sprinklers, fire alarms and other safety devices.

fusible plug A hollowed threaded plug having the hollowed portion filled with a low melting point material, usually located at the lowest permissible water level.

fusion welding Any welding process that involves melting of a portion of the base metal.

fusion zone In a weldment, the area of base metal melted, as determined on a cross section through the weld.

future alarm point See alarm point.

FWA See first word address.

G

g See gram.

G See specific gravity.

gage Also spelled gauge. 1. The thickness of metal sheet, or the diameter of rod or wire. 2. A device for determining dimensions such as thickness or length. 3. A visual inspection aid that helps an inspector to reliably determine whether size or contour of a formed, stamped or machined part meets tolerances.

gage block A rectangular chromium steel block having two flat parallel surfaces, with flatness and parallelism guaranteed within a few millionths of an inch; they are usually manufactured and sold in sets for use as standards in linear measurement. Also known as Johanssen block; Jo block; precision block; size block.

gage cock A valve attached to a water column or drum for checking water level.

gage factor A measure of the ratio of the relative change of resistance to the relative change in length of a resistive strain transducer (strain gage) [S37.1].

gage glass A glass or plastic tube for measuring liquid level in a tank or pressure vessel, usually by direct sight; it is usually connected directly to the vessel through suitable fittings and shutoff valves.

gage length In materials testing, the original length of an elongated specimen over which measurements of strain, thermal expansion or other properties are taken.

gage point A specific location used to position a part in a jig, fixture or qualifying gage.

gage pressure 1. Pressure measured relative to ambient pressure. 2. The difference between the local absolute pressure of the system and the atmospheric pressure at the place of the measurement. 3. Static pressure as indicated on a gage.

gain 1. Ratio of output signal magnitude to input signal magnitude; when less than one this is usually called attenuation [S26]. 2. The relative degree of amplification in an electronic circuit. 3. The ratio of the change in output to the change in input which caused the change. 4. In a controller, the reciprocal of proportional band—for example, if the proportional band is set at 25%, the controller gain is .25. Proportional band can be expressed as a dimensionless number (gain) or as a percent.

gain, antenna By common definition, the difference in signal strengths between a given antenna and an isotropic antenna.

gain, closed loop In process instrumentation, the gain of a closed loop system, expressed as the ratio of the output change to the input change at a specified frequency [S51.1].

gain, crossover frequency See frequency, gain crossover.

gain, derivative action (rate gain) The ratio of maximum gain resulting from proportional plus derivative control action to the gain due to proportional control action alone [S51.1].

gain, dynamic The magnitude ratio of the steady-state amplitude of the output signal from an element or system to the amplitude of the input signal to that element or system, for a sinusoidal signal [S51.1].

gain, loop In process instrumentation, the ratio of the absolute magnitude of the change in the feedback signal to the change in its corresponding error signal at a specified frequency. Note: The gain of the loop elements is frequently measured by opening the loop, with appropriate termination. The gain so measured is often called the open loop gain [S51.1].

gain (magnitude ratio) For a linear system or element, the ratio of the magnitude (amplitude) of a steady-state sinusoidal output relative to the causal input. The quantity may be separated

171

into two factors: (a) A proportional amplification often denoted as K, which is frequency independent and is associated with a dimensioned scale factor relating to the units of input and output. (b) A dimensionless factor often denoted as G (jω) which is frequency-dependent. Frequency, conditions of operation, and conditions of measurement must be specified. A loop-gain characteristic is a plot of log gain versus log frequency. In nonlinear systems, gains are often amplitude-dependent [S51.1].

gain margin The reciprocal of the open-loop gain for a stable feedback system at the frequency at which the phase angle reaches -180°.

gain, open loop See gain, loop.

gain, proportional The ratio of the change in output due to proportional control action to the change in input. See proportional band [S51.1].

gain, static (zero-frequency gain) Of gain of an element, or loop gain of a system, the value approached as a limit as frequency approaches zero. Note: Its value is the ratio of change of steady-state output to a step change in input provided the output does not saturate [S51.1].

gain, zero frequency See gain, static (zero-frequency gain).

gal See gallon.

Gal A unit of acceleration equal to 1 cm/s². The milligal is frequently used because it is about 0.001 times the earth's gravity.

galling Localized adhesive welding with subsequent spalling and roughening of rubbing metal surfaces as a result of excessive friction and metal-to-metal contact at high spots.

gallon A unit of capacity (volume) usually referring to liquid measure in the British or U.S. Customary system of units. The capacity defined by the British (Imperial) gallon equals 1.20095 U.S. gallons; one U.S. gallon equals four quarts or 3.785×10^{-3} m³.

galvanic corrosion Electrochemical corrosion associated with current in a galvanic cell, which is set up when two dissimilar metals (or the same metal in two different metallurgical conditions) are in electrical contact and are immersed in an electrolytic solution.

galvanizing Coating a metal with zinc, using any of several processes, the most common being hot dipping and electroplating.

galvanometer An instrument for measuring small electric currents using electromagnetic or electrodynamic forces to create mechanical motion, such as changing the position of a suspended moving coil.

galvanometer recorder A sensitive moving-coil instrument having a small mirror mounted on the coil; a small signal voltage applied to the coil causes a light beam reflected from the mirror to move along the length of a slit, producing a trace on a light-sensitive recording medium that moves transverse to the slit at constant speed.

game theory A mathematical process of selecting an optimum strategy in the face of an opponent who has a strategy of his own.

gamma A measure of the contrast properties of a photographic or radiographic emulsion which equals the slope of the straight-line portion of its HD curve.

gamma counter An instrument for detecting gamma radiation—either by measuring integrated intensity over a period of time or by detecting each photon separately.

gamma ray 1. Electromagnetic radiation emitted by the nucleus of an atom, each photon resulting from the quantum transition between two energy levels of the nucleus. 2. A term sometimes used to describe any high-energy electromagnetic radiation, such as x-rays exceeding about 1 MeV or photons of annihilation radiation.

gamma-ray spectrometer An instrument for measuring the energy distribution in a beam of gamma rays.

Gantt chart A style of bar chart used in production planning and control to display both work planned and work done in relation to time.

gap 1. An interval of space or time that is used as an automatic sentinel to indicate the end of a word, record, or file of data on a tape; for example, a word gap at the end of a word, a record or item gap at the end of a group of words, or a file gap at the end of a group of records or items. 2. The absence of information for a specified length of time, or space on a recording medium, as contrasted with marks and sentinels that indicate the presence of specific information to achieve a similar purpose. 3. The space between the reading or recording head and the recording medium, such as tape, drum, or disk; related to gap, head. 4. In a weldment, the space between members, prior to welding, at the point of closest approach for opposing faces.

gap scanning In ultrasonic examination, projecting the sound beam through a short column of fluid produced by pumping couplant through a nozzle in the ultrasonic search unit.

garbage In data processing, meaningless or incorrect data.

garter spring A closed ring made by welding the ends of a closely wound helical spring together.

gas amplification A counter-tube or ionization-chamber characteristic equal to the charge collected divided by the charge produced in the active volume by a given ionizing event.

gas analysis The determination of the constituents of a gaseous mixture.

gas bearing A journal or thrust bearing that uses a film of gas to lubricate the running surfaces. Also known as gas-lubricated bearing.

gas burner A burner for use with gaseous fuel.

gas carburizing A surface hardening process in which steel or an alloy of suitable alternative composition is exposed at elevated temperature to a gaseous atmosphere with a high carbon potential; hardening of the resulting carbon-rich surface layers is done by quenching the part from the carburizing temperature or by reheating and quenching.

gas counter A type of counter tube in which a gaseous sample whose radiation is to be measured is introduced directly into the counter tube itself.

gas current A current of positive ions flowing to a negatively biased electrode, the positive ions being produced when electrons flowing between two other electrodes collide with residual gas molecules.

gas detection instrument An assembly of electrical and mechanical components (either a single integrated unit, or a system comprising two or more physically separate but interconnected component parts) which senses the presence of such combustible gas and provides an indication, alarm, or other output function [S12.13] [S12.15].

gasdynamic pumping The production of a population inversion by a gasdynamic process, in which a hot, dense gas is expanded into a near vacuum, causing the gas to cool rapidly. If the gas cools faster than energy can be redistributed, a population inversion is generated.

gas etching Removing material from a semiconductor material by reacting it with a gas to form a volatile compound.

gas house tar By-product from the distillation of coal for illuminating gas.

gasification The process of converting solid or liquid fuel into a gaseous fuel such as the gasification of coal.

gasket A sealing member, usually made by stamping from a sheet of cork, rubber, metal or impregnated synthetic material and clamped between two essentially flat surfaces to prevent pressurized fluid from leaking through the crevice; typical applications include flanged joints in piping, head seals in a reciprocating engine or compressor, casing seals in a pump, or virtually anywhere a pressure-tight joint is needed between stationary members. Also known as static seal.

gas lift The technique of raising a liquid in a vertical flow line by injecting a gas below a portion of the liquid column causing upward flow.

gas metal-arc welding (GMAW) A form of electric arc welding in which the electrode is a continuous filler metal wire and in which the welding arc is shielded by supplying a gas such as argon, helium or CO_2 through a nozzle in the torch or welding head; the term GMAW includes the methods known as MIG welding.

gas meter An instrument for measuring and recording the volume or mass of a gaseous fluid that flows past a given point in a piping system.

gasometer A piece of apparatus typically used in analytical chemistry to hold and measure the quantity of gas evolved in a reaction; similar equipment is used in some industrial applications.

gas plasma display A data display screen used on some laptop computers. Characters are easier to read than those on liquid crystal display screens, but the unit is more expensive.

gas pliers A pinchers-type tool for grasping round objects such as pipes, tubes and rods.

gas pocket A cavity within a solid or liquid body that is filled with gas.

gas seal A type of shaft seal that prevents gas from leaking axially along a shaft where it penetrates a machine casing.

gas-sensing element (sensor) 1. The primary element in the gas detection system which responds to the presence of a combustible gas—including any reference of compensating unit, where applicable [S12.13]. 2. The particular subassembly or element in the gas detection instrument that, in the presence of gas, produces a change in its electrical, chemical, or physical characteristics [S12.15].

gas-shielded arc welding An all-inclusive term for any arc welding process that utilizes a gas stream to prevent direct contact between the ambient atmosphere and the welding arc and weld puddle.

gassing 1. Absorption of gas by a material. 2. Formation of gas pockets in a material. 3. Evolution of gas during a process—for example, evolution of hydrogen at the cathode during electroplating, gas evolution from a metal during melting or solidification, or desorption of gas from internal surfaces during evacuation of a vacuum system; the last is sometimes referred to as outgassing.

gas specific gravity balance A weighing device consisting of a tall gas column with a floating bottom; a pointer mechanically linked to the floating bottom indicates density or specific gravity directly, depending on scale calibration.

gas thermometer A temperature transducer that converts temperature to pressure of gas in a closed system. The relation between temperature and pressure is based on the gas laws at constant volume.

gas tube An electron tube whose operating characteristics are substantially affected by the presence of gas or vapor within the tube envelope.

gas tungsten-arc welding (GTAW) A form of metal-arc welding in which the electrode is a nonconsumable pointed tungsten rod; shielding is provided by a stream of inert gas, usually helium or argon; filler metal wire may or may not be fed into the weld puddle, and pressure may or may not be applied to the joint; the term GTAW includes the method known as Heliarc or TIG welding.

gas welding See preferred term oxyfuel-gas welding.

gate 1. A flat or wedge-shaped sliding element that modifies flow rate with linear motion across the flow path [S75.05]. 2. A movable barrier. 3. A device such as a valve or door which controls the rate at which materials are admitted to a conduit, pipe or conveyor. 4. A device for positioning film in a movie camera, printer or projector. 5. The passage in a casting mold that connects the sprue to the mold cavity. Also known as ingate. 6. An electronic component that allows only signals of predetermined amplitudes, frequencies or phases to pass.

gate valve 1. A valve with a linear motion clo-

sure member that is a flat or wedge shaped gate which may be moved in or out of the flow stream. It has a straight-through flow path [S75.05]. 2. A type of valve whose flow-control element is a disc or plate that undergoes translational motion in a plane transverse to the flow passage through the valve body.

gateway A network device that interconnects two networks that may have different protocols.

gauge See gage.

gauss The CGS unit of magnetic flux density or magnetic induction; the SI unit, the tesla, is preferred.

Gaussian beam A laser beam in which the intensity has its peak at the center of the beam, then drops off gradually toward the edges. The intensity profile measured across the center of the beam is a classical Gaussian curve.

gaussmeter A magnetometer for measuring only the intensity, not the direction, of a magnetic field; its scale is graduated in gauss or kilogauss.

gauze 1. A sheer, loosely woven textile fabric; one of its widest uses is for surgical dressings, but it also has some industrial uses such as for filter media. 2. Plastic or wire cloth of fine to medium mesh size.

gear 1. A toothed machine element for transmitting power and motion between rotating shafts whose axes are relatively close to each other or are intersecting. 2. A collective term for equipment that performs a specific functions—lifting gear, for example. 3. A collective term for the portion of a machine that transmits motion from one mechanism to another. 4. A specific combination of gears in a transmission or adjustable gear train that determines mechanical advantage, speed and direction of rotation.

gear down To arrange a gear train so that the driven shaft rotates at a slower speed than the driving shaft.

gear drive A mechanism for transmitting power (torque) and motion from one shaft to another by means of direct contact between toothed wheels.

gear level To arrange a gear train so that the driving and driven shafts rotate at the same speed.

gear meter A positive-displacement fluid meter in which two meshing gear wheels provide the metering action.

gear motor 1. A device consisting of an electric motor and a direct-coupled gear train; the arrangement allows the motor to run at optimum speed—usually 1800 or 3600 rpm—while delivering rotational motion at a substantially lower speed.

gear pump A pump in which fluid is fed to one side of a set of meshing gears, which entrain the fluid and discharge it on the other side. 2. A gear pump supplied with pressurized fluid which converts fluid flow to rotary motion.

gear train A combination of two or more gears, arranged to transmit power and motion between two rotating shafts or between a rotating shaft and a member that moves linearly.

gear up To arrange a gear train so that the driven shaft rotates at a higher speed than the driving shaft.

gear wheel A wheel with integral gear teeth that mesh with another gear, a rack, or a worm.

Geiger-Muller counter A radiation-measuring instrument whose active element is a gas-filled chamber usually consisting of a hollow cathode with a fine-wire anode along its axis; in operation, the voltage between anode and cathode is high enough that the discharge caused by a primary ionizing event spreads over the entire anode until stopped when the space charge reduces the electric-field magnitude. Also known as Geiger counter.

Geiger threshold The lowest voltage applied to a counter tube which results in output pulses of essentially equal amplitude, regardless of the magnitude of the ionizing event.

gel coat A resin gelled on the internal surface of a plastics mold prior to filling it with a molding material; the finished part is a two-layer laminate, with the gel coat providing improved surface quality.

general processor In numerical control, a computer program which carries out computations on the part program and prepares the cutter location data (CL data) for a particular part without reference to the machine on which it might be made.

general-purpose computer A computer designed to solve a large variety of problems, e.g., a stored program computer which may be adapted to any of a very large class of applications.

general-purpose simulation system A generic class of discrete, transaction-oriented simulation languages based on a block (diagramming) approach to problem statement. Abbreviated GPSS.

generating electric field meter An instrument for measuring electric field strength in which a flat conductor is alternately exposed to the field and shielded from it; potential gradient of the field is determined by measuring the rectified current through the conductor.

generating magnetometer An instrument for measuring magnetic field strength by means of the electromotive force generated in a rotating coil immersed in the field being measured.

geotechnology Application of science and engineering to problems involved in utilizing natural resources.

gesso A mixture of chalk and either gelatine or casein glue, which is painted on panels to provide a suitable surface for tempera work or for polymer-based paints.

getter A material exposed to the interior of a vacuum system in order to reduce the concentration of residual gas by absorption or adsorption.

getter-ion pump A type of vacuum pump that produces and maintains high vacuum by continuously or intermittently depositing chemically active metal layers on the wall of the pump, where they trap and hold inert gas atoms which have been ionized by an electric discharge and drawn to the activated pump wall. Also known as sputter-ion pump.

gewel hinge A hinge consisting of a hook inserted in a loop.

ghost point A term used in boiler water testing with soap solution. A lather appears to form but will disappear upon the addition of more soap solution. This point represents total calcium hardness and the final lather total hardness.

gib A removable plate that holds other parts or that acts as a bearing or wear surface.

gilbert The CGS unit of magnetomotive force; the SI unit, the ampere (or ampere-turn) is preferred.

gimbal 1. A cage or frame with two mutually perpendicular, intersecting axes of rotation, which gives free angular movement in two directions to any device or mechanism mounted within the frame. 2. A gyro support that gives the spin axis a degree of freedom.

gimbal lock A position in a gyro having two degrees of freedom where the spin axis becomes aligned with an axis of freedom, thus depriving it of a degree of freedom and therefore depriving it of its useful properties.

gimbal mount An optical mount which allows

position of a component to be adjusted by rotating it independently around two orthogonal axes.

gimlet A small tool for boring holes in wood, leather and similar materials; it consists of a threaded point, spiral-fluted shank and cross handle; a tool without the handle and adapted for use in a drill is known as a gimlet bit.

gin A hoisting machine consisting of a windlass, pulleys and ropes in a tripod frame.

gland 1. A device for preventing a pressurized fluid from leaking out of a casing at a machine joint, such as at a shaft penetration. Also known as gland seal. 2. A movable part that compresses the packing in a stuffing box. See packing follower; see also lantern ring.

glass A hard, brittle, amorphous, inorganic material, often transparent or translucent, made by fusing silicates (and sometimes borates and phosphates) with certain basic oxides and cooling rapidly to prevent crystallization.

glass fiber A glass thread less than 0.001 in. (0.025 mm) thick; it is used in loose, matted or woven form to make thermal, acoustical or electrical insulation; in matted, woven or filament-wound form to make fiber-reinforced composites; or in loose, chopped form to make glass-filled plastics parts.

glassine A thin, dense, transparent, super-calendered paper made from highly refined sulfite pulp; it is used industrially as insulation between layers of iron-core transformer windings.

glassmaker's soap A substance such as MnO_2, which is added to glass to eliminate the green color imparted by the presence of iron salts.

glass paper 1. An abrasive material made by bonding a layer of pulverized glass to a paper backing. 2. Paper made of glass fibers.

glass sand The raw material for glassmaking; it normally consists of high-quartz sand containing small amounts of the oxides of Al, Ca, Fe and Mg.

glassware Laboratory containers, vessels, graduated cylinders, tubing and the like which are made from glass.

glass wool A relatively loose mass of glass fibers used chiefly in insulating, packing and filtering applications.

glassy alloy A metallic material having an amorphous or glassy structure. Also known as metallic glass.

glaze A glossy, highly reflective, glasslike, inorganic fused coating. See enamel.

glazing 1. Cutting and fitting glass panes into frames. 2. Smoothing the exposed solder of a wiped pipe joint with a hot iron.

glazing compound A caulking compound, such as putty, used to seal the edges of a pane of glass where it fits into its frame.

glitch Undesirable electronic pulses that cause processing errors.

glitter Decorative flaked powder having a particle size large enough so that the individual flakes produce a visible reflection or sparkle; used in certain decorative paints and in some compounded plastics stock.

global 1. Any name that is declared global has as its scope the entire system in which it resides. 2. A computer instruction that causes the computer to locate all occurrences of specific data. 3. A value defined in one program module and used in others; globals are often referred to as entry points in the module in which they are defined, and externals in the other modules that use them.

global array A set of data listings that can be referenced by other parts of the software.

global common An un-named data area that is accessible by all programs in the system. Sometimes referred to as blank common.

global variable Any variable available to all programs in the system. Contrast with reserved variable.

globe valve 1. A valve with a linear motion closure member, one or more ports and a body distinguished by a globular shaped cavity around the port region [S75.05]. 2. A type of flow-regulating valve consisting of a movable disc and a stationary-ring seat in a generally spherical body. In the general design, the fluid enters below the valve seat and leaves from the cavity above the seat.

globe valve plug guides The means by which the plug is aligned with the seat and held stable through out its travel. The guide is held rigidly in the body or bonnet [S75.05].

globe valve trim The internal parts of a valve which are in flowing contact with the controlled fluid. Examples are the plug, seat ring, cage, stem and the parts used to attach the stem to the plug. The body, bonnet, bottom flange, guide means and gaskets are not considered as part of the trim [S75.05].

glossimeter An instrument for measuring the 'glossiness' of a surface—that is, the ratio of light reflected in a specific direction to light reflected in all directions—usually by means of a photoelectric device. Also known as glossmeter.

glow discharge A discharge of electrical energy through a gas, in which the space potential near the cathode is substantially higher than the ionization potential of the gas.

glue 1. Generally, a term often used improperly to describe an adhesive. 2. Specifically, a crude, impure form of commercial gelatine that softens to a gel consistency when wetted with water and dries to form a strong adhesive layer.

glued A mixed signal simulation system that combines existing analog and digital simulation software into a hybrid analog or digital simulation system.

GMAW See gas metal-arc welding.

Golay cell An infrared detector in which the incident radiation is absorbed in a gas cell, thereby heating the gas. The temperature induced expansion of the gas deflects a diaphragm, and a measurement of this deflection indicates the amount of incident radiation.

goniometer 1. Generally, any instrument for measuring angles. 2. Specifically, an instrument used in crystallography to determine angles between crystal planes, using x-ray diffraction or other means. 3. An instrument used to measure refractive index and other optical properties of transparent optical materials or optically scattering in materials at UV, VIZ or IR wavelengths.

go/no-go gage A composite gaging device that enables an inspector to quickly judge whether specific dimensions or contours are within specified tolerances; in many instances, the device is so constructed that the part being inspected will fit one part of the gage easily and will not fit another part if it is within tolerance, and will pass both parts or pass neither if it is not within tolerance.

go/no-go test A test in which one or more parameters are determined, but which can only result in acceptance or rejection of the test object, depending on the value(s) measured.

gouging Forming a groove in an object by electrically, mechanically, thermomechanically or manually removing material; the process is typically used to remove shallow defects prior to repair welding.

governor A device for automatically regulating the speed or power of a prime mover—especially a device that relies on centrifugal force in whirl-ing weights opposed by springs or gravity to actuate the controlling element.

GPSS See general-purpose simulation system.

grab-sample point The point in the sample line where the flow of sample fluid can be directed to a portable container. It may be referred to as "sample point." [S67.10]

grab sampling A method of sampling bulk materials for analysis, which consists of taking one or more small portions (usually only imprecisely measured) at random from a pile, tank, hopper, railcar, truck or other point of accumulation.

graceful degradation A system attribute wherein when a piece of equipment fails, the system falls back to a degraded mode of operation rather than failing catastrophically and giving no response to its users.

grade 1. To move earth, making a land surface of uniform slope. 2. A classification of materials, alloys, ores, units of product or characteristic according to some attribute or level of quality. 3. To sort and classify according to attributes or quality levels. 4. Oil classification according to quality.

graded index fiber (GRIN) An optical fiber in which the refractive index changes gradually between the core and cladding, in a way designed to refract light so it stays in the fiber core. Such fibers have lower dispersion and broader bandwidth than step-index fibers.

graded refractive index lens A lens in which the refractive index of the glass is not uniform. Typically the index will differ with distance from the center of the lens.

gradient The rate of change of some variable with respect to another, especially a regular uniform or stepwise rate of change.

graduation Any of the major or minor index marks on an instrument scale.

grain 1. The appearance or texture of wood, or a woodlike appearance or texture of another material. 2. In paper or matted fibers, the predominant direction most fibers lie in, which corresponds to directionality imparted during manufacture. 3. In metals and other crystalline substances, an individual crystallite in a polycrystalline mass. 4. In crumbled or pulverized solids, a single particle too large to be called powder.

grain boundary The plane of mismatch between adjacent crystallites in a polycrystalline mass, as revealed on a polished and etched cross section through the material.

grain flow Fibrous appearance on a polished and etched section through a forging, which is caused by orientation of impurities and inhomogeneities along the direction of working during the forging process.

grain growth An increase in the average grain size in a metal, usually as a result of exposure to high temperature.

graininess Visible coarseness in a photographic or radiographic emulsion, which is caused by countless small grains of silver clumping together into relatively large masses visible to the naked eye or with slight magnification.

graining Working a translucent stain while still wet to simulate the grain in wood or marble; tools such as special brushes, combs and rags are used by hand to create the desired irregular patterns.

grain size 1. For metals, the size of crystallites in a polycrystalline solid, which may be expressed as a diameter, number of grains per unit area, or standard grain size number determined by comparison with a chart such as those published by ASTM; in most instances the grain size is given as an average, unless there are substantial proportions which can be given as two distinct sizes; if two or more phases are present, grain size of the matrix is given. 2. For abrasives, see preferred term grit size.

grains per cu ft The term for expressing dust loading in weight per unit of gas volume (7000 grains equals one pound).

grains (water) A unit of measure commonly used in water analysis for the measurement of impurities in water (17.1 grains=1 part per million—ppm).

gram The CGS unit of mass; it equals 0.001 kilogram, which has been adopted as the SI unit of mass.

granular fracture A rough, irregular fracture surface, which can be either transcrystalline or intercrystalline, and which often indicates that fracture took place in a relatively brittle mode, even though the material involved is inherently ductile.

granular structure Nonuniform appearance of molded or compressed material due to the presence of particles of varying composition.

graphic Pertaining to representational or pictorial material, usually legible to humans and applied to the printed or written form of data such as curves, alphabetic characters, and radar scope displays.

graphical display unit An electronics device that can display both text and pictorial representations.

graphic character See graphic.

graphic panel A master control panel which, pictorially and usually colorfully, traces the relationship of control equipment and the process operation. It permits an operator, at a glance, to check on the operation of a far-flung control system by noting dials, valves, scales, and lights.

graphics A computer technique using lines and symbols to display information rather than letters and numbers.

graphic symbol An easily recognized pictorial representation [S5.5].

graphite flake A form of graphite present in gray cast iron which appears in the microstructure as an elongated, curved inclusion.

graphite rosette A form of graphite present in gray cast iron which appears in the microstructure as graphite flakes extending radially outward from a center of crystallization.

graphitic carbon Free carbon present in the microstructure of steel or cast iron; it is an essential feature of most cast irons, but is almost always undesirable in steel.

graphitic corrosion Corrosion of gray cast iron in which the iron matrix is slowly leached away, leaving a porous structure behind that is largely graphite but that may also be held together by corrosion products; this form of corrosion occurs in relatively mild aqueous solutions and on buried pipe and fittings.

graphitic steel Alloy steel in which some of the carbon is present in the form of graphite.

graphitization Formation of graphite in iron or steel; it is termed primary graphitization if it forms during solidification, and secondary graphitization if it forms during subsequent heat treatment or extended service at high temperature.

graphitizing Annealing a ferrous alloy in such a way that at least some of the carbon present is converted to graphite.

gravimeter A device for measuring the relative force of gravity by detecting small differences in weight of a constant mass at different points on the earth's surface. Also known as gravity meter.

gravimetric A descriptive term used to designate an instrument or procedure in which gravitational forces are utilized. However, the results or indications of such procedures are not necessarily influenced by the magnitude of the acceleration of gravity [RP31.1].

gravitational constant A dimensionless conversion factor in English units which arises from Newton's second law (F=ma) when mass is expressed in pounds-mass (lb_m).

gravitometer See densimeter.

gravity Weight index of fuels; liquid, petroleum products expressed either as specific, Baumé or A.P.I. (American Petroleum Institute) gravity; weight index of gaseous fuels as specific gravity related to air under specified conditions; or weight index of solid fuels as specific gravity related to water under specified conditions.

gravity meter 1. A device that uses a U-tube manometer to determine specific gravities of solutions by direct reading. 2. An electrical device for measuring variations in gravitational forces through different geological formations. 3. A gravimeter.

gray Metric unit for absorbed dose.

gray body An object having the same spectral emissivity at every wavelength, or one whose spectral emissivity equals its total emissivity.

gray code A generic name for a family of binary codes which have the property that a change from one number to the next sequential number can be accomplished by changing only one bit in a code for the original number. This type of code is commonly used in rotary shaft encoders to avoid ambiguous readings when moving from one position to the next. See also cyclic code and shaft encoder.

gray iron Cast iron containing free graphite in flake form; so named because a freshly broken bar of the alloy appears gray.

grease 1. Rendered, inedible animal fat. 2. A semisolid to solid lubricant consisting of a thickening agent, such as metallic soap, dispersed in a fluid lubricant, such as petroleum oil.

grease seal ring See lantern ring.

green Unfired, uncured or unsintered.

green strength The mechanical strength of a ceramic or powder metallurgy part after molding or compacting but before firing or sintering; it represents the quality needed to maintain sharpness of contour and physical integrity during handling and mechanical operations to prepare it for firing or sintering.

greenware Unfired ceramic ware.

grid 1. A network of lines, typically forming squares, used in layout work or in creating charts and graphs. 2. A criss-cross network of conductors used for shielding or controlling a beam of electrons.

grid circuit An electronic circuit which includes the grid-cathode path of an electron tube in series with other circuit elements.

grid control A method of controlling anode current in an electron tube by varying the potential of the grid electrode with respect to the cathode.

grid emission Emission of electrons or ions from the grid electrode of an electron tube.

grid nephoscope A device for determining the direction of cloud motion by sighting through a grid work of bars and adjusting the angular position of the grid until some feature of the cloud in the field of view appears to move along the major axis of the grid.

GRIN See graded index fiber.

grinding 1. Removing material from the surface of a workpiece using an abrasive wheel or belt. 2. Reducing the particle size of a powder or granular solid.

grinding aid Something added to the charge in a rod or ball mill to accelerate the grinding process.

grinding burn Localized overheating of a workpiece surface due to excessive grinding pressures, or inadequate supply of coolant or both.

grinding cracks Shallow cracks in the surface of a ground workpiece; they appear most often in relatively hard materials due to excessive grinding friction or high material sensitivity.

grinding fluid A cutting fluid used in grinding operations, primarily to cool the workpiece but also to lubricate the contacting surfaces and carry away grinding debris.

grinding medium Any material—including balls, rods, and quartz or chert pebbles—used in a grinding mill.

grindstone A stone disc mounted on a revolving axle and used for grinding or tool sharpening.

grit A particulate abrasive consisting of angular grains.

grit blasting Abrasively cleaning metal surfaces by blowing steel grit, sand or other hard particulate against them to remove soil, rust and scale. Also known as sandblasting.

grommet 1. A metal washer or eyelet, often used to reinforce a hole in cloth or leather. 2. A rubber or soft plastic eyelet inserted in a hole through sheet metal, such as an electronic equipment chassis or enclosure, to prevent a wire from chafing against the side of the hole, damaging its insulation or shorting out to the chassis. 3. A circular piece of fibrous packing material used under a bolt head or nut to seal the bolthole.

grommet nut A blind nut with a round head that is sometimes used with a screw to attach a hinge to a door.

groove 1. A long narrow channel or furrow in a solid surface. 2. In a weldment, a straight-sided, angled, or curved gap between joint members prior to welding which helps confine the weld puddle and ensure full joint penetration to produce a sound weld.

grooved drum A windlass drum whose face has been grooved, usually in a helical fashion, to support and guide the rope or cable wound on it.

grooved tube-seat A tube seat having one or more shallow groves into which the tube may be forced by the expander.

gross porosity In weld metal or castings, large or numerous gas holes, pores or voids that are indicative of substandard quality or poor technique.

ground 1. A conducting connection, whether intentional or accidental, between an electrical circuit or electrical equipment and either the earth or some other conducting body that serves in place of the earth [S82.01]. 2. A (neutral) reference level for electrical potential, equivalent to the level of electrical potential of the earth's crust. 3. A secure connection to earth which is used to reference an entire system. Usually the connection is in the form of a rod driven or buried in the soil or a series of rods connected into a grid buried in the soil.

grounded Refers to the presence or absence of an electrical connection between the "low" side of the transducer element and the portion of the transducer intended to be in contact with the test structure. Method of ungrounding should be stated as "internally ungrounded" or "by means of separate stud." [RP37.2]

grounding The act of establishing a conductive connection, whether intentional or accidental, between an electrical circuit or electrical equipment and either the earth or some other conducting body that serves in place of the earth [S82.01].

ground lead See work lead.

grouping Combining two or more computer records into one block of information to conserve storage space or disk or tape. Also known as

blocking.

group velocity The velocity corresponding to the rate of change of average position of a wave packet as it travels through a medium.

grouting Placing or injecting a fluid mixture of cement and water (or of cement, sand and water) into a grout hole, crevice, seam or joint for the purpose of forming a seepage barrier, consolidating surrounding earth or rock, repairing concrete structures, or sealing the joint where an equipment base rests on a concrete floor.

grub screw A headless screw that is slotted at one end to receive a screw driver.

GTAW See gas tungsten-arc welding.

guard A shield or cowling that surrounds moving parts to prevent workers from being injured or to prevent incidental equipment damage from foreign objects.

guard bit A bit contained in each word or groups of words of memory which indicates to computer hardware or software whether the content of that memory location may be altered by a program. See protected location.

guard ring An auxiliary ring-shaped electrode in a counter tube or ionization chamber whose chief functions are to control potential gradients, reduce insulation leakage or define the active region of the tube.

guard vacuum An enclosed evacuated space between a primary vacuum system and the atmosphere whose primary purpose is to reduce seal leakage into the primary system.

guide 1. A pulley, idler roll or channel member that keeps a rope, cable or belt traveling in a predetermined path. 2. A runway in which a conveyor travels. 3. A stationary machine element—a beam, bushing, rod or pin, for instance—whose primary function is to maintain one or more moving elements confined to a specific path of travel.

guide bearing A plain bushing used to prevent lateral movement of a machine element while allowing free axial translation, with or without (usually without) simultaneous rotation. Also known as guide bushing.

guide bushing See bushing.

guided bend test A bend test in which the specimen is bent to a predetermined shape in a jig or around a grooved mandrel.

guided missile An unmanned airborne vehicle whose flight path or trajectory can be altered by some mechanism within or attached to the vehicle in response to either a preprogrammed control sequence or a control sequence transmitted to the vehicle while in flight.

guided wave A wave whose energy is confined by one or more extended boundary surfaces, and whose direction of propagation is effectively parallel to the boundary.

guides, closure component The means by which the closure component is aligned with the seat and held stable through its travel. The guide is held rigidly in the control valve body, bonnet, and/or bottom plate [S75.05].

gutter 1. A drainage trough or trench, usually surrounding a raised surface. 2. A groove around the cavity of a forging or casting die to receive excess flash.

guy A wire, rope or rod used to secure a pole, derrick, truss or temporary structure in an upright position, or to hold it securely against the wind.

guyed-steel stack A steel stack of insufficient strength to be self-supporting which is laterally stayed by guys.

gyratory screen A sieving machine having a series of nested screens whose mesh sizes are progressively smaller from top to bottom of the stack; the mechanism shakes the stacked screens in a nearly circular fashion, which causes fines to sift through each screen until an entire sample or batch has been classified.

gyromagnetic ratio The magnetic moment of a system divided by its angular momentum.

gyroscope 1. A transducer which makes use of a self-contained spatial directional reference [S37.1]. 2. An instrument that maintains a stable, angular reference direction by virtue of the application of Newton's second law of motion to a mechanism whose chief component is a rapidly spinning heavy mass.

gyroscopic couple The turning moment generated by a gyroscope to oppose any change in the position of its axis of rotation.

gyroscopic horizon A gyroscopic instrument that simulates the position of the natural horizon and indicates the attitude of an aircraft with respect to this horizon.

gyro wheel The heavy rotating element of a gyroscope, which consists of a wheel whose rather large mass is distributed uniformly around its rim; in precision gyroscopes, the gyro wheel is specially constructed to have nearly perfect balance.

H

H 1. See henry.

hair-line cracks Fine, random cracks in a coating such as paint or any rigid surface.

half-adder A logic circuit that accepts two binary input signals and produces corresponding sum and carry outputs; two half-adders and an OR gate can be combined to realize a full-adder. See also full adder.

half-adjust To round a number so that the least significant digit(s) determines whether or not a "one" is to be added to the digit next higher in significance that the digit(s) used as criterion for the determination. After the adjustment is made, if required, the digit(s) used as criterion will be dropped. e.g. 432.784 using the terminal 4 as criterion yields 432.78 as the half-adjusted value. The number 432.785 half-adjusts to 432.79, since the terminal digit is "one half, or more."

half-and-half solder A lead-tin alloy (50Pb-50Sn) used primarily to join copper tubing and fittings.

half cycle In alternating circuits, the time to complete one-half of a full cycle at the operating frequency.

half duplex Communications in both directions (transmit and receive), but in only one direction at a given instant in time. See duplex, half; see also full duplex.

half-height drive A 5 1/4-inch disk drive that is 1 5/8-inches wide when installed.

half-life The time span necessary for the atoms of a nuclide to disintegrate by one-half.

half-thickness The thickness of an absorbing medium that will depreciate the intensity of radiation beam by one-half.

half-wave plate A polarization retarder which causes light of one linear polarization to be retarded by a half wavelength 180° relative to the phase of the orthogonal polarization.

half-wave rectifier 1. An electronic circuit that converts an a-c input signal to a d-c output signal, with current flowing in the output circuit during only one half of each cycle of the input signal. 2. A rectifier that feeds current during the half cycle when the alternating current voltage is in the polarity at which the rectifier has low resistance. Whereas during the other half cycle the rectifier passes no current.

halide Compound containing fluorine, bromine, chlorine, or iodine [S71.04].

Hall effect An electromotive force developed as a result of interaction when a steady-state current flows in a steady-state magnetic field; the direction of the emf is at right angles to both the direction of the current and the magnetic field vector, and the magnitude of the emf is proportional to the product of current intensity, magnetic force, and sine of the angle between current direction and magnetic field vector.

halogen Any one of the four chemical elements; chlorine, fluorine, bromine or iodine [S71.04].

hammer 1. A hand tool used for striking a workpiece to shape it or to drive it into another object. 2. A machine element consisting of an arm and a striking head, such as for ringing a bell, or consisting of a guided striking head, often carrying one-half of a die set, for shaping metals by forging.

Hamming code An error-correcting code, with or without parity, that allows a data device to detect and correct single-bit errors in coded digital data.

Hamming distance A characteristic of any given data code that indicates the ability to detect single-bit errors; it equals the number of bits in any given character that must be changed to produce another legitimate character.

H and D curve The measurement of photographic emulsion shown as a curve in which density is expressed as a function of the logarithm of exposure.

hand-held equipment Any piece of equipment either designed to be or indicated by the manufacturer that it can be (that is, in advertising literature or in the operating instructions) held in one hand during any phase of normal operation, regardless of its weight [S82.01].

handhole An opening in a pressure part for access, usually not exceeding 6 in. in longest dimension.

handhole cover A handhole closure.

hand jack A manual override device, using a lever, to stroke a value or to limit its travel [S75.05].

hand lance A manually manipulated length of pipe carrying air, steam, or water for blowing ash and slag accumulations from heat absorbing surfaces.

handshake The recognition between two computers that they are able to communicate.

handwheel A manual override device to stroke a valve or limit its travel. (a) Top mounted. The handwheel is mounted on top of the valve actuator case. This type of handwheel does not have a clutch and is usually used to restrict the motion of the valve stem in one direction only. (b) Side-mounted. Bellcrank lever types are externally mounted on the control valve yoke. They can provide a limit to the extent a valve stem will travel in either direction, but not in both directions. (c) In-yoke mounted. In-yoke gear types are designed with a worm gear drive which is contained in a lubricated housing. The gear box is integral with the yoke which is usually elongated to provide space for the worm gear assembly. With this type of handwheel, stops may be set in either or both directions to limit the travel of the valve stem. This type of handwheel is declutchable. (d) Shaft mounted, declutchable. A shaft mounted worm gear drive that can be declutched from the power actuator [S75.05].

hard card A type of computer hard disk on a card rather than a spinning disk.

hard clad silica fibers Silica optical fibers which are coated with hard plastic material, not with the soft materials typically used in plastic clad silica.

hard copy Readable data from the computer to a printer which is produced on paper.

hard disk A computer storage medium with a large storage capacity as compared to floppy disks. 80 megabytes of storage is now common.

hard-disk management Since hard disk life is limited, there are four basic things that will enhance disk life and use: (A) Use subdirectories rather than have all work files in one directory; (B) Delete files that are no longer needed; (C) Run CHKDSK periodically to check for lost clusters; (D) Run a defragmentation program every three months.

hard-drawn wire Heavily cold-drawn metal wire of relatively high tensile strength and low ductility.

hardening Producing increased hardness in a metal by quenching from high temperature, such as hardening steel, or by precipitation-hardening (aging) a dilute alloy, such as hardening certain aluminum or other nonferrous alloys.

hardfacing A material harder than the surface to which it is applied. Used to resist fluid erosion and/or to reduce the chance of galling between moving parts, particularly at high temperature [S75.05].

hard lead Any of a series of lead-antimony alloys of low ductility; typically, hard lead contains 1 to 12% Sb.

hardness A measure of the amount of calcium and magnesium salts in a boiler water. Usually expressed as grains per gallon or p.p.m. as $CaCO_3$.

hard-plating A thin metal deposit, sometimes electroplated, used to induce surface hardening. Hard plating is many orders of magnitude thinner than hard facing [S75.05].

hardware 1. Physical equipment directly involved in performing industrial process measuring and controlling functions [S51.1]. 2. In data processing, hardware refers to the physical equipment associated with the computer. 3. The electrical, mechanical and electromechanical equipment and parts associated with a computing system, as opposed to its firmware and software.

hardware priority interrupt See priority interrupt and software priority interrupt.

hard water Water which contains calcium or magnesium in amounts which require an excessive amount of soap to form a lather.

harmonic Having a frequency that is a multiple of the basic cyclical quantity to which it is related.

harmonic analyzer 1. An instrument for measuring the magnitude and phase of harmonic segments of a cyclical function from a graph. 2. An electronic instrument which measures the amplitude and frequency of an a.c. signal; including those of its harmonics.

harmonic content The distortion in a transducer's sinusoidal output, in the form of harmonics other than the fundamental component. It is usually expressed as a percentage of rms output [S37.1].

harmonic conversion transducer A transducer in which the output frequency is a multiple of the input frequency.

harmonic distortion 1. Distortion characterized by the appearance in the output of harmonics other than the fundamental component when the input wave is sinusoidal [RP55.1]. 2. Distortion caused by the presence of harmonics of a desired signal.

harmonic generation 1. The multiplication of the frequency of a lightwave by nonlinear interactions of the lightwave with certain materials. Generating the second harmonic is equivalent to dividing the wavelength in half. 2. Electronic means of multiplying frequency, usually accomplished with the assistance of nonlinear devices.

Hartley information unit In information theory, a unit of logarithmic measurement of the decision content of a set of 10 mutually exclusive events, expressed as the logarithm to the base 10; for example, the decision content of an eight-character set equals log 8, or 0.903 Hartley.

hashing The generation of a meaningless number from a group of records that can be used as a location address.

Hastelloy B An International Nickel Co. alloy having a nominal composition of nickel (Ni) 66.7%; iron (Fe) 5%; molybdenum (Mo) 28%; vanadium (V) 0.3%.

Hastelloy C An International Nickel Co. alloy having a nominal composition of nickel (Ni) 59%; iron (Fe) 5%; molybdenum (Mo) 16%; tungsten (W) 4%; chromium (Cr) 16%.

Hay bridge A general-purpose a-c bridge circuit in which two opposing sides of the bridge are fixed resistances; the unknown leg is a combination of resistance and inductance, and the remaining side consists of a variable resistor and a variable capacitor.

hazardous area An area in which explosive gas/air mixtures are, or may be expected to be, present in quantities such as to require special precautions for the construction and use of electrical apparatus.

hazardous area classifications 1. Division 1

(hazardous). Where concentrations of flammable gases or vapors exist (a) continuously or periodically during normal operations; (b) frequently during repair or maintenance or because of leakage; or (c) due to equipment breakdown or faulty operation which could cause simultaneous failure of electrical equipment. (See National Electrical Code, Paragraph 500-4(a) for detailed definition.) 2. Division 2 (normally non-hazardous). Locations in which the atmosphere is normally non-hazardous and may become hazardous only through the failure of the ventilating system, opening of pipe lines, or other unusual situations. (See National Electrical Code, Paragraph 500-4(b) for detailed definition.) 3. Non-hazardous. Areas not classified as Division 1 or Division 2 are considered non-hazardous. Note: It is safe to have open flames or other continuous sources of ignition in non-hazardous areas [S12.4].

hazardous atmosphere 1. A combustible mixture of gases and/or vapors. 2. An explosive mixture of dust in air.

hazardous (classified) location 1. A location where fire or explosion hazards may exist due to the presence of flammable gases or vapors, flammable liquids, combustible dust, or easily ignitable fibers or flyings [RP12.6]. See location, hazardous (classified).

hazardous dust layer Any accumulation of combustible dust that will propagate or cause a fire [S12.11].

hazardous location A space of limited and definable extent in which may be found a hazardous atmosphere or a hazardous dust layer [S12.11].

hazardous material Any substance that requires special handling to avoid endangering human life, health or well being. Such substances include poisons, corrosives, and flammable, explosive or radioactive chemicals.

hazemeter See transmissometer.

HDDR See high density digital recording.

HDLC See high-level data link control.

head The portion of a computer disk drive that reads, writes, or erases any magnetic storage medium.

head crash In data processing, the malfunction of the read and write head in a disk drive.

header 1. A conduit or chamber that receives fluid flow from a series of smaller conduits connected to it, or that distributes fluid flow among a series of smaller conduits. 2. In data processing, data placed at the beginning of a file for identification. 3. The portion of a batch recipe that contains information about the purpose, source and version of the recipe such as recipe and product identification, originator, issue date and so on.

head gap 1. The space between the reading or recording head and the recording medium, such as tape, drum, or disk. 2. The space or gap intentionally inserted into the magnetic circuit of the head in order to force or direct the recording flux into the recording medium.

head loss Pressure loss in terms of a length parameter such as inches of water or millimeters of mercury.

head pressure Expression of a pressure in terms of the height of fluid. $P = y\rho g$, where ρ is fluid density and y is the fluid column height.

head pulley The pulley at the discharge end of the belt conveyor. The power drive for the belt is generally applied to the end pulley [RP74.01].

health physics The technology associated with the measurement and control of radiation dose in humans.

heat Energy that flows between bodies because of a difference in temperature; same as thermal energy.

heat absorbing filter A glass filter which transmits most visible light, but strongly absorbs infrared light.

heat available The thermal energy above a fixed datum that is capable of being absorbed for useful work. In boiler practice, the heat available in a furnace is usually taken to be the higher heating value of the fuel corrected by subtracting radiation losses, unburned combustible, latent heat of the water in the fuel formed by the burning of hydrogen, and adding the sensible heat in the air for combustion, all above ambient temperatures.

heat balance An accounting of the distribution of the heat input and output.

heat content The amount of heat per unit mass that can be released when a substance undergoes a drop in temperature, a change in state or a chemical reaction.

heat exchanger A vessel in which heat is transferred from one medium to another.

heating surface Heating surface shall be expressed in square feet and shall include those surfaces which are exposed to products of combustion on one side and water on the other. This surface shall be measured on the side receiving the heat, which is as provided in the ASME Power Test Code.

heat rate The ratio of heat input to work output of a thermal power plant. It is a measure of power plant efficiency.

heat release The total quantity of thermal energy above a fixed datum introduced into a furnace by the fuel, considered to be the product of the hourly fuel rate and its high heat value, expressed in Btu per hour per cubic foot of furnace volume.

heatsink Any device, usually a static device, used primarily to absorb heat and thereby protect another component from damage due to excessive heat.

heat transfer, coefficient of Heat flow per unit time across a unit area of a specified surface under the driving force of a unit temperature difference between two specified points along the direction of heat flow. Also known as over-all coefficient of heat transfer.

heat tracing The technique of adding heat to a process or instrument measurement line by placing a steam line or electric heating element adjacent to the line.

heat treatment Controlled heating and cooling to alter the properties or structure of a metal, alloy or glass-like material.

heavier-than-heavy key The remaining components in the bottoms stream other than the heavy key.

Heaviside bridge A type of a-c bridge for making mutual inductance measurements when the inductance of the primary winding is already known.

heavy duty cable Generally a type of fiber optic or electrical cable designed to withstand unfriendly conditions, such as those encountered outdoors. Some varieties are armored to withstand hostile conditions.

heavy ends The fraction of a petroleum mixture having the highest boiling point.

heavy fraction The final products retrieved by distilling crude oil.

heavy key The component in multicomponent distillation that is removable in the bottoms stream and that has the highest vapor pressure of the components at the bottoms. If more reboiler heat is added, the heavy key component is the first component to be put in the overhead product.

heavy oil   A viscous fraction of petroleum or coal-tar oil having a high boiling point.

heavy water (deuterium)   A liquid compound D_2O whose chemical properties are similar to H_2O (light water), and occurs in a ratio of 1 part in 6000 in fresh water.

hectare   A metric unit of land measure equal to 10,000 m^2, or approximately 2.5 acres.

hectare-meter   A metric unit of volume, commonly used in irrigation work, which equals 10,000 m^3; it represents the amount of water needed to cover an area of one hectare to a depth of one metre.

heel block   A block or plate attached to a die that keeps the punch from deflecting too much.

height gage   A mechanical device, usually having a vernier or micrometer scale, used for measuring precise distances above a reference plane.

heliarc welding   See gas tungsten-arc welding.

Hellige turbidimeter   A variable-depth instrument for visually determining the cloudiness of a liquid caused by the presence of finely divided suspended matter.

help   In data processing, an on-screen information resource that a user can activate to answer questions.

henry   Metric unit for inductance.

Henry's law   A principle of physical chemistry that relates equilibrium partial pressure of a substance in the atmosphere above a liquid solution to the concentration of the same substance in the liquid; the ratio of concentration to equilibrium partial pressure equals the Henry's-law constant, which is a temperature-sensitive characteristic; Henry's law generally applies only at low liquid concentrations of a volatile component.

hermetically sealed device   A device which is sealed against the entrance of an external atmosphere and in which the seal is made by fusion, e.g., soldering, brazing, welding, or the fusion of glass to metal [S12.12].

hertz   Unit for frequency of a periodic phenomenon measured in cycles per second.

heterodyne   A combination of a.c. signals of two different frequencies, coupled so as to produce beats whose frequency is the sum or difference of the frequencies of the original signals.

heterodyne conversion transducer   A transducer in which output frequency is the sum and difference of the input frequency and the local oscillator frequency.

heterogeneous radiation   A beam of radiation containing rays of several different wavelengths or particles of different energies or different types.

heterojunction   A junction between semiconductors that differ in doping levels and also in their atomic compositions. Example: a junction between layers of GaAs and GaAIAs—a double heterojunction laser contains two such junctions, a single heterojunction laser contains only one.

heuristic   Pertaining to a method or problem solving in which solutions are discovered by evaluation of the progress made toward the final solution, such as a controlled trial and error method. An exploratory method of tackling a problem, or sequencing of investigation, experimentation, and trial solution in closed loops, gradually closing in on the solution. A heuristic approach usually implies or encourages further investigation, and makes use of intuitive decisions and inductive logic in the absence of direct proof known to the user. Thus, heuristic methods lead to solutions of problems or inventions through continuous analysis of results obtained thus far, permitting a determination of the next step. A stochastic method assumes a solution on the basis of intuitive conjecture or speculation and testing the solution against known evidence, observations, or measurements. The stochastic approach tends to omit intervening or intermediate steps toward a solution. Contrast with stochastic and algorithmic.

heuristic program   A program that monitors its performance with the objective of improved performance.

hex   A number of representation system of base 16. The hex number system is very useful in cases where computer words are composed of multiples of four bits (that is, 4-bit words, 8-bit words, 16-bit words, and so on).

hexadecimal   Number system using base 16 and the digit symbols from 0 to 9 and A to F. See hex.

hexadecimal notation   A numbering system using 0, 9, A, B, C, D, E, and F with 16 as a base.

hexagonal-head bolt   A standard threaded fastener with an integral hexagon-shaped head.

hexagonal nut   A hexagon-shaped fastener with internal threads used with a mating, externally threaded bolt, stud or machine screw.

hex code   A low-level code in which the machine code is represented by numbers using a base of 16.

187

Heydweiller bridge A type of a-c bridge circuit suitable for determining the mutual inductance between two interacting windings, both having unknown inductances.

hierarchical distributed control A hierarchy of computer systems in which one computer acting as supervisor controls several lower level computers.

hierarchy Specified rank or order of items, thus, a series of items classified by rank or order.

high-alloy steel An iron-carbon alloy containing at least 5% by weight of additional elements.

high brass A commercial wrought brass containing 65% copper and 35% zinc.

high-carbon steel A plain carbon steel with a carbon content of at least 0.6%.

high density digital recording (HDDR) The technique which combines the good features of NRZ and biphase codes, to achieve a packing density of up to 33,000 bits per inch per track in instrumentation tape recording.

high-frequency bias A sinusoidal signal that is mixed with the data signal during the magnetic tape direct recording process for the purpose of increasing the linearity and dynamic range of the recording medium. Bias frequency is usually three to four times the highest data frequency to be recorded.

high-frequency heating See electronic heating.

high gas pressure switch A switch to stop the burner if the gas pressure is too high.

high-heat value See calorific value.

high-level computing device (HLCD) A microprocessor-based device used to perform computer-like functions.

high-level data link control (HDLC) A type of data link protocol.

high-level human interference (HLHI) A device that allows a human to interact with the total distributed control system over the shared communications facility.

high-level language A programming language whose statements are translated into more than one machine-language instruction. Examples of high-level languages are BASIC, FORTRAN, COBOL, and TELEVENT.

high-level operator interface (HLOI) A type of HLHI designed for use by a process operator.

highlighting A term encompassing various attention-getting techniques, such as blinking, intensifying, underscoring, and color coding [S5.5].

high limiting control See control, high limiting.

high-low bias test Same as marginal check.

high order Pertaining to the weight or significance assigned to the digits of a number, e.g., in the number 123456, the highest order digit is 1, the lowest order digit is 6. One may refer to the three higher-order bits of a binary word, as another example.

highpass filter A filter which passes high frequencies above the cut-off frequency, with little attenuation.

high-pressure boiler A boiler furnishing steam at pressure in excess of 15 pounds per square inch or hot water at temperatures in excess of 250 °F or at pressures in excess of 160 pounds per square inch.

high resolution graphics A finely defined graphical display on a computer monitor screen.

high-strength alloy A metallic material having a strength considerably above that of most other alloys of the same type or classification.

high-temperature alloy A metallic material suitable for use at 500 °C (930 °F) or above. This classification includes iron-base, nickle-base and cobalt-base superalloys, and the refractory metals and their alloys, which retain enough strength at elevated temperature to be structurally useful and generally resist undergoing metallurgical changes that weaken or embrittle the material.

high-temperature hot-water boiler A water heating boiler operating at pressure exceeding 160 psig or temperatures exceeding 250 °F.

hinge A mechanical device that connects two members across a joint yet allows one member to pivot about an axis that runs along the joint.

hit In data processing, the isolation of a matching record.

hitch pin See cotter pin.

hit rate The number of successful matches in a computer search.

HLCD See high-level computing device.

HLHI See high-level human interference.

HLOI See high-level operator interface.

holding beam An electron beam for reactivating the charge on the surface of an electronic device.

hold time In any process cycle, an interval during which no changes are imposed on the system. Hold time is usually used to allow a chemical or metallurgical reaction to reach completion, or to allow a physical or chemical condition to stabilize before proceeding to the next step.

Hollerith card A punched card used in digital computing; it is named for Herman Hollerith, who developed a computing method using punched cards to compile the 1890 U.S. census.

Hollerith code A widely used system of encoding alphanumeric information onto cards, hence Hollerith cards are the same as punch cards. Such cards were first used in 1890 for the U.S. Census and were named after Herman Hollerith, their originator. See code, Hollerith.

holographic diffraction grating Diffraction grating in which the pattern of light diffracting lines was recorded holographically rather than mechanically ruled into the surface.

holographic optical elements Holograms which have been made to diffract light in the same pattern as other optical components. It is possible to produce (usually by computer synthesis) a hologram which mimics the function of the lens. In some applications, such holographic optical elements are less costly than conventional optics.

home In personal computers, a key which places the cursor at the upper left-hand position on the screen or the upper left-hand position of the entire file.

homogeneous radiation A beam of radiation containing rays whose wavelengths all fall within a narrow band of wavelengths, or containing particles of a single type having about the same energy.

homojunction A junction between semiconductors that differ in doping levels but not in atomic composition. Example: A junction between n-type and p-type GaAs.

homologous pair In optical spectroscopy, two lines so chosen that the ratio of their radiant powers has minimal change with variations in the input conditions.

hone To remove a small amount of material using fine-grit abrasive stones and thereby obtain an exceptionally smooth surface finish or very close dimensional tolerances.

Hookean behavior A condition in liquid expansion when the fractional change in volume is proportional to the hydrostatic stress, if under such stress it evidences ideal elastic behavior.

hook gage An instrument consisting of a pointed metal hook mounted on a micrometer slide that is used to measure the level of a liquid in an evaporation pan. The level with respect to a ref-erence height is determined when the point of the hook just breaks the liquid surface.

hopper, card A device that holds cards and makes them available to a card feed mechanism. Synonymous with input magazine [RP55.1].

hopper scale A weighing device consisting of a bulk container or hopper suspended on load cells or a lever system and used to batch-weigh bulk solids, often in connection with automated batch processing or with continuous receiving or shipping operations.

horizontal boiler A water tube boiler in which the main bank of tubes are straight and on a slope of 5 to 15 degrees from the horizontal.

horizontal return-tubular boiler A fire-tube boiler consisting of a cylindrical shell, with tubes inside the shell attached to both end closures. The products of combustion pass under the bottom half of the shell and return through the tubes.

horn A device for directing and intensifying sound waves that consists of a tube whose cross section increases from one end to the other.

horn antenna The flared end of a radar waveguide, whose dimensions are chosen to give efficient radiation of electromagnetic energy into the surrounding environment.

horn (antenna) A moderate-gain wide-beamwidth antenna, generally limited to use in manually steerable applications.

horn relay contact See auxiliary output.

host (computer) The primary computer in a multielement system; the system that issues commands, has access to the most important data, and is the most versatile processing element in a system. Compare with target computer or object computer.

hot dip galvanizing A process for rust-proofing iron and steel products by the application of a coating of metallic zinc.

hot dipping A process for coating parts by briefly immersing them in a molten metal bath, then withdrawing them and allowing the metal to solidify and cool.

hot-wire instrument A measuring device that depends on the heating reaction of a wire carrying a current for its operation.

housekeeping Administrative or overhead operations or functions that are necessary in order to maintain control of a situation; for example, for a computer program, housekeeping involves the setting up of constants and variables to be used in the program; synonymous with "red tape."

housing A protective enclosure or case.

HPLC See high-performance liquid chromatography.

hue The characteristic of color that determines whether the color is basically yellow, blue, red, etc.

Huggenberger tensometer A magnifying extensometer that employs a compound lever system to intensify changes taking place in a 10 to 20 mm gage length about 1200 times.

hum An undesirable by-product in an alternating current power supply.

human-factors engineering A branch of engineering in which the capabilities and limitations of human beings are integrated into design models to obtain enhanced overall performance of a system that uses both humans and machines.

humidification Artificially increasing the moisture content of a gas.

humidistat An instrument for measuring and controlling relative humidity.

humidity, absolute The moisture content of air on a mass or volumetric basis.

humidity element The part of a hygrometer that senses the amount of water vapor in the atmosphere.

humidity, relative The moisture content of air relative to the maximum that the air can contain at the same pressure and temperature.

humidity test A corrosion test for comparing relative resistance of specimens to a high humidity environment at constant temperature.

hunting 1. A continuing cyclic motion caused by friction, with the positioner attempting to find the set position [RP75.18]. 2. An undesirable oscillation of appreciable magnitude, prolonged after external stimuli disappear. Note: In a linear system, hunting is evidence of operation at or near the stability limit; non-linearities may cause hunting of well-defined amplitude and frequency [S51.1]. In data processing, the system usually contains a standard, a method of determining deviation from this standard and a method of influencing the system, such that the difference between standard and the state of the system is brought to zero. In automatic control, hunting is generally caused by the gain or reset of the controller being set too high. See dither.

hybrid computer 1. A computer for data processing using both analog representation and discrete representation of data. 2. A computing system using an analog computer and a digital computer working together.

hybrid T Series T and shunt T junctions located at the same point in a waveguide and designed to restrict energy flow to specified channels.

hydraulic Referring to any device, operation or effect that uses pressure or flow of oil, water or any other liquid of low viscosity.

hydraulic circuit A fluid-flow circuit that operates somewhat like an electric circuit.

hydraulic engineering A branch of civil engineering that deals with the design and construction of such structures as dams and other flood-control devices, sewers and sewage-disposal plants, water-driven electric power stations, and water treatment and distribution systems.

hydraulic fluid A light oil or other low-viscosity liquid used in a hydraulic circuit.

hydraulic gage A gage designed for service at extremely high pressure.

hydraulic valve A fluid powered device which converts the energy of an incompressible fluid into motion [S75.05]. See also actuator, hydraulic.

hydrocarbon A chemical compound of hydrogen and carbon.

hydrocracker A chemical reactor in which large hydrocarbon molecules are fractured in the presence of hydrogen.

hydroelectric plant An electric power generating station where the power is produced by generators driven by hydraulic turbines.

hydrogen damage Any of several forms of metal failure caused by dissolved hydrogen, including blistering, internal void formation, and hydrogen-induced delayed cracking.

hydrokineter A device for recirculating or causing flow of water by the use of a jet of steam or water at higher pressure than the water caused to flow.

hydrometer An instrument for directly indicating density or specific gravity of a liquid.

hydrophone A transducer that reacts to water-borne sound waves.

hydropneumatic Referring to a device operated by both liquid and gas power.

hydroscopic Refers to any material that easily absorbs and retains moisture.

hydrostatic head The pressure created by a height of liquid above a given point.

hydrostatic-head gage A pressure gage that is unique from others in the graduation of scale;

usually in feet.

hydrostatic test Determining the burst resistance or leak tightness of a fluid component or system by imposing internal pressure.

hygrometer An instrument for directly indicating humidity.

hygrometry Any process for determining the amount of moisture present in air or another gas.

hygroscopic Having a tendency to absorb water [S71.04].

hygrothermograph An instrument that records both temperature and humidity on the same chart.

hypergolic A term related to spontaneous ignition upon contact.

hypermodel A behavioral model of the analog-digital interface in a mixed-mode simulator.

hypsometer An instrument that determines elevation above a reference plane (such as sea level) by measuring the boiling point of a liquid and from that measurement finding atmospheric pressure.

hysteresimeter A device for measuring a lagging effect related to physical change, such as the relationship between magnetizing force and magnetic induction.

hysteresis 1. That property of an element evidenced by the dependence of the value of the output, for a given excursion of the input, upon the history of prior excursions and the direction of the current traverse [S51.1] [S67.04]. Note: (a) It is usually determined by subtracting the value of dead band from the maximum measured separation between upscale going and downscale going indications of the measured variable (during a full range traverse, unless otherwise specified) after transients have decayed. This measurement is sometimes called hysteresis error or hysteretic error. (b) Some reversal of output may be expected for any small reversal of input; this distinguishes hysteresis from dead band. See test procedure [S51.1]. The maximum difference in output value for any single input value during a calibration cycle, excluding errors due to dead band [S75.05]. 2. The maximum difference in output, at any measurand value within the specified range, when the value is approached first with increasing and then with decreasing measurand. Hysteresis is expressed in percent of full scale output, during any one calibration cycle. Friction error is included with hysteresis unless dithering is specified [S37.1]. 3. A phenomenon demonstrated by materials which make their behavior a function of the history of the environment to which they have been subjected. 4. The tendency of an instrument to give a different output for a given input, depending on whether the input resulted from an increase or decrease from the previous value. 5. The lagging in the response of a unit of a system behind an increase or a decrease in the strength of signal.

Hz See hertz.

I

I See current; also moment (definition 2.)

IAE See integral absolute error.

IC See integrated circuit.

I&C Technician Instrumentation and Control System Technician.

icon In data processing, a picture that represents a particular command that is used with a mouse.

I controller See controller, integral (reset) (I).

ICP See integrated circuit piezoelectric.

ideal elastic behavior A material characteristic, under given conditions, when the strain is a unique straight-line function of stress and is independent of previous stress history.

ideal gas A hypothetical gas characterized by its obeying precisely the equation for a perfect gas, PV=nRT.

idealized system See system, idealized.

ideal transducer A hypothetical passive transducer which produces the maximum possible output for a given input.

ideal value See value, ideal.

identification The sequence of letters or digits, or both, used to designate an individual instrument or loop [S5.1].

identification plate See data plate.

identifier A symbol used in data processing whose purpose is to identify, indicate or name a body of data.

idle characters Control characters interchanged by a synchronized transmitter and receiver to maintain synchronization during non-data periods.

idlers (idler rollers) Freely-turning cylinders mounted on a frame to support the conveyor belt. For a flat belt, the idlers may consist of one or more horizontal cylinders transverse to the direction of belt travel. For a troughed belt, the idlers may consist of one or more horizontal cylinders with one or more additional cylinders at an angle that lifts the sides of the belt to form a trough [RP74.01].

idler spacing The center-to-center distance between consecutive idler roller, measured parallel to the belt [RP74.01].

idle time 1. That part of available time during which computer hardware is not being used. Contrast with operating time. 2. That part of uptime in which no job can run because all jobs are halted or waiting for some external action such as I/O data transfer.

ID synchronization A count contained in one word of a telemetry frame to indicate which subframe is being sampled at any given time.

ID synchronizer A method of PCM telemetry subframe recognition in which a specific word in the format activates a counter that identifies the number of the subframe word being received.

ID (Time Code) A three-numeral identification that can be inserted into time code manually in the place of the "day of the year" information.

IEC See International Electrotechnical Commission.

IEE 488 A parallel transmission standard for connecting instruments to a computer. An industry standard byte serial, bit parallel system handling 8-bit words.

IF See intermediate frequency.

IF amplifier An intermediate-frequency stage in a typical superhetrodyne radio receiver.

if and only if (IFF) A conditional statement implying that an action is to be taken or that a result is true if, and only if, stated prerequisite conditions are satisfied.

IFIP See International Federation for Information Processing.

I format In FORTRAN Iw indicates that w characters are to be converted as a decimal integer, e.g., 17 yields –24680 as input, +24680 internally, and –24680 as output.

if-then See inclusion.

igniter A device for initiating an explosion or combustion in a fuel-air mixture.

ignition The initiation of combustion.

ignition capable Equipment or wiring which in its normal operating condition releases sufficient electrical or thermal energy to cause ignition of a specific hazardous atmosphere or hazardous dust layer [S12.11].

ignition capable equipment and wiring Equipment and wiring which in its normal operating condition releases sufficient electrical or thermal energy to cause ignition of a specific hazardous atmosphere, under normal operating conditions.

ignition lag The time interval between spark discharge and fuel ignition in an internal combustion engine. Also known as ignition delay.

ignition period See trial for ignition.

ignition system That portion of the electrical subsystems of an internal combustion engine that produces a spark to ignite the fuel.

ignition temperature Lowest temperature of a fuel at which combustion becomes self-sustaining.

ignitor A flame or high energy spark which is utilized to ignite the fuel at the main burner.

ignitor intermittent An electric-ignited pilot which is automatically lighted each time there is a call for heat. It burns during the entire period that the main burner is firing.

ignitor interrupted An electric-ignited pilot which is automatically lighted each time there is a call for heat. The pilot fuel is cut off automatically at the end of the trial-for-ignition period of the main burner.

ILI (In Limits) algorithm See compressor.

illuminance Luminous flux per unit area over a uniformly illuminated surface.

illuminants Light oil or coal compounds that readily burn with a luminous flame such as ethylene, propylene and benzene.

illuminated dial A transparent, semitransparent or nontransparent circular scale that is artificially illuminated.

image converter camera A camera which converts images from one wavelength region to another, typically from the infrared to the visible.

image digitizer A device which measures light intensity at each point in an image and generates a corresponding digital signal which indicates that intensity. It converts an analog image to a digital data set.

image impedances Of a transducer, the impedances that will simultaneously produce equal impedances in both directions at each of its inputs and outputs.

image intensifier A viewing system which functions as a light amplifier, taking a faint image and amplifying it so that it can be viewed more easily.

image inverter A fused fiber optic bundle which is permanently twisted during manufacture to turn the image it transmits upside down. The same can be done with conventional optics, but a fiber optic image inverter can do it in a distance of less than an inch.

image orthicon A camera tube whose output is generated using a low-velocity electron beam to scan the reverse side of a storage target containing an image produced by focusing the electron image from a photoemitting surface on it.

IMC See Institute of Measurement and Control.

immediate-access storage A device, usually consisting of an array of storage elements, in which stored information can be read in one microsecond or less.

immediate address Incorporating an operand, instead of merely the address of an operand, in the address portion of a digital computer instruction.

immediately dangerous to life and health (IDLH) Represents the maximum level from which one could escape within 30 minutes without escape-impairing symptoms or any irreversible effects [S12.15].

immediate mode In data processing, the ability to interrupt a program sequence to perform another function.

immersion length Of a thermometer, the distance along the thermometer body from the boundary of the medium whose temperature is being determined to the free end of the well, bulb or element, if unprotected.

immunity An inherent or induced electrochemical condition that enables a metal to resist attack by a corrosive solution.

impact extrusion See cold extrusion.

impact idler A belt idler incorporating resilient roll coverings to absorb large amounts of shock at the loading point [RP74.01].

impact pressure The pressure a moving stream of fluid produces against a surface which brings part of the moving stream abruptly to rest; it is approximately equal to the stagnation pressure for subsonic flow in the fluid medium.

impact strength A material property that indicates its ability to resist breaking under extremely rapid loading, usually expressed as energy absorbed during fracture.

impact temperature The temperature of a gas, after impact with a solid body which converts some of the kinetic energy of the gas to heat and thus raises the gas temperature above ambient.

impact tube A small diameter tube, immersed in a fluid, and oriented so that the fluid stream impinges normally on its open end.

impedance The complex ratio of a forcelike parameter to a related velocity-like parameter—for instance, force to velocity, pressure volume velocity, electric voltage to current, temperature to heat flow, or electric field strength to magnetic field strength.

impedance bridge A four-arm bridge circuit in which one or more of the arms have reactive components instead of purely resistive components; an impedance bridge must be excited by an a-c signal to yield complete analysis of the unknown bridge element.

impedance, input 1. Impedance presented by a device to the source [S51.1]. 2. The impedance presented by a device or system output element to the input [S26].

impedance, load Impedance presented to the output of a device by the load [S51.1].

impedance, output 1. Impedance presented to the input of a device by the source [S51.1]. 2. The internal impedance of an output element which limits that element's ability to deliver power [S26].

impedance, source Impedance presented to the input of a device by the source [S51.1].

impeller 1. As applied to pulverized coal burners, a round metal device located at the discharge of the coal nozzle in circular type burners, to deflect the fuel and primary air into the secondary air stream. As applied to oil burners, same as diffuser. 2. The driven portion of a centrifugal pump or blower.

impingement 1. The striking of moving matter, such as the flow of steam, water, gas or solids, against similar or other matter. 2. A method of removing entrained liquid droplets from a gas stream by allowing the stream to collide with a baffle plate.

impingement attack A form of accelerated corrosion in which a moving corrosive liquid erodes a protective surface layer, thus exposing the underlying metal to renewed attack.

implication See inclusion.

impregnated bit A diamond cutting tool made of fragmented bort or screened whole diamonds in a sintered powder-metal matrix.

improvement maintenance Efforts to reduce or eliminate the need for maintenance. Reliability engineering efforts should emphasize elimination of failures that require maintenance. Includes modification, retrofit, redesign, or change-order.

impulse excitation A method of producing oscillations in which the duration of stimulus is relatively short in relation to the duration of oscillation.

impulse line 1. Piping or tubing connecting the process to the sensor [S67.06]. 2. The conduit that transfers the pressure signal from the process to the measuring instrument.

impulse-type telemetering Employing intermittent electrical impulses to transmit instrument readings to remote locations.

inaccessible area 1. An area for which the radiation level, as defined by the Architect Engineer, precludes personnel entry during power operations and other operational situations. These areas are typically indicated by "zones," which depict accessibility based on various plant evolutions [S67.10]. 2. An area is considered inaccessible if entry may be dangerous without special controls to enter.

inaccuracy See error.

incandescence Spontaneous radiation of light energy from a hot object.

inches water gage (″w.g.) Usual term for expressing a measurement of relatively low pressures or differentials by means of a U-tube. One inch w.g. equals 5.2 lb per sq ft or 0.036 lb per sq in.

incident wave A wave in a given medium that impinges on a discontinuity or a medium of different propagation characteristics.

in-circuit emulation (ICE) A development aid for testing the software in computer hardware. It involves an umbilical link between a development system and the target hardware being plugged into the microprocessor socket.

inclination error See error, inclination.

inclined-tube manometer A glass-tube manometer having one leg inclined from the vertical to give more precise readings.

inclinometer 1. An instrument for determining the angle of the earth's magnetic field vector from the horizontal. 2. A device for finding the direction of the earth's magnetic field with respect to the horizon. 3. An instrument on a ship which indicates the angular deviation of the ships attitude to the true vertical.

inclusion A logic operator having the property that if P is a statement and Q is a statement, then P inclusion Q is false if P is true and Q is false, true if P is false, and true if both statements are true. P inclusion Q is often represented by P>Q. Synonymous with if-then and implication.

inclusive OR See OR.

incoherent fiber optics A bundle of fibers in which the fibers are randomly arranged at each end. The pattern may be truly random to achieve uniform illumination, or the manufacturer may simply not bother to align individual fibers. In either case, the fiber bundle cannot transmit an image along its length.

incomplete combustion The partial oxidation of the combustible constituents of a fuel.

incompressible Liquids are referred to as being incompressible since their change in volume due to pressure is negligible.

incompressible flow Fluid flow under conditions of constant density.

Inconel A series of International Nickel Co. high-nickel, chromium and iron alloys characterized by inertness to certain corrosive fluids.

increased safety A type of protection by which measures are applied so as to prevent with a higher degree of security the possibility of excessive temperatures and of the occurrence of sparks in the interior and on the external parts of electrical apparatus which does not produce them in normal service and which is intended for use in hazardous locations defined by the IEC as Zone 1.

increaser A pipe fitting identical to a reducer except specifically referred to for enlargements in the direction of flow.

increment The specific amount in which a variable is changed.

incremental See incremental representation.

incremental backup A computer routine that copies only those files that have not yet been backed up.

incremental compiler Computer software that compiles programs as they are entered into a computer rather than compiling a program upon completion.

incremental cost The cost of the next increment of output from a process.

incremental encoder An electronic or electromechanical device which produces a coded digital output based on the amount of movement from an arbitrary starting position; the output for any given position with respect to a fixed point of reference is not unique.

incremental feedback In numerical control, assignment of a value for any given position of machine slide or actuating member based on its last previous stationary position.

incremental plotter A discrete X-Y plotter.

incremental representation A method of representing a variable in which changes in the value of the variables are represented, rather than the values themselves.

independent conformity See conformity, independent.

independent linearity See linearity, independent.

independent variable 1. A process or control-system parameter that can change only due to external stimulus. 2. A parameter whose variations, intentional or unintentional, induce changes in other parameters according to predetermined relationships.

index 1. An ordered reference list of the contents of a computer file or document, together with keys or reference notations for identification or location of those contents. 2. To prepare a list as in 1. 3. A symbol or number used to identify a particular quantity in an array of similar quantities, for example, the terms of an array represented by X(1), X(2),..., 100 respectively. 4. Pertaining to an index register. 5. To move a machine part to a predetermined position, or by a predetermined amount, on a quantized scale.

index address modification (indexing) See address modification.

indexed address 1. An address in a computer instruction that indicates a location where the address of the reference operand is to be found. In some computers, the machine address indicated can itself be indirect. Such multiple levels of addressing are terminated either by prior control or by a termination symbol. Synonymous with second-level address. 2. An address that is to be modified or has been modified by an index register or similar device. Synonymous with variable address.

indexed addressing A method of addressing computer data whereby the address is obtained by adding the instruction operand to the address in the index register.

indexed sequential files Collection of related computer records stored on discs. The records are arranged in the same sequence as the key number and an index or table is used to define the actual location of these records on the disc.

index graduations The heaviest or longest division marks on a graduated scale, opposite the scale numerals.

indexing A technique of address modification often implemented by means of index registers.

index matching fluid A liquid with refractive index that matches that of the core or cladding of an optical fiber. It is used in coupling light into or out of optical fibers, and can help in suppressing reflections at glass surfaces.

index register A register which contains a quantity which may be used to modify addresses.

index-word A computer storage position or register, the contents of which may be used to modify automatically the effective address of any given instruction.

indicating extensions, direct A device that provides flow rate indication by means of viewing the position of the extension of the metering float within a glass extension tube [RP16.4].

indicating extensions, magnetic A device that provides flow rate indication by means of a magnetic coupling between the extension of the metering float and an external indicator follower surrounding the extension tube [RP16.4].

indicating gage Any measuring device whose output can be read visually but is not automatically transcribed on a chart or other permanent record.

indicating instrument See instrument, indicating.

indicating measuring equipment Equipment which indicates the value of the measured quantity [S82.03].

indicating scale On a recording instrument, a scale that allows a recorded quantity to be simultaneously observed.

indication In nondestructive testing, any visible sign or instrument reading that must be interpreted to determine whether or not a flaw exists.

indicator 1. An instrument which graphically shows a value of a variable. 2. The pointer on a dial or scale that provides a visual readout of a measurement. 3. An instrument for diagramming pressure-volume changes during the working cycle of a positive-displacement compressor, engine or pump.

indicator card A chart for recording an indicator diagram.

indicator diagram A graphic representation of work done by or on the working fluid in a positive-displacement device such as a reciprocating engine.

indicator travel The length of the path described by the indicating means or the tip of the pointer in moving from one end of the scale to the other. NOTE: 1: The path may be an arc or a straight line. 2: In the case of knife-edge pointers and others extending beyond the scale division marks, the pointer shall be considered as ending at the outer end of the shortest scale division marks [S51.1].

indicator tube An electron-beam tube in which useful information is conveyed by variations in beam cross section at a luminescent target.

indirect-acting recording instrument An instrument in which the output level of the primary detector is raised through intermediate mechanical, electric, electronic or photoelectric means to actuate the writing or marking device.

indirect address An address that specifies a computer storage location that contains either a direct address or another indirect address. Synonymous with multi-level address.

indirect addressing A method of addressing computer data in which the operand of the instruction is a location address which contains the address of the data.

indirect commands In data processing, commands to the system from previously recorded inputs, rather than from the operator terminal; the operator can call a sequence of indirect commands by file name.

indirect file In data processing, a file that contains commands that are processed sequentially, yet could have been entered interactively at a terminal.

indirectly controlled system See system, indirectly controlled.

indirectly controlled variable See variable, indirectly controlled.

indirectly heated cathode A cathode in a thermionic tube that is heated by an independent heating element.

indirect test A test that measures a quantity other than response time. The actual response time is determined using this quantity and previous measurements of this quantity which have a known relationship to the actual response times [S67.06].

induced draft Airflow through a device such as a firebox or drying unit which is produced by placing a fan or suction jets in the exit duct.

induced draft fan A fan exhausting hot gases from the heat absorbing equipment.

inductance 1. In an electrical circuit, the property that tends to oppose changes in current magnitude or direction. 2. In electromagnetic devices, generating electromotive force in a conductor by means of relative motion between the conductor and a magnetic field such that the conductor cuts magnetic lines of force.

inductance-type pressure transducer Any of several designs of pressure sensor where motion of the primary sensor element, such as a bourdon tube or diaphragm, is detected and measured by a variable-inductance element and measuring circuit.

induction heating Raising the temperature of an electrically conductive material by electromagnetically inducing eddy currents in the material.

induction instrument A type of meter whose indicated output is determined by the reaction between magnetic flux in fixed windings and flux in a moving coil where the two fluxes are induced by electric currents from different sources.

induction motor meter A type of meter resembling an induction motor, in which the rotor moves in direct relation to the reaction force between a magnetic field and currents induced in the rotor.

inductive Converting a change of measurand into a change of the self-inductance of a single coil [S37.1].

inductive bridge position transducer A device for measuring linear position by means of induction between a fixed member slightly longer than the limits of motion and a movable member approximately half as long; position is determined by selecting appropriate taps from the longer member that are connected in a successive decade with external inductors to form a bridge circuit, and relating the configuration that balances the bridge with actual position of the movable member; the chief advantage is the relatively high output voltage developed for a relatively small change in position.

inductive coupling Using common or mutual inductance to cause signals in one circuit to vary in accordance with signals in another.

inductive plate position transducer A device for measuring rotary position by means of induction between a stationary and rotary plate, each having an etched winding projected onto a nonconductive surface, or for measuring linear motion by means of induction between a stator plate and a sliding member, each also having etched windings; advantages include eliminating wear and backlash as well as providing good resolution, often within 0.001 in. or less.

inductor A wire coil that will store energy in the form of a magnetic field.

industrial computer A computer used on-line in various areas of manufacturing including process industries (chemical, petroleum, etc.), numerical control, production lines, etc. See process computer and numerical control.

industrial computer language A computer language for industrial computers. A language used for programming computer control applications and system development, e.g., assembly language, FORTRAN, RTL, PROSPRO, BICEPS, and AUTRAN.

industrial controls A collective term for control instrumentation used in industry.

industrial engineering A branch of engineering that deals with the design and operation of integrated systems of personnel, equipment, materials and facilities.

Industrial Technology Institute (ITI) A nonprofit organization founded by the University of Michigan dedicated to computer integrated manufacturing. ITI offers MAP conformance testing and certification.

inelastic collision A collision between two or more bodies in which there is a net change in internal energy of at least one of the participating bodies and a net change in the sum of their kinetic energies.

inert gaseous constituents Incombustible gases such as nitrogen which may be present in a fuel.

inertia Inherent resistance of a body to changes in its state of motion.

inertia-type timer Any of several types of relay devices that incorporate extra weights or flywheels to achieve brief time delay in normal relay action by providing additional inertia to be overcome; delays are usually on the order of 80 to 120 milliseconds.

infiltration Casing molten metal to be drawn into void spaces in a powder-metal compact, foamed-metal shape, or fiber-metal layup.

infinite loop In data processing, a routine that can be ended only by terminating the program.

influence The change in an instrument's indicated value caused solely by a difference in value of a specified variable or condition from its reference value or condition when all other variables are held constant.

information processing The organization and manipulation of data usually by a computer. See data processing.

information theory The mathematical theory concerned with information rate, channels, channel width, noise, and other factors that affect information transmission; initially developed for electrical communications, it is now applied to business systems and other phenomena that deal with information units and flow of information in networks.

infrared Any electromagnetic wave whose wavelength is 0.78 to 300 μm.

infrared absorption moisture detector An instrument for determining moisture content of a material such as sheet paper; moisture content can be read directly by determining the ratio of two beam intensities, one at a wavelength within the resonant-absorption band for water and the other at a wavelength just outside the band.

infrared imaging device Any device that receives infrared rays from an object and displays a visible image of the object.

infrared spectroscopy A technique for determining the molecular species present in a material, and measuring their concentrations, by detecting the characteristic wavelengths at which the material absorbs infrared energy and measuring the relative drop in intensity associated with each absorption band.

infrasonic frequency A sound-wave frequency lower than the audio-frequency range.

infrequently Referring to the frequency of a hazardous event connotes "not normally occurring" and "not be likely during normal operations of the facility." [S12.11]

inherent damping Using mechanical hysteresis of materials such as cork or rubber to reduce vibrational amplitude.

inherent error The error in quantities that serve as initial conditions at the beginning of a step in a step-by-step set of operations. Thus, the error carried over from the previous operation from whatever source or cause.

inherent flow characteristic The relationship between the flow rate through a valve and the travel of the closure member as the closure member is moved from the closed position to rated travel with constant pressure drop across the valve [S75.11].

inherent rangeability The ratio of the largest flow coefficient (Cv) to the smallest flow coefficient (Cv) within which the deviation from the specified inherent flow characteristic does not exceed the limits stated [S75.11].

inherent regulation See self-regulation.

inhibitor A substance which selectively retards a chemical action. An example in boiler work is the use of an inhibitor, when using acid to remove scale, to prevent the acid from attacking the boiler metals.

in-house maintenance 1. Maintenance performed by plant maintenance personnel. 2. Not contract maintenance.

initialize In data processing, to send a rest command to clear all previous or extraneous information, as when starting a new operating sequence.

initial set The start of a hardening reaction following water addition to a powdery material such as plaster or portland cement.

injection laser diode A semiconductor device in which lasing takes place within the P-N junction. Light is emitted from the diode edge.

injector Any nozzle or nozzle-like device through which a fluid is forced into a chamber or passage.

ink A liquid or semisolid material consisting of a pigment or dye and a carrier, and used to produce a design or mark on a material such as paper or cloth, after which the carrier evaporates leaving behind a colored residue of pigment or dye.

ink-jet printer A printer that forms characters by shooting tiny dots of ink onto paper. See also dot matrix printer.

ink-vapor recording A type of electromechanical recording in which the trace is produced by depositing vaporized particles of ink directly on the chart paper.

inlet 1. The body end opening through which

fluid enters the valve [S75.05]. 2. A passage or opening where fluid enters a conduit or chamber.

inlet box An enclosure at or near the entrance to a chamber or duct system for attaching a fan to the system.

inlet valve A valve for admitting the working fluid to the cylinder of a positive-displacement device such as a reciprocating pump or engine.

in line 1. Centered on an axis. 2. Having several features, components or units aligned with each other. 3. In a motor-driven device, having the motor shaft parallel to the device's driven shaft and approximately centered on each other.

in-line valve A valve having a piston actuated closure member shaped like a globe valve plug which moves to seat axially in the direction of the flow path. In-line valves are normally operated by a fluid energy source but may be operated mechanically [S75.05].

in-plant system 1. A system whose parts, including remote terminals, are all situated in one building or localized area. 2. The term is also used for communication systems spanning several buildings and sometimes covering a large distance, but in which no common carrier facilities are used.

input 1. The data to be processed [RP55.1]. 2. The state or sequence of states occurring on a specified input channel [RP55.1]. 3. The device or collective set of devices used for bringing data into another device [RP55.1]. 4. A channel for impressing a state on a device or logic element [RP55.1]. 5. The process of transferring data from an external storage to an internal storage [RP55.1]. 6. Signals taken in by an input interface as indicators of the condition of the process being controlled. 7. Data keyed into a computer or computer peripherals. See excitation or measurand.

input, analog Information or data in analog form transferred or to be transferred from an external device into the computer system [RP55.1].

input area An area of computer storage reserved for input. Synonymous with input block.

input block See input area.

input channel A channel for impressing a state on a device or logic element. See channel, input.

input, contact A digital input generated by operating an external contact [RP55.1].

input counter See counter, input.

input device In data processing, the device or collective set of devices used for conveying data

into another device.

input, digital Information or data in digital form transferred or to be transferred from an external device into the computer system [RP55.1].

input impedance The impedance (presented to the excitation source) measured across the excitation terminals of a transducer. Unless otherwise specified, input impedance is measured at room conditions, with no measurand applied, and with the output terminals open-circuited [S37.1]. See impedance, input.

input interface Any device that connects computer hardware or other equipment for the input of data.

input-output (I/O) A general term for the equipment used to communicate with a computer and the data involved in the communications. Synonymous with I/O.

input-output control system (IOCS) A set of flexible routines that supervises the input and output operations of a computer at the detailed machine-language level.

input-output, data processing 1. A general term for the equipment used to communicate with a computer [RP55.1]. 2. The data involved in such communication [RP55.1]. 3. The media carrying the data for input-output operations on data [RP55.1].

input-output limited Pertaining to a computer system or condition in which the time for input and output operation exceeds other operations.

input-output (I/O) software That portion of the operational software which organizes efficient flow of data and messages to and from external equipment.

input-output (I/O) statement A statement that controls the transmission of information between the computer and the input/output units.

input ports In computer hardware, terminals for connection in external devices which input data to the computer.

input resistance See resistance, input.

input signal A signal applied to a device, element, or system. See signal, input.

input state The state occurring on a specified computer input channel.

input strobe (INSTRB) A signal that enters set-up data into registers.

input work queue A list of summary information of job-control statements maintained by the job scheduler, from which it selects the jobs and job steps to be processed.

inquiry A technique whereby the interrogation of the contents of a computer's storage may be initiated at a keyboard.

insensitive time See dead time.

insert 1. Any design feature of a cast or molded component that is made separately and placed in the mold cavity prior to the casting or molding step. 2. A removable part of a die, mold or cutting tool.

insertion gain The ratio of the power delivered to the portion of a transmission system following a transducer to the power delivered to the same portion without the transducer in place.

insertion point Usually indicated by a computer cursor, the place where characters will appear when an operator starts typing.

inside caliper A caliper having outward-turned feet on each leg for measuring inside dimensions.

inside diameter The maximum dimension across a cylindrical or spherical cavity. Ideally, this is a line passing through the exact center of the cavity and perpendicular to the cavity's inner surface.

inside gage 1. A fixed-dimension device for checking inside diameters. 2. The inside diameter of a bit, measured between opposing cutting points.

inside micrometer A micrometer caliper designed for measuring inside diameters and similar inside dimensions between opposing surfaces.

inspection A deliberate critical examination to determine whether or not an item meets established standards. Inspection may involve measuring dimensions, observing visible characteristics, or determining inherent properties of an object, but usually does not involve determining operating characteristics. The last is more properly termed testing.

inspection door A small door in the outer enclosure so that certain parts of the interior of the apparatus may be observed.

instability See stability.

installation Putting equipment or software in place prior to commencing operation.

instantaneous frequency In an angle-modulated wave, the derivative of the angle with respect to time.

instantaneous sampling Taking a series of readings of the instantaneous values of one or more wave parameters.

Institute of Measurement and Control A British professional organization.

instruction In data processing, a statement that specifies an operation and the values or locations of its operands. In that context, the term instruction is preferable to the terms command or order, which are sometimes used synonymously. Command should be reserved for electronic signals, and order should be reserved for sequence, interpolation and related usage.

instruction address An instruction's computer memory address. An asterisk is frequently used to designate this address.

instruction area 1. A part of computer storage allocated to receive and store the group of instructions to be executed. 2. The storage locations used to store the program.

instruction buffer An eight-bit byte buffer in the computer processor that is used to contain bytes of the instruction currently being decoded and to prefetch instructions in the instruction system.

instruction code See operation code.

instruction counter A counter that indicates the location of the next computer instruction to be interpreted.

instruction format The bits or characters of a computer instruction allocated to specific functions.

instruction register In data processing, a storage register which contains the address of the instruction.

instruction repertory 1. The set of instructions which a computing or data processing system is capable of performing. 2. The set of instructions which an automatic coding system assembles.

instruction set In computer software, the particular set of instructions that are implemented on a microcomputer.

instruction time The portion of an instruction cycle during which the computer control unit is analyzing the instruction and setting up to perform the indicated operation.

instrument 1. A device used directly or indirectly to measure and/or control a variable. The term includes primary elements, final control elements, computing devices, and electrical devices such as annunciators, switches, and pushbuttons. The term does not apply to parts (e.g., a receiver bellows or a resistor) that are internal components of an instrument [S5.1]. 2. A device that performs some analysis of the sample fluid and for which a sample line is required and con-

201

nected. Also referred to as "analyzer" or "monitor" [S67.10]. 3. A device for measuring the value of an observable attribute; the device may merely indicate the observed value, or it may also record or control the value. 4. Measuring, recording, controlling, and similar apparatus requiring the use of small to moderate amounts of electrical energy in normal operation.

instrument air (IA) Clean dry air that meets ISA standard RP7.7. See plant air.

instrumental analysis Any analytical procedure that uses an instrument to measure a value, detect the presence or absence of an attribute, or signal a change or end point in a process.

instrumentation 1. Any system of instruments and associated devices used for detecting, signaling, observing, measuring, controlling or communicating attributes of a physical object or process [S51.1]. 2. A collection of instruments or their application for the purpose of observation, measurement, control, or any combination of these [S5.1].

instrumentation amplifiers High precision amplifiers with high noise rejection capabilities.

instrumentation tape Analog magnetic tape, ungapped, for continuous data (as PCM or FM telemetry).

instrument channel An arrangement of components and modules as required to generate a single protective action signal when required by a generating station condition. A channel loses its identity where single protective action signals are combined [S67.04].

instrument channel, response time The time interval from the time when the monitored variable exceeds its trip setpoint until the time when a protective action is initiated [S67.06].

instrument computing A device in which the output is related to the input or inputs by a mathematical function such as addition, averaging, division, integration, lead/lag, signal limiting, squaring, square root extraction, subtraction, etc. [S51.1]

instrument correction A quantity added to, subtracted from, or multiplied into an instrument reading to compensate for inherent inaccuracy or degradation of instrument function.

Instrument Engineer 1. Applies standard engineering standards and practices to the specification, sizing, and functional design of instrumentation hardware or control systems. Involves a clear understanding of the manufacturing or

scientific process to be controlled. Serves as the key person on the instrumentation design and operation team, often supervising and reviewing the team's efforts. 2. Under supervision, participates in the design and planning of control and instrument systems as required by the project assignment, including: a) collecting background information. b) preparing drawings and calculations. c) designing or modifying systems. d) assisting in selection and procurement of equipment. e) ensuring compliance with applicable standards and codes. f) completing assigned tasks on schedule. g) assisting technicians and designers as needed. h) possible specializing in specific engineering discipline.

Instrument Engineering Technician Helps engineers in the design of control and instrumentation systems by providing semi-professional technical assistance, including: a) collecting background information. b) performing calculations. c) transmitting information to project team members. d) preparing design specification. e) checking design documents to ensure compliance with applicable standards and codes. f) preparing diagrams. g) preparing requisitions. h) executing necessary test, collecting data and making analyses. i) maintaining equipment. j) performing miscellaneous administrative work. k) assisting in testing, field start-ups and training.

Instrument Field Engineer Provides field engineering services at installation and start-up sites, ensuring specification fulfillment and operating ability. Also defines any potential problems that may arise, including: a) managing installation and initial servicing. b) maintaining own technical abilities and awareness of new methodologies. c) checking out operation of panels and instruments. d) providing instruction in the maintenance and repair of equipment to company service personnel. e) consulting on service problems in the field.

Instrument Field Service Representative Fulfills customer service requirements on company-produced instruments and equipment, including: a) analyzing and correcting instrument operating problems. b) contacting supervisor or field engineer for assistance when necessary. c) promoting scheduled maintenance agreements and sales of replacement or spare parts. d) training new service personnel. e) maintaining tools and equipment in good operating order. f) reporting time, activities and expenses in accordance with company requirements. g) keeping up-to-date on new products and product applications.

instrument, indicating A measuring instrument in which only the present value of the measured variable is visually indicated [S51.1].

instrument loop diagram 1. A loop diagram contains the information useful to engineering, construction, commissioning and startup, and maintenance work. Loop diagrams are extension of P&ID's. A loop diagram generally contains only one loop. General layout of a loop diagram is divided into sections for relative location of devices. Symbols used are from ISA-S5.1 Instrument Symbols and Identification and ISA-S5.3 Graphic Symbols for Distributed Control/Shared Display Instrumentation. 2. A loop diagram contains the information needed to understand the operation of the loop and also show all connections to facilitate instrument startup and maintenance of the instruments. The loop diagram must show the components and accessories of the instrument loop, highlighting special safety and other requirements.

Instrument Maintenance Technician Assemble, install, maintain troubleshoot, and repair various components of measurement and control systems. A high school diploma or the equivalent is required, but with today's technology changing so rapidly, vocational/technical school or community college training is strongly recommended. Most employers also provide on-the-job training.

instrument, measuring A device for ascertaining the magnitude of a quantity or condition presented to it [S51.1].

Instrument Mechanic Installs, calibrates, inspects tests, and repairs instruments and control system devices. Mechanics must be able to work well with their hands, be willing to improve their skills through training, and be interested in learning new technologies.

instrument oil A special grade of lubricating oil for instruments and other delicate mechanisms. It is formulated to resist oxidation and gumming, to be compatible with electric insulation, and to inhibit metals from tarnishing.

instrument range The region between the limits within which a quantity is measured, received, or transmitted, expressed by stating the lower and upper range values [S67.04].

instrument reading time The time lag between an actual change in an attribute and stable indication of that change on a continuous-reading instrument.

instrument, recording A measuring instrument in which the values of the measured variable are recorded. The record may be either analog or digital and may or may not be visually indicated [S51.1].

instruments Measuring, indicating, recording, computing, controlling, and similar apparatus requiring the use of small to moderate amounts of electrical energy in normal operation [S12.11].

Instrument Service Specialist Maintains and/or troubleshoots control and instrumentation equipment and calibrates instrument hardware, including: a) maintains thorough knowledge of equipment. b) possibly specializing in specific technology or complex equipment. c) performing analyses to ensure proper functioning of instruments. d) surveying available equipment. e) installing and repairing equipment.

instrument shutoff valve The valve or valve manifold of the sample line located nearest the instrument [S67.02]. Also referred to as "component isolation valve" [S67.10].

Instrument Society of America (ISA) A U.S. society of instrument and controls professionals.

instrument specification A detailed and exact statement of particulars, especially a statement prescribing performance, dimensions, construction, tolerances, bill of materials, features, and operating conditions.

Instrument Supervisor Usually an experienced instrument technician who supervises the work of a team of instrument specialists. In addition to mechanical ability, the supervisor must understand the entire system or process operations for which the team is responsible. Leadership ability is also important.

instrument system See instrumentation.

Instrument Technician 1. Usually requires certification or graduation from a technical college program. The technician works with theoretical and analytical problems, helping engineers find ways to improve the performance of an instrument or a system, as well as helping mechanics troubleshoot system components. 2. An "engineering technician" is one who, in support of engineers or scientists, can carry out in a responsible manner either proven techniques, known to those who are technically expert in a particular technology (instrumentation and control systems), or those techniques especially prescribed by engineers. 3. Performance as an engineering technician requires the application of principles, methods, and techniques especially prescribed by engineers. Performance as an engineering technician requires the application of principles, methods, and techniques appropriate to a field of technology (instrumentation and control systems), combined with practical knowledge of the construction, application, properties, operation, and limitations of engineering systems, processes, structures, machinery, devices or materials, and, as required, related manual crafts, instrumental, mathematical, or graphic skills. Under professional direction, an engineering technician analyzes and solves technological problems, prepares formal reports on experiments, tests, and other projects, or carries out functions such as drafting, surveying, designing, technical sales, advising consumers, technical writing, teaching, or training. 4. The education of an engineering technician places great emphasis on mathematics and applied physics with intensive laboratory work in which the technician develops practical knowledge and skills. 5. Technicians differ from craftsmen in the extent of their knowledge of engineering theory and methods, and they differ from engineers by reason of their more specialized technical background and skills.

Instrument Technologist 1. The "engineering technologist" is qualified to practice engineering technology by reason of having the knowledge and the ability to apply well-established mathematical, physical science, and engineering principles and methods of technological problem-solving which were acquired by engineering technology education and engineering technology experience. 2. The engineering technologist will usually have earned a baccalaureate degree in engineering technology or gained considerable technical experience on the job. 3. The technologist is a member of the engineering team which will normally include technicians and engineers and, for special projects, may include scientists, craftsmen, and other specialists. The configuration of technical personnel possessing complementary capabilities that facilitate the engineering process is, by necessity, peculiar to each situation. The technologist is expected to have a thorough knowledge of the equipment,

applications, and established state-of-the-art design and problem-solving methods in a particular field (instrumentation and control systems).

instrument torque The turning moment on an instrument's moving element produced directly or indirectly by the quantity being measured.

instrument transformer A precision transformer capable of reproducing a signal in a secondary circuit that is suitable for use in measuring, control or protective devices.

insulation 1. A material of low thermal conductivity used to reduce heat loss. 2. A material of specific electrical properties used to cover wire and electrical cable.

insulation resistance The resistance measured between specified insulated portions of a device when a specified direct current voltage is applied at reference operating conditions unless otherwise stated. The objective is to determine whether the leakage current would be excessive under operating conditions [S51.1] [S37.1].

insulation voltage breakdown The voltage at which a disruptive discharge takes place through or over the surface of the insulation [S51.1].

insulator A material through which electrical current cannot flow.

intake 1. An opening where a fluid enters a chamber or conduit; an inlet. 2. The amount of fluid entering through the opening.

InTech ISA journal previously known as Instrumentation Technology.

integer A whole number signified by a binary "word."

integer programming 1. In operations research, a class of procedures for locating the maximum or minimum of a function subject to constraints, where some or all variables must have integer values. Contrast with convex programming, dynamic programming, linear programming, mathematical programming. 2. Loosely discrete programming.

integral 1. This control action will cause the output signal to change according to the summation of the input signal values sampled at regular intervals up to the present time. 2. Mathematically it is the reciprocal of reset.

integral absolute error (IAE) A measure of controller error defined by the integral of the absolute value of a time-dependent error function; used in tuning automatic controllers to respond properly to process transients. See also integral time absolute error.

integral action A type of controller function where the output (control) signal or action is a time integral of the input (sensor) signal.

integral action limiter A device which limits the value of the output signal due to integral control action, to a predetermined value [S51.1].

integral action rate (reset rate) 1. In proportional plus integral, or proportional plus integral plus, derivative control action devices, step input, the ratio of the initial rate of change of output due to integral control action to the change in steady-state output due to proportional control action. Integral action rate is often expressed as the number of repeats per minute because it is equal to the number of times per minute that the proportional response to a step input is repeated by the initial integral response [S51.1]. 2. In integral control action devices, step input, the ratio of the initial rate of change of output to the input change [S51.1].

integral action time constant See time constant, integral action.

integral blower A blower built as an integral part of a device to supply air thereto.

integral-blower burner A burner of which the blower is an integral part.

integral control Form of control action that returns the value of the controlled variable to the set point when sustained offset occurs without this action. Also called reset control.

integral control action (reset) Control action in which the output is proportional to the time integral of the error input, i.e., the rate of change of output is proportional to the error input. See control action, integral.

integral controller See controller, integral (reset).

integral control mode A controller mode in which the controller output increases at a rate proportional to the controlled variable error. Thus, the controller output is the integral of the error over time with a gain factor called the integral gain.

integral flange A flange on a length of pipe, a nozzle or a pressure vessel which is cast or forged with the item itself, or is permanently attached to it by welding.

integral logic annunciator An annunciator that includes visual displays and sequence logic circuits in one assembly [S18.1].

integral orifice A differential pressure measuring technique for small flow rates in which the fluid flows through a miniature orifice plate integral with a special flow fitting.

integral seat A flow control orifice and seat that is an integral part of the body or cage material or may be constructed from material added to the body or cage [S75.05].

integral stem A design in which the stem is either physically a part of the ball or mechanically made part of the ball. Some integral stems are designed to perform a turning and then lifting action [S75.05].

integral time absolute error (ITAE) A measure of controller error defined by the integral of the product of time and the absolute value of a time-dependent error function; whereas the absolute value prevents opposite excursions in the process variable from canceling each other, the multiplication by time places a more severe penalty on sustained transients. See also integral absolute error.

integrated circuit (IC) A complete electronic circuit containing active and passive elements fabricated and assembled as a single unit, usually as a single piece of semiconducting material, resulting in an assembly that cannot be disassembled without destroying it.

integrated circuit piezoelectric (ICP) A type of pressure-sensitive sensor that combines a piezoelectric element with isolation amplifier and signal conditioning microelectronics inside the sensor housing so that the output signal can be transmitted over ordinary two-wire cable instead of special low-noise cable.

integrated software A computer program that combines several functions for ease of use.

Integrated Systems Digital Network (ISDN) A suite of protocols being defined by CCITT to provide voice and data services over wide area networks (WANs).

integrating Providing an output which is a time integral function of the measurand [S37.1].

integrating accelerometer A device that measures acceleration of an object, and converts the measurement to an output signal proportional to speed or distance traveled.

integrating ADC A type of analog-to-digital converter where the analog input is integrated over a specific time with the advantages of high resolution, noise rejection and linearity.

integrating extensions An integrator which derives its input from the motion of the float can be installed within the extension housing.

integrating frequency meter A master frequency meter for an electric power system that measures the actual number of cycles of alternating voltage for comparison with the theoretical number of cycles for the same time at the prescribed frequency.

integrating meter 1. A totalizing meter, such as for electric energy consumed. 2. An instrument whose output is proportional to the single (or higher order) integral of the quantity measured.

integrating network A transducer circuit whose output waveform is a time integral of its input waveform.

integrating sphere A sphere used in optical measurements which is intended to integrate the input light over the output aperture to provide uniform illumination.

integrator 1. A device which continually totalizes or adds up the value of a quantity for a given time. 2. A device whose output is proportional to the integral of the input variable with respect to time.

integrity In data processing, a word to describe data that has not been corrupted.

intelligence In data processing, the processing capability of a computer.

intelligent terminal A computer terminal with some local processing capability.

intensifying screen A sheet of material placed in contact with radiographic film that undergoes secondary fluorescence when struck with x-rays or gamma rays, thereby increasing image density for a given exposure.

intensity 1. The lumination level (i.e., brightness) of the pixels of a VDU [S5.5]. 2. The amount of light incident per unit area. For human viewing of visible light, the usual term is illuminance; for electromagnetic radiation in general, the term is radiant flux.

intensity level The amplitude of a sound wave, commonly measured in decibels.

interaction A phenomena, characteristic of a multivariable process, in which the effect of a manipulative variable change in one control loop not only affects its own controlled variable, but also the controlled variable in another loop. In distillation the primary consideration is interaction between the overhead composition control loop and the bottoms composition control loop.

interaction analysis A technique used in determining the pairing of manipulative and controlled variables in a control loop.

interactive In data processing, a technique of user/system communication in which the operating system immediately acknowledges and sets upon a request entered by the user at a terminal; compare with batch.

interactive computing See conversational mode.

interblock gap Blank space on a computer storage medium between two adjacent blocks of data.

intercept method A method for estimating the quantity of particles or number of grains within a unit area of a microscopic image by counting the number intercepted by a series of straight lines through the image. This is one of the standard methods of determining grain size of a polycrystalline metal.

intercooler A heat exchanger in the path of fluid flow between stages of a compressor to cool the fluid and allow it to be further compressed at lower power demand.

interelectrode capacitance 1. The capacitance between electrodes of a vacuum tube. 2. A capacitance determined by measuring the short-circuit transfer admittance between two electrodes.

interface 1. A common boundary between automatic data processing systems or parts of a single system [RP55.1]. 2. A specific electronic circuit that is a boundary between other circuits or devices.

interference 1. The waveform resulting from superimposing one wave train on another. 2. In signal transmission, spurious or extraneous signals which prevent accurate reception of desired signals. 3. A disturbance in a useful signal resulting from spurious or extraneous signals in the circuit or in the transmission system.

interference, common mode A form of interference which appears between measuring circuit terminals and ground.

interference, differential mode See interference, normal mode [S51.1].

interference (electrical) Any spurious voltage or current rising from external sources and appearing in the circuits of a device. See noise.

interference, electromagnetic Any spurious effect produced in the circuits or elements of a device by external electromagnetic fields. NOTE: A special case of interference from radio transmitters is known as "Radio Frequency Interference (RFI)" [S51.1].

interference, electrostatic field See interference, electromagnetic.

interference filter An optical filter which selectively transmits specific wavelengths of light because of interference resulting from dielectric coatings on the surface of the material. Multilayer interference coatings may include metallic layers.

interference fit Any combination of pin or shaft diameter and mating hole diameter where the tolerance envelope of the hole overlaps or is smaller than the tolerance envelope of the pin.

interference, longitudinal See interference, common mode.

interference, magnetic field See interference, electromagnetic.

interference, normal mode A form of interference which appears between measuring circuit terminals [S51.1].

interference pattern The pattern of some characteristic of a stationary wave produced by superimposing one wave train on another—it may be the distribution in space of energy density, energy flux, particle velocity, pressure or some other characteristic.

interference, transverse See interference, normal mode.

interferometer 1. An instrument so designed that the variance of wavelengths and light path lengths within the mechanism allows very accurate measurement of distances. 2. A device which divides a single beam of light into two (or sometimes more) components, then recombines them to produce interference. In general, the path lengths light travels along the different arms will differ; the difference in distance is proportional to the wavelength of the light times the number of interference rings.

interferometric pressure transducer A type of pressure sensor developed to read pressure differentials on the order of 200 Pa (.030 psi) with a resolution of 1 Pa (0.00015 psi) by detecting very small deflections of a fragile diaphragm through optical interferometry.

interleaving 1. The act of accessing two or more bytes or streams of data from distinct computer memory banks simultaneously. 2. The alternating of two or more operations or functions through the overlapped use of a computer facility.

interlock 1. To arrange the control of machines or devices so that their operation is interdependent in order to assure their proper coordination [RP55.1]. 2. Instrument which will not allow one part of a process to function unless another part is functioning. 3. A device such as a switch that prevents a piece of equipment from operating when a hazard exists. 4. To join two parts together in such a way that they remain rigidly attached to each other solely by physical interference. 5. A device to prove the physical state of a required condition, and to furnish that proof to the primary safety control circuit.

interlock, motor start A connection made through contacts on the motor controller which is wired in series with the safety circuit so that the motor must be energized before the system is allowed to proceed.

intermediate addressing Method of addressing data stored in a computer memory. The instruction operand is the data to be used with the instruction.

intermediate band A mode of recording and playback in which the frequency response at a given tape speed is "intermediate."

intermediate frequency In a superheterodyne receiver, the stage where a down-converted carrier is passed through a bandpass filter and amplifier.

intermediate means In an instrumentation or control system, all system elements between the primary detector and the end device which transmit or modify the output of the former to make it compatible with input requirements of the latter.

intermediate mode Method of operating a computer, with interpretive languages such as BASIC, whereby an individual instruction or a small number of instructions, not forming part of a program, are executed.

intermediate phase A distinct compound or solid solution in an alloy system whose composition limits do not extend to any of the pure constituents.

intermediate position A specified position that is greater than zero and less than 100 percent open.

intermediate zone See zone, intermediate.

intermittency effect In photography or radiography, a departure from the reciprocity law when the emulsion is exposed in a series of discrete increments, compared to the response when it is exposed continuously to the same total energy level.

intermittent blowdown   The blowing down of boiler water at intervals.

intermittent duty   An operating cycle that consists of alternating periods of use and idle time—for example, on and off, load and no-load, load and rest, or load, no-load and rest; in most instances, successive periods of use or idle time vary widely in length, although some intermittent-duty cycles follow well-defined patterns.

intermittent firing   A method of firing by which fuel and air are introduced into and burned in a furnace for a short period, after which the flow is stopped, this succession occurring in a sequence of frequent cycles.

intermittent rating   The rating applicable to specified operation over a specified number of time intervals of specified duration; the length of time between these time intervals must also be specified [S37.1].

intermodulation   The modulation of the components of a complex wave by each other, producing new waves whose frequencies are equal to the sums and differences of integral multiples of the component frequencies of the original complex wave.

intermodulation distortion (IMD)   Defined as 20 log (rms sum of the sum and difference distortion products)/(rms amplitude of the fundamental).

internal   In PC ladder programs, a coil or contact whose reference is a logical element in the program and not directly concerned with I/O. May also refer to the storage location used for the logical status of such an element.

internal combustion engine   A mechanical prime mover that uses exhaust gases resulting from the burning of fuel within the engine as the thermodynamic working fluid.

internal energy   Ability of a working fluid to do its work based on the arrangement and motion of its molecules.

internal furnace   A furnace within a boiler consisting of a straight or corrugated flue, surrounded with water.

internal gear   Any ring-type or annular gear whose teeth are on the inner surface of the rim.

internally fired boiler   A fire tube boiler having an internal furnace such as a Scotch, Locomotive Fire-Box, Vertical Tubular, or other type having a water-cooled plate-type furnace.

internal-mix oil burner   A burner having a mixing chamber in which high velocity steam or air impinges on jets of incoming liquid fuel which is then discharged in a completely atomized form.

internal oxidation   A form of degradation of a material involving absorption of oxygen at the surface and diffusion of oxygen to the interior, where it forms subsurface scale or oxide inclusions.

internal pressure   See burst pressure, proof pressure, or reference pressure.

internal standard   In chemical analysis, especially instrumental analysis, a material present in or added to a sample in known amounts to serve as a reference in determining composition.

internal storage   Addressable storage directly controlled by the central processing unit of a digital computer.

internal treatment   The treatment of boiler water by introducing chemicals directly into the boiler.

International Electrotechnical Commission   An international standards development and certification group in the area of electronics and electrical engineering.

International Federation for Information Processing   An international group of technical societies.

International Standard (IS)   The third (and highest) stage of the ISO standard process. Prospective ISO standards are balloted three times. The first stage is a Draft Proposal (DP). After a Draft Proposal has been in use a period of time (typically six months to a year) the standard, frequently with corrections and changes, is reballoted as a Draft International Standard (DIS). After the Draft International Standard has been in use for a period of time (typically one to two years) it is reballoted as an International Standard (IS).

interpass temperature   The lowest temperature reached by weld metal before the next pass is deposited in a multiple-pass weld.

interpreter   A system program which allows the execution of computer programs using a step by step translation of individual instructions instead of translating the complete program before execution.

Inter-Range Instrumentation Group (IRIG)   The telemetry working group of IRIG is responsible for specifying the industrywide standards and practices of telemetry.

inter-record gap (IRG)   On magnetic tape, the

blank gap between records; the tape can stop and start within this gap.

interrupt (INT) 1. Suspension of the execution of a routine as a result of a hardware or program generated signal [RP55.1]. 2. In data processing, a signal that, when activated, causes a transfer of control to a specific location in memory, thereby breaking the normal flow of control of the routine being executed. An interrupt is normally caused by an external event such as a "done" condition in a peripheral. It is distinguished from a trap, which is caused by the execution of a processor instruction.

interrupt, process Those interrupts available for connection to the user supplied equipment. Synonymous with external interrupt [RP55.1].

interrupt service routine In data processing, a unique address that points to two consecutive memory locations containing the start address of the interrupt service routine and priority at which the interrupt is to be serviced.

interrupt vector In data processing, an address generated by an interrupt. It points to the start of the interrupt service routine.

interrupt vector register In data processing, a register for storing the interrupt vector.

interval The number of word times that occur between successive repetitive samples of the same channel; synonymous with supercommutation and strapping interval.

interval timer A device that provides an interrupt signal upon completion of a predetermined or programmed interval of time. See timer.

intrinsically safe apparatus Apparatus in which all circuits are intrinsically safe [RP12.6].

intrinsically safe circuit A circuit in which no spark nor any thermal effect produced under prescribed test conditions (which include normal operation and specified fault conditions) is capable of causing ignition of a given explosive atmosphere [RP12.6].

intrinsically safe equipment and wiring Equipment and wiring which are incapable of releasing sufficient electrical or thermal energy under normal or abnormal conditions to cause ignition of a specific hazardous atmospheric mixture in its most easily ignited concentration [S51.1] [S12.11].

intrinsically safe system An assembly of interconnected intrinsically safe apparatus, associated apparatus, other apparatus, and interconnecting cables in which those parts of the system which may be used in hazardous (classified) locations are intrinsically safe circuits [RP12.6].

intrinsic joint loss A loss intrinsic to the fiber that is caused by fiber parameter mismatches when joining two nonidentical fibers.

intrinsic safety 1. A method to provide safe operation of electric process control instrumentation where hazardous atmospheres exist. The method keeps the available electrical energy so low that ignition of the hazardous atmosphere cannot occur. 2. A protection technique based upon the restriction of electrical energy within apparatus and of interconnecting wiring, exposed to a potentially explosive atmosphere, to a level below that which can cause ignition by either sparking or heating effects. Because of the method by which intrinsic safety is achieved, it is necessary to ensure that not only the electrical apparatus exposed to the potentially explosive atmosphere but also other electrical apparatus with which it is interconnected is suitably constructed.

intrinsic safety barrier 1. A network designed to limit the energy (voltage and current) available to the protected circuit in the hazardous (classified) location, under specified fault conditions [RP12.6]. 2. A device inserted in wire between process control instrumentation and the point where the wire passes into the hazardous area. It limits the voltage and current on the wire to safe levels.

intrinsic safety ground bus A grounding system which has a dedicated conductor separate from the power system so that ground currents will not normally flow and which is reliably connected to a ground electrode in accordance with Article 250 of the NEC or Section 10 of CEC Part I, CSA C22.1 [RP12.6].

inverse response The dynamic characteristic of a process by which its output responds to an input change by moving initially in one direction but finally in the other.

inversion temperature In a thermocouple, the temperature of the "hot" junction when the thermoelectric emf of the circuit is equal to zero.

inverter A NOT element. The output signal is the reverse of the input signal.

I/O See input/output.

I/O-bound A state of program execution in which all operations are dependent on the activity of an I/O device; for example, when a program is waiting for input from a terminal. See also CPU-bound.

I/O hardware Computer hardware used to carry signals into and out of the processing hardware.

I/O isolation Usually refers to the electrical separation of field circuits from computer internal circuits. Accomplished by optoelectronic devices. Occasionally refers to the ability to have input or output field wiring on isolated circuits, i.e., with one return for each.

I/O limited See input-output limited.

I/O module Basic set of I/O interfaces sharing a common computer unit housing. Can be a set of discrete I/O or a smart control I/O.

ion A charged atom or radical which may be positive or negative.

ion exchange A chemical process for removing unwanted dissolved ions from water by inducing an ion-exchange reaction (either cation or anion) as the water passes through a bed of special resin containing the substitute ion.

ion-exchange resin A synthetic organic compound (resin) that can remove unwanted ions from a dilute solution by combining with them or by exchanging them for ions that produce desirable or neutral effects.

ionic strength Effective strength of all ions in a solution that is equal to the sum of one half of the product of the individual ion concentration and their ion valence or charge squared for dilute solutions.

ion implantation A process for enhancing surface properties of a solid by bombarding it with a beam of high-energy ions, which are absorbed into the material's surface layer.

ionization The process of splitting a neutral molecule into positive and negative ions, or of detaching one or more electrons from a neutral atom.

ionization chamber An enclosure filled with gas that is ionized when radiation enters the chamber; it contains two or more electrodes that sustain an electric field and collect the charge resulting from ionization.

ionization constant A measure of the degree of dissociation of a polar compound in dilute solution at equilibrium; it equals the product of the concentrations of the dissociated compound (ions) divided by the concentration of the undissociated compound.

ionization gage A pressure transducer based on conduction of electric current through ionized gas of the system whose pressure is to be measured. Useful only for very low pressures (for example, below 10^{-3} atm).

ionization time In a gas tube, the interval between the time when conduction conditions are established and the time when conduction actually begins at some stated value of tube potential.

ionization vacuum gage An instrument for measuring very low pressures (high vacuums) by means of a current of positive ions produced in the gas by electrons emitted from a hot cathode and accelerated across a portion of the evacuated space toward another electrode.

ionizing Converting a change of measurand into a change in ionization current, such as through a gas between two electrodes [S37.1].

ionizing event Any interaction between an atom or molecule and an energy beam, particle, atom or molecule that causes one or more ions to be generated.

ionizing radiation Any electromagnetic or particulate radiation that can produce ions, either directly or indirectly, when it interacts with matter.

ion laser A laser in which the active medium is an ionized gas, typically one of the rare gases, argon or krypton, or a mixture of the two.

ionosphere That portion of the earth's atmosphere where ionization takes place due to ultraviolet radiation of the sun or from bombardment by hydrogen bursts from sunspots. The various layers, identified as B, C, D, E and F, have characteristics that reflect and refract radio waves according to their frequency, time of day, sunspot cycle and earth weather.

ion pair The combination of a positive ion and a negative ion having the same magnitude of charge, and formed from a neutral molecule due to absorption of the energy in radiation.

I/O page That portion of computer memory in which specific storage locations are associated directly with I/O devices.

I/O rack Chassis for mounting computer I/O modules. May be local or remote from CPU/memory unit.

I/P converter A device that linearly converts electric current into gas pressure (for example, 4-20 mA into 3-15 psi).

IRIG See Inter-Range Instrumentation Group.

irradiance The power per unit area incident upon a surface. Also called radiant flux density.

irradiation 1. Exposing an object or person to penetrating ionizing radiation such as x-rays or gamma rays. 2. Exposing an object or person to ultraviolet, visible, or infrared energy.

IS See International Standard.

ISA See Instrument Society of America.

ISDN See Integrated Systems Digital Network.

isentropic Proceeding at constant entropy.

isentropic exponent A ratio defined by the specific heat at constant pressure divided by the specific heat at constant volume.

isobaric Proceeding at constant pressure.

isochronous governor A device that maintains rotational speed of an engine constant, regardless of load.

isolated circuit A circuit in which the current, with the equipment at reference-test conditions, to any other circuit or conductive part does not exceed the limit for leakage current [S82.01].

isolating element A movable membrane, usually of metal, that physically separates the measured fluid from the sensing element. Usually this membrane is considerably more flexible than the sensing element and is coupled to the sensing element using a transfer fluid. Its purpose is to provide material compatibility with the measured fluid while maintaining the performance integrity of the sensing element [S37.6].

isolation valve The isolation valve nearest the instrument, grab-sample point, or in-line component which is available to personnel during normal plant operation. The root valve may or may not perform the function of the isolation valve, depending on its location [S67.10].

isopotential point Point on the millivolt versus pH plot at which a change in temperature has no effect. It is at 7 pH and zero millivolts unless shifted by the standardization and meter zero adjustments or an electrode asymmetry potential.

isothermal Proceeding at constant temperature.

isotope Any of two or more nuclides that have the same number of protons in their nuclei but different numbers of neutrons; such atoms are of the same element, and thus cannot be separated from each other by chemical means, but because they have different masses can be separated by physical means.

isotope effect The effect of nuclear properties other than the number of protons on the nonnuclear physical and chemical behavior of the nuclides.

ITAE See integral time absolute error.

iterate To repeatedly execute a loop or series of steps. For example, a loop in a routine.

iterative Describing a procedure or process which repeatedly executes a series of operations until some condition is satisfied. An iterative procedure can be implemented by a loop in a routine.

ITI See Industrial Technology Institute.

IVD See Integrated Voice Data LAN.

J

J See joule.

jack A connecting device to which a wire or wires of a circuit may be attached and which is arranged for the insertion of a plug.

jacket 1. A plastic layer applied over the coating of an optical fiber, or sometimes over the bare fiber. Used for color coding in optical cables, to make handling easier, or for protection of the fiber against mechanical stress and strain. 2. Stiff plastic protective material that encases a floppy disk with slots for a disk drive to access the data. 3. The layer of plastic, fiber or metal surrounding insulated electrical wires to form a cable. This outer cover may be for mechanical or environmental protection of the wires contained therein.

jacketed valves A valve body cast with a double wall or provided with a double wall by welding material around the body so as to form a passage for a heating or cooling medium. Also refers to valves which are enclosed in split metal jackets having internal heat passageways or electric heaters. Also referred to as steam jacketed or vacuum jacketed. In a vacuum jacketed valve, a vacuum is created in the space between the body and secondary outer wall to reduce the transfer of heat by convection from the atmosphere to the internal process fluid, usually cryogenic [S75.05].

jackscrew 1. A portable device for lifting a heavy load a short distance by means of a screw mechanism. 2. The screw of such a device.

jackshaft A countershaft, especially an auxiliary shaft used between two other shafts.

jamming Intentional transmission of radio-frequency waves for the purpose of interfering with transmissions from another station.

Japanese Map Users Group See World Federation.

JCL See job control language.

jerk 1. The time rate of change of acceleration. Expressed in $feet/s^3$, cm/s^3, gn/s [S37.1]. 2. A sudden, abrupt motion.

jet Rapid flow of fluid from a nozzle or orifice.

jet pump A type of pump that uses a jet of fluid to induce flow in another fluid.

jewels Recessed bearings of glass, sapphire or diamond that support the ends of a pivot pin in an instrument or a fine mechanical watch or clock.

jitter A computer signal instability.

job A group of computer data and control statements that do a unit of work, such as a program and all its related subroutines, data, and control statements; also a batch control file.

job control language (JCL) A language for identifying a job and requesting action from a computer operating system.

Johansson curved crystal spectrometer A type of spectrometer having a reflecting crystal whose face is concave so that x-rays that diverge slightly after passing through the primary slit are refocused at the detector slit.

joint A separable or inseparable juncture between two or more materials.

joule Metric unit for energy, quantity of heat, or work.

JOVIAL See Jules' Own Version of International Algorithmic Language.

joy stick A device by which an individual can communicate with an electronic information system through a cathode ray tube.

Jules' Own Version of International Algorithmic Language (JOVIAL) A compiler language based on the International Algorithmic Language, ALGOL.

jump An instruction which causes a new address to be entered into a computer program counter; the program continues execution from the new program counter address.

213

jumper A temporary wire used to bypass a portion of an electrical circuit or to attach an instrument or other device during testing or troubleshooting.

jumper tube A short tube connection for bypassing, routing, or directing the flow of fluid as desired.

junction The portion of a transistor where opposite types of semiconductor material meet.

junction box A protective enclosure around connections between electric wires or cables.

jury rig A makeshift or temporary assembly.

justification 1. The act of adjusting, arranging, or shifting digits to the left or right, to fit a prescribed pattern. 2. The use of space between words in printed material that causes each line to be of equal length.

justify To align computer data about a specified reference.

K

K Symbol used to indicate a kilobyte which is a measure of a computer memory. One Kilobyte of memory can store 1024 characters. See also kelvin; see also ratio of specific heats.

Kalman filter A technique for calculating the optimum estimates of process variables in the presence of noise; the technique, which generates recursion formulas suitable for computer solutions, also can be used to design an optimal controller.

Karl Fischer technique A titration method for accurately determining moisture content of solid, liquid or gas samples using Karl Fischer reagent —a solution of iodine, sulfur dioxide and pyridine in methanol or methyl Cellusolve; the titration is highly suitable for automation, and has high sensitivity (5 ppm) and good accuracy (+ or – 1%) over a wide range of moisture content (10 ppm to 100%).

Karnaugh map A tubular arrangement that facilitates the combination and elimination of logical functions by listing similar logical expressions, thereby taking advantage of the human brain's ability to recognize visual patterns to perform the minimization.

K$_b$ See Bernoulli coefficient.

KByte $1024(2^{10})$bytes.

Kelvin Metric unit for thermodynamic temperature. An absolute temperature scale where the zero point is defined as absolute zero (the point where all spontaneous molecular motion ceases) and the scale divisions are equal to the scale divisions in the Celsius system; in the Kelvin system, the scale divisions are not referred to as degrees as they are in other temperature-measurement systems; 0 °C equals approximately 273.16 K.

Kelvin bridge A type of d-c bridge circuit similar to a Wheatstone bridge, but incorporating two extra resistances in parallel with two of the known resistances to minimize the inaccuracies introduced because of finite lead and contact resistances in the circuit.

Kelvin-Varney voltage divider(KVVD) A resistive-type voltage divider used in some d-c bridge circuits to provide greater sensitivity at low values of the unknown resistance.

Kennison nozzle A specially shaped nozzle designed for measuring flow through partially filled pipes; because of its self-scouring, nonclogging design, it is especially useful for measuring flow of raw sewage, raw and digested sludge, final effluent, trade wastes, and other liquids containing suspended solids or debris; it also functions well at low flow rates or when flow rates vary widely.

Kern counter See dust counter.

Kerr cell A device in which the Kerr effect is used to modulate light passing through the material. The modulation depends on rotation of beam polarization caused by the application of an electric field to the material. The degree of rotation determines how much of the beam can pass through a polarizing filter.

key 1. A machine part inserted into a groove (or keyway) to lock two parts together, such as a shaft and a gear or pulley. 2. One of a set of control levers used to operate a machine such as a typewriter or computer processing unit. 3. A device that moves or pivots to secure or tighten components in an assembly. 4. A component, usually having notches or grooves in its working face, that is inserted into a lock to engage or disengage the locking mechanism. 5. In data processing, characters that identify a record.

keyboard An orderly arrangement of keys for operating a machine such as a typewriter.

keyboard entry 1. An element of information inserted manually, usually via a set of switches or marked punch levers, called keys, into an au-

215

tomatic data processing system. 2. A medium for achieving access to or entrance into an automatic data processing system.

keyboard lockout An interlock feature which prevents sending from the keyboard while the tape transmitter of another station is sending on the same circuit.

keyboard monitor A computer program that provides and supervises communication between the user at the system console and an operating system.

keyboard perforator A machine that operates somewhat like a typewriter and produces punched paper tape for automatically operating computers or communications equipment.

key disk A disk required to start certain programs.

keypunch 1. A special device to record information on cards or tape by punching holes in the cards or tape to represent letters, digits, and special characters. 2. To operate a device for punching holes in cards or tape.

key-to-disk device Input equipment designed to accept keyboard entry directly on magnetic disks.

keyword One of the significant and informative words in a title or document which describe the content of that document.

kg See kilogram.

kiln An oven or similar heated chamber for drying, curing or firing materials or parts.

kilo A decimal prefix denoting 1,000.

kilogram Metric unit of mass.

kinematic viscosity Absolute viscosity of a fluid divided by its density.

kinetic energy 1. The energy of a working fluid caused by its motion. 2. Energy related to the fluid of dynamic pressure, $1/2 \, p \, V^2$.

kinetic vacuum system A vacuum system capable of attaining and sustaining limiting pressures of 5×10^{-5} to 5×10^{-7} torr despite a relatively high outgassing load or the presence of small leaks.

kink A tight loop in wire or wire rope that results in permanent damage due to deformation.

Kirchhoff's Law The sum of the voltage across a device in a circuit series is equal to the total voltage applied to the circuit.

knockout pin See ejector pin.

knot A unit of speed commonly used to measure the speed of ships or aircraft; it equals one nautical mile (6080 ft) per hour, or 1.151 statute miles per hour.

knowledge Familiarity with theory and concepts, and detailed understanding of job-related topics.

Knudsen flow Gas flow in a long tube at pressures such that the mean free path of a gas molecule is significantly greater than the tube radius.

kohm See kilo-ohm.

konimeter A device for determining dust concentration by drawing in a measured volume of air, directing the air jet against a coated glass surface thereby depositing dust particles for subsequent counting under a microscope.

koniscope An indicating instrument for detecting dust in the air.

Kurtosis number Figure of merit, K, used for monitoring impulsive type vibrations of ball bearings.

KVVD See Kelvin-Varney voltage divider.

Kynar Tradename of Polyvinylidene fluoride, by Pennwalt Corp.

L

L See litre; also see length.

La Chemical symbol for lanthanum.

label In data processing, a set of symbols used to identify or describe an item, record, message, or file. Occasionally it may be the same as the address in storage.

labeled common Named data areas that are accessible to all computer programs declaring the named area.

labeled molecule A molecule of a specific chemical substance in which one or more of its component atoms is an abnormal nuclide—that is a nuclide that is radioactive when the molecules normally are composed of stable isotopes, or vice versa.

lactometer A hydrometer designed for measuring the specific gravity of milk.

ladder diagram Symbolic representation of a control scheme. The power lines form the two sides of a ladder-like structure, with the program elements arranged to form the rungs. The basic program elements are contacts and coils as in electromechanical logic systems.

lag 1. A relative measure of the time delay between two events, states, or mechanisms. 2. In control theory, a transfer function term in the form, $1/(Ts+1)$. 3. The dynamic characteristic of a process giving exponential approach to equilibrium.

lagging 1. In an a-c circuit, a condition where peak current occurs at a later time in each cycle than does peak voltage. 2. A thermal insulation, usually made of rock wood and magnesia plaster, that is used to prevent heat transfer through the walls of process equipment, pressure vessels or piping systems.

lag time An interval of time between the initiation of a discrete sample (particle, molecule, atom) at the sample tap to termination at a specific volumetric flow rate through the sample line [S67.10].

lambert A unit of luminance; it equals the uniform luminance of a perfectly diffusing surface emitting or reflecting light at one lumen per square centimeter.

Lambert's cosine law The radiance of certain surfaces, known as Lambertian reflectors, Lambertian radiators, or Lambertian sources, is independent of the angle from which the surface is viewed.

laminar boundary layer A layer of a moving turbulent stream adjacent to the wall of a pipe or other conduit, where the motion approximates streamline flow.

laminar flow 1. A type of streamline flow most often observed in viscous fluids near solid boundaries, which is characterized by the tendency for fluid to remain in thin, parallel layers to maintain uniform velocity. 2. A nonturbulent flow regime in which the stream filaments glide along the pipe axially with essentially no transverse mixing. Also known as viscous or streamline flow. 3. Flow under conditions in which forces due to viscosity are more significant than forces due to inertia.

lamp Any device for producing light, usually one that converts electric energy to light.

lamp cabinet A cabinet containing visual displays only [S18.1].

lamp follower See auxiliary output.

lamp test See test.

LAN See Local Area Network.

lance door A door through which a hand lance may be inserted for cleaning heating surfaces.

language In data processing, a set of representations, conventions and rules used to convey information. See algorithmic language, artificial language, machine language, natural language, object language, problem-oriented language, procedure-oriented language, programming lan-

217

guage, source language, and target language.

language extendibility The ability to change a programming language through source statements written in that language.

language translator A general term for any assembler, compiler, or other routine that accepts statements in one language and produces equivalent statements in another language.

lantern ring A rigid spacer assembled in the packing box with packing normally above and below it and designed to allow lubrication of the packing or access for a leak-off connection [S75.05].

lap joint A connection between two parts made by overlapping members at the junction and welding, riveting or bolting them together.

Laplace transform In control theory, a mathematical method for solution of differential equations.

lapped-in Mating contact surfaces that have been refined by grinding and/or polishing together or separately in appropriate fixtures [S75.05].

lapping Smoothing or polishing a surface by rubbing it with a tool made of cloth, leather, plastic, wood or metal in the presence of a fine abrasive.

lapping in A process of mating contact surfaces by grinding and/or polishing.

laptop A small, portable computer usually with a flip-up screen.

lap weld A lap joint made by welding.

large core fiber An optical fiber with a comparatively large core, usually a step index type. There is no standard definition of "large" but for the purposes here, diameters of 400 micrometers or more are designated as "large".

large scale integration (LSI) 1. A computer chip containing a large number of digital circuits in a small area. 2. Integrated circuits with more than 100 logic gates.

LASER Light Amplification by Stimulated Emission of Radiation. It is a source of EM radiation generally in the IR, visible, or UV bands and is characterized by small divergence, coherence, and monochromaticity.

laser diode array A device in which the output of several diode lasers is brought together in one beam. The lasers may be integrated on the same substrate, or discrete devices coupled optically and electronically.

laser doppler flowmeter An apparatus for de-

termining flow velocity and velocity profile by measuring the Doppler shift in laser radiation scattered from particles in the moving fluid stream; contaminants such as smoke may have to be introduced into a gas stream to provide scattering centers; the technique can be used to measure velocities of 0.01 to 5000 in./s (0.25 mm/s to 125 m/s).

laser glass An optical glass doped with a small concentration of a laser material. When the impurity atoms are excited by light, they are stimulated to emit laser light.

laser interferometer A type of optical interferometer using a laser as the source of monochromatic light; accuracies of better than 20 microinch (1.25 μm) are achieved when measuring lengths up to 200 in. (5.08 m).

laser line filter A filter which transmits light in a narrow range of wavelengths centered on the wavelength of a laser. Light at other wavelengths is reflected. Such filters are used to remove light from non-laser sources, which could interfere with operation of a laser system.

laser printer A print-quality printer that uses a laser beam to electrostatically transfer an image to paper.

laser simulator A light source which simulates the output of a laser. In practice, the light source is a 1.06 micrometer LED which simulates the output of a neodymium laser at much lower power levels.

last-in, first-out (LIFO) In an ordered pushdown stack, a discipline wherein the last transaction to enter a stack is also the first to leave it. Contrast with first-in, first-out.

latching digital output A contact closure output that holds its condition (set or reset) until changed by later execution of a computer program. See momentary digital output.

latching relay Real device or program element that retains a changed state without power. In a computer program element, the power removal is only in terms of the logic power expressed in the diagram. Real power removal will affect the PC outputs according to some scheme provided by the manufacturer.

latch switch A control to prevent fuel valve opening if the burner is not secured in the firing position.

latency In data processing, the time between the completion of the interpretation of an address and the start of the actual transfer from the addressed location. Latency includes the delay associated with access to storage devices such as drums and delay lines.

latent heat Heat that does not cause a temperature change.

latitude Of a photographic emulsion, the ratio of the exposure limits between which the film density curve (Hurter and Driffield curve, or H & D curve) is essentially linear.

lattice network An electronic network composed of four branches connected end-to-end to form a mesh, and in which two nonadjacent junctions are the input terminals and the two remaining nonadjacent junctions are the output terminals.

lattice parameter In crystallography, the length of any side of the unit cell in a given space lattice; if the sides are unequal, all unequal lengths must be specified.

Lauritsen electroscope An electroscope in which the sensitive element is a metallized quartz fiber.

layer A subdivision of the OSI architecture.

lb See pound.

L-band In telemetry, the radio spectrum which is available for manned vehicles; 1435-1540 MHz.

LCD See liquid crystal display.

LCU See Local Control Unit.

L/D (reflux-to-distillate ratio) A quantity used in analyzing column operations. See reflux ratio.

lead 1. In control theory, a transfer function term in the form, (Ts+1). 2. The distance a screw mechanism will advance along its axis in a single rotation.

lead angle 1. In welding, the angle between the axis of the electrode and the axis of the weld. 2. The angle between the tangent to a helix and a plane perpendicular to the axis of the helix.

lead equivalent The radiation-absorption rating of a specific material expressed in terms of the thickness of lead that reduces radiation dose an equal amount under given conditions.

leader 1. A blank section of tape at the beginning of a reel of magnetic tape or at the beginning of a paper tape. 2. A system program which enables other programs to be loaded into the computer.

leading 1. In an a-c circuit, a condition where peak current occurs earlier in each cycle than does peak voltage. 2. In printing, the insertion of additional space between lines.

leading edge The first transition of a pulse going in either a positive (high) or a negative (low) direction.

lead-lag compensator A dynamic compensator combining lead action (the inverse of lag) with lag.

lead time In industrial engineering, the amount of time required to design and develop a piece of equipment before it is ready for use.

lead wire Any wire connecting two points in an electrical circuit, but especially a wire connecting an electric device to a source of power or connecting an indicating or controlling instrument to a sensor.

leak An opening, however minute, that allows undesirable passage of a fluid from its containing boundaries [S67.03].

leakage 1. The quantity of fluid passing through a valve when the valve is in the full closed position under stated closure forces, with the pressure differential and temperature as specified. Leakage is usually expressed as a percentage of the valve capacity at full rated travel [S75.05]. 2. Undesirable loss or entry across the boundary of a system. The term is usually applied to slow passage of a fluid through a crack or fissure, but may also be used to describe passage of small quantities of particles, radiation, electricity or magnetic lines of force beyond desired boundaries.

leakage, class Classification established by ANC1/FC1 70-2 to categorize seat leakage of control valve trim [S75.05].

leakage, packing The quantity of process fluid that escapes through the valve packing [S75.05].

leakage rate 1. The maximum rate at which a fluid is permitted or determined to leak through a seal. The type of fluid, the differential pressure across the seal, the direction of leakage and the location of the seal must be specified [S37.1]. 2. Leakage expressed in volumetric units per unit of time at 20 °C and one atmosphere pressure [S67.03]. 3. The amount of leakage across a defined boundary per unit time.

leakage, seat The quantity of fluid passing through a valve when the valve is in the fully closed position with pressure differential and temperature as specified [S75.05].

leak detector An instrument such as a helium mass spectrometer used for detecting small cracks or fissures in a vessel wall.

leak-off gland A packing box with packing above and below the lantern ring so as to pro-

219

vide a sealed low pressure leak collection point for fluid leaking past the primary seal (lower packing) [S75.05].

leak pressure See pressure, leak.

least significant bit (LSB) The smallest bit in a string of bits, usually at the extreme right.

least significant digit (LSD) The right-most digit of a number.

least-squares line The straight line for which the sum of the squares of the residuals (deviations) is minimized [S37.1].

LED A semiconductor diode, the junction of which emits light when passing a current in the forward (junction on) direction. See also light emitting diode.

Ledoux bell meter A type of manometer whose reading is directly proportional to flow rate sensed by a head producing measuring device such as a pitot tube.

left justified A field of numbers (decimal, binary, etc.) which exists in a memory cell, location or register, possessing no zeroes to its left.

leg 1. One of the members of a branched object or system. 2. The distance between the root of a fillet weld and the toe. 3. Any structural member that supports an object above the horizontal.

length A fundamental measurement of the distance between two points, measured along a straight or curved path.

lens joint ends Valves with the ends prepared for lens ring gaskets [S75.05].

level 1. A measure of the logarithm of the ratio of some quantity to a reference quantity of the same kind. The reference quantity must be identified [S37.10]. 2. Any bubble-tube device used to establish a horizontal line or plane. 3. To make the earth's surface even and roughly horizontal.

level indicator 1. An indicating instrument for determining the position of a liquid surface within a vessel. 2. An instrument containing a meter, neon lamp or cathode-ray tube which shows audio voltage level in an operating sound-recording system.

leveling saddle A pipe clamp anchoring a swivel joint 2 in. threaded socket allowing a 2 in. pipe-stand to be properly positioned.

level (logic) A signal that remains at the "O" or "I" level for long amounts of time.

lexical analysis In data processing, a stage in the compilation of a program in which statements, such as IF, AND, END, etc. are replaced by codes.

L/F (reflux-to-feed ratio) A quantity used in analyzing column operations.

Liapunov's second method A method analogous to the rate-of-change-of energy method for mechanical systems whereby stability or instability or a process control system can be determined. Also referred to as the indirect method.

liberation See heat release.

library 1. A collection of information and standard programs available to a computer, usually on auxiliary storage. 2. A file that contains one or more relocatable binary modules which are routines that can be incorporated into other programs.

Lichtenberg figure camera A device for indicating the polarity and approximate crest value of a voltage surge; it consists of a photographic film or plate backed by an extended plane electrode, with its emulsion contacting a small electrode that is connected to the circuit in which a surge occurs.

life, cycling The specified minimum number of full range excursions or specified partial range excursions over which a transducer will operate as specified without changing its performance beyond specified tolerances [S37.1].

life, operating The specified minimum length of time over which the specified continuous and intermittent rating of a transducer applies without change in transducer performance beyond specified tolerances [S37.1].

life, storage The specified minimum length of time over which a transducer can be exposed to specified storage conditions without changing its performance beyond specified tolerances [S37.1].

life test A destructive test in which a device is operated under conditions that simulate a lifetime of use.

LIFO See last-in, first-out.

lift See travel.

ligament The minimum cross section of solid metal in a header, shell or tube sheet between two adjacent holes.

light 1. An electromagnetic radiation whose wavelength is between approximately 10^{-2} and 10^{-6} cm. NOTE: By strict definition only visible radiation (4×10^{-5} to 7×10^{-5} cm) can be considered as light [S37.1]. 2. Electromagnetic radiation having a wavelength in the range over which it can be detected with the unaided human eye.

light-beam galvanometer A type of sensitive galvanometer whose null-balance point is indicated by the position of a beam of light reflected from a mirror carried in the moving coil of the instrument. Also known as d'Arsonval galvanometer.

light-beam instrument A measurement device that indicates measured values by means of the position of a beam of light on a scale.

light-coupled switch A switch in which the switching signal is transmitted to the activating device by means of a light beam.

light crude Crude oil rich in low-viscosity hydrocarbons of low molecular weight.

light curtain An arrangement whereby a wide, thin beam of invisible modulated light is used to detect passage of objects through a plane up to about 8 by 78 in. (200 mm by 2 m).

light duty cable Generally a type of fiber optic cable designed to withstand conditions encountered in a building—not outdoor conditions.

lighted dial A dial or indicating scale and pointer which has a small lamp within the assembly so that the scale and pointer are illuminated for viewing in darkness. Compare with luminous dial.

light emitting diode (LED) A semiconductor diode which emits visible or infrared light. Light from an LED is incoherent spontaneous emission, as distinct from the coherent stimulated emission produced by diode lasers and other types of lasers. The indicator lights on most I/O modules are LED's.

light ends The fraction of a petroleum mixture having the lowest boiling point.

lighter-than-light key The remaining components in the overhead stream other than the light key.

lighting off torch A torch used for igniting fuel from a burner. The torch may consist of asbestos wrapped around an iron rod and saturated with oil or may be a small oil or gas burner.

light key The component in multicomponent distillation that is removed in the overhead stream and that has the lowest vapor pressure of the components in the overhead. If the reboiler head is decreased or the reflux flow increased, the light key component is the first component to fall into the bottoms product.

light meter A small, hand-held instrument for measuring intensity of illumination.

light modulator An apparatus that produces a sound track by means of a source of light, an appropriate optical system, and a device for inducing controlled variations in light-beam characteristics.

light oil Any oil whose boiling point is in the temperature range 110 to 210 °C, especially a coal tar fraction obtained by distillation.

light pen A device by which an individual can communicate with an information system through a cathode ray tube.

light valve A device whose ability to transmit light can be made to vary by applying an external electrical quantity such as a current, voltage, electric field, electron beam or magnetic field.

limit-check The comparison of data from a specific source with pre-established allowable limits for that source.

limit checking Internal program checks for high, low, rate-of-change, and deviation from a reference. These checks are to detect signals indicating undesirable or unsafe plant operation.

limit control A sensing device that shuts down an operation or terminates a process step when a prescribed limiting condition is reached.

limit cycle A sustained oscillation of finite amplitude.

limiter A device which applies limits to a signal.

limiting The action which causes a transducer output to become constant even though its input continues to rise above a certain value.

limiting safety system setting (LSSS) Limiting safety system settings for nuclear reactors are settings for automatic protective devices related to those variables having significant safety functions [S67.04].

limit of detection In any instrument or measurement system, the smallest value of the measured quantity that produces discernible movement of the indicator.

limit of error In an instrument or control device, the maximum error over the entire scale or range of use under specific conditions.

limit of measurement In any instrument or measurement system, the smallest value of the measured quantity that can be accurately indicated or recorded.

limit priority A priority specification associated with every task in a multitask operation, representing the highest dispatching priority that the task may assign to itself or to any of its subtasks.

limits The prescribed maximum and minimum values of a dimension or other attribute.

limit switch An electromechanical device that is operated by some moving part of a power-driven machine to alter an electrical circuit associated with the machine. See also position switch.

lin See linear.

Lindemann electrometer An electrometer in which a metallized quartz fiber mounted on a quartz torsion fiber perpendicular to its axis is positioned within a system of electrodes to produce a visual indication of electric potential.

line 1. In word processing, a string of characters that terminates with a vertical tab, form-feed, line-feed, or carriage return. 2. In communications, a line provides a data transmission link. 3. In process plants, a collection of one or more associated units and equipment modules, arranged in serial and/or parallel paths, used to make a complete batch of material or finished product. Also "production line" or "train."

lineal scale length The distance from one end of an instrument scale to the other, measured along the arc if the scale is curved or circular.

linear The type of relationship that exists between two variables when the ratio of the value of one variable to the corresponding value of the other is constant over the entire range of possible values.

linear actuator A device for converting power into linear motion.

linear control system A control system in which the transfer function between the controlled condition and the command signal is independent of the amplitude of the command signal.

linearity 1. The closeness to which the curve relating two variables approximates a straight line [RP55.1] [S75.05]. NOTE: A. It is usually measured as a nonlinearity and expressed as linearity; e.g., a maximum deviation between an average curve and a straight line. The average curve is determined after making two or more full range traverses in each direction. The value of linearity is referred to the output unless otherwise stated. B. As a performance specification linearity should be expressed as independent linearity, terminal-based linearity or zero-based linearity. When expressed simply as linearity it is assumed to be independent linearity [S51.1]. 2. The closeness of a calibration curve to a specified straight line. Linearity is expressed as the maximum deviation of any calibration point on a specified straight line, during any one calibration cycle. It is expressed as "within ± _____ percent of full scale output." [S37.1] 3. Characteristic of a device or system which can be described by a linear differential equation with constant coefficients [S26]. 4. The linearity error is generally the greatest departure from the best straight line that can be drawn through the measured calibration points. See also conformity.

linearity, differential Any two adjacent digital codes should result in measured output values that are exactly 1 LSB apart. Any deviation of the measured "step" from the ideal difference is called differential nonlinearity expressed in multiples of 1 LSB.

linearity, end point Linearity referred to the end-point line [S37.1].

linearity, independent 1. The maximum deviation of the calibration curve (average of upscale and downscale readings) from a straight line so positioned as to minimize the maximum deviation [S51.1]. 2. Linearity referred to the best straight line [S37.1]. See test procedure.

linearity, least squares Linearity referred to the least-squares line [S37.1].

linearity of a turbine flowmeter The maximum percentage deviation from the average sensitivity (K) across the linear range [RP31.1].

linearity range of a turbine flowmeter The flow range over which the output frequency is proportional to flow (constant K factor) within the limits of linearity specified [RP31.1].

linearity, terminal Linearity referred to the terminal line [S37.1].

linearity, terminal-based The maximum deviation of the calibration curve (average of upscale and downscale readings) from a straight line coinciding with a calibration curve at upper and lower range-values [S51.1]. See test procedure.

linearity, theoretical slope Linearity referred to the theoretical slope [S37.1].

linearity, zero-based The maximum deviation of the calibration curve (average of upscale and downscale readings) from a straight line so positioned as to coincide with the calibration curve at the lower range-value and to minimize the maximum deviation [S51.1]. See test procedure.

linearization The process of converting a nonlinear (nonstraight-line) response into a linear response.

linear meter An instrument whose indicated

output is proportional to the quantity measured.

linear optimization See linear programming.

linear polarization Light in which the electric field vector points in only a single direction.

linear position sensing detector An optical detector which can measure the position of a light spot along its length.

linear potentiometer A variable resistance device whose effective resistance is a linear function of the position of a control arm or other adjustment; most often, the device is constructed so that a single length of straight or coiled wire whose resistance varies uniformly along its length is in contact with a shoe or similar sliding member; effective resistance is varied by connecting the circuit to one end of the wire and to the shoe, and then varying the position of the shoe; use of a wire-wound resistor, thin film or printed circuit element allows greater voltage drop per unit length along the potentiometer, and therefore stronger and more useful output signals.

linear programming (LP) A method of solution for problems in which a linear function of a number of variables is subject to a number of constraints in the form of linear inequalities.

linear system See system, linear.

linear transducer A type of transducer for which a plot of input signal level versus output signal level is a straight line.

linear variable differential transformer (LVDT) A type of position sensor consisting of a central primary coil and two secondary coils wound on the same core; a moving-iron element linked to a mechanical member induces changes in self induction that are directly proportional to movement of the member.

linear variable reluctance transducer (LVRT) A type of position sensor consisting of a center-tapped coil and an opposing moving coil attached to a linear probe; the winding is continuous over the length of the core, instead of being segmented as in an LVDT; the chief disadvantage of an LVRT is that the overall length must be at least double the stroke, whereas the chief advantage is its excellent linearity over an effective stroke up to 24 in. (610 mm).

linear velocity A vector quantity whose magnitude is expressed in units of length per unit time and whose direction is invariant; if the direction varies in circular fashion with time, the quantity is known as angular or rotational ve-

locity, and if it varies along a fluctuating or noncircular path the quantity is known as curvilinear velocity.

line-class valve A valve qualified by its design characteristics to be used as the first valve off the process line.

lined body A body having a lining which makes an interference fit with the disk in the closed position thus establishing a seal [S75.05].

lined valve body A valve body to which a protective coating or liner has been applied to internal surfaces of pressure containing parts or to the surfaces exposed to the fluid [S75.05].

line mixer See flow mixer.

line pressure See reference pressure.

line printer A computer printer that operates on a line-by-line (rather than character-by-character) basis for high-speed systems.

liner, slip-in An annular shaped liner which makes a slight interference fit with the body bore and which may be readily forced into position through the body end. May be plain or reinforced [S75.05].

line spectrum The spectrum of a complex wave consisting of several components having discrete frequencies.

lining The material used on the furnace side of a furnace wall. It is usually of high grade refractory tile or brick or plastic refractory material.

link See communication link.

linkage 1. A technique for providing interconnections between routines. 2. A mechanism consisting of bars, slides, pivots and rotating members, which transfers motion from one part of a machine to another.

linkage editor A computer program that produces a load module by transforming object modules into a format that is acceptable to fetch, combining separately produced object modules and previously processed load modules into a single load module, resolving symbolic cross references among them, replacing, deleting, and adding control sections automatically on request, and providing overlay facilities for modules requesting them.

linked list In data processing, a method of organizing data so that it is retrievable in an order that is not always the same order as the data is stored.

linker A computer program that binds together independently assembled programs. The program is developed in modules which are then linked together to form the whole.

link library A generally accessible partitioned computer data set which, unless otherwise specified, is used in fetching load modules referred to in execute (EXEC) statements and in attach, link, load, and transfer control (XCTL) macro instructions.

liquid barometer A simple device for measuring atmospheric pressure, which can be constructed by filling a glass tube having one closed end with a liquid such as mercury, then temporarily plugging the open end, inverting the tube into a container partly filled with the liquid, and unplugging the open end; if the liquid is mercury the tube must be at least 30 in. (76.2 mm) long; liquids of different densities require tubes of different lengths.

liquid cooled dissipator See cold plate.

liquid crystal display A type of digital display device.

liquid crystal light valve A device used in optical processing to convert an incoherent light image into a coherent light image.

liquid-filled thermometer Any of several designs of temperature-measurement devices that depend for their operation on predictable change in volume with temperature of a liquid medium confined in a closed system.

liquid knockout See impingement.

liquid level control A device for sensing and regulating the position of a liquid surface within a vessel.

liquid-level manometer A differential pressure gage in which the reading is obtained by viewing the change in level of one or both of the free surfaces of a liquid column spanning both gage legs.

liquid-metal embrittlement A decrease in strength or ductility of a solid metal caused by contact with a liquid metal.

liquid pressure recovery factor The ratio (F_1) of the valve flow coefficient (C_v) based on the pressure drop at the vena contracta, to the usual valve flow coefficient (C_v) which is based on the overall pressure drop across the valve in non-vaporizing liquid service. These coefficients compare with the orifice metering coefficients of discharge for vena contracta taps and pipe taps, respectively [S75.05].

Lissajous figure Pattern on an oscilloscope screen which indicates relative phase and magnitude of sinusoidal signals [S26].

list An ordered set of items contained within an electronic memory in such a way that only two items are readily program addressable. These items are the earliest appended (beginning item) and the most recently appended (ending item). Items stored into the list are "appended" following the ending item. Items read from the list are "removed." Same as push-up list.

listing The hard copy generated by a line printer; may also refer to a visual CRT display generated in lieu of hard copy.

list processing A method of processing data in the form of lists. Usually, chained lists are used so that the logical order of items can be changed without altering their physical location.

literal An element of a programming language that permits the explicit representation of character strings in expressions and command and function elements; in most languages, a literal is enclosed in either single or double quotes to denote that the enclosed string is to be taken "literally" and not evaluated.

litmus A blue, water-soluble powder derived from lichens and used as an acid-base indicator; it is blue at pH 8.3 and above, and is red at pH 4.5 and below.

litre Also spelled liter. Abbreviated L. The SI unit of volume; it equals 0.001 m^3 or 1.057 quarts.

live center A lathe center held in the headstock that rotates with the headstock and part being turned.

live front An assembly arrangement which has all moving or energized parts exposed on the front of the panel, framework or cabinet.

live part A part which is considered capable of rendering an electric shock.

live room An enclosed space characterized by unusually small capacity for absorbing sound.

live steam Steam which has not performed any of the work for which it was generated.

live zone See zone, live.

LLC See logical link control.

LLEI See low-level engineering interface.

LLHI See low-level human interface.

LLOI See low-level operator interface.

lm See lumen.

L network An electronic network composed of two branches in series, with the junction and the free end of one branch being connected to one pair of terminals and the free ends of both branches being connected to another pair of terminals.

load 1. An electrical device connected to the output terminals [RP55.1]. 2. To connect a signal receiving device to the output terminals of a signal source [RP55.1]. 3. To store a computer program or data into memory. 4. To mount a magnetic tape on a device so that the read point is at the beginning of the tape. 5. To place a removable disk in a disk drive and start the drive. 6. The amount of force applied to a structural member in service. 7. The quantity of parts placed in a furnace, oven or other piece of process equipment. 8. The quantity or mass of bulk material placed in a hopper, railcar or truck. 9. The power demand on an electrical distribution system. 10. The amount of power needed to start or maintain motion in a power-driven machine. 11. The process load is a term to denote the nominal values of all variables in a process that affect the controlled variable. 12. In an electric power circuit, the resistive and reactive components which comprise the device being powered by the circuit. 13. In a physical structure the externally applied force, or the sum of external forces and the weight of the structure borne by a single member or by the entire structure. See also load impedance.

load-and-go In data processing, an automatic coding procedure that not only compiles the program, creating machine language, but also proceeds to execute the created program; load-and-go procedures are usually part of a monitor.

load cell A transducer for the measurement of force or weight. Action is based on strain gages mounted within the cell on a force beam.

load circuit A circuit or a branch of a network which carries the main portion of current flow.

load factor The ratio of the average load in a given period to the maximum load carried during that period.

load impedance The impedance presented to the output terminals of a transducer by the associated external circuitry [S37.1]. See also impedance, load.

loading 1. That system connected to the output of a device, including the transmission network [S26]. 2. Buildup of material along the cutting edge of a bit or other tool. Similarly, buildup of grinding debris on the working face of a grinding wheel or abrasive disc.

loading error An error due to the effect of the load impedance on the transducer output. In the case of force transducers the term loading has been applied to application of force [S37.1].

loading point The location at which material to be conveyed is applied to the conveyor [RP74.01].

load module A program prepared in a format that is ready for loading and executing.

load point (mag tape) The point, near the beginning of the tape, at which the computer can start to record data.

load reactor A device that generates a signal proportional to the force imposed upon it by the load sensor [RP74.01].

load regulation The change in output (usually speed or voltage) from no-load to full-load (or other specified load limits). It may be expressed as the percentage ratio of the change from no-load to full load divided by the no-load value [S51.1]. See offset.

load resistance The load resistance is the sum of the resistances of all connected receivers and the connection lines [S50.1].

load sensor See weigh carriage. Also called load receiving element.

local The location of an instrument that is neither in nor on a panel or console, nor is it mounted in a control room. Local instruments are commonly in the vicinity of a primary element or a final control element. The word field is often used synonymously with local [S5.1].

Local Area Network (LAN) 1. A communications mechanism by which computers and peripherals in a limited geographical area can be connected. They provide a physical channel of moderate to high data rate (1 to 20 Mbit) which has a consistently low error rate (typically 10^{-9}). 2. The connecting of several data processing machines that can share programs, data files and printers.

Local Control Unit (LCU) A control device that performs closed-loop control and interfaces directly with the process.

local oscillator An oscillator whose output is combined with another frequency to generate a sum or difference frequency either of which may be easier to amplify and use, as in a superheterodyne receiver.

local panel A panel that is not a central or main panel. Local panels are commonly in the vicinity of plant subsystems or sub-areas. The term local panel instrument should not be confused with local instrument [S5.1].

local processing unit Field station with input/output circuitry and the main processor. These devices measure analog and discrete inputs, convert these inputs to engineering units, perform analog and logical calculations (including control calculations) on these inputs and provide both analog and discrete (digital) outputs.

local reference A copper bar mounted on the cabinets of a subsystem which become the signal reference point for the entire subsystem. All power commons and signal commons of a subsystem are tied to the local reference. Each local reference is tied to the master reference, by a separate wire.

location An address in computer storage or memory where a unit of data or an instruction can be stored.

location counter 1. In data processing, the control-section register which contains the address of the instruction currently being executed. 2. A register in which the address of the current instruction is recorded. Synonymous with instruction counter and program address counter.

location, hazardous (classified) That portion of a plant where flammable or combustible liquids, vapors, gases or dusts may be present in the air in quantities sufficient to produce explosive or ignitable mixtures [S51.1].

lock 1. In a forging, having the flash line in more than one plane. 2. A device for securing a door, drawer or hatch that features a movable bolt operated by a key. 3. To prevent a movable part from moving; to seize.

locked-in liner In a butterfly valve body, a liner retained in the body bore by a key ring or other means.

lock-in A sequence feature that retains the alarm state until acknowledged when the abnormal process condition is momentary [S18.1].

lock-in amplifier An amplifier which selects signals at one prespecified frequency and amplifies them, while discriminating against signals at other frequencies.

locking Pertaining to code extension characters that change the interpretation of an unspecified number of following characters. Contrast with nonlocking.

lockout Any condition which prevents any or all senders or receivers from communicating.

lock-step A method to synchronize a mixed-signal simulation system where each simulator progresses one time step and passes all interacting signals to the other simulator.

locomotive boiler A horizontal fire-tube boiler with an internal furnace the rear of which is a tube sheet directly attached to a shell containing tubes through which the products of combustion leave the furnace.

log 1. A record of everything pertinent to a machine run including: identification of the machine run, record of alteration switch settings, identification of input and output tapes, copy of manual key-ins, identification of all stops, and a record of action taken on all stops. 2. To record occurrences in a chronological sequence.

logarithmic amplifier An amplifier whose output is a logarithmic function of its input.

logarithmic decrement In an exponentially damped oscillation, the natural logarithm of the ratio of one peak value to the next successive peak value in the same direction.

logger A device which automatically records physical processes and events, usually chronologically.

logic 1. A means of solving complex problems through the repeated use of simple functions which define basic concepts. Basic logic functions are "AND", "OR", "NOT", etc. 2. The science dealing with the criteria or formal principles of reasoning and thought. 3. The systematic scheme which defines the interactions of signals in the design of an automatic data processing system. 4. The basic principles and application of truth tables and interconnection between logical elements required for arithmetic computation in an automatic data processing system. Related to symbolic logic.

logical block An arbitrarily-defined, fixed number of contiguous bytes used as the standard I/O transfer unit throughout a computer operating system. For example, the commonly used logical block in PDP-11 systems is 512 bytes long. An I/O device is treated as if its block length is 512 bytes, although a device's actual (physical) block length may be different. Logical blocks on a device are numbered from block 0 consecutively up to the last block on the volume.

logical connectives The computer operators or words, such as and, or, or else, if then, neither, nor, and except, which make new expressions from given expressions and which have the property that the truth or falsity of the new expressions can be calculated from the truth or falsity of the given expressions and the logical meaning of the operator.

logical decision 1. The choice or ability to choose

between alternatives. Basically, this amounts to an ability to answer yes or no with respect to certain fundamental questions involving equality and relative magnitude. 2. The utilization of a logic instruction.

logical device name An alphanumeric name assigned by a user to represent a physical device; the name can be used synonymously with the physical device name in all references to the device. Logical device names are used in device-independent systems to enable a program to refer to a logical device name that can be assigned to a physical device at run time.

logical difference All elements belonging to Class A but not to Class B, when two classes of elements, Class A and Class B, are given.

logical element The smallest building block in a computer or data processing system, which can be represented by logical operators in an appropriate system of symbolic logic. Typical logical elements are the and-gate and the or-gate, which can be represented as operators in a suitable symbolic logic.

logical expression A logical expression consists of logical constants, variables, array elements, function references, and combinations of those operands, separated by logical operators and parentheses.

logical link control (LLC) The upper sublayer of the Data Link Layer (Layer 2) used by all types of IEEE 802 Local Area Networks. LLC provides a common set of services and interfaces to higher layer protocols. Three types of services are specified: a) Type 1 - Connectionless: A set of services that permit peer entities to transmit data to each other without the establishment of connections. Type 1 service is used by both MAP and TOP; b) Type 2 - Connection Oriented: A set of services that permit peer entities to establish, use and terminate connections with each other in order to transmit data; c) Type 3 - Acknowledged Connectionless: A set of services that permit a peer entity to send messages requiring immediate response to another peer entity. This class of service can also be used for polled (master-slave) operation.

logical operation 1. An operation in which logical (yes or no) quantities form the elements being operated on, e.g., AND, OR. 2. The operations of logical shifting, masking, and other nonarithmetic operations of a computer. Contrasted with arithmetic operation.

logical operator See logical connectives.

logical product Same as AND.

logical record A logical unit of data within a file whose length is defined by the user and whose contents have significance to the user; a group of related fields treated as a unit.

logical sum A result, similar to an arithmetic sum, obtained in the process of ordinary addition, except that the rules are such that a result of one is obtained when either one or both input variables is a one, and an output of zero is obtained when the input variables are both zero. The logical sum is the name given the result produced by the inclusive OR operator.

logical unit number A number associated with a physical device unit during a task's I/O operations; each task in the system can establish its own correspondence between logical unit numbers and physical device units.

logical variable A variable that may have only the value true or false. Also called Boolean variable.

logic analyzer A device used to analyze the logical operation of the microcomputer. It is a test device used for debugging systems.

logic cabinet A cabinet containing logic circuits and no visual displays [S18.1].

logic design The specification of the working relations between the parts of a computer system in terms of symbolic logic and without primary regard for hardware implementation.

logic diagram 1. In data processing, a diagram that represents a logic design and sometimes the hardware implementation. 2. Graphic method of representing a logic operation or set of operations.

logic gate A device that takes binary bits as input and produces an output bit to some specification.

logic instruction A computer instruction that executes an operation that is defined in symbolic logic, such as AND, OR, NOR.

logic levels Electrical convention for representing logic states. For TTL systems, the logic levels are nominally 5V for logic 1, and 0V for logic 0.

logic network In data processing, an arrangement of logic gates designed to achieve specific outputs.

log-in See log-on.

log-on In data processing, to enter into a system or network.

log-out   In data processing, to exit from a system or network.

long flame burner   A burner in which the fuel emerges in such a condition, or one in which the air for combustion is admitted in such a manner, that the two do not readily mix, resulting in a comparatively long flame.

longitudinal drum boiler   A sectional header or box header boiler in which the axis on the horizontal drum or drums is parallel to the tubes in a vertical plane.

longitudinal interference   See common mode interference.

longitudinal redundancy check (LRC)   A system of error control based on the formulation of a block check following preset rules. The check formation rule is applied in the same manner to each character.

longitudinal wave   A wave in which the medium is displaced in a direction perpendicular to the wave front at all points along the wave.

look-up table   Same as table, and not to be confused with the verb form, table look up.

loop   1. A combination of two or more instruments or control functions arranged so that signals pass from one to another for the purpose of measurement and/or control of a process variable [S5.1][S67.02]. 2. A sequence of instructions that is executed repeatedly until a terminal condition prevails. 3. Synonymous with control loop. See closed loop and open loop. 4. The doubled part of a cord, wire, rope or cable; a bight or noose. 5. A complete hydraulic, electric, magnetic or pneumatic circuit. 6. A length of magnetic tape or motion picture film that has been spliced together, end-to-end, so it can be played repeatedly without interruption. 7. In data processing, a closed sequence of instructions that are repeated. 8. All the parts of a control system: process, or, sensor, any transmitters, controller, and final control element. 9. In a computing program: A sequence of instructions that is written only once but executes many times (iterates) until some predefined condition is met.

loop, closed (feedback loop)   A signal path which includes a forward path, a feedback path and a summing point, and forms a closed circuit [S51.1].

loop (computing)   Instructions that actually perform the primary function of a loop, as distinguished from loop initialization, modification, and testing.

loop diagram   A schematic representation of a complete hydraulic, electric, magnetic or pneumatic circuit.

loop, feedback   See loop, closed (feedback loop).

loop gain   The product of the gains of all the elements in a loop. See also gain, loop.

loop gain characteristics   In process instrumentation, of a closed loop, the characteristic curve of the ratio of the change in the return signal to the change in the corresponding error signal for all real frequencies [S51.1].

loop identification   1. Consists of a first-letter and a number of an instrument loop. Each instrument within a loop has assigned to it the same loop number and, in the case of parallel numbering, the same first letter. 2. Each instrument loop has a unique loop identification.

loop (initialization)   Instructions that immediately precede the loop proper, that set addresses, counters, or data to initial values.

loop-modification   Instructions of a loop that alter instruction addresses, counters, or data.

loop, open   A signal path without feedback [S51.1].

loop-testing   Instructions of a loop that determine whether the loop is complete.

loop transfer function   Of a closed loop, the transfer function obtained by taking the ratio of the Laplace transform of the return signal to the Laplace transform of its corresponding error signal [S51.1].

loose stem   A design in which the stem is not physically or mechanically attached to the ball, but drives the ball through intimate contact of surfaces. Typical loose stem drives are: A. tang, B. pin, C. splined [S75.05].

loran   A navigation aid consisting of long-range pulsed radio waves; positions are determined by measuring the time of arrival of synchronized pulses and finding the intersection of position lines determined from signals transmitted by two or more fixed transmitters.

loss   1. Dissipation of power, which reduces the efficiency of a machine or system. 2. Dissipation of material or energy due to leakage.

lost cluster   A group of one or more disk sectors that are not available for storage use.

loudness   The relative auditory intensity of a sound wave.

loudness level   A measurement of sound intensity numerically equal to the sound pressure, in decibels, relative to 0.0002 microbar, of a simple tone whose frequency is 1000 Hz and which is judged by the listeners to be equivalent in loudness; the units of measure determined in this way are called phons.

loudspeaker An electroacoustic transducer usually constructed to effectively radiate sound of varying frequencies into the air.

low-alloy steel An iron-carbon alloy which contains up to about 1% C, and less than 5% by weight of additional elements.

low brass A binary copper-zinc alloy containing about 20% zinc.

low-carbon steel An iron-carbon alloy containing about 0.05 to 0.25% C, and up to about 0.7% Mn.

low draft switch A control to prevent burner operation if the draft is too low. Used primarily with mechanical draft.

lower limit 1. The lower limit of the signal current is the current corresponding to the minimum value of the dc current signal [S50.1] [S7.4]. 2. The signal corresponding to the minimum value of the transmitted input.

lower range-limit See range-limit, lower.

lower range-value See range-value, lower.

low-fire start The firing of a burner with controls in a low-fire position to provide safe operating condition during light-off.

low gas pressure switch A control to stop the burner if gas pressure is too low.

low head boiler A bent tube boiler having three drums with relatively short tubes in a vertical plane.

low-heat value The high heating value minus the latent heat of vaporization of the water formed by burning the hydrogen in the fuel.

low-hydrogen electrode A covered welding electrode that provides an atmosphere around a welding arc which is low in hydrogen.

low-level engineering interface (LLEI) A type of LLHI designed for use by an instrumentation engineer.

low-level human interface (LLHI) A device that allows a human to interact with a local control unit.

low-level language In data processing, a program instruction that usually has a single machine instruction.

low-level operator interface (LLOI) A type of LLHI designed for use by a process operator.

low limiting control See control, low limiting.

low oil temperature switch (cold oil switch)
A control to prevent burner operation if the temperature of the oil is too low.

low order Pertaining to the weight or significance assigned to the digits of a number, e.g. in the number, 123456, the low order digit is six. One may refer to the three low order bits of a binary word, as another example. See order.

lowpass filter A filter which passes frequencies below its cut-off frequency with little attenuation.

low-pass output filter (LPOF) In a subcarrier discriminator, the filter which rejects subcarrier components and all extraneous noise while passing the frequencies which are known to contain data.

low-pressure hot-water and low-pressure steam boiler A boiler furnishing hot water at pressures not exceeding 160 pounds per square inch or at temperatures not more than 250 °F or steam at pressures not more than 15 pounds per square inch.

low resolution graphics In data processing, the ability of a dot-matrix printer to reproduce simple forms or pictures.

low-temperature hygrometry The measurement of water vapor at low temperatures; requires special techniques because of the small amounts of moisture typically present and because of unusual instrument operating characteristics at such temperatures.

low water cutoff A device to stop the burner on unsafe water conditions in the boiler.

LP See linear programming.

LPOF See low-pass output filter.

L_{po} Sound pressure level at a point four feet downstream of a valve and three feet from the surface of the pipe.

LSB See least significant bit.

LSI See large-scale integration.

lubricant ring See lantern ring.

lubricated packing box A packing arrangement consisting of a lantern ring with packing rings above and below with provision to lubricate the packing [S75.05].

lubricator A device for automatically applying lubricant.

lubricator isolating valve In a control valve, an isolating valve is a small hand operated valve located between the packing lubricator assembly and the packing box assembly. It shuts off the fluid pressure from the lubricator assembly [S75.05].

lug   Any projection, like an ear, used for supporting or grasping.

lugged body   See body, wafer, lugged.

lumen   Metric unit for measuring the flux or power of light visible to the human eye; the photometric equivalent of the watt.

luminance   The luminous intensity of any surface in a given direction per unit of projected area in a plane perpendicular to that direction.

luminosity   Emissive power with respect to visible radiation.

luminosity coefficients   The constant multipliers for the respective tristimulus values of any color such that the sum of the three products is the luminance of the color.

luminous   Emitting radiation in the form of visible light.

luminous dial   A dial or indicating scale and pointer whose scale divisions, numerals and pointer are made of or coated with a light-emitting substance such as luminous paint so that they can be seen in the dark. Compare with lighted dial.

luminous efficiency   Luminous flux divided by radiant flux.

luminous flux   The amount of light passing a given point per unit time.

lumped-constant wavemeter   A device for determining frequency using a tunable resonant LC circuit coupled to a crystal detector; the circuit generally utilizes plug-in coils of various inductances and a continuously variable capacitor with a dial calibrated in frequency.

lux   Metric unit of illuminance.

LVDT   See linear variable differential transformer.

LVRT   See linear variable reluctance transducer.

Lx   See lux.

M

m See metre.

MAC See media access control.

mach See machine.

Mach angle The angle between the path of a body moving with supersonic velocity and a corresponding Mach line; the speed of sound divided by the body's velocity equals the sine of the Mach angle.

machine Any device capable of performing useful work, especially a device for producing controlled motion or for regulating the effect of a given force.

machine address An absolute, direct unindexed address expressed as such, or resulting after indexing and other processing has been completed.

machine code The lowest level of computer language in the form of the digital code that can be directly executed by the computer.

machine code instruct A code that defines a particular computer operation that can be used without further translation.

machine-dependent program A program that operates on only one type of computer.

machine element Any standard mechanical part used in constructing a machine, such as a bearing, fastener, cam, gear, lever, link, pin or spring.

machine error A deviation from correctness in computer data resulting from an equipment failure.

machine-independent Pertaining to procedures or programs created without regard for the actual devices which will be used to express them.

machine-independent program A program that operates on a variety of different computers.

machine instruction An instruction that a machine can recognize and execute.

machine language 1. A language that is used directly by a machine. 2. Binary words (on the PDP-11 family, sixteen-bit) that are required to make the computer perform. 3. In software, the language which a computer understands; ones and zeros.

machine-language code Same as computer code and contrasted with symbolic code.

machine-language programming The term basically means programming using machine language. See machine-language code.

machine operator The person who manipulates the computer controls, places information media into the input devices, removes the output and performs other related functions.

machine-oriented language 1. A language designed for interpretation and use by a machine without translation. 2. A system for expressing information which is intelligible to a specific machine, e.g., a computer or class of computers. Such a language may include instructions which define and direct machine operations, and information to be recorded by or acted upon by these machine operations. 3. The set of instructions expressed in the number system basic to a computer, together with symbolic operation codes with absolute addresses, relative addresses, or symbolic addresses. Synonymous with machine language. Clarified by language. Related to object language and contrasted with problem-oriented language.

machine program 1. A program that is to be loaded in a computer and executed by it. 2. In numerical control, an ordered set of instructions in automatic control language and format and based on the part program, recorded on appropriate input media and sufficiently complete to effect the direct operation of an automatic control.

machine readable Data that will be accepted by a computer through an input device.

machine-readable medium A medium that can

231

convey data to a given sensing device.

machinery One or a group of machines; an apparatus or system constructed of machines.

machine word A unit of information of a standard number of bits or characters which a machine regularly handles in each transfer, e.g. a machine may regularly handle numbers or instructions in units of 36 binary digits.

machining center A versatile CNC machine tool with multi-axis control and usually automatic tool loading. The machining centers are designed to carry out a range of operations.

Mach number The ratio of the fluid velocity to the velocity of sound in the fluid, at the same temperature and pressure.

macro Directions for expanding abbreviated text; a boilerplate that generates a known set of instructions, data, or symbols. A macro is used to eliminate the need to write a set of instructions that are used repeatedly; for example, an assembly-language macro instruction enables the programmer to request the assembler to generate a predefined set of machine instructions.

macro-assembler An assembler which allows the use of macros and converts them to machine code.

macro instruction The more powerful instructions which allow a programmer to refer to several instructions as though they were a single instruction. When a programmer uses the name of a macro instruction, all of the instructions are inserted at that point in his coding by the macroprocessor.

macro modeling The representation of a component or device in terms of a net-list description of an equivalent circuit. Standard components, such as resistors or capacitors, are typically employed.

macroprocessor 1. A program which translates a single symbolic statement into one or more assembly language statements. 2. A phase of an assembler that has the capability of translating selected mnemonic or symbolic instructions into multiple machine-language instructions.

macro-program A program containing macros.

macroprogramming Programming with macro instructions.

macroscopic stress Load per unit area distributed over an entire structure or over a visible region of the structure.

macrostructure The features of a polycrystalline metal revealed by etching and visible at magnifications of 10 diameters or less.

magnetically actuated extensions A device attached to the meter body which contains an electrical switch and which is magnetically actuated by the metering float extension to signal a high or low flow. The switch is adjustable with respect to the float position over a range equal to the travel of the metering float. Standard switch ratings are usually 0.3 amperes for 110 volt, 60 cycle a-c supply (five amperes or more if relays are used).

magnetic amplifier An electronic amplification or control device that functions through the use of saturable reactors, either alone or in combination with other circuit elements.

magnetic bearing The angle between the line of sight to an object and the direction from the observer to magnetic North, measured in a plane parallel to the earth's surface.

magnetic biasing Simultaneous conditioning of a magnetic recording medium by superimposing a second magnetic field on the magnetic signal being recorded.

magnetic blowout switch A special type of switch designed to switch high d-c loads; a small permanent magnet contained in the switch housing deflects the arc to quench it when the contacts open.

magnetic bubble memory A high-density information storage device composed of a magnetic film only a few micrometres thick deposited on a garnet substrate; information is stored in small magnetized regions (bubbles) whose magnetic polarity is opposite to that of the surrounding region.

magnetic card A card with a magnetic surface on which data can be stored by selective magnetization of portions of the flat surface.

magnetic compass Any of several devices for indicating the direction of the horizontal component of a magnetic field, but especially for indicating magnetic North in the earth's magnetic field.

magnetic contactor A device for opening and closing one or more sets of electrical contacts which is actuated by either energizing or deenergizing an electromagnet within the device.

magnetic core 1. A configuration of magnetic material that is, or is intended to be, placed in a spatial relationship to current-carrying conductors and whose magnetic properties are essential to its use. It may be used to concentrate an induced magnetic field, as in transformer, induction coil, or armature, to retain a magnetic polarization for the purpose of storing data, or for its nonlinear properties as in a logic element. It may be made of such material as iron, iron oxide, or ferrite and in such shapes as wires, tapes, toroids, or thin film. 2. A storage device in which binary data is represented by the direction of magnetization in each unit of an array of magnetic material, usually in the shape of toroidal rings, but also forms such as wraps on bobbins. Synonymous with core.

magnetic damping Progressive reduction of oscillation amplitude by means of current induced in electrical conductors due to changes in magnetic flux.

magnetic disk A flat, circular plate with a magnetic surface on which data can be stored by selective magnetization of portions of the flat surface.

magnetic drum A right circular cylinder with a magnetic surface on which computer data can be stored by selective magnetization of portions of the curved surface.

magnetic extensions A device that provides flow rate indication by means of a magnetic coupling between the extension of the metering float and an external indicator follower surrounding the extension tube.

magnetic field interference A form of interference induced in the circuits of a device due to the presence of a magnetic field. It may appear as common mode or normal mode interference in the measuring circuits. See also interference, electromagnetic.

magnetic float gage Any of several designs of liquid-level indicator that use a magnetic float to position a pointer or change the orientation of bicolor wafers.

magnetic float switch A device for operating a mercury switch by repositioning a magnetic piston with respect to a small permanent magnet attached to the pivoting mercury switch capsule; in the usual configuration, a float attached to the piston positions it near the small magnet when liquid level is high, and drops the piston out of proximity when the level is low, allowing a light spring to retract the magnet and pivot the mercury capsule.

magnetic focusing Causing an electron beam to become diverging or converging to position an image or beam on an object—usually CRT screen—by interacting with a magnetic or electromagnetic field.

magnetic hardness comparator A device for determining hardness of a steel part by comparing its response to electromagnetic induction with the response of a similar part of known hardness.

magnetic head A transducer for converting electrical signals into magnetic signals suitable for storing on magnetic recording media, for converting stored magnetic signals into electrical output signals, or for erasing stored magnetic signals.

magnetic ink An ink which contains magnetic particles. Characters printed in magnetic ink can be read both by humans and by machines designed to read the magnetic pattern.

magnetic lens Electric coils electromagnets or permanent magnets assembled into a configuration that can accomplish magnetic focusing.

magnetic printing Permanently transferring a recorded signal from one magnetic recording medium to another magnetic recording medium (or to another portion of the same medium) by bringing the two sections into close proximity.

magnetic proximity sensor Any of several devices that are activated when a magnetized or ferromagnetic object passes within a defined distance of the active element; there are four types—variable-reluctance sensors, hermetically sealed dry-reed switches, Hall-effect switches, and Weigand-effect sensors.

magnetic recorder A device for producing a stored record of a variable electrical signal as a variable magnetic field in a ferromagnetic recording medium.

magnetic separator A machine that uses strong magnetic fields to remove pieces of magnetic material from a mixture of magnetic material and nonmagnetic or less strongly magnetic material.

magnetic shield A metal shield which insulates the contents from external magnetic fields. Such shields are often used with photomultiplier tubes.

magnetic storage A device or devices which utilize the magnetic properties of materials to store information.

magnetic tape 1. A tape with a magnetic surface on which computer data can be stored by selective polarization of portions of the surface. 2. A tape of magnetic material used as the constituent in some forms of magnetic cores.

magnetic tape encoder An electronic device that will accept data from a keyboard and write it to magnetic tape.

magnetic test coil A coil used in conjunction with a suitable indicating or recording instrument to measure variations or changes in magnetic flux when the coil is linked with a magnetic field.

magnetic variometer An instrument for measuring variations in magnetic field strength with respect to space or time.

magnet meter An instrument for measuring the magnetic flux of a permanent magnet under specified conditions; it usually incorporates a torque coil or a moving-magnet magnetometer with a unique arrangement of pole pieces.

magnetometer An instrument for measuring the magnitude of a magnetic field, and sometimes for also determining its direction.

magnetostriction A characteristic of some ferromagnetic materials whereby their physical dimensions vary with the intensity of an applied magnetic or electromagnetic field.

magnetostrictive effect An inherent property of some ferromagnetic materials to deform elastically, thereby generating mechanical force, when subjected to a magnetic field.

magnetostrictive resonator A ferromagnetic rod so constructed that an alternating magnetic field can excite it into resonance at one or more frequencies.

magnification 1. The ratio of output to input signal magnitudes. 2. Attaining a change in magnitude without a change in power. 3. Producing an enlarged visual image. 4. The ratio of a specific dimension on a virtual image to the corresponding dimension on the physical object being viewed.

magnitude ratio Ratio of output signal magnitude to input signal magnitude [S26]. See also gain; see also gain, magnitude ratio.

magnitude signal Peak-to-peak value of signal [S26].

mag tape In computer systems, the nine-track (seven-track in some cases) 1/2-inch medium on which data can be stored and transferred to other computers.

main fractionator A large, multiproduct distillation column used to separate the effluent from a reacting system. Some of these product streams will go on to further distillation processing.

mainframe 1. The central processor of the computer system. It contains the main storage, arithmetic unit, and special register groups. Synonymous with CPU and central processing unit. 2. All that portion of a computer exclusive of the input, output, peripheral, and in some instances, storage units.

main-line class Refers to the pressure and temperature ratings, the material from which the pipe is constructed, and the appropriate code, such as ANSI B31.1 [S67.10].

main memory The set of storage locations connected directly to the central processing unit; also called (generically) core memory.

main program The module of a computer program that contains the instructions at which program execution begins; normally, the main program exercises primary control over the operations performed and calls subroutines or subprograms to perform specific functions.

main storage 1. Usually the fastest storage device of a computer and the one from which instructions are executed. Contrasted with storage, auxiliary [RP55.1]. 2. The fastest general-purpose storage of a computer.

maintain 1. To keep in continuance or in a certain state, as of repair. 2. To preserve or keep in a given existing condition, as of efficiency or repair.

maintainability 1. The relative ability of a device or system to remain in operation, requiring only routine scheduled maintenance and occasional unscheduled maintenance without extensive periods of downtime for major repairs. 2. The probability that a device will be restored to operating condition within a specified period of time when maintenance is done with prescribed resources and procedures. 3. The inherent characteristic of a design or installation that determines the ease, economy, safety, and accuracy with which maintenance actions can be performed. 4. The ability to restore a product to service or to perform preventive maintenance within required limits.

maintained 1. An alarm that returns to normal after being acknowledged. 2. Remains maintained after removal of pressure or signal.

maintained alarm See alarm.

maintenance 1. Any act that either prevents equipment failure or malfunction, or restores operating capability following a failure or malfunction. 2. Any activity intended to eliminate faults or to keep computer hardware or programs in satisfactory working condition, including test, measurements, replacements, adjustments, and repairs.

maintenance conditions Conditions under which maintenance is performed.

Maintenance Engineer 1. Assist maintenance supervisors, technicians, and mechanics in maintaining the facilities/systems by providing in-depth engineering assistance on high maintenance equipment. 2. Assist design and project engineers in plant start-up, retrofits and modifications to ensure that the design is engineered for maintenance. Ensures standardization of equipment and systems. 3. Provides assistance in determining the frequency of preventive maintenance. 4. Determine service life of equipment by cost to repair and frequency of repair. 5. Ensures that the equipment continues to conform to the technical specifications. 6. Ensures that equipment is kept in calibration. 7. Work with management to ensure customer/production/maintenance cooperation by providing systems and equipment so that the end item meets all specifications and production schedules. 8. Ensure that new systems are designed to be maintainable. 9. Ensure proper documentation is provided. 10. Suggest and provide training when required to ensure that the maintenance personnel have the knowledge to maintain the equipment or system. 11. Help determine if in-house maintenance or contract maintenance is more feasible. 12. Ensure operational maintenance is being performed. 13. Ensure maintenance mechanics, technicians, technologists and engineers are involved in a project from the very beginning, looking at instrument/wiring/equipment standardization, new equipment training, test equipment, and equipment calibration.

maintenance engineering Developing concepts, criteria, and technical requirements for maintenance during the conceptual and acquisition phases of a project. Providing policy guidance for maintenance activities and exercising technical and management direction and review of maintenance programs.

Maintenance Management System A part of the Management Information System (MIS) that is useful for maintaining the companies equipment. It accesses equipment information, spare parts availability and location, maintenance work order systems, preventive maintenance systems, maintenance personnel qualifications, equipment maintenance history, and any other information that will help the maintenance engineer, supervisor, technician or mechanic be more proficient in his job.

maintenance time Time used for equipment maintenance. It includes preventive maintenance time and corrective maintenance time.

major diameter The largest diameter of a screw thread; it is measured at the crest of an external thread and at the root of an internal thread.

major frame With reference to telemetry formats, the time period where all data of a multiplex are sampled at least once; includes one or more minor frames. Major frame length is determined as (N)(Z) words, where N = the number of words per minor (prime) frame and Z = the number of words in the longest submultiple frame.

major graduations Intermediate graduation marks on a scale which are heavier or longer than other graduation marks but which are not index graduations.

majority A logic operator having the property that if P is a statement, Q is a statement, and R is a statement, then the majority of P, Q, R is true if more than half the statements are true, false if half or less are true.

major time In telemetry computer systems, two sixteen-bit words: minutes/seconds, and hours/days.

make/break components Components having contacts which can interrupt a circuit (even if the interruption is transient in nature). Examples of make/break components are relays, circuit breakers, servo potentiometers, adjustable resistors, switches, connectors, and motor brushes [S12.12].

make-up The water added to boiler feed to compensate for that lost through exhaust, blowdown, leakage, etc.

male fitting An element of a connection in pipe, tubing, electrical conductors or mechanical assemblies that fits into the mating (female) element; for example, the externally threaded end of a pipe fitting is termed male.

malfunction The effect of a fault.

malfunction routine Same as diagnostic routine.

malleable iron A somewhat ductile form of cast iron made by heat treating white cast iron to convert the carbon-containing phase from iron carbide to nodular graphite.

man See manual.

Management Information System (MIS) A computerized system using a large database containing information on: a)customers, b) equipment, c) supplies, d) spare parts, e) personnel, f) process, g) sales forecast, h) history, i) costs, j) profits, etc. Selected information is available to those persons making decisions.

Manager, Control Systems and Instrumentation Responsible for all departmental/sectional activities, including: a) defining and reporting activities and needs to management. b) maintaining liaisons with other departments/sections to coordinate work assignments. c) providing administrative and technical support to other department/sections. d) assigning of projects to personnel and manpower scheduling, recruitment, evaluation, and salary review. e) managing development of training program. f) making decisions on crucial or complex project activities. g) supervising all departmental/sectional personnel. h) preparing departmental/sectional budgets, forecasts and goals. i) promoting safety.

manhead The head of a boiler drum or other pressure vessel having a manhole.

manhole The opening in a pressure vessel of sufficient size to permit a man to enter.

manifold 1. A pipe or header for collecting a fluid from, or the distributing of a fluid to a number of pipes or tubes. 2. A branch pipe which distributes intake or exhaust fluids to a series of valve ports, as in a multicylinder engine such as an automobile engine.

manifold equalizing line The conduit within a manifold which connects the high and low differential pressure impulse lines.

manifold (instrumentation) Any configuration of valves which can be manipulated to create zero differential pressure at the measuring instrument.

manifold pressure The fluid pressure in the intake manifold of an internal combustion engine.

manifold variable A quantity or condition which is varied so as to change the value of the controlled variable.

manipulated variable 1. In a process that is desired to regulate some condition, a quantity or a condition that is altered by the control in order to initiate a change in the value of the regulated condition. 2. The part of the process which is adjusted to close the gap between the set point and the controlled variable. See also variable, manipulated.

manipulative variable In a control loop, the variable that is used by the controller to regulate the controlled variable.

manipulators Mechanical devices for the remote handling of hazardous materials; they are usually hand operated, often from behind a shield, and may or may not be power assisted.

manometer A gage for measuring pressure or a pressure difference between two fluid chambers. A U-tube manometer consists of two legs, each containing a liquid of known specific gravity.

manometric equivalent The length of a vertical column of a given liquid at standard room temperature which indicates a pressure differential equal to that indicated by a 1-mm-long column of mercury at 0 °C.

mantissa See floating point.

manual backup An alternate method of process control by means of manual adjustment of final control elements in the event of a failure in the computer system.

manual control The operation of a process by means of manual adjustment of final control elements.

manual controller A control device whose output signals, power or motions are all varied by hand.

manual data entry module A device which monitors a number of manual input devices from one or more operator consoles and/or remote data entry devices and transmits information from them to the computer.

manual input 1. The entry of data by hand into a device at the time of processing. 2. The data entered as in definition 1.

manual loading station A device or function having a manually adjustable output that is used to actuate one or more remote devices. The station does not provide switching between manual and automatic control modes of a control loop (see controller and control station). The station may have integral indicators, lights, or other features. It is also known as a manual station or a manual loader [S5.1].

manual operation Processing of data in a system by direct manual techniques.

manual override A device to manually impart motion in either one or two directions to the valve stem. It may be used as a limit stop [S75.05]. See also handjack and/or handwheel.

manual rest See reset.

manual station 1. Synonymous with manual loading station. 2. A single loop hard manual control to operate the final control devices in case of control system failure. 3. Provides for bypassing normal controller operation to manually vary an analog output signal in a controller. Used primarily in an emergency, or possibly during a maintenance shutdown of the controller.

manufactured gas Fuel gas manufactured from coal, oil, etc., as differentiated from natural gas.

manufacturer software A complex program package that develops the user's application and organizes computer procedures to obtain efficient response to the application program. Often this software is referred to as an operating system.

Manufacturing Automation Protocol (MAP) A specification for a suite of communication standards for use in manufacturing automation developed under the auspices of General Motors Corporation. The development of this specification is being taken over by the MAP/TOP Users Group under the auspices of CASA/SME.

Manufacturing Messaging Format Standard (MMFS) One of the application protocols specified by MAP.

map 1. To establish a correspondence or relationship between the members of one set and the members of another set and perform a transformation from one set to another, for example, to form a set of truth tables from a set of Boolean expressions. Information should not be lost or added when transforming the map from one to another. 2. See memory map.

MAP See Manufacturing Automation Protocol.

MAP/EPA Part of the EPA architecture, a MAP/EPA node contains both the MAP protocols and the protocols required for communication to MINI-MAP. It can communicate with both MINI-MAP nodes on the same segment and full MAP nodes anywhere in the network.

mapped system A system that uses the computer hardware memory management unit to relocate virtual memory addresses.

MAP/TOP Users Group United States and Canada's MAP/TOP Users Group. See CASA/SME.

marginal check A preventive-maintenance procedure in which certain operating conditions are varied about their normal values in order to detect and locate incipient defective units, e.g., supply voltage or frequency may be varied.

Synonymous with marginal test and high-low bias test, and related to check.

marginal test Same as marginal check.

margin of safety The ratio between maximum service load (allowable design load) for a structure and the load that would cause the structure to deform, collapse or break.

marine engineering A branch of engineering that deals with the design, construction and operation of shipboard propulsion systems and associated auxiliary machinery.

mark A sign or symbol used to signify or indicate an event in time or space, *e*.g., end of word or message mark, a file mark, a drum mark, or an end of tape mark.

marking pointer An adjustable stationary pointer, usually of a color different from that of the indicating pointer, that can be positioned opposite any location on the scale of interest to the user.

markings Information shown on the transducer itself, will normally include manufacturer, model number and serial number [RP37.2].

Markov chain A probabilistic model of events, in which the probability of an event is dependent only on the event that precedes it.

mark-sense To mark a position on a punch card with an electrically conductive pencil, for later conversion to machine punching.

mark-sense device An electronic machine that will read mark-sensed forms.

mark sensing A technique for detecting special pencil marks entered in special places on a punch card and automatically translating the marks into punched holes. See sensing, mark.

Marx generators A high voltage electrical pulse generator in which capacitors are charged in parallel, then discharged in series to generate a voltage much higher than the charging voltage.

MASER See Microwave Amplification by the Stimulated Emission of Radiation.

mask 1. A protective face covering which usually provides for filtration of breathing air or for attachment to an external supply of breathing air. 2. A frame or similar device to prevent certain areas of a workpiece surface from being coated, as with paint. 3. A frame that conceals the edges of a cathode-ray tube, such as a television screen. 4. A machine word or register that specifies which parts of another machine word or register are to be operated on.

masking 1. The process of extracting a nonword

group or a field of characters from a word or a string of words. 2. The process of setting internal program controls to prevent transfers which otherwise would occur upon setting of internal machine latches.

mask programmed memory Computer memory dedicated to the storage of a particular set of data. A mask containing the particular pattern of bits is used in the manufacture of the memory.

mass Amount of matter an object contains.

MASSBUSS The thirty-two-bit direct-memory-access bus on the PDP-11/70 and VAX-11 computers.

mass feedwater flow rate The mass flow rate of all water delivered to the boiler, derived either from direct process measurements and/or calculations from other parameters. When volumetric feedwater flow rate measurement techniques are employed and the feedwater temperature at the flow-measuring element varies 100 °F (37.8 °C), the measured (indicated) flow shall be compensated for flowing feedwater density to determine the true mass feedwater flow rate [S77.42].

mass flow The amount of fluid, measured in mass units, that passes a given location or reference plane per unit time.

mass-flow bin A bin with steep, smooth sides which allow its contents to flow, without stagnant regions, whenever some of the contents are withdrawn.

mass flowmeter An instrument for measuring the rate of flow in a pipe, duct or channel in terms of mass per unit time.

mass flow rate The mass of fluid moving through a pipe or channel within a given period of time.

mass number The sum of the number of protons and the number of neutrons in the nucleus of a specific nuclide.

mass spectrograph A mass spectroscope which records intensity distributions on a photographic plate.

mass spectrometer A mass spectroscope which uses an electronic instrument to indicate intensity distribution in the separated ion beam.

mass spectroscope An instrument for determining the masses of atoms or molecules, or the mass distribution of an ion mixture, by deflecting them with a combination of electric and magnetic fields which act on the particles according to their relative masses.

mass spectrum In a mixture of ions, the statistical distribution by mass or by mass-to-charge ratio.

mass steam flow rate The mass flow rate of steam from the boiler, derived either from direct process measurements and/or calculations from other parameters. If volumetric steam flow-rate measuring techniques are employed, the measured (indicated) flow shall be compensated for flowing steam density to determine the true mass steam flow rate [S77.42].

mass storage Pertains to a computer device that can store large amounts of data so that they are readily accessible to the central processing unit; for example, disks, DEC tape, or magnetic tape.

mass velocity Mass flow per unit cross-sectional area.

master 1. A device which controls other devices in a system. 2. A precise pattern for making replicate workpieces, as in certain types of casting processes.

master clock A device which functions as the primary source of timing signals.

master file directory The system-maintained file on a volume that contains the names and addresses of all the files stored on the volume.

master flowmeter Flowmeter used as an interlaboratory standard in correlation checks of calibration systems [RP31.1].

master gage A device with fixed locations for positioning parts or holes in three dimensions.

master recipe A basic recipe which has been made site-specific.

master reference A signal point which is the signal reference point for an entire system. Usually a ground rod or grid. All local references are tied back to the master reference point.

master-slave A mode of operation where one data station (the master) controls the network access of one or more data stations (the slaves).

master/slave manipulator A remote manipulator which mechanically, hydromechanically or electromechanically reproduces hand or arm motions of an operator.

material balance 1. The procedure of accounting for the mass of material going into a process versus the mass leaving the process. 2. The balance relating the material in and material out of a distillation column. Material-balance manipulative variables are overhead flow, bottoms flow, sidestreams flow, and feed flow.

material dispersion Light pulse broadening due to differential delay of various wavelengths of light in a waveguide material. This group delay is aggravated by broad bandwidth light sources.

materials handling Transporting or conveying materials, parts or assemblies, including all aspects of loading, unloading, moving, storing and shipping them, both within a facility and between facilities.

materials science The study of materials used in research, construction and manufacturing; includes the fields of metallurgy, ceramics, plastics, rubber and composites.

mathematical check A check which uses mathematical identities or other properties, occasionally with some degree of discrepancy being acceptable, e.g., checking multiplication by verifying that AxB=BxA. Synonymous with arithmetic check.

mathematical logic Same as symbolic logic.

mathematical model The general characterization of a process, object, or concept, in terms of mathematics, which enables the relatively simple manipulation of variables to be accomplished in order to determine how the process, object, or concept would behave in different situations.

mathematical programming In operations research, a procedure for locating the maximum or minimum of a function subject to constraints. Contrast with convex programming, dynamic programming, integer programming, linear programming, nonlinear programming, and quadratic programming.

matrix 1. In mathematics, an n dimensional rectangular array of quantities. Matrices are manipulated in accordance with the rules of matrix algebra. 2. In computers, a logic network in the form of an array of input leads and output leads with logic elements connected at some of their intersections. 3. The principal microstructural constituent of an alloy. 4. The binding agent in a composite or agglomerated mass.

matrix printer A type of computer device that forms letters and symbols by printing a pattern of dots.

matte 1. A smooth but relatively nonreflective surface finish. 2. An intermediate product in the refining of sulfide ores by smelting.

max See maximum.

maximum allowable working pressure The highest gage pressure that can safely be applied to an internally pressurized system under normal operating conditions. It is usually well below the design bursting pressure and the hydrostatic test pressure for the system, and is the pressure at which relief valves are set to lift.

maximum continuous load The maximum load which can be maintained for a specified period.

maximum, error See error, maximum.

maximum excitation 1. The maximum value of excitation voltage or current that can be applied to the transducer at room conditions without causing damage or performance degradation beyond specified tolerances [S37.1]. 2. The maximum allowable voltage (current) applied to the potentiometric element at room conditions while maintaining all other performance characteristics within their limits. The excitation value is particularly associated with temperature [S37.12] [S37.6]. See excitation, maximum.

maximum instantaneous demand The sudden load demand on a boiler beyond which an unbalanced condition may be established in the boiler's internal flow pattern and/or surface release conditions.

maximum (minimum) ambient temperature The value of the highest (lowest ambient temperature that a transducer can be exposed to, with or without excitation applied, without being damaged or subsequently showing a performance degradation beyond specified tolerances [S37.1].

maximum (minimum) fluid temperature The value of the highest (lowest) measured-fluid temperature that a transducer can be exposed to, with or without excitation applied, without being damaged or subsequently showing a performance degradation beyond specified tolerances. NOTE: When a maximum or minimum fluid temperature is not separately specified it is intended to be the same as any specified maximum or minimum ambient temperature [S37.1].

maximum pointer A movable pointer that is repositioned as the indicating pointer of an instrument moves upscale, but remains stationary at the highest point reached when the indicating pointer moves downscale.

maximum thermometer A thermometer that indicates maximum temperature reached during a given interval of time; a clinical thermometer used to determine a patient's body temperature is one type of maximum thermometer.

maximum working pressure See pressure, maximum working (MWP).

Maxwell The CGS unit of magnetic flux.

239

Maxwell bridge A type of a-c bridge circuit in which the impedance of an unknown inductor is measured in terms of an adjustable resistor and adjustable inductor; since the latter may be difficult to obtain, an alternative bridge arrangement uses an adjustable resistor and capacitor in parallel with the unknown inductor.

Maxwellian distribution The velocity distribution of the moving molecules of a gas in thermal equilibrium, as determined by applying the kinetic theory of gases.

MBit Million bits per second.

MBM See magnetic bubble memory.

MByte 1,048,576(2^{20}) bytes.

McLeod vacuum gage A common type of mercury filled pressure gage whose design is a special case of a liquid manometer used as a pressure amplifier; the design enables use of a manometer-type instrument for measuring vacuum on the order of 10^{-6} torr instead of the 10^{-2} torr usually achieved with precision manometers.

mean accuracy See accuracy, mean.

mean effective pressure The average net pressure difference across a piston in a positive displacement machine such as a compressor, engine or pump. It is commonly used to evaluate performance of such a machine.

mean error (E) (data processing) The mean error is defined as the deviation between the mean value of a statistically significant number of output readings and the true value of the input signal. The mean error is expressed as a percentage of the full range (F.R.) [RP55.1].

mean free path In a gas, liquid or colloid, the average distance traveled by an individual atom, molecule or particle between successive collisions with other particles.

mean output curve The curve through the mean values of output during any one calibration cycle or a different specified number of calibration cycles [S37.1].

mean-time-between-failures (MTBF) The limit of the ratio of operating time of equipment to the number of observed failures as the number of failures approaches infinity. The total operating time divided by the quantity (n+1), where n is the number of failures during the time considered.

mean-time-to-failure (MTTF) The average or mean-time between initial operation and the first occurrence of a failure or malfunction, as the number of measurements of such time on many pieces of identical equipment approaches infinity.

measurand A physical quantity, property or condition which is measured. NOTE: The term measurand is preferred to input, parameter to be measured, physical phenomenon, stimulus, and variable [S37.1]. See also measured variable; see also variable, measured.

measured accuracy See accuracy, measured.

measured fluid The fluid which comes in contact with the sensing element. NOTE: The chemical and/or physical properties of this fluid may be specified to insure proper transducer operation [S37.1].

measured signal See signal, measured.

measured value See value, measured.

measured value of an analog dc current signal The measured value of an analog dc current signal is its specified mean value during the stated duration [S50.1].

measured value of a pneumatic transmission signal The indicated value during a stated duration [S7.4].

measured variable 1. The physical quantity, property, or condition which is to be measured. Common measured variables are temperature, pressure, rate of flow, thickness, speed, etc. 2. The part of the process that is monitored to determine the actual condition of the controlled variable. See also variable, measured.

measured variable modifier The second letter when first-letter is used in combination with modifying letters D (differential), F (ratio), M (momentary), K (time rate of change), Q integrate or totalize), (A could be used for absolute), or any combination of these is intended to represent a new and separate measured variable, and the combination is treated as a first-letter entity.

measurement 1. The determination of the existence or the magnitude of a variable [S5.1]. 2. A data point which is or can be converted into a suitable signal for telemetry transmission.

measurement component A general term indicating the components or subassemblies in a specific device that together determine the value of a quantity and produce the indicated or recorded output.

measurement device A self-contained assembly comprised of all the necessary components to perform one or more measuring operations.

measurement energy The energy, usually obtained from the measurand or the primary detector required to operate a measurement device or system.

measurement equipment A general term used to describe components, devices, assemblies or systems capable of performing measuring operations.

measurement mechanism A mechanical device that performs one or more operations in a measuring sequence.

measurement range The portion of the total response range of an instrument over which specific standards of accuracy are met.

measurement system Any set of interconnected components, including one or more measurement devices, that performs a complete measuring function, from initial detection to final indication, recording or control-signal output.

measuring-circuit voltage The voltage between two terminals of a measuring circuit or between one of these terminals and ground [S82.01].

measuring instrument See instrument, measuring.

measuring junction The electrical connection between the two legs of a thermocouple which is attached to the body, or immersed in the medium, whose temperature is to be measured.

measuring means The components of an automatic controller which determine the value of a controlled variable and communicate that value to the controlling means.

measuring modulator A component in a measuring system which modulates a direct-current or low-frequency alternating-current input signal to produce an alternating-current output signal whose amplitude is related to the measured value, usually as a preliminary step to producing an amplified output signal.

measuring range The extreme values of the measured variable within which measurements can be made within the specified accuracy. The difference between these extreme values is called "span."

measuring vessel The container in which the liquid metered by the turbine flowmeter during calibration interval is collected and measured . In a direct-gravimetric calibration system, this is a tank on a weigh scale and the exact dimensions are not significant. In indirect-gravimetric systems and volumetric systems the cross-sectional area or actual volume, respectively, must be known to a precision compatible with the desired accuracy of calibration [RP31.1].

mechanical Referring to tools or machinery.

mechanical atomizing oil burner A burner which uses the pressure of the oil for atomization.

mechanical chart drive A spring-driven clock mechanism which feeds continuous chart paper past a recorder head at a predetermined speed.

mechanical classification Any of several methods for separating mixtures of particles or aggregates according to size or density, usually involving the action of a stream of water.

mechanical compliance Displacement of a mechanical element per unit force; it is the mechanical equivalent of capacitance in an electrical circuit.

mechanical damping Attenuating a vibrational amplitude by absorption of mechanical energy.

mechanical draft The negative pressure created by mechanical means.

mechanical efficiency The ratio of power output to power input.

mechanical engineering A branch of engineering that deals with the generation and use of thermal and mechanical energy, and with the design, manufacture and use of tools and machinery.

mechanical hygrometer A hygrometer that uses an organic material, such as a bundle of human hair, to sense changes in humidity. In operation, the organic material expands and contracts with changes in moisture content in the air, and the change in length alters the position of a pointer through a spring-loaded mechanical linkage.

mechanical impedance The complex quotient of alternating force applied to a system divided by the resulting alternating linear velocity in the direction of the force at its point of application.

mechanical isolation of transduction element Internal construction of transducer which allows forces (particularly bending forces and external pressures) to be applied to the transducer case with negligible resulting forces on the transducer element [RP37.2].

mechanical limit stop A mechanical device to limit the valve stem travel [S75.05].

mechanical linkage A set of rigid bars, or links, that are joined together at pivot points and used to transmit motion. Frequently they are used in a mechanism along with a crank and slide to convert rotary motion to linear motion.

mechanical properties The properties of a

241

material that can be determined through application of a force and measuring the material's response.

mechanical reactance The imaginary component of mechanical impedance.

mechanical register A mechanical or electro-mechanical recording or indicating counter.

mechanical resistance The real component of mechanical impedance.

mechanical transmission An assembly of mechanical components suitable for transmitting mechanical power and motion.

mechanical scale A weighing device in which objects are balanced through a system of levers against a counterweight or counterpoise.

mechanical shock The momentary application of an acceleration force to a device. It is usually expressed in units of acceleration of gravity (g) [S51.1].

mechanism 1. Generally, an arrangement of two or more mechanical parts in which motion of one part compels the motion of the others. 2. Specifically the arrangement of parts in an indicating instrument which control motion of the pointer or other indicating means, excluding those parts which form the enclosure, scale, or support structure, or which adapt the instrument to the quantity being measured. 3. In a recording instrument, the arrangement of parts which control motion of the marking device, the marking device itself, the device for driving the chart, and the parts which carry the chart.

mechanized dew-point meter See dew-point recorder.

media 1. The physical interconnection between devices attached to the LAN. Typical LAN media are twisted pair, baseband coax, broadband coax, and fiber optics. 2. The plural of medium. 3. A name for the various materials used to hold or store electronic data, such as printer paper, disks, magnetic tape or punched cards.

media access control (MAC) The lower sub-layer of the Data Link Layer (Layer 2) unique to each type of IEEE 802 Local Area Networks. MAC provides a mechanism by which users access (share) the network.

medium In data processing, the material on which data is recorded and stored.

medium-carbon steel An alloy of iron and carbon containing about 0.25 to 0.6% C, and up to about 0.7% Mn.

medium scale integration (MSI) 1. A medium

level of chip density, lower than for LSI circuits but more than small scale integration. 2. An integrated circuit with 10 to 100 logic gates.

mega Prefix denoting 1,000,000.

megabit One million bits.

megabyte (MByte) A unit of computer memory size. One million bytes. See MByte.

megahertz One million hertz or cycles per second.

melting point The temperature at which a solid substance becomes liquid; for pure substances and some mixtures it is a single unique temperature; for impure substances, solutions and most mixtures it is a temperature range.

membrane 1. A thin tissue which covers organs, lines cavities, and forms canal walls in the body of an animal. 2. A thin sheet of metal, rubber or treated fabric used to line cavities or ducts, or to act as a semirigid separator between two fluid chambers.

memory Any form of computer data storage, including main memory and mass storage, in which data can be read and written; in its strictest sense, memory refers to main memory.

memory access time See access time.

memory address The address in computer memory of the location containing an instruction or operand.

memory addressability A measure of capability and ease of programming used in evaluating computers. The maximum number of locations specifiable by a nonindexed instruction using the instruction's minimum execution time.

memory bus The computer bus (or buses) which interconnects the processor, memory, and peripherals on a high-speed data processing highway.

memory capacity Same as storage capacity.

memory chip An electronic device that accepts data for computer use or storage.

memory cycle time The minimum time between two successive data accesses from a memory.

memory dump A listing of the contents of a storage device, or selected parts of it.

memory image A replication of the contents of a portion of memory.

memory latency time See latency.

memory management A function of a PDP-11 computer that enables it to operate with a larger memory than 32k words.

memory map Graphic representation of the general functional assignments of various areas in memory. Areas are defined by ranges of addresses.

memory mapping A map showing the usable and unusable (or protected) areas of memory.

memory protect A technique of protecting the contents of sections of memory from alteration by inhibiting the execution of any memory modification instruction upon detection of the presence of a guard bit associated with the accessed memory location. Memory modification instructions accessing protected memory are usually executed as a no-operation and a memory protect violation program interrupt is generated.

memory protection A scheme for preventing read and/or write access to certain areas of memory.

memory resident A program that remains in RAM memory even when other programs are operating, and be called up by interrupting the currently running program.

meniscus The concave or convex surface, caused by surface tension, at the top of a liquid column, as in a manometer tube.

meniscus lens A lens with one concave surface and one convex surface.

menu In data processing, a list from which an operator can select the tasks to be done.

mercury meter A differential pressure measuring device utilizing mercury as the seal between the high and low chambers.

mercury switch A type of switch consisting of two wires sealed into the end of a glass capsule containing a bead of mercury; if the capsule is tipped one way, the mercury covers the exposed ends of the wires and completes the circuit; if tipped the other way, the mercury exposes the wires and breaks the circuit.

mercury vapor lamp A type of ion discharge lamp widely used in ultraviolet analyzers because it emits several strong monochromatic lines with characteristic wavelengths such as 254, 313, 360, 405, etc.; lamp emission can be made almost completely monochromatic by using special filters.

mercury-vapor tube A gas tube in which the active gas is mercury vapor.

mercury-wetted relay A device using mercury as the relay contact closure substance.

merge In data processing, to combine two or more groups of records into a single file.

merge sort In data processing, an operation of combining data and then sorting it in some prescribed manner.

meridian plane Any plane which contains the optical axis.

mesh 1. A measure of screen size equal to the number of openings per inch along the principal direction of the weave. 2. The size classification of particles that pass through a sieve of the stated screen size. 3. Engagement of a gear with its mating pinion or rack. 4. A closed path through ductwork in a ventilation survey.

message An arbitrary amount of information whose beginning and end are defined or implied.

message exchange A device placed between a communication line and a computer to take care of certain communication functions and thereby free the computer for other work.

message routing The function performed at a central message processor of selecting the route, or alternate route if required, by which a message will proceed to the next point in reaching its destination.

message switching The technique of receiving a message, storing it until the proper outgoing circuit is available, and then retransmitting it.

metal A chemical element that is crystalline in the solid state, exhibits relatively high thermal and electrical conductivity, and has a generally lustrous or reflective surface appearance.

metallic Exhibiting characteristics of a metal.

metallic coating A thin layer of metal applied to an optical surface to enhance reflectivity.

metallic glass See glassy alloy.

metallography The study of the structure of metals—the most common techniques are optical microscopy, electron microscopy and x-ray diffraction analysis.

metal-nitride-oxide semiconductor One type of computer semiconductor memory used in EAROMs.

metal piston type seal A self-expandable metal seal ring installed in a groove on the disk circumference to block the clearance between the disk outer diameter and the liner bore with the disk in closed position [S75.05].

metal units Concentration units defined as the number of gm-moles per 1000 gm of solvent.

meteorograph A recording instrument for measuring meteorological data—temperature, barometric pressure and humidity, for example.

meteorological instrumentation Equipment for measuring weather data.

meter 1. A device for measuring and indicating the value of an observed quantity. 2. An interna-

243

tional metric standard for measuring length, equivalent to approximately 39.37 in. in the U.S. customary system of units. Spelled metre in the International Standard of Units (SI).

meter factor 1. A constant used to multiply the actual reading on a scale or chart to produce the measured value in actual units. 2. A correction factor applied to a meter's indicated value to compensate for variations in ambient conditions such as a temperature correction applied to a pressure indication.

metering 1. Regulating the flow of a fluid so that only a measured amount is permitted to flow past a given point in the system. 2. Measuring any variable (flow rate, electrical power, etc.).

meter prover A device for checking the accuracy of a gas meter.

meter proving tank See calibrating tank.

meter run A flowmeter installed and calibrated in a section of pipe having adequate upstream and downstream length to satisfy standards of flowmeter installation. See also orifice run.

meter sensitivity The accuracy with which a meter can measure a value; it is usually expressed as percent of the meter's full scale reading.

metre Metric unit of length (SI). See meter.

MeV Mega-electron-volts; a unit of energy equivalent to the kinetic energy of a single electron accelerated through an electric potential of 1-million volts.

mho A customary unit of conductance and admittance generally defined as the reciprocal of one ohm, or the conductance of an element whose resistance is one ohm; the equivalent SI unit Siemen is preferred.

micro A common term meaning very small.

microbalance A small analytical balance for weighing masses of 0.1 g or less to the nearest μg.

microbar A unit of pressure equal to one dyne per square centimeter.

microchannel plate A glass device with many tiny, parallel holes passing through it. It is used, with suitable biasing, as an electron amplifier, primarily for use in imaging detectors.

micro code 1. A system of coding making use of suboperations not ordinarily accessible in programming, e.g., coding that makes use of parts of multiplication or division operations. 2. A list of small program steps; combinations of these steps, performed automatically in a prescribed sequence from a macro-operation like multiply,

divide, and square root. See multiprocessor.

microcomputer 1. A computer based on the use of a microprocessor integrated circuit. The entire computer often fits on a small printed circuit board and works with a data word of 4, 8, or 16 bits. 2. A complete computer in which the CPU is a microprocessor.

microcurie A unit of radioactivity equal to one millionth (10^{-6}) curie.

microdensitometer A device for measuring the density of photographic films or plates on a microscopic scale; the small scale version of a densitometer.

microfaradmeter A capacitance meter calibrated in microfarads.

micro-floppy disks 3 1/2-inch disks that have greater storage capacity than a 5 1/4-inch floppy disk.

microinstruction Controls the operations of the various primitive resources of a computer: main and local store registers (both general and special purpose), arithmetic and logic units (ALUs), data paths, and so on. Microinstructions are stored as words in a control store that is traditionally (but not necessarily) separate from the main storage.

micromanipulator A positioning device for making small adjustments to the position of an optical component or other device.

micrometer 1. A metric measure with a value of 10^{-6} meters or 0.000001 meter, previously referred to as "micron" 2. Any device incorporating a screw thread for precisely measuring distances or angles, such as is sometimes attached to a telescope or microscope. 3. A type of calipers that incorporates a precision screw thread and is capable of measuring distance between two opposing surfaces to the nearest 0.001 or 0.0001 in.

micrometre A metric measure with a value of 10^{-6} metres or 0.000001 metre, previously referred to as micron [S7.3].

micron One millionth of a meter, or 0.000039 in. The diameter of dust particles is often expressed in microns.

microphone An electroacoustic transducer which transmits an electrical output signal that is directly related to the loudness and frequency distribution of sound waves that strike the active element.

microphonism 1. In an electron tube, modulation of one or more electrode currents as a direct result of mechanical vibrations of a tube element.

2. An undesirable electrical output signal in response to mechanical or acoustic vibration of an electronic or electrical devise.

microprocessor 1. A usually monolithic, large-scale-integrated (LS) central processing unit (CPU) on a single chip of semi-conductor material; memory, input/output circuits, power supply, etc. are needed to turn a microprocessor into a microcomputer. 2. A large-scale integrated circuit that has all the functions of a computer, except memory and input/output systems. The IC thus includes the instruction set, ALU, registers and control functions. 3. Sometimes abbreviated as MPU, uP, etc.

microprogramming A method of operating the control unit of a computer, wherein each instruction initiates or calls for the execution of a sequence of more elementary instructions. The microprogram is generally a permanently stored section of nonvolatile storage. The instruction repertory of the microprogrammed system can thus be changed by replacing the microprogrammed section of storage without otherwise affecting the construction of the computer.

microradiography Production of a magnified radiographic image.

microradiometer A device for detecting radiant power which consists of a thermopile supported on and directly connected to the moving coil of a galvanometer.

microscopic stress Load per unit area over a very short distance, on the order of the diameter of a metal grain, or smaller. A term usually reserved for characterizing residual stress patterns.

microwave Electromagnetic radiation having a wavelength of 1 to 300 mm.

Microwave Amplification by the Stimulated Emission of Radiation (MASER) The microwave equivalent and predecessor of the laser. It produces coherent microwaves.

microwave spectrum The portion of the electromagnetic spectrum of frequencies lying between infrared waves and radio waves.

migration The movement of ions from an area of the same charge to an area of opposite charge.

MIG welding Metal inert-gas welding; see gas metal-arc welding.

mil 1. A unit of linear measurement equal to 0.001 in. 2. A unit of angular measurement commonly used in the military for setting artillery elevations.

mile A British and U.S. unit of length commonly used to specify distances between widely separated points on the earth's surface; a statute mile, used for distances over land, is defined as 5280 ft; a nautical mile, used for distances along the surface of the oceans, is defined as one minute of arc measured along the equator, which equals 6080.27 ft or 1.1516 statute miles.

millimeter Also spelled millimetre. 1. A unit of length equal to 0.001 meter. 2. A millimeter of mercury, abbreviated mm Hg, is a unit of pressure equivalent to the pressure exerted by a column of pure liquid mercury one mm high at 0 °C under a standard gravity of 980.665 cm/s²; it is roughly equivalent to 1/760th of standard atmospheric pressure.

MIL-STD-1533 The military standard that defines serial data communications protocol on modern military vehicles, especially aircraft.

min See minute.

mini In data processing, a term used to describe a smaller computer of 12- to 32-bit word length and memory sizes of 16K-8M bytes.

miniature boiler Fired pressure vessels which do not exceed the following limits: 16 in. inside diameter of shell; 42 in. over-all length to outside of heads at center; 20 sq ft water heating surface; or 100 psi maximum allowable working pressure.

miniaturization The design and production of a scaled-down version of a device or mechanism that is capable of performing all of the same functions as the larger-sized original.

minicomputer A medium size computer for more dedicated applications than mainframe computers. It generally has a larger instruction set, wider range of languages and better support than microcomputers.

MINI-MAP A subset of the MAP protocols extended to provide higher performance for application whose communications are limited to a single LAN. A MINI-MAP node contains only the lower two layers (physical and link) of the MAP protocols. It can only communicate directly with MAP/EPA or MINI-MAP nodes on the same segment.

mini-micro In data processing, a very small microcomputer containing a CPU, memory, and I/O interfaces for data exchange and timing circuits to control the flow of data.

minimum bend radius The smallest radius around which a piece of sheet metal, wire, bar stock or tubing can be bent without fracture, or in the case of tubing, without collapse.

minimum cloud ignition temperature The minimum temperature at which a combustible dust atmosphere will autoignite and propagate an explosion [S12.10] [S12.11].

minimum dust layer ignition temperature The minimum temperature of a surface that will ignite dust lying on it after a long time (theoretically, until infinity). In most dusts, free moisture has been vaporized before ignition [S12.10] [S12.11].

minimum explosible concentration The minimum concentration of combustible dust that when ignited produces light explosive force [S12.11].

minimum explosion concentration The minimum concentration of a dust cloud that, when ignited, will propagate flame away from the source of ignition. NOTE: The measurable combustible properties of dusts depend not only on the chemical structure of the dust, but on test conditions, dust particle size, weight, density, and other particle characteristics [S12.10].

minimum reflux The quantity of reflux required to perform a specified separation in a column that has an unlimited number of trays. At minimum reflux, no products are withdrawn.

minimum thermometer A thermometer that indicates the lowest temperature reached during a given interval of time.

mining engineering A branch of engineering that deals with the discovery, extraction and initial processing of minerals—usually metal ores or coal—found in the earth's crust.

minor frame The period between frame synchronization words that include one complete cycle of a commutator having the highest rate; normally does not exceed 8192 bit internals; synonymous with prime frame.

minor graduations The shortest or lightest division marks on a graduated scale, which indicate subdivisions lying between successive major graduations or between an index graduation and an adjacent major graduation.

minor time In these systems, one sixteen-bit word: binary milliseconds.

minute 1. A measure of angle equal to 1/60th of one degree. 2. A measure of time equal to 60 s.

mips (million instructions per second) The measure of computer machine code instructions per second.

mirror scale An instrument scale and a mirror, so arranged that the indicating pointer and its reflection are aligned when the observer's eye is in the correct position to read the instrument without parallax error.

mismatch Lateral offset between two halves of a casting mold or forging die, which produces distortion in shape across the parting line.

miter valve A valve in which the disc is at an angle of approximately 45° to the axis of the valve body.

mixed level A simulation system combining both low-level transistor and gate circuit descriptions with high-level behavioral circuit representations.

mixed mode See mixed signal.

mixed radix Pertaining to a numeration system that uses more than one radix, such as the biquinary system.

mixed signal A simulation system combining both analog and digital circuit representations.

mixer In sound recording or reproduction equipment, a device capable of combining two or more input signals into a single linearly-proportioned output signal, usually with the additional capability of adjusting the levels of any of the inputs.

mixing valve A valve having more than one inlet but only one outlet port; it is used to blend two or more fluids to give a mixture of predetermined composition.

MMA (maximum-minimum) algorithm See compressor.

MMFS See Manufacturing Messaging Format Standard.

MMS (Manufacturing Message Specification) ISO/IEC 9506 A set of international standards developed to facilitate the interconnection of information processing systems. The first part is to define the service provided by the MMS. The second part specifies the protocol that supports the MMS.

mnemonic Pertaining to assisting, or intending to assist, human memory; thus a mnemonic term, usually an abbreviation, that is easy to remember, e.g., MPY for multiply and ACC for accumulator.

mnemonic operation code An operation code in which the names of operations are abbreviated and expressed mnemonically to facilitate remembering the operations they represent. A mnemonic code normally needs to be converted to an actual operation code by an assembler before execution by the computer. Examples of mnemonic codes are ADD for addition, CLR for clear storage, and SQR for square root.

mnemonics An assembly language instruction, defined by a symbol, that has some resemblance to the operations carried out. Mnemonics are easier to remember and use than the equivalent Hex code or machine code.

MNOS See metal-nitride-oxide semiconductor (one type of computer semiconductor memory used in EAROMs).

mobile A continuous-monitoring instrument mounted on a vehicle such as, but not limited to, a mining machine or industrial truck [S12.13] [S12.15].

mobile telemetering Any arrangement for transmitting instrument readings from a movable data-acquisition station to a remote stationary or movable indicating or recording station without the use of interconnecting wire.

mobility 1. The average drift velocity of a charged particle induced by a unit electrical potential gradient. 2. In gases, liquids, solids or colloids, the relative ease with which atoms, molecules or particles can move from one location to another without external stimulus.

mockup A model of a piece of equipment or a system, frequently full size, used for experiments, performance testing or training.

mode 1. A computer system of data representation, e.g., the binary mode [RP55.1]. 2. A selected method of computer operation [RP55.1]. 3. A single component in a computer network. 4. Real or complex (number system). 5. A stable condition of oscillation in a laser. A laser can operate in one mode (single-mode) or in many modes (multimode).

mode changer A device for changing the characteristics of a guided wave from one mode of propagation to another.

mode filter In a waveguide circuit, an arrangement of waveguide elements which pass waves that are being propagated in certain mode(s) and exclude waves being propagated in other modes.

model See mathematical model.

model basin A large tank of water for design experiments and performance studies of ship hulls using scale models. Also known as model tank, towing tank.

model dispersion That component of pulse

spreading caused by differential optical path lengths in a multimode fiber.

modem 1. A term used in reference to a device used in data transmission; a contraction of modulator-demodulator. The term may be used with two different meanings: A. The modulator and the demodulator of a modem are associated at the same end of a circuit. B. The modulator and the demodulator of a modem are associated at the opposite ends to form a channel [RP55.1]. 2. A device that provides both combining (modulation) and separation (demodulation) of data. Typically used to connect a node to a broadband network. See also modulator-demodulator; see also transceiver. 3. An electronic device for serial transmission of digital data in the audio frequency spectrum over a voice-grade telephone line. 4. A device that converts signals in one form to another form compatible with another kind of equipment. In particular, a circuit board that changes digital data being transmitted within a particular device into a form suitable to be transmitted over a data highway and vice versa. (From MOdulator + DEModulator).

moderation Reducing the kinetic energy of neutrons, usually by means of successive collisions with hydrogen, carbon or other light atoms.

modes The PID algorithms operate in several modes. These are determined by the operator and/or by states of other instructions inside the controller. 2. AUTO, CASCADE, MANUAL, etc.

modularity The degree to which a system of programs is developed in relatively independent components, some of which may be eliminated if a reduced version of the program is acceptable.

modularization Designing a series of components, subassemblies or devices for interchangeability of physical location, so that different assemblies can be easily constructed on a standard frame or mounted in standard enclosures.

modular programming Programming in which tasks are programmed in distinct sections or sub-sections resulting in the ability to modify one section without reference to other sections.

modulated wave A radio-frequency wave in which amplitude, phase or frequency is varied in accordance with the waveform of a modulating signal.

modulating The actions to keep a quantity or quality in proper measure or proportion. See also throttling [S75.05].

modulation 1. The process or the result of the

process by which some parameter of one wave is varied in accordance with some parameter of another wave [S51.1]. 2. The process of impressing information on a carrier for transmission (AM, amplitude modulation; PM, phase modulation; FM, frequency modulation). 3. Regulation of the fuel-air mixture to a burner in response to fluctuations of load on a boiler. 4. The action of a control valve to regulate fluid flow by varying the position of the closure component [S75.05].

modulation factor The ratio of peak variation actually used in a given type of modulation to the maximum design variation possible.

modulation index In frequency modulation with a sinusoidal waveform, the modulation index is the ratio of the peak (not peak-to-peak) frequency deviation to the frequency of the modulating wave.

modulation meter An instrument for measuring modulation factor of a wave train, usually expressed in percent.

modulation noise The noise in an electronic or acoustic circuit caused by the presence of a signal, but not including the waveform of the signal itself.

module 1. An assembly of interconnected components which constitutes an identifiable device, instrument, or piece of equipment. A module can be disconnected, removed as a unit, and replaced with a spare. It has definable performance characteristics which permit it to be tested as a unit. NOTE: A module could be a card or other subassembly of a larger device, provided it meets the requirements of this definition [S51.1]. 2. A program unit that is discrete and identifiable with respect to compiling, combining with other units, and loading, for example the input to, or output from, an assembler, compiler, linkage editor, or executive routine.

modulo A mathematical operation that yields the remainder function of division; thus, 39 modulo 6 equals 3.

modulo N check 1. A check that makes use of a check number that is equal to the remainder of the desired number when divided by n, e.g., in a modulo 4 check, the check number will be 0, 1, 2, or 3 and the remainder of the desired number when divided by 4 must equal the reported check number, otherwise an equipment malfunction has occurred. 2. A method of verification by congruences, e.g., casting out nines.

modulus of elasticity In any solid, the slope of the stress-strain curve within the elastic region; for most materials, the value is nearly constant up to some limiting value of stress known as the elastic limit; modulus of elasticity can be measured in tension, compression, torsion or shear; the tension modulus is often referred to as Young's modulus.

moisture Water in the liquid or vapor phase.

moisture barrier A material or coating that retards the passage of moisture through a wall made of more permeable materials.

moisture-free See bone dry.

moisture in steam Particles of water carried in steam usually expressed as the percentage by weight.

moisture loss The loss representing the difference in the heat content of the moisture in the exit gases and that at the temperature of the ambient air.

mol See mole.

molar units Concentration units defined as the number of gm-moles of the component per liter of solution.

mole Metric unit for amount of a substance.

molecular attrition See fretting.

molecular beam A unidirectional beam of neutral-charge molecules passing through a vacuum.

molecular flow Gas flow in a tube at a pressure low enough that the mean free path of the molecules is greater than the inside diameter of the tube.

molecule The smallest division of a unique chemical substance which maintains its unique chemical identity.

moles Number of molecular weights which is the weight of the component divided by its molecular weight.

Moll thermopile A type of thermopile used in some radiation-measuring instruments. It consists of multiple manganan-constantan thermocouples connected in series. Alternate junctions are embedded in a shielded nonconductive plate of large heat capacity; the remaining junctions are blackened and exposed directly to the radiation. The voltage across the thermopile is directly proportional to the intensity of radiation.

moment 1. Of force, the effectiveness of a force in producing rotation about an axis; it equals the product of the radius perpendicular to the axis of rotation that passes through the point of force application and the tangential component of force perpendicular to the plane defined by the radius and axis of rotation. 2. Of inertia, the resistance of a body at rest or in motion to changes in its angular velocity.

momentary 1. An alarm that returns to normal before being acknowledged. 2. Returns to normal state when pressure or signal is removed.

momentary alarm See alarm.

momentary digital output A contact closure, operated by a computer, that holds its condition (set or reset) for only a short time. See latching digital output.

momentary switch A spring-loaded switch whose contacts complete a circuit only while an actuating force is applied; for a typical momentary pushbutton, electric current flows through the switch only while the operator has a finger on the button.

momentum The product of a body's mass and its linear velocity.

Monel A series of International Nickel Co. high-nickel, high-copper alloys used for their corrosion resistant properties to certain conditions.

monitor 1. A general term for an instrument or instrument system used to measure or sense the status or magnitude of one or more variables for the purpose of deriving useful information. The term monitor is very unspecific—sometimes meaning analyzer, indicator, or alarm. Monitor can also be used as a verb [S5.1]. 2. To measure a quantity continuously or at regular intervals so that corrections to a process or condition may be made without delay if the quantity varies outside of prescribed limits. 3. Software or hardware that observes, supervises, controls, or verifies the operations of a system. 4. In data processing, a high-resolution viewing screen.

monitor command An instruction issued directly to a monitor from a user.

monitor console The system control terminal.

monitor light See pilot light.

monitor routine See executive program.

monitor software That portion of the operational software which controls on-line and off-line events, develops new on-line applications and assists in their debugging. This software is also known as a batch monitor.

monitor system Same as operating system.

monochromatic A single wavelength or frequency. In reality, light cannot be purely monochromatic, and actually extends over a range of wavelengths. The breadth of the range deter-

mines how monochromatic the light is.

monochromatic radiation Any electromagnetic radiation having an essentially single wavelength, or in which the photons all have essentially the same energy.

monochromator An optical device which uses a prism or diffraction grating to spread out the spectrum, then pass a narrow portion of that spectrum through a slit; it generates monochromatic light from a non-monochromatic source.

monostable Pertaining to a device that has one stable state.

monotonic A digital-to-analog converter is monotonic if the output either increases or remains constant as the digital input increases.

Monte Carlo method A trial-and-error method of repeated calculations to discover the best solution of a problem. Often used when a great number of variables are present, with interrelationships so extremely complex as to forestall straightforward analytical handling.

most significant bit (MSB) The bit in a digital sequence which defines the largest value. It is usually at the extreme left.

most significant digit The leftmost non-zero digit.

mother board A circuit board that includes the primary components of a microcomputer.

motion balance instrument An instrument design technique utilizing the motion of the measuring element against a spring to reach a balance of forces representing the magnitude of the measured variable.

motion conversion mechanism 1. Device needed on some, but not all, assemblies, to convert linear action to rotary valve operation [S75.05]. 2. A mechanism between the valve and the power unit of the actuator to convert between linear and rotary motion. The conversion can be from linear actuator action to rotary valve operation or from rotary actuator action to linear valve operation [S75.05].

motor meter A type of integrating meter; it consists of a rotor, one or more stators, a retarding device which makes the speed of the rotor directly proportional to the integral of the quantity measured (usually power or electric current), and a counter or set of dials that indicates the number of rotor revolutions.

motor operator The electric or hydraulic power mechanism that receives a control signal and repositions a valve or other final control element.

mounting effects The effects (errors) introduced into transducer performance during installation caused by fastening of the unit or its mounting hardware or by irregularities of the surface on (or to) which the transducer is mounted [S37.12] [S37.6].

mounting error The error resulting from mechanical deformation of the transducer caused by mounting the transducer and making all measureand and electrical connections [S37.1].

mounting position 1. The position of a device relative to physical surroundings [S51.1]. 2. The location and orientation of an actuator or auxiliary component relative to the control valve. This can apply to the control valve itself relative to the piping [S75.05].

mounting strain error See error, mounting strain.

mouse In data processing, an input device capable of moving a cursor on a computer screen.

moving-coil instrument An instrument whose output is related to the reaction between the magnetic field set up by current flow in one or more movable coils and the magnetic field of a fixed-position permanent magnet. Also known as permanent-magnet moving-coil instrument.

moving-dial indicator A type of indicator where a flat, circular scale (dial) is attached to the moving element and the instrument scale is continually repositioned with respect to a fixed pointer to indicate changing values of a measured variable.

moving-drum indicator A type of indicator where a circular member (drum) with a scale along its periphery revolves in relation to a fixed pointer to indicate changing values of a measured variable.

moving element Of an instrument, the parts which move as a direct result of a variation in the quantity being measured.

moving-iron instrument An instrument which depends on the reaction between one or more pieces of magnetically soft material, at least one of which moves, and the magnetic field set up by electric current flowing in one or more fixed coils, to produce its output indication or signal.

moving-magnet instrument An instrument which depends on the reaction between a movable permanent magnet that aligns itself with the resultant field produced by another permanent magnet interacting with one or more current-carrying coils, or by two or more coils interacting with each other, to produce its output indication or signal.

moving-scale indicator Any of several designs of instrument indicator where the scale moves in relation to a fixed pointer to indicate changing values of a measured variable.

MPL A high-level language suitable for the development of microprocessor application software.

MSB See most significant bit.

MS-DOS The most widely used operating system for microcomputers.

MSI See medium scale integration.

M synchronization A type of linking between a camera shutter and a flash unit that gives a 15 millisecond delay so that the metal foil flash lamp reaches peak brightness before the shutter actually trips.

MTBF See mean-time-between-failures.

MTTF See mean-time-to-failure.

multi-access A multiprogramming system which permits a number of users to simultaneously make on-line program changes.

multi-address Same as multiple address.

multi-bus Intel's proprietary link between single board systems used for industrial systems.

multicast A message addressed to a group of stations connected to a LAN.

multichannel spectral analyzer A measurement system which sorts signals into a number of different channels, then counts and analyzes the signals channel by channel.

multicoupler A device for coupling several receivers or transmitters to one antenna; it also enables the proper impedance match between same.

multicraft 1. Maintenance personnel who are proficient in more than one craft such as Instrument Technician, Electrician and Instrument Mechanic. 2. Responsible to maintain a variety of equipment used in control systems.

MULTICS The time-sharing system developed at Project MAC (Man and Computer Project at M.I.T.). See task, definition 3.

multi-element control system A control system utilizing input signals derived from two or more process variables for the purpose of jointly affecting the action of the control system. Examples are input signals from pressure and temperature or from speed and flow, etc. See also multi-variable control; see also control system, multi-element (multi-variable).

multifiber cable   A fiber optic cable containing many fibers which transmit signals independently and are housed in separate substructures within the cable.

multifuel burner   A burner by means of which more than one fuel can be burned, either separately or simultaneously, such as pulverized fuel, oil or gas.

multifunction multiloop controller   A type of microprocessor based controller that combines the process control functions of a dedicated loop controller with many of the logic functions of a programmable logic controller to provide the control strategy of an entire unit operation.

multilayer   A type of printed circuit board which has several layers of circuit etch or pattern, one over the other and interconnected by electroplated holes. Since these holes can also receive component leads, a given component lead can connect to several circuit points, reducing the required dimensions of a printed circuit board.

multilayer coating   Optical coatings in which several layers of different thicknesses of different materials are applied to an optical surface. Interference affects light passing through the layers. Reflection and transmission are influenced differently at different angles of incidence and different wavelengths.

multilevel address   See indirect address.

multimeter   See volt-ohm-milliammeter.

multimode fiber   An optical fiber capable of carrying more than one mode of light in its core.

multiple action   A control-system action that is a composite of the actions of two or more individual controllers.

multiple address   A type of instruction which specifies the addresses of two or more items which may be the addresses of locations of inputs or outputs of the calculating unit or the addresses of locations of instructions for the control unit. The term multi-address is also used in characterizing computers, e.g., two-, three-, or four-address machines. Synonymous with multiaddress.

multiple input   See reflash.

multiple orifice   A flow control orifice consisting of a moving member (gate) which slides reciprocally against a stationary member (plate). Both elements contain several matching orifices and the flow area is changed as the gate slides [S75.05].

multiple-output system   A system which ma-nipulates a plurality of variables to achieve control of a single variable.

multiple-purpose meter   See volt-ohm-milliammeter.

multiple sampling   A type of statistical quality control in which several samples, each consisting of a specified number of items, are withdrawn from a lot and inspected. The lot is accepted, rejected, or resampled, depending on the number of unacceptable items found.

multiplexer   1. A device which interleaves or simultaneously transmits two or more messages or signals on a single channel [RP55.1]. Optical multiplexers combine signals at different wavelengths. Electronic multiplexers combine signals electronically before they are converted into optical form. 2. A device for combining two or more signals, as for multiplex, or for creating the composite color video signal from its components in color television. 3. A device which samples input and/or output channels and interleaves signals in frequency or time. 4. A device that allows selection of one of many input channels of analog data under computer control. The device is often an integral part of a DAS. 5. A device which mixes several measurements for transmission and/or tape recording; time-division (PAM or PCM) or frequency-division (FM). 6. A device or circuit that samples many data lines in a time ordered sequence, one at a time, and puts all sampled data onto a single bus. (A demultiplexer does the reverse job).

multiplexer channel   1. An input/output channel which serves several input/output units. 2. A single path that is capable of transferring data from multiple sources or to multiple destinations through the use of time multiplexing.

multiplexing   1. The transmission of a number of different messages simultaneously over a single circuit. 2. Utilizing a single device for several similar purposes or using several devices for the same purpose, e.g., a duplexed communications channel carrying two messages simultaneously. 3. Technique of selecting from many inputs to provide a specified output. Allows a single A/D to serve several voltage sources by selecting them one at a time.

multiplier tube   A phototube in which secondary emission from auxiliary electrodes produces an internally amplified output signal. Also known as multiplier phototube; photomultiplier tube.

multiport burner A burner having a number of nozzles from which fuel and air are discharged.

multiposition action A type of controller action in which the final control element is positioned in one of three or more preset configurations, each corresponding to a definite range of values for the controlled variable.

multi-position controller See controller, multi-position.

multiprocessing 1. Pertaining to the simultaneous execution of two or more programs or sequences of instructions by a computer or computer network. 2. Loosely, parallel processing.

multiprocessor A machine with multiple arithmetic and logic units for simultaneous use.

multi-processors A number of independent central processing units each having access to a common memory. One unit is usually an information interchange controller while others carry out distinct defined parts of a task.

multiprogramming A computer processing method in which more than one task is in an executable state at any one time.

multirange Having two or more specific ranges of values over which an instrument or control device can be used; changing from one range to another usually involves simply repositioning a switch and does not require removing or replacing any internal parts.

multiskilled Maintenance personnel who are skilled in more than one craft.

multi-speed floating controller See controller, multiple-speed floating.

multistage Occurring in a sequence of separate steps; in a multistage pump or compressor, for instance, pressure is raised by passing the working fluid through a series of impellers or pistons, each of which raise the pressure above the outlet pressure from the previous stage.

multitasking The facility that allows the programmer to make use of the multiprogramming capability of a computer system.

multivariable control A control system involving several measured and controlled variables where the interdependences are considered in the calculation of the output variables.

multi-variable control system See control system, multi-element (multi-variable).

Munsell chroma The dimension of the Munsell system of color corresponding most closely to saturation.

MUX See multiplexer.

N

N See newton.

nameplate A plate attached to a control valve bearing the name of the manufacturer. It may also contain specification and limitation information [S75.05]. See also data plate; see also window.

NAND 1. A logical operator having the property that if P is a statement, Q is a statement, R is a statement, ..., then the NAND of P, Q, R, ... is true if at least one statement is false; false if all statements are true. Synonymous with NOT-AND. 2. Logical negation of AND. Supplies a logic 0 when all inputs are at logic 1.

nano A prefix which means one billionth.

nappe A sheet of liquid passing through the notch and falling over the weir crest.

narrowband radiation thermometer A type of temperature-measuring instrument that responds accurately only over a given, relatively narrow band of wavelengths, often a band chosen to meet a special requirement of the intended application.

National Electric Code (NEC) A set of regulations governing construction and installation of electrical wiring and apparatus, established by the National Fire Protection Association. It is widely used by state and local authorities within the United States.

National Institute for Certification in Engineering Technologies (NICET) Provides certification in "Industrial Instrumentation Engineering Technology Technician".

natural circulation The circulation of water in a boiler caused by differences in density.

natural draft Convective flow of a gas—as in a boiler, stack or cooling tower—due to differences in density. Warm gas in the chamber rises toward the outlet, drawing in colder, more dense gas through inlets near the bottom of the chamber.

natural frequency See frequency, undamped; see also frequency, natural and frequency, resonant.

natural gas A mixture of gaseous hydrocarbons trapped in rock formations below the earth's surface. The mixture consists chiefly of methane and ethane, with smaller amounts of other low-molecular-weight combustible gases, and sometimes noncombustible gases such as nitrogen, carbon dioxide, helium and H_2S called "sour gas."

natural language A language, the rules of which reflect and describe current usage rather than prescribed usage. Contrast with artificial language.

natural radioactivity Spontaneous radioactive decay of a naturally occurring nuclide.

near letter quality With computer printers, a dot-matrix character formation that resembles the print of earlier cloth-ribbon typewriters.

neck A reduced section of pipe or tubing between sections of larger diameter or between a pipe and a chamber.

needle point valve A type of valve having a needle point plug [S75.05].

needle valve A type of metering valve used chiefly for precisely controlling flow. Its essential design feature is a slender tapered rodlike control element which fits into a circular or conoidal seat. Operating the valve causes the rod to move into or out of the seat, gradually changing the effective cross-sectional area of the gap between the rod and its seat.

neg See negative.

negative feedback Returning part of an output signal and using it to reduce the value of an input signal.

negative-going edge The edge of a pulse going from a high to a low level.

negatron A negatively charged beta particle.

253

NEMA standard Consensus standards for electrical equipment approved by the majority of the members of the National Electrical Manufacturers Association.

neopheloscope An apparatus for making clouds in the laboratory by expanding moist air or by condensing water vapor.

neoprene A synthetic rubber made by polymerization of chloroprene (2-chlorobutadiene-1,3). Its color varies from amber to silver to cream. It exhibits excellent resistance to weathering, ozone, flames, various chemicals and oils.

neper A unit of measure determined by taking the natural logarithm of the scalar ratio of two voltages or two currents.

nephelometry The application of photometry to the measurement of the concentration of very dilute suspensions.

nephelometer A general term for instruments that measure the degree of cloudiness or turbidity.

nephoscope An instrument for determining the direction in which clouds move.

nest 1. To embed a subroutine or block of data into a larger routine or block of data. 2. To evaluate as nth degree polynomial by a particular algorithm which uses (n–1) multiply operations and (n–1) addition operations in succession.

nested Used in relation to subroutines used or called with another subroutine.

nested DO loop A FORTRAN statement which directs the computer to perform a given sequence repeatedly.

nesting In computer software, a program that has loops within loops.

net fan requirements The calculated operating conditions for a fan excluding tolerances.

net positive suction head The minimum difference between the static pressure at the inlet to a pump and vapor pressure of the liquid being pumped. Below that pressure, fluid is not forced far enough into the pump inlet to be acted upon by the impeller.

network 1. In data processing, any system consisting of an interconnection of computers and peripherals. Information is transferred between the devices in the network. 2. LAN (Local Area Network) is a system at one location linked by cables; WAN (Wide Area Network) is a widely dispersed system usually connected by telephone lines. 3. In an electric or hydraulic circuit, any combination of circuit elements.

network analyzer An apparatus which contains numerous electric-circuit elements that can be readily combined to form models of electric networks.

network management The facility by which network communication and devices are monitored and controlled.

network structure A type of alloy microstructure in which one phase occurs predominantly at grain boundaries, enveloping grains of a second phase.

neutral atmosphere An atmosphere which tends neither to oxidize nor reduce immersed materials.

neutral density filter A filter which has uniform transmission throughout the part of the spectrum where it is used.

neutral filter A light-beam filter which exhibits constant transmittance at all wavelengths within a specified range.

neutral point Point on the titration curve where the hydrogen ion concentration equals the hydroxyl ion concentration.

neutral zone See zone, neutral.

neutron A nuclear particle with a mass number of one and exhibiting zero (neutral) charge.

newton Metric unit for force.

Newtonian flow Fluid characteristics adhering to the linear relation between shear stress, viscosity and velocity distribution.

nexus The point in a computer system where interconnections occur.

nibble A word with four bits, or one-half a byte.

nibbling Contour cutting of sheet metal by a rapidly reciprocating punch which makes numerous, successive small cuts.

nine-light indicator A remote indicator used in conjunction with a contact anemometer and a wind vane. It consists of a center lamp surrounded by eight lamps, equally spaced and labeled to indicate compass points. Wind speed is indicated by the number of flashes the center lamp makes in a certain time interval; wind direction, by the position of an illuminated lamp in the outer ring.

nipple A short piece of pipe or tube, usually with an external thread at each end.

noble metal thermocouple A thermocouple whose elements are made of platinum (Pt) or platinum-rhodium (Pt-Rh alloys), and that resist oxidation and corrosion at temperatures up to about 1550 °C (2800 °F); three standard alloy pairs are in common use—Pt vs Pt-10%Rh, Pt vs Pt-13%Rh, and Pt-6%Rh vs Pt-30%Rh.

node In data processing, one component of a computer network where interconnections occur.

nodular iron See ductile iron.

noise 1. In process instrumentation, an unwanted component of a signal or variable which obscures the information content [RP55.1]. It may be expressed in units of the output or in percent of output span. See interference, electromagnetic [S51.1]. 2. Any spurious variation in the electrical output not present in the input. Noise is defined quantitatively in terms of an equivalent parasitic transient resistance appearing between the wiper and the resistance element while the input shaft is being moved. The equivalent noise resistance is established independently of the functional characteristics, in the noise test circuit. The wiper is required to be excited by a specified dc constant current source. The noise test circuit output measuring system is an oscilloscope with defined frequency bandwidth or time constant. The magnitude of the equivalent noise resistance is measured as ohms variation while the input shaft is moved at a specified speed, and observed as peak-to-peak deflection on the oscilloscope. NOTE: Noise may be characterized as generally reproducible, exhibited as a local nonlinearity, or it may be the classical sporadic type. Manufacturing cleanliness and improved quality control on processing may significantly reduce noise problems [S37.12]. 3. Random variations of one or more characteristics of any entity, such as voltage, current, or data [RP55.1]. 4. A random signal of known statistical properties of amplitude, distribution, and spectral density [RP55.1]. 5. Loosely, any disturbance tending to interfere with the normal operation of a device or system [RP55.1]. 6. Meaningless stray signals in a control system similar to radio static. Some types of noise interfere with the correctness of an output signal.

noise equivalent power The amount of optical power which must be incident on a detector to produce an electrical signal equal to the r.m.s. level of noise inherent in the detector. It is the measure of the sensitivity of the sensor.

noise factor In an electronic circuit, the ratio of total noise in the output signal to the portion thereof in the input signal under the following conditions—a selected input frequency and its corresponding output frequency, an input termination whose noise temperature is a standard 290 K at all frequencies, a linear system, and noise expressed as power per unit bandwidth.

noise figure A calculated or measured mathematical figure that denotes the inherent noise in a unit, system, or link.

noise immunity A device's ability to discern valid data in the presence of noise.

noise quantization Inherent noise that results from the quantization process.

noise temperature At a pair of terminals and at a specified frequency, the temperature of a passive system exhibiting the same noise power per unit bandwidth as the actual terminals.

nominal bandwidth The difference between upper and lower nominal cutoff frequencies of an acoustic, electric or optical filter.

nominal size 1. The standard dimension closest to the central value of a toleranced dimension. 2. A size used for general identification.

nominal stress The stress calculated by dividing nominal load by nominal cross sectional area, ignoring the effect of stress raisers but taking into account localized variations due to general part design.

nominal voltage That given by manufacturers as the recommended operating voltage of their gas detection equipment. If a range (versus a specific voltage) is given, the nominal voltage shall be considered as the midpoint of the range, unless otherwise specified [S12.15].

non-blackbody A term used to describe the thermal emittance of real objects, which emit less radiation than blackbodies at the same temperature, which may reflect radiant energy from other sources, and which may have their emitted radiation modified by passing through the medium between the body and a temperature-measuring instrument.

noncondensable gas The portion of a gas mixture (such as vapor from a chemical processing unit or exhaust steam from a turbine) that is not easily condensed by cooling. It normally consists of elements or compounds that have very low, often subzero, boiling points and vapor pressures.

noncontact gaging A method of determining physical dimensions without actual contact between the measuring device and the object.

noncontacting tachometer Any of several devices for measuring rotational speed without physical contact between a sensor and the rotating element—for example, stroboscopes or eddy-current tachometers.

noncritical dimension Any dimension which can be altered without affecting the basic function of a device.

non-destructive read out A method of reading from memory where the stored value is left intact by the reading process; plated wire and modern semiconductor random-access memory (RAM) are examples of NDRO memory.

nondestructive testing Any testing method which does not damage or destroy the sample. Usually, it consists of stimulating the sample with electricity, magnetism, electromagnetic radiation or ultrasound, and measuring the sample's response.

non-hazardous area An area in which explosive gas/air mixtures are not expected to be present so that special precautions for the construction and use of electrical apparatus are not required.

nonhazardous location 1. A location not designated as hazardous (classified) [RP12.6]. 2. A location where neither a hazardous atmosphere nor a hazardous dust layer is to be expected [S12.11].

nonincendive Equipment and wiring which in its normal operating condition is incapable of igniting a specific hazardous atmosphere or hazardous dust layer. Equipment and wiring having exposed blanketed surface temperatures above 80 percent of the ignition temperature in degrees centigrade of the specific hazardous dust layer shall NOT be classed as non-incendive. The blanketed surface temperature shall be determined at the outside surface of the enclosure beneath the surface of a dust accumulation 0.2 inch or more thickness [S12.11].

nonincendive circuit A circuit in which any arc or thermal effect produced under normal operating conditions of the equipment is not capable, under the conditions prescribed in this standard, of igniting the specified flammable gas or vapor-in-air mixture [S12.12].

nonincendive circuit field wiring Wiring which enters or leaves the equipment enclosure and which, under normal operating conditions of the equipment, is not capable, due to arcing or thermal effects, of igniting the specified flammable gas or vapor-in-air mixture by opening, shorting, or grounding the field wiring [S12.12].

nonincendive component A component with contacts for making or breaking a specified incendive circuit where either the contacting mechanism or the enclosure in which the contacts are housed is so constructed that the component is not capable of propagating ignition of the specified flammable gas or vapor-in-air mixture when tested according to Section 9. The housing of a nonincendive component is not intended to exclude the flammable atmosphere [S12.12].

non-incendive equipment Equipment which in its normal operating condition would not ignite a specific hazardous atmosphere in its most easily ignited concentration. The electrical circuits may include sliding or make-and-break contacts releasing insufficient energy to cause ignition [S51.1] [S12.4]. Wiring which under normal conditions cannot release sufficient energy to ignite a specific hazardous atmospheric mixture by opening, shorting or grounding, shall be permitted using any of the methods suitable for wiring in ordinary locations [S51.1]. NOTE: Equipment having exposed surface temperatures above 80% of the ignition temperature in °C of the specific hazardous atmosphere shall not be classed as non-incendive [S12.4].

noninteracting control system A multi-element control system designed to avoid disturbances to other controlled variables due to the process input adjustments which are made for the purpose of controlling a particular process variable. See control system, non-interacting.

nonlinear distortion A departure from a desired linear relationship between corresponding input and output signals of a system.

nonlinear effects Optical interactions which are proportional to the square or higher powers of electromagnetic field intensities. Nonlinear effects generate harmonics of optical frequencies, and sum and difference frequencies when two lightwaves are mixed.

nonlinearity A type of error in an FM system, where the input to a device does not relate to the output in a linear manner. See linearity.

nonlinear optimization See nonlinear programming.

nonlinear programming 1. In operations research a procedure for locating the maximum or minimum of a function of variables which are subject to constraints, when either the function or the constraints, or both, are nonlinear. Contrast with convex programming and dynamic programming. 2. Synonymous with nonlinear optimization.

nonlinear system Any system whose operation cannot be represented by a finite set of linear differential equations.

nonlocking Pertaining to code extension characters that change the interpretation of one or a specified number of characters. Contrast with locking.

non-nuclear safety (NNS) Instrumentation not included in nuclear safety related (NSR) [S67.03].

non-operating conditions See environmental conditions, non-operating.

nonprocessor request The system for accomplishing data transfers between two devices without involving the CPU.

nonreclosing pressure relief device A device for relieving internal pressure which remains open when actuated and must be replaced or reset before it can actuate again.

non-repeatability See repeatability.

non-return-to-zero (NRZ) Coding of digital data for serial transmission or storage whereby a logic ONE is represented by one signal level and a logic ZERO is represented by a different signal level.

non-scheduled maintenance 1. Unscheduled maintenance specifically intended to eliminate an existing fault. 2. An urgent need for repair or upkeep that was unpredicted or not previously planned, and must be added to or substituted for previously planned work.

nontransferred arc In arc welding and cutting, an arc sustained between the electrode and a constricting nozzle rather than between the electrode and the work.

nonvolatile memory Computer memory that retains data when power is removed.

NOR 1. A logic operator having the property that if P is a statement, Q is a statement, R is a statement, ..., then the NOR of P, Q, R, ... is true if all statements are false, false if at least one statement is true. P NOR Q is often represented by a combination of OR and NOT symbols. P NOR Q is also called NEITHER P NOR Q. Synonymous with NOT-OR. 2. Logical negation of OR. Supplies a logic 0 when any input is at logic 1.

Nordel 1070 An ethylene propylene rubber by E. I. du Pont de Nemours Co.

normal capacity Normal capacity is 80 percent of design capacity [RP74.01].

normality Concentration units defined as the number of gram-ions of replaceable hydrogen or hydroxyl groups per liter of solution. A shorter notation of gram-equivalents per liter is frequently used.

normalize 1. In programming, to adjust the exponent and fraction of a floating-point quantity such that the fraction lies in a prescribed normal standard range. 2. In mathematical operations, to reduce a set of symbols or numbers to a normal or standard form; synonymous with standardize. 3. In heat treating ferrous alloys, to heat 50 to 100 °F above the upper transformation temperature, then cool in still air.

normally closed (NC) 1. A switch position where the usual arrangement of contacts permits the flow of electricity in the circuit. 2. In a solenoid valve, an arrangement whereby the disk or plug is seated when the solenoid is deenergized. 3. A field contact that is closed for a normal process condition and open when the process condition is abnormal. 4. A valve with means provided to move to and/or hold in its closed position without actuator energy supply. See also field contact.

normally closed valve A valve with means provided to move to and/or hold in its closed position without actuator energy supply. See failclose [S75.05].

normally open (NO) 1. A switch position where the usual arrangement of contacts provides an open circuit (no current flowing). 2. In a solenoid valve, an arrangement whereby the disk or plug is seated when the solenoid is energized. 3. A field contact that is open for normal process condition and closed when the process conditions are abnormal. 4. A valve with means provided to move to and/or hold in its wide-open position without actuator energy supply. See also field contact.

normally open valve A valve with means provided to move to and/or hold in its wide open position without actuator energy supply. See failopen [S75.05].

normal mode interference See interference, normal mode. It may be expressed as a dimensionless ratio, a scalar ratio, or in decibels as 20 times the $\log_{10}$ of that ratio [S51.1].

normal mode rejection The ability of a circuit to discriminate against normal mode voltage usually expressed as a ratio or in decibels [RP55.1].

normal mode voltage An extraneous voltage induced across the circuit path (transverse mode voltage). See also voltage, normal mode.

normal operating conditions See operating conditions, normal.

normal operation Intrinsically safe electrical apparatus or associated electrical apparatus is in normal operation when it complies electrically and mechanically with the requirements of its design specification and is used within the limits specified by the manufacturer.

normal operational conditions Equipment is in normal operational conditions when it conforms electrically and mechanically with its design specifications and is used within the limits specified by the manufacturer. This includes: a) supply voltage, current, and frequency, b) environmental conditions (including process interface), c) all tool-removable parts in place, (e.g., covers), d) all operator-accessible adjustments at their most unfavorable settings, and e) opening, shorting, or grounding of nonincendive field wiring [S12.12].

Normal Thermometric Scale The first international standard temperature scale, adopted in 1887, which was based on the fundamental interval of 100° between the ice point of pure water and the condensing point of pure water vapor.

NOT 1. A logic operator having the property that if P is a statement, then the NOT of P is true if P is false, false if P is true. 2. Logical negation symbol. Supplies the complement of any input.

NOT-AND Same as NAND.

not-at-intermediate position A position that is either above or below the specified intermediate position.

notation 1. The act, process, or method of representing facts or quantities by a system or set of marks, signs, figures, or characters. 2. A system of such symbols or abbreviations used to express technical facts or quantities, as mathematical notation. 3. An annotation.

notch 1. A V-shaped indentation in an edge or surface. 2. An indentation of any shape that acts as a severe stress raiser.

notching Cutting out various shapes from the edge of a metal strip, blank or part.

notch width The horizontal distance between opposite sides of the weir notch.

not-closed position A position that is more than zero-percent open. A device that is not closed may or may not be open.

NOT-IF-THEN Same as exclusive OR.

not-open position A position that is less than 100 percent open. A device that is not open may or may not be closed.

NOT-OR Same as NOR.

NOVRAM Nonvolatile random-access memory (one type of nonvolatile semiconductor computer memory).

nozzle 1. A short flanged or welded neck connection on a drum or shell for the outlet or inlet of fluids; also a projecting spout through which a fluid flows. 2. A streamlined device for accelerating and directing fluid flow into a region of lower fluid pressure. 3. A particular type of restriction used in flow system to facilitate flow measurement by pressure drop across a restriction.

nozzle efficiency The efficiency of a nozzle in converting potential energy to kinetic, commonly expressed as the ratio of actual to ideal change in kinetic energy at a specific pressure ratio.

nozzle/flapper A fundamental part of pneumatic signal processing and pneumatic control operations. Basically, the device converts a displacement of the flapper to a pressure signal.

NPSH See net positive suction head.

NRZ See non-return-to-zero.

NSE (nth sequential) algorithm See compressor.

nuclear emulsion A photographic emulsion specially designed to record the tracks of ionizing particles.

nuclear fluorescence thickness gage A device for determining the weight of an applied coating by exciting the coated material with gamma rays and measuring low-energy fluorescent radiation that results.

nuclear radiation The emission of charged and uncharged particles and of electromagnetic radiation from atomic nuclei [S37.1].

nuclear safety-related (NSR) Activation or control of systems or components that are essential to emergency reactor shutdown, containment isolation, reactor core-cooling, and containment and reactor heat removal, or are otherwise essential in preventing or mitigating a significant release of radioactive material to the environment, or are otherwise used to provide reasonable assurance that a nuclear power plant can be operated without any risk to the health and safety of the public [S67.10] [S67.03] [S67.02].

nuclear safety-related instrumentation That which is essential to: a) provide emergency reactor shutdown, b) provide containment isolation, c) provide reactor core cooling, d) provide for containment or reactor heat removal, e) prevent or mitigate a significant release of radioactive material to the environment; or is otherwise essential to provide reasonable assurance that a nuclear power plant can be operated without undue risk to the health and safety of the public [S67.04].

nucleonics Technology involving atomic nuclei; includes nuclear reactors, particle accelerators, radiation detectors and radioisotope applications.

nucleus 1. The positively charged core of an atom; it contains almost all of the mass of the atom but occupies only a small fraction of its volume. 2. A number of atoms or molecules bound together with interatomic forces sufficiently strong to make a small particle of a new phase stable in a mass otherwise consisting of another phase; creating a stable nucleus is the first step in phase transformation by a nucleation-and-growth process. 3. That portion of the control program that must always be present in main storage. Also, the main storage area used by the nucleus and other transient control program routines.

nucleus counter An instrument that measures the number of condensation or ice nuclei in a sample volume of air.

nuclide A species of atom characterized by a unique combination of charge, mass number and quantum state of its nucleus.

nude vacuum gage A hot-filament ionization gage mounted entirely within the vacuum system whose pressure is being measured.

null A condition, such as of balance, which results in a minimum absolute value of output.

null-balance recorder An instrument that records a measured value by means of a pen or printer attached to a motor-driven slide, where the position of the slide is determined by continuously balancing current or voltage in the measuring circuit against current or voltage from a sensing element.

null indicator An indicating device such as a galvanometer used to determine when voltage or current in a circuit is zero; used chiefly in balancing bridge circuits. Also known as null detector.

number 1. A mathematical entity that may indicate quantity or amount of units. 2. Loosely, a numeral. 3. See binary number and random numbers.

number system 1. A systematic method for representing numerical quantities in which any quantity is represented as the sequence of coefficients of the successive powers of a particular base with an appropriate point. Each succeeding coefficient from right to left is associated with and usually multiplies the next higher power of the base. 2. The following are names of the number systems with bases 2 through 20: 2, binary; 3, ternary; 4, quaternary; 5, quinary; 6, senary; 7, septenary; 8, octal or octonary; 9, novenary; 10, decimal; 11, undecimal; 12, duodecimal; 13, terdenary; 14, quaterdenary; 15, quindenary; 16, sexadecimal—or hexadecimal; 17, septendecimal; 18, octodenary; 19, novemdenary; 20, vicenary. Also 32, duosexadecimal—or duotricinary; and 60, sexagenary. The binary, octal, decimal, and sexadecimal systems are widely used in computers. See decimal number and binary number and related to positional notation and clarified by octal digit and binary digit.

numerical analysis The study of methods of obtaining useful quantitative solutions to mathematical problems, regardless of whether an analytic solution exists or not, and the study of the errors and bounds on errors in obtaining such solutions.

numerical aperture The sine of the half-angle over which an optical fiber or optical system can accept light rays, multiplied by the index of refraction of the medium containing the rays.

numerical control Automatic control of a process performed by a device that makes use of all or part of numerical data generally introduced as the operation is in process.

numerical keypad Typical of a computer keyboard, a separate set of 0 through 9 keys arranged like numerical keys on a 10-key adding machine.

numeric word A word consisting of digits and possibly space characters and special characters.

nutating-disk flowmeter A type of positive-displacement flowmeter in which the advancing volume of fluid causes a measuring disk to wobble (nutate), thereby passing a precise volume of fluid through the meter with each revolution of the disk.

nutation Rocking back and forth, or periodically repeating a circular, elliptical, conical or spiral path, usually involving relatively small degrees of motion.

N-value The exponent in the power function $V(T)=KT^N$, which is the calibration function for a ratio thermometer. The N-value and mean effective wavelength can be used to express operating characteristics of a given ratio thermometer.

nylon A plastics material used to make filaments, fibers, fabric, sheet and extrusions; a generic name for a type of long-chain polymer containing recurring amide groups within the main chain.

Nyquist frequency One-half of the sampling frequency in a sampled data system.

O

object code 1. The machine code that can be directly executed by the computer. It is produced as a result of the translation of the source code. 2. A relocatable machine-language code.

objective variable A quantity or condition that is not measured directly for the purpose of controlling it, but rather is controlled through its relation to another, controlled variable.

object language A language which is the output of an automatic coding routine. Usually object language and machine language are the same, however, a series of steps in an automatic coding system may involve the object language of one step serving as a source language for the next step and so forth.

object machine The computer on which the object program is to be executed. Same as target computer.

object module The primary output of an assembler or compiler, which can be linked with other object modules and loaded into memory as a program. The object module is composed of the relocatable machine-language code, relocation information, and the corresponding symbol table defining the use of symbols within the module.

object program A fully compiled or assembled program that is ready to be loaded into the computer. See also target program.

object time system The collection of modules called by the compiled code to perform various utility or supervisory operations; for example, an object time system usually includes I/O and trap-handling routines.

obsolescent Lower in physical or functional value due to changes in technology rather than to deterioration.

obsolete No longer suitable for the intended use because of changes in technology or requirements.

octal Pertaining to eight; usually describing a number system of base or radix eight, e.g., in octal notation, octal 214 is 2 times 64, plus 1 times 8, plus 4 times 1, and equals decimal 140.

octal digit The symbol 0, 1, 2, 3, 4, 5, 6, or 7 used as a digit in the system of notation which uses 8 as the base or radix. Clarified by number systems.

octal number A number of one or more figures, representing a sum in which the quantity represented by each figure is based on a radix of eight. The figures used as 0, 1, 2, 3, 4, 5, 6, and 7. Clarified by octal.

octave 1. Any group or series of eight [S26]. 2. The interval between two frequencies with a ratio of 2:1.

octave-band analyzer A portable sound analyzer which amplifies a microphone signal, feeds it into one of several band-pass filters selected by a switch, and indicates signal amplitude on a logarithmic scale; except for the highest and lowest band, each band spans an octave in frequency.

octave-band filter A band-pass filter in which the upper and lower cutoff frequencies are in a fixed ratio of 2:1.

octet A group of eight bits treated as a unit. See byte.

OD See outside diameter.

odd-even check Same as parity check.

odograph An instrument mounted in a vehicle to automatically plot its course and distance traveled on a map.

odometer An instrument for measuring and indicating distance traveled.

Oersted The CGS unit of magnetic field strength; the SI unit, ampere-turn per metre, is preferred.

off Describing the nonoperating state of a device or circuit.

off-axis mirrors Mirrors in which the mechanical center of the mirror does not correspond to the axis of the optical figure of the mirror.

off-line 1. Not being in continuous, direct communication with the computer. 2. Done independent of the computer (as in off-line storage). 3. Describing the state of a subsystem or piece of computer equipment which is operable, but currently bypassed or disconnected from the main system. 4. Pertaining to a computer that is not actively monitoring or controlling a process or operation, and pertaining to a computer operation performed while the computer is not monitoring or controlling a process or operation. 5. Describing lateral or angular deviation from the intended axis of a drilled or bored hole.

off-line diagnostics 1. Describing the state of a control system, subsystem or piece of computer equipment which is operable, but currently not actively monitoring or controlling the process. 2. A program to check out systems and subsystems, providing error codes if an error is detected. This diagnostic program is run while the system is off line.

off-line equipment The peripheral equipment or devices not in direct communication with the central processing unit of a computer.

off-line memory Any media, capable of being stored remotely from the computer, which can be read by the computer when placed into a suitable reading device. Also see external storage.

off-line system That kind of system in which human operations are required between the original recording functions and the ultimate data processing function. This includes conversion operations as well as the necessary loading and unloading operations incident to the use of point-to-point or data-gathering systems. Compare on-line system.

offset 1. A sustained deviation of the controlled variable from set point. This characteristic is inherent in proportional controllers that do not incorporate reset action. 2. Offset is caused by load changes. 3. The steady-state deviation when the set point is fixed. The offset resulting from a no-load to a full-load change (or other specified limits) is often called "droop" of "load regulation". See also deviation, steady-state [S51.1]. 4. The count value output from an A/D converter resulting from a zero input analog voltage. Used to convert subsequent nonzero measurements. 5. A short distance measured perpendicular to a principal line of measurement in order to locate a point with respect to the line. 6. A constant and steady state of deviation of the measured variable from the set point. 7. A printing process in which ink is transferred from the printing plate or master to a rubber covered roller, which in turn transfers the ink to the paper.

offset (programming) The difference between a base location and the location of an element related to the base location; the number of locations relative to the base of an array, string, or block.

ohm The metric unit for electrical resistance; it is the resistance (or impedance) of a conductor such that an electrical potential of one volt exists across the ends of the conductor when it carries a current of one ampere.

ohmmeter A device for measuring electrical resistance.

ohms per volt A standard rating of instrument sensitivity determined by dividing the instrument's electrical resistance by its full-scale voltage.

oil Any of various viscous organic liquids that are soluble in certain organic solvents such as naphtha or ether but are not soluble in water; may be of animal, vegetable, mineral or synthetic origin.

oil bath 1. Oil, in a container or chamber, which a part or mechanism is submerged, or into which it dips, during operation or manufacture. 2. Oil poured on a cutting tool or in which it is submerged during a machining operation.

oil burner A burner for firing oil.

oil cone The cone of finely atomized oil discharged from an oil atomizer.

oil gas A heating gas made by reacting petroleum oil vapors and steam.

oil heating and pumping set A group of apparatus consisting of a heater for raising the temperature of the oil to produce the desired viscosity, and a pump for delivering the oil at the desired pressure.

olemeter 1. A device for measuring the specific gravity of oil. 2. A device for measuring the proportion of oil in a mixture.

OLI (Out of Limits) Algorithm See compressor.

Olsen ductility test A method for determining relative formability of metal sheet. A sheet metal sample is deformed at the center by a steel ball until fracture occurs; the cup height at fracture indicates relative ease of forming deep-drawn or stamped parts.

ombroscope An instrument for indicating when precipitation occurs. A heated water-sensitive surface is exposed to the weather; when it rains or snows, an electrical or mechanical output trips an alarm or records the occurrence on a time chart.

omnidirectional (antenna) An antenna having equal gains in all directions.

omnigraph An automatic acetylene flame-cutting device that cuts several blanks simultaneously, duplicating the pattern traced by a mechanical pointer.

on Describing the operating state of a device or circuit.

on-condition maintenance 1. On-condition maintenance is done when equipment needs it. 2. Inspect critical components, regard safety as paramount, repair defects, it works, don't fix it.

one-piece-element clamp A one-piece-element clamp or pinch valve is a valve consisting of a one-piece flexible element or liner installed in a body with the element or liner extending over the flange faces and acting as gaskets between the valve and connecting piping [S75.08].

ones complement The radix-minus-one complement in binary notation. The ones complement of an octal 3516 is 4261. See also complement.

on-line 1. Describing the state of a subsystem or piece of computer equipment which is operable and currently connected to the main system. 2. Pertaining to a computer that is actively monitoring or controlling a process or operation, or pertaining to a computer operation performed while the computer is monitoring or controlling a process or operation. 3. Describing coincidence of the axis of a drilled or bored hole and its intended axis, without measurable lateral or angular deviation. 4. Directly controlled by, or in continuous communication with, the computer (on-line storage). 5. Done in real time.

on-line data-reduction The processing of information as rapidly as information is received by the computing system or as rapidly as it is generated by the source.

on-line debugging The act of debugging a program while time sharing its execution with an on-line process program.

on-line diagnostics 1. Describing the state of a control system, subsystem or piece of computer equipment which is operable and actively monitoring or controlling the process. 2. A program to check out systems and subsystems, providing error codes and alarms if errors are detected. This diagnostic program runs in background while the control system is in the operating mode.

on-line equipment A computer system and the peripheral equipment or devices in the system in which the operation of such equipment is under control of the central processing unit, and in which information reflecting current activity is introduced into the data processing system as soon as it occurs. Thus, directly in-line with the main flow of transaction processing. Clarified by on-line.

on-line memory Any media directly accessible by the computer system. Also see internal storage.

on-line processing Same as on-line.

on-line system 1. Synonymous with on-line. 2. A system in which the input data enters the computer directly from the point of origin and/or in which output data is transmitted directly to where it is used. Compare off-line.

ON-OFF control A simple form of control whereby the control variable is switched fully ON or fully OFF in response to the process variable rising above the set-point or falling below the set-point respectively. Cycling always occurs with this form of control.

on-off controller See controller, on-off.

opacity The reciprocal of optical transmissivity.

opcode The pattern of bits in an instruction that indicates the addressing mode.

open circuit 1. An interruption in an electrical or hydraulic circuit, usually due to a failure or disconnection, which renders the circuit inoperable. 2. A nonrecirculating (once-through) system or process.

open-end protecting tube A tube extending from a physical boundary into the body of a medium to surround and protect a thermocouple yet allowing direct contact between the thermocouple's measuring junction and the medium.

open-flow nozzle See Kennison nozzle.

opening pressure The static inlet pressure which initiates a discharge.

open loop Pertaining to a control system in which there is no self-correcting action for misses of the desired operational condition, as there is in a closed-loop system. See feed-forward control action.

open loop control 1. A control system which does not take any account of the error between the desired and actual values of the controlled variables. 2. An operation in which computer-evaluated control action is applied by an operator. See open loop and closed loop. 3. A system in which no comparison is made between the actual value and the desired value of a process variable.

open-loop numerical control A type of numerical control system in which the drive motor provides both actuation and measurement with no feedback to the control console.

openness of scale With respect to measuring instruments, the amount of change in a measured quantity that causes the pointer to move 1 mm (or in some instances, 1 in.) on the instrument scale.

open position A position that is 100 percent open.

open seal An impulse line filled with a seal fluid open to the process.

open system A system that complies with the requirements of the OSI reference model in its communication with other open systems.

open system interconnection (OSI) A connection between one communication system and another using a standard protocol.

operand The address of an instruction to be executed by the processor.

Operating Basis Earthquake (OBE) That earthquake which ". . . could reasonably be expected to affect the plant site during the operating life of the plant; it is that earthquake which produces the vibratory ground motion for which those features of the nuclear power plant necessary for continued operation without undue risk to the health and safety of the public are designed to remain functional." [S67.03]

operating conditions Conditions to which a device is subjected, not including the variable measured by the device. Examples of operating conditions include: ambient pressure, ambient temperature, electromagnetic fields, gravitational force, inclination, power supply variation (voltage, frequency, harmonics), radiation, shock and vibration. Both static and dynamic variations in these conditions should be considered [S51.1].

operating conditions, normal The range of operating conditions within which a device is designed to operate and for which operating influences are stated [S51.1].

operating conditions, reference The range of operating conditions of a device within which

operating influences are negligible. NOTE: A. The range is usually narrow. B. They are the conditions under which reference performance is stated and the base from which the values of operating influences are determined [S51.1]. 2. Conditions to which a device is subjected, not including the variable measured by the device. See also environmental condition.

operating control A control to start and stop the burner—must be in addition to the high limit control.

operating influence The change in a performance characteristic caused by a change in a specified operating condition from reference operating condition, all other conditions being held within the limits of reference operating conditions. The specified operating conditions are usually the limits of the normal operating conditions. Operating influence may be stated in either of two ways: a) As the total change in performance characteristics from reference operating condition to another specified operating condition. b) As a coefficient expressing the change in a performance characteristics corresponding to unit change of the operating condition, from reference operating condition to another specified operating condition. NOTE: If the relation between operating influence and change in operating condition is linear, one coefficient will suffice. If it is non-linear, it may be desirable to state more than one coefficient such as 0.05% per volt from 120 to 125V to 130V [S51.1].

operating level The nominal position or output at which a system or process operates. Typical examples are water level in a boiler, production rate of a manufacturing process, or acoustical output (volume) of a loudspeaker system.

operating pressure 1. The nominal pressure or pressure limits at which a system or process operates. 2. In a pneumatic or hydraulic system, the high and low values (range) of pressure that will produce full-range operation of an output device such as a motor operator, positioning relay or data-transmission device. See pressure, operating.

operating system 1. An integrated collection of service routines for supervising the sequencing of programs by a computer. Operating systems may perform debugging, input-output, accounting, compilation, and storage assignment tasks. Synonymous with monitor system and executive system. 2. A group of programming systems operating under control of a data processing monitor program.

operating temperature range The range in extremes of ambient temperature within which the transducer must perform to the requirements of the temperature error or temperature error band [S37.12] [S37.6].

operating time That part of available time during which the hardware is operating and assumed to be yielding correct results. It includes development time, production time, and makeup time. Contrast with idle time.

operation A set of tasks or processes, usually performed at one location.

operational Describing a state of readiness for immediate use, as may be said of equipment or vehicles.

operational maintenance Any maintenance activity, other than corrective maintenance, intended to be performed by the operator and which is required in order for the equipment to serve its intended purpose. Such activities typically include the correcting of "zero" on a panel instrument, changing charts, making records, adding ink, or the like [S12.12] [S82.01]. Such activities are expected to be performed by a person(s) not familiar with the risks of electrical shock, likelihood of fire, or personal injury [S82.01].

operational test See test.

operation analysis An evaluation process in industrial engineering that assesses design, materials, equipment, tools, working conditions, methods and inspection standards, usually for the purpose of improving production output or decreasing cost.

operation code The part of a computer instruction word which specified, in coded form, the operation to be performed.

operations analysis See operations research.

operations research The use of analytic methods adopted from mathematics for solving operational problems. The objective is to provide management with a more logical basis for making sound predictions and decisions. Among the common scientific techniques used in operations research are the following: linear programming, probability theory, information theory, game theory, Monte Carlo method, and queuing theory.

operative limits The range of operating conditions to which a device may be subjected without permanent impairment of operating characteristics. NOTE 1: In general, performance characteristics are not stated for the region between the limits of normal operating conditions and the operative limits. 2: Upon returning within the limits of normal operating conditions, a device may require adjustments to restore normal performance [S51.1].

operator 1. A person using or operating the equipment [S82.01]. 2. The person who initiates and monitors the operation of a computer. 3. The person who initiates and monitors the operation of a process. 4. A mathematical symbol which represents a mathematical process to be performed on an associated operand. 5. The portion of an instruction which tells the machine what to do.

operator command A statement to the control program, issued via a console device, which causes the control program to provide requested information, alter normal operations, initiate new operations, or terminate existing operations.

operator control An operator-accessible control, usually a knob, push button, lever, or the like, provided to enable the operator to cause the equipment to perform its intended function and to serve its intended purpose [S82.01].

operator's console A device which enables the operator to communicate with the computer. It can be used to enter information into the computer, to request and display stored data, to actuate various preprogrammed command routines, etc. See also process engineer's console and programmer's console.

operator station 1. Serves as the interface between the operator and other devices on the data highway. 2. The operator can observe and control several devices.

opisometer An instrument incorporating a tracing wheel used for measuring the length of curved lines, such as those on a map.

optical ammeter An electrothermic instrument commonly employing a photoelectric cell and indicating device to determine the magnitude of electric current by measuring the light emitted by a lamp filament carrying the current; the instrument is calibrated by determining the amount of light emitted when known currents are carried by the same filament.

optical amplifier A type of amplifier in which an electric input signal is converted to light, amplified as light, then converted back to an electric output signal.

optical attenuation meter A device which measures the loss or attenuation of an optical fiber, fiber optic cable, or fiber optic system. Measurements are usually made in decibels.

optical bench A rigid horizontal bar or track for holding and supporting optical devices in fixed positions, yet allowing the positions to be changed or adjusted quickly and easily.

optical character reader A scanning device that can recognize some typewritten characters.

optical comparator 1. Any comparator in which movement of a measuring plunger tilts a small mirror, which in turn reflects light in an optical system. 2. A type of comparator in which the silhouette of a part is projected on a graduated screen and the dimensions or contour evaluated from the image.

optical density A measurement of transmission equal to the base 10 logarithm of the reciprocal of transmittance. An object with optical density of zero is transparent; an optical density of one corresponds to 10% transmission.

optical disk A large electronic storage device that uses laser beam patterns to store data.

optical-emission spectrometry Measurement of the wavelength(s) and intensities of visible light emitted by a substance following stimulation.

optical encoder tachometer A type of instrument that combines a sensor (optical encoder) with a microprocessor to convert sensor impulses into a measurement of rotational velocity.

optical fiber Fine glass stands that transmit data using light signals.

optical filter A semitransparent device that selectively passes rays of light having predetermined wavelengths.

optical flat A transparent disk, usually made of fused quartz, having precisely parallel faces, one face polished for clear vision and the other face ground optically flat; when placed on a surface and illuminated under proper conditions, interference bands can be observed and used to either assess surface contour (relative flatness) or determine differences between a reference gage or gage block and a highly accurate part or inspection gage.

optical fluid-flow measurement Any method for measuring the density of a fluid in motion which depends on measuring refraction and phase shift among different rays of light as they pass through a flow field of varying density.

optical gage A gage that measures the image of an object without touching the object itself.

optical glass Glass free of imperfections, such as bubbles, chemical inhomogeneity or unmelted particles, which degrade its ability to transmit light.

optical grating 1. Diffraction grating usually employed with other appropriate optics to fabricate a monochromator. These gratings consist of a series of parallel grooves carefully and uniformly shaped in an optical surface either flat or concave depending upon the application. The number of grooves form and the shape of the grooves (its profile) determines in what region of the spectrum it is applicable. 2. Commonly referred to as a Ronchi grating. 3. A highly accurate device used in precision dimensional measurement which consists of a polished surface, commonly aluminum coating on a glass substrate, onto which close, equidistant and parallel grooves have been ruled; the distribution of grooves ranging from several hundred to many thousands of grooves per inch; gratings are used in conjunction with monochromatic light to produce interference patterns sometimes referred to as moiré patterns. They are used in optical testing as well as generating the dot matrix for picture reproduction from a photographic negative.

optical indicator An instrument which plots pressure variations as a function of time by making use of magnification in an optical system coupled with photographic recording.

optical mark reader Using light sensing, a device that reads marks made on special forms.

optical material Any material which is transparent to visible light or to x-ray, ultraviolet or infrared radiation.

optical plastic Any plastics material which is transparent to light and can be used in optical devices and instruments to take advantage of physical or mechanical properties where the plastics material is superior to glass, or to take advantage of the lower cost of the plastics material.

optical pressure transducer Any of several devices that use optical methods to accurately measure the position of the sensitive element of the pressure transducer.

optical pyrometer An instrument that determines the temperature of an object by comparing its incandescent brightness with that of an electrically heated wire; the current through the wire is adjusted until the visual image of the wire blends into the image of the hot surface, and temperature is read directly from a calibrated dial attached to the current adjustment.

optical rangefinder An optical instrument for measuring distance, usually from the instrument's location to a target some distance away, by measuring the angle between rays of light from the target to separate windows on the rangefinder body.

optical recording Making a record of an instrument reading by focusing a tiny beam of light on photosensitive paper, the position of the light along one axis of the resulting orthogonal plot being directly related to the value of the quantity being measured.

optical rotation Rotation of the plane of polarization about the axis of a beam of polarized light.

optical storage disk A computer storage medium using lasers to form surface patterns that represent data. CD-ROM (Compact Disk Read-Only Memory) is an optical storage disk that stores data in digital form.

optical time domain reflectometer A device that sends a very short pulse of light down a fiber optic communication system and measures the time history of the pulse reflection. The reflection indicates fiber dispersion and discontinuities in the fiber path, such as breaks and connectors. The time it takes for the light pulse to travel to and from the discontinuity indicates how far it is from the test set.

optimization 1. Theoretical analysis of a system, including all of the characteristics of the process, such as thermal lags, capacity of tanks or towers, length and size of pipes, etc. This analysis is made, sometimes with the aid of frequency response curves, to obtain the most desirable instrumentation and control. 2. Making a design, process or system as nearly perfect in function or effectiveness as possible. 3. Using a structured decision-making technique to select the best way of achieving a defined goal from a set of alternatives.

optimize 1. To establish control parameters so as to make control as effective as possible. 2. To rearrange the instructions or data in storage so that the program can be run in minimum time.

optimizing control See control, optimizing, steady-state optimization and dynamic optimization.

option module Any additional device that expands a computer's capability.

optoelectronic amplifier An amplifier whose input and output signals and method of amplification may be either optical or electronic.

OR A logic operator having the property that if P is an expression, Q is an expression, R is an expression ..., then the OR of P, Q, R ... is true if at least one expression is true, false if all expressions are false. P OR Q is often represented by P + Q, PVQ. Synonymous with inclusive OR. Contrast with exclusive OR.

organic matter Compounds containing carbon often derived from living organisms.

orient To place an instrument, particularly one for making optical measurements, so that its physical axis is aligned with a specific direction or reference line.

orientation Alignment with a specific direction or reference line.

orifice 1. The opening from the whirling chamber of a mechanical atomizer or the mixing chamber of a steam atomizer through which the liquid fuel is discharged. 2. A calibrated opening in a plate, inserted in a gas stream for measuring velocity of flow.

orifice fitting A specially designed orifice plate holding device.

orifice flange taps The 1/2 in. or 3/4 in. pipe taps in the edge of an orifice flange union.

orifice flange union Two unique flanges used to hold an orifice plate primary element with specific design dimensions established by the American Gas Association.

orifice meter A general term used to describe any recording differential pressure measuring instrument.

orifice mixer A piece of equipment for mixing two or more liquids by simultaneously directing them, under pressure, through a constriction where the resulting turbulence blends them together.

orifice plate A disc or platelike member, with a sharp-edged hole in it, used in a pipe to measure flow or reduce static pressure.

orifice run The differential pressure producing arrangement consisting of selected pipe, orifice flange union and orifice plate. An orifice run has rigid specifications defined by the American Gas Association.

orifice-type variable-area flowmeter A flow-measurement device consisting of a tube section containing an orifice and a guided conically tapered float that rides within the orifice; flow of a fluid through the meter positions the float in relation to flow rate, with float position being determined magnetically or by other indirect means.

O ring A toroidal sealing ring made of synthetic rubber or similar material. The cross section through the torus is usually round or oval, but may be rectangular or some other shape.

orometer A barometer for measuring elevation above sea level.

orsat A gas-analysis apparatus in which certain gaseous constituents are measured by absorption in separate chemical solutions.

orthicon A camera tube that utilizes a low-velocity electron beam to scan an image stored electrically on a photoactive mosaic panel.

orthometric correction A systematic correction that must be applied to a measured difference in elevation to compensate for the fact that level surfaces at different elevations are not exactly parallel.

OS See operating system.

oscillating-piston flowmeter A flow measurement device similar to a nutating-disk flowmeter but in which motion of the piston takes place in one plane only; rotational speed of the piston is directly related to the volume of fluid passing through the meter.

oscillation Fluctuation around the set point.

oscillator A nonrotating device for producing alternating current; the output frequency is determined by characteristics of the device. In some cases the frequency is fixed, but in others it can be varied.

oscillator crystal A piezoelectric crystal device used chiefly to determine the frequency of an oscillator.

oscillatory circuit A circuit that produces a periodically reversing current when energized by a direct-current voltage; the circuit contains R, L and C elements, which may be varied to change the characteristics of the resultant a-c output.

oscillogram The permanent record created by an oscillograph. Alternatively, a permanent record of the trace on an oscilloscope, such as might be recorded photographically.

oscillograph A device for determining waveform by plotting instantaneous values of a quantity such as voltage as a function of time.

oscilloscope A CRT device that can display instantaneous values of alternating-current voltages or currents with respect to time or with respect to other alternating-current voltages or currents; it also can be used to display instantaneous values of other quantities that vary rapidly with time (not necessarily oscillatory values) and which can be converted to suitable electrical signals by means of a transducer; the display is a graphical representation of electrical signals produced by varying the position of the focused spot where an electron beam strikes the fluorescent coating on the inside surface of the CRT face.

OSI See open system interconnection.

OSI reference model A seven layered model of communications networks defined by ISO. The seven layers are: Layer 7—Application: provides the interface for application to access the OSI environment. Layer 6—Presentation: provides for data conversion to preserve the meaning of the data. Layer 5—Session: provides user-to-user connections. Layer 4—Transport: provides end-to-end reliability. Layer 3—Network: provides routing of data through the network. Layer 2—Data Link: provides link access control and reliability. Layer 1—Physical: provides an interface to the physical medium.

ounce A U.S. unit of weight; one ounce (avoirdupois) equals 1/16 pound, and is used for most commercial products; one ounce (troy) equals 1/12 pound, and is used for precious metals.

outdoor area See area, outdoor.

outdoor location Location where neither air temperature nor humidity are controlled and the equipment is exposed to outdoor atmospheric conditions such as direct sunshine, wind, rain, hail, sleet, snow and icing [S82.03].

outgassing The release of adsorbed or occluded gases and water vapor, usually during evacuation or subsequent heating of an evacuated chamber.

out-of-round A dimensional condition where diameters taken in different directions across a nominally circular object are unequal, the difference between them being the amount of out-of-roundness.

output 1. The electrical quantity, produced by a transducer, which is a function of the applied measurand [S37.1]. 2. The information transferred from the internal storage of a computer to secondary or external storage or to any device outside of the computer. 3. The routines which direct 2.4. The device or collective set of devices necessary for 2.5. To transfer from internal storage on to external media.

output, analog Nominally pertains to output of data in the form of continuously variable physical quantities as contrasted with digital output. Most analog output subsystems utilize digital to analog converters which provide a finite number of output levels and only approximate a continuous variable [RP55.1].

output area An area of storage reserved for output.

output block 1. A block of computer words considered as a unit and intended or destined to be transferred from an internal storage medium to an external destination. 2. A section of internal storage reserved for storing data which are to be transferred out of the computer. Synonymous with output area. 3. A block used as an output buffer. See buffer.

output, contact A digital output generated by operating a contact [RP55.1].

output device The part of a machine which translates the electrical impulses representing data processed by the machine into permanent results such as printed forms, punched cards, magnetic writing on tape or into control signals for a process.

output, digital Pertaining to the output of data in the form of digits. Contrast with analog output [RP55.11].

output impedance The impedance across the output terminals of a transducer presented by the transducer to the associated external circuitry [S37.1]. See impedance, output.

output indicator A device connected to a radio receiver to indicate variations in output signal without indicating a specific signal value; usually used for alignment or tuning.

output noise The rms, peak, or peak-to-peak (as specified) a-c component of a transducer's d-c output in the absence of measurand variations. NOTE: Unless otherwise specified, output impedance is measured at room conditions and with the excitation terminals open/circuited, except that nominal excitation and measurand between 80 and 100 percent-of-span is applied when the transducer contains integral active output-conditioning circuitry [S37.1].

output regulation The change in output due to a change in excitation. NOTE: Unless otherwise

specified, output regulation is measured at room conditions and with the measurand applied at its upper range limit [S37.1].

output signal A signal delivered by a device, element, or system. See also signal, output.

output variable A variable delivered by a control algorithm, e.g., the signal going to a steam valve in a temperature control loop. See controlled variable.

outside caliper A caliper used to measure distances across two external opposing surfaces.

outside diameter The outer dimension of a circular member such as a rod, pipe or tube.

oval-shaped gear flowmeter A type of positive-displacement flowmeter that operates by trapping a precise volume of fluid between an oval, toothed rotor and the meter housing as the rotor revolves in mesh with a second rotor; volume flow of an incompressible fluid is indicated directly by determining rotor speed.

oven A heated enclosure for baking, drying or heating.

oven dry A term often used by papermakers to indicate paper from which all moisture has been removed by artificial evaporation using heat. See bone dry.

overdamped See damping.

overflow 1. The condition which arises when the result of an arithmetic operation exceeds the capacity of the storage space allotted in a digital computer. 2. The digit arising from this condition if a mechanical or programmed indicator is included, otherwise the digit may be lost.

overflow pipe A pipe with its open end protruding above the liquid level in a tank; it limits the height of liquid in the tank by carrying away any liquid entering the open end, usually to a drain or sewage system.

overfractionation Operation of a distillation column to produce a purer product than required.

overheat To raise the temperature above a desired or safe limit; in metal heat treating, to reach a temperature that results in degraded mechanical or physical properties.

overlay The technique of repeatedly using the same blocks of internal storage during different stages of a problem. When one routine is no longer needed in storage, another routine can replace all or part of it.

overload The maximum magnitude of measurand that can be applied to a transducer without causing a change in performance beyond

269

specified tolerance [S37.1].

overload capacity The force weight, power, pressure or other capacity factor, usually higher than the rated capacity, beyond which permanent damage occurs to a device or structure.

overload recovery Refers to an effect caused by an analog signal input greater than that for which the feedback capacity of an amplifier can compensate. The result is a loss of feedback control by the amplifier and thereby requiring some recovery time after the overload is removed [RP55.1].

overrange In process instrumentation, of a system or element, any excess value of the input signal above its upper range value or below its lower range-value [S51.1].

overrange limit The maximum input that can be applied to a device without causing damage or permanent change in performance [S51.1]. See also overload.

override control 1. Generally, two control loops connected to a common final control element—one control loop being normally in control with the second being switched in by some logic element when an abnormal condition occurs so that constant control is maintained. 2. A technique in which more than one controller manipulates a final control element. The technique is used when constraint control is important.

overshoot 1. The amount of output measured beyond the final steady output value, in response to a step change in the measurand. Expressed in percent of the equivalent step change in output [S37.1]. 2. A transient response to a step change in an input signal which exceeds the normal or expected steady-state response. 3. The maximum difference between the transient response and the steady-state response. See also transient overshoot.

overvoltage protection A protective device that interrupts power or reduces voltage supplied to an operating device in the event the incoming voltage exceeds a preset value.

overwriting In data processing, the elimination of data by writing new data over it.

Owen bridge A type of a-c bridge circuit in which one leg contains a fixed capacitor, the opposite leg contains an unknown inductance and resistance, the third leg contains a fixed resistor and the fourth leg contains a variable resistor and a variable capacitor; this type of bridge is especially useful for measuring wide ranges of inductances using reasonable ranges of standard capacitances, and can be used to measure permeability or core loss.

oxidation 1. Loss of electrons by a constituent of a chemical reaction [S71.04]. 2. Chemical combination with oxygen.

oxide Chemical compound of an element, usually metal, with oxygen [S71.04].

oxidizing atmosphere An atmosphere which tends to promote the oxidation of immersed materials.

oxygen attack Corrosion or pitting in a boiler caused by oxygen.

P

p See pressure.

P See poise.

P&ID See piping and instrumentation drawing.

Pa See pascal.

pachymeter An instrument used to measure the thickness of material such as paper.

pack 1. In data processing, a method to condense data in order to increase storage capacity. 2. A removable disk.

packaged boiler A packaged steam or hot water firetube boiler is defined as a modified Scotch unit engineered, built, fire tested before shipment, with material, workmanship, and performance warranteed by manufacturer as stated in the manufacturer's standard conditions of sale. Components include, but are not limited to, burner, boiler and controls.

packaged steam generator See packaged boiler.

packed column A distillation column filled with packing (commonly Raschig rings) to mix the descending liquid with the ascending vapors. Packing is often used instead of trays in columns for certain applications (such as gas adsorption) or very-low-pressure drop systems.

packed decimal A method of representing a decimal number by storing a pair of decimal digits in one eight-bit byte, which takes advantage of the fact that the numbers zero through nine can be represented by four bits.

packet Block of data assembled with other control bits as a basic chunk of information to be transmitted.

packet switching system (PSS) In a wide area network, a method of sending data between computers.

packing 1. A sealing system consisting of deformable material of one or more mating and deformable elements contained in a packing box which may have an adjustable compression means to obtain or maintain an effective pressure seal [S75.05]. 2. A method of sealing a mechanical joint in a fluid system. A material such as oakum or treated asbestos is compressed into the sealing area (known as a packing box or stuffing box) by a threaded seal ring. 3. In data processing, the compression of data to save storage space.

packing box The chamber, in the bonnet, surrounding the stem and containing packing and other stem sealing parts [S75.05].

packing box, purged A packing arrangement consisting of a lantern ring inside the packing rings to permit introduction of a purge fluid to continually flush the space between the stem and body [S75.05].

packing density The number of units of useful information contained within a given linear dimension, usually expressed in units per inch; for example, the number of binary digit magnetic pulses or numbers of characters stored on tape or drum per-linear-inch on a single track by a single head.

packing flange A device that transfers the deforming mechanical load to the packing follower [S75.05].

packing follower A part which transfers mechanical load to the packing from the packing flange or nut [S75.05].

packing gland See packing follower.

packing lubricator assembly A device for injection of lubricant/sealer into a lubricator packing box [S75.05].

packing nut See packing flange.

pad 1. A pad is larger than a boss and is attached to a pressure vessel to reinforce an opening. 2. A fixed-value attenuator.

paddle-wheel level detector A device for detecting the presence or absence of bulk solids at the device location; it consists of a motor that slowly rotates a paddle in the absence of material, and rotates itself against a momentary switch when material is at or above the paddle location.

page A block of information that can be stored as a complete unit in the computer memory.

palette In data processing, the range of display colors that will show on a screen.

PAM The process (or the results of the process) in which a series of pulses is generated having amplitudes proportional to the measured signal samples.

PAM/FM Frequency modulation of a carrier by pulse amplitude modulated information.

PAM/FM/FM Frequency modulation of a carrier by subcarriers that are frequency modulated by pulse amplitude modulated (PAM) information.

panel 1. A structure that has a group of instruments mounted on it, houses the operator-process interface, and is chosen to have a unique designation. The panel may consist of one or more sections, cubicles, consoles, or desks. Synonym for board [S5.1]. 2. A sheet of material held in a frame. 3. A section of an equipment cabinet or enclosure, or a metallic or nonmetallic sheet, on which operating controls, dials, instruments or subassemblies of an electronic device or other equipment are mounted.

panel-mounted A term applied to an instrument that is mounted on a panel or console and is accessable for an operator's normal use. A function that is normally accessible to an operator in a shared-display system is the equivalent of a discrete panel-mounted device [S5.1].

paper Felted or matted sheets of cellulose fibers, bonded together and used for various purposes, especially involving printed language, artwork or diagrams.

paper machine A synchronized series of mechanical devices such as screens and heated rolls that transforms a dilute suspension of cellulose fibers (digested pulp) into a dry sheet of paper.

paper tape punch A hardware device that punches digital data into a paper tape.

paper tape reader (PTR) A hardware device for accepting punched hole paper tapes and transmitting their information content to the computer in digital form.

parallax The apparent differences in spatial relations when objects in different planes are viewed from different directions; in making instrument readings, for instance, parallax will cause an error in the observed value unless the observer's eye is directly in line with the pointer.

parallel 1. Pertaining to the simultaneity of two or more processes [RP55.1]. 2. Pertaining to the simultaneity of two or more similar or identical processes [RP55.1]. 3. Pertaining to the simultaneous processing of the individual parts of a whole, such as the bits of a character and the characters of a word, using separate facilities for the various parts [RP55.1]. 4. In data transfer operations, a procedure that handles a multiple-bit code, working with all bits simultaneously, usually one word at a time.

parallel computer 1. A computer having multiple arithmetic or logic units that are used to accomplish parallel operations or parallel processing. Contrast with serial computer. 2. Historically, a computer, some specified characteristic of which is parallel, for example, a computer that manipulates all bits of a word in parallel.

parallel elements In an electric circuit, two or more two-terminal elements connected between the same pair of nodes.

parallel I/O The simultaneous input/output of all the bits. Eight lines are required for the simultaneous transmission of eight bits.

parallel linkage A linkage mechanism that amplifies reciprocating motion. A parallel linkage depending on the geometry of the drive crank, driven crank and connecting link can amplify, attenuate as well as characterize the relationship of output driven crank to the input driven crank.

parallel operation The performance of several actions, usually of a similar nature, simultaneously through provision of individual similar or identical devices for each such action. Particularly flow or processing of information. Parallel operation is performed to save time.

parallel output To send data simultaneously between interconnecting devices.

parallel processing Pertaining to the concurrent or simultaneous execution of two or more processes in multiple devices such as channels or processing units. Contrast with serial processing.

parallels Spacers or pressure pods used in molding equipment to regulate height and prevent mold parts from being crushed.

parallel search storage A storage device in which one or more parts of all storage locations are queried simultaneously. Contrast with associative storage.

parallel task execution   Concurrent execution of two or more programs. Also, simultaneous execution of one program and I/O.

parallel transfer   A method of data transfer in which the characters of an element of information are transferred simultaneously over a set of paths.

parallel transmission   A method of transmitting digitally coded data in which a separate channel is used to transmit each bit making up a coded word. See also serial transmission.

parameter   1. A quantity or property treated as a constant but which may sometimes vary or be adjusted [S51.1]. 2. A quantity in a subroutine whose value specifies or partly specifies the process to be performed; it may be given different values when the subroutine is used in different main routines, or in different parts of one main routine, but usually remains unchanged throughout any one such use. 3. A quantity used in a generator to specify machine configuration, designate subroutines to be included, or otherwise to describe the desired routine to be generated. 4. A constant or a variable in mathematics that remains constant during some calculation. 5. A definable characteristic of an item, device, or system. See also measurand.

parameterize   To set up for variable execution depending on run-time parameter.

parametric analysis   Analysis of the impact on circuit performance of changes in the individual parameters, such as component values, process parameters, temperature, etc.

parametric oscillator   A nonlinear device which, when pumped by light from a laser, can generate tunable output. The beam produced by the parametric oscillator relies on oscillation within the nonlinear material.

parametric variation   A change in system properties—magnification, resistance or area, for example—which may affect performance of a control system that incorporates a feedback loop.

parasitic oscillations   Unintended self-sustaining oscillations or transient pulsations.

parity   A code that is used to uncover data errors by making the sum of the "1" bits in a data unit either an odd or even number.

parity bit   A binary digit appended to a group of bits to make the sum of all the bits always odd

(odd parity) or always even (even parity); used to verify data storage and transmission.

parity check   A check that tests whether the number of ones or zeroes in an array of binary digits is odd or even. Synonymous with odd-even check. See also check, parity.

park   A computer routine that will disengage a hard disk as protection from possible damage.

Parr turbidimeter   A device for determining the cloudiness of a liquid by measuring the depth of the turbid suspension necessary to extinguish the image of a lamp filament of fixed intensity.

parse   To break a command string into its elemental components for the purpose of interpretation.

Parshall flume   A venturi-type device for measuring flow in an open channel at flow rates up to 1.5-billion gal/day (5.7-million m^3/day); it consists of a converging upstream section, a downward sloping throat and an upward sloping discharge section, and may be made of any suitable structural material, usually concrete.

part   An element of an assembly or subassembly that normally is of little use by itself and cannot be disassembled further for repair or maintenance.

partial node   A point, line or plane in a standing wave field where some attribute of the wave has a nonzero minimum value.

partial pressure   The portion of total pressure in a closed system containing a gas mixture that is due to a single element or compound.

particle accelerator   Any of several different types of devices for imparting motion to charged atomic particles.

particle size   A measure of dust size, expressed in microns or per cent passing through a standard mesh screen.

parting tool   See cutoff tool.

partition   A contiguous area of computer memory within which tasks are loaded and executed.

part program   In numerical control, an ordered set of instructions in a language and format required to cause operations to be effected under automatic control which then is either written in the form of a machine program on an input medium or stored as input data for processing in a computer to obtain a machine program.

parts per million (ppm)   Represents parts per million and should be given on a weight basis. The abbreviation shall be ppm (w/w). If inconvenient to present data on a weight bases (w/w), it may be given in a volume basis; (v/v) must be stated after the term ppm, e.g., 5ppm (v/v) or 7 ppm (w/w) [S7.3].

pascal Metric unit for pressure or stress.

PASCAL A programming language developed by Nicholas Wirth and named for the mathematician Blaise Pascal.

pass 1. A single circuit through a process, such as gases through a boiler, metal between forging rolls, or a welding electrode along a joint. 2. In data processing, the single execution of a loop. 3. The shaped open space between rolls in a metal-rolling stand. 4. A confined passageway, containing heating surface, through which a fluid flows in essentially one direction. 5. A single circuit of an orbiting satellite around the earth. 6. A transit of a metal-cutting tool across the surface of a workpiece with a single tool setting.

passivating A process for the treatment of stainless steel in which the material is subjected to the action of an oxidizing solution which augments and strengthens the normal protective oxide film providing added resistance to corrosive attack.

passivation of metal The chemical treatment of a metal to improve its resistance to corrosion.

passive AND gate An electronic or fluidic device which generates an output signal only when both of two control signals appear simultaneously.

passive metal A metal which has a natural or artificially produced surface film that makes it resistant to electrochemical corrosion.

passive transducer A transducer that produces output waves without any direct interaction with the source of power that produces the actuating waves.

password In data processing, a series of characters needed to access a computer known only to those authorized to access the data stored in the computer or diskettes.

paste solder Finely divided solder alloy combined with a semisolid flux.

pasteurizing column A column that purges either a lighter-than-light key impurity through a purge stream at the top of the column or heavier-than-heavy key impurity through a purge stream at the bottom of the column.

patch 1. A section of coding inserted into a routine to correct a mistake or alter the routine. Often it is not inserted into the actual sequence of the routine being corrected, but placed somewhere else, with an exit to the patch and a return to the routine provided. 2. To insert corrected coding.

path In MS-DOS, the instructions to the computer as to how to locate a particular file.

path loss (radio) The signal loss between transmitting and receiving antennas.

pattern recognition The recognition of shapes or other patterns by a machine system.

PAW See plasma arc welding.

P band In telemetry, the portion of the radio frequency spectrum from 215 to 260 MHz; generally a narrow section of that band near 225 MHz is available for telemetry application.

PBW See proportional bandwidth.

p chart A type of data display in quality control which charts the fraction defective in a sample or over a production period against time or number of units of production.

PCM See pulse code modulation

PCM serial recording The technique of recording a train of bits on a single track of magnetic tape.

P controller See controller, proportional.

PD control Proportional plus derivative control, used in processes where the controlled variable is affected by several different lag times. See both proportional control and derivative control.

PD controller See controller, proportional plus derivative.

PDM See pulse duration modulation.

PDP-11 A family of sixteen-bit minicomputers manufactured by DEC.

PDU See protocol data unit.

peak-to-peak Pertains to the maximum amplitude excursion of a signal; for example, in the case of a pure sine wave, the maximum value between the $90°$ and $270°$ excursion points. See also double amplitude.

peak-to-peak amplitude In an oscillating or alternating function, the difference between maximum and minimum instantaneous values of the function.

pedestal In PAM, an arbitrary minimum signal value assigned to provide for channel synchronization and decommutation.

pedometer 1. A device for determining the distance traveled by walking. 2. A device for determining birth weight of an infant.

peep door A small door usually provided with a shielded glass opening through which combustion may be observed.

peep hole   A small hole in a door covered by a movable cover.

peer entities   Entities within the same layer.

peer-to-peer protocol   Communication protocol between peer entities.

peg count meter   A meter that counts the number of trunks tested, the number of circuits passed busy, the number of tests failed, or the number of repeat tests completed.

pellicle   An extremely thin, tough membrane which is stretched over a frame. Because of its thinness, it transmits some light and reflects other light, and hence can serve as a beamsplitter. Its thinness avoids the problem of ghost reflections sometimes produced by other beamsplitters. Usually found as beam splitters in interferometers.

Peltier effect   The principle in solid-state physics that forms the basis of thermocouples—if two dissimilar metals are brought into electrical contact at one point, the difference in electrical potential at some other point depends on the temperature difference between the two points.

pen   1. A device for writing with ink. 2. An ink-filled device for drawing a graphical record of an instrument reading.

pencil   An implement for making marks with graphite, carbon or a colored solid substance.

pendulum scale   A type of weighing device in which the weight of the load is counterbalanced by rotation of a bent lever with a fixed weight at the free end.

penetrameter   A stepped piece of metal used to assess density of exposed and developed radiographic film, and to determine relative ability of the radiographic technique to detect flaws in a workpiece.

penetration   1. Distance from the original base metal surface to the point where weld fusion ends. 2. A surface defect on a casting where molten metal filled surface voids in the sand mold.

penetration number   A measure of the consistency of materials such as waxes and greases expressed as the distance that a standard needle penetrates a sample under specified ASTM test conditions.

penetration rate   The distance per unit time that a drill cuts into a material, measured along the drill axis.

penetrometer   An instrument for determining penetration number.

pen-motor recorder   A data-versus-time strip chart recorder where each trace is written by a motor-driven pen.

pen recorder   See pen.

pentode   An electron tube containing five electrodes—an anode, a cathode, a control electrode, and two others which usually are grids.

pen travel   The length of the path described by the pen in moving from one end of the chart scale to the other. The path may be an arc or a straight line [S51.1].

percent defective   The number of defective pieces in a lot or sample, expressed as a percent.

percent of actual   Same accuracy value applies over the entire flow rate range.

percent of span   Accuracy value applies only at the maximum rated flow.

perfect combustion   The complete oxidation of all the combustible constituents of a fuel, utilizing all the oxygen supplied.

perfect vacuum   A reference datum analogous to a temperature of absolute zero that is used to establish scales for expressing absolute pressures.

performance characteristic   A qualitative or quantitative measurement unique to a piece of equipment or a system that is evident only during its test or operation.

performance chart   A graphic representation of some aspect of operation of a piece of equipment or a system.

performance curves   Plots of the abilities of rotating equipment under various operating conditions.

performance data   Information on the way a material or device behaves during actual use.

performance evaluation   A comparison of performance data, usually taken by an automatic data logging system, with predetermined standards or estimates, for assessing operating experience or identifying any need for corrective action.

performance index   In industrial engineering, the ratio of standard hours to hours of work actually used to produce a given output. A ratio greater than 1.00 (100%) indicates standard output is being exceeded.

performance number   Any of a series of numbers used to rate aviation gasolines with octane values greater than 100; the PN (performance number) compares fuel antiknock values with those of a standard reference fuel in terms of an index which indicates relative engine performance.

period 1. Of a periodic function, the smallest increment of the independent variable that can be repeated to generate the function. 2. Of an undamped instrument, the time between two successive transits of the pointer through the rest position in the same direction following a step change in the measured quantity.

periodic duty A type of intermittent duty involving regularly repeating load conditions.

periodic function An oscillating quantity whose values repeatedly recur for equal increments of the independent variable.

peripheral 1. A supplementary piece of equipment that puts data into, or accepts data from the computer (printers, floppy disc memory devices, videocopiers). 2. Any device, distinct from the central processor, that can provide input or accept output from the computer.

peripheral speed See cutting speed.

permanently connected equipment Equipment connected to a supply circuit by field wiring terminals or by means of separate installed leads [S82.01].

permanent magnet A shaped piece of ferromagnetic material that retains its magnetic field strength for a prolonged period of time following removal of the initial magnetizing force.

permanent-magnet moving-coil instrument See moving-coil instrument.

permanent pressure drop The unrecoverable reduction in pressure that occurs when a fluid passes through a nozzle, orifice or other throttling device.

permeameter 1. A device for determining the average size or surface area of small particles; it consists of a powder bed of known dimensions and degree of packing through which the particles are forced under pressure. Particle size is determined from flow rate and pressure drop across the bed; surface area, from pressure drop. 2. A device for determining the coefficient of permeability by measuring the gravitational flow of fluid across a sample whose permeability is to be determined. 3. An instrument for determining magnetic permeability of a ferromagnetic material by measuring the magnetic flux or flux density in a specimen exposed to a magnetic field of a given intensity.

permissible dose The amount of ionizing radi-

ation that a human being can absorb over a given period of time without harmful result.

personal computer (PC) 1. A personal computer is generally used at an office desk for word processing, data bases and spreadsheets. A personal computer may be used to aid in the configuration of programmable logic controllers and distributed control systems or may be used for data acquisition and control of small processes. Ruggedized PC's have been used on the process floor for control and data acquisition. 2. The letters PC are sometimes used to signify a programmable (logic) controller.

PERT See Program Evaluation and Review Technique.

perturbation generator An instrument that simulates typical data link perturbations such as blanking, noise, bit rate jitter, baseline offset, and wow.

petroleum Naturally occurring mineral oil consisting predominately of hydrocarbons.

petroleum engineering A branch of engineering that deals with drilling for and producing oil, natural gas and liquifiable hydrocarbons.

pH The symbol for the measurement of acidity or alkalinity. Solutions with a pH reading of less than 7 are acid; solutions with a pH reading of more than 7 are alkaline on the pH scale of 0 to 14, where the midpoint of 7 is neutral.

phase 1. The relationship between voltage and current waveforms in a-c electrical circuits. 2. A microstructural constituent of an alloy that is physically distinct and homogeneous. 3. For a particular value of the dependent variable in a periodic function, the fractional part of a period that the independent variable differs from some arbitrary origin. 4. In batch processing, an independent process-oriented action within the procedural part of a recipe. The phase is defined by boundaries that constitute safe and logical points where processing can be interrupted.

phase angle 1. The difference between the phase of current and the phase of voltage in an alternating-current signal, usually determined as the angle between current and voltage vectors plotted on polar coordinates. 2. A measure of the propagation of a sinusoidal wave in time or space from some reference instant or position on the wave.

phase angle firing A method of operation for a SCR stepless controller in which power is turned on for the proportion of each half cycle in the a-c power supply necessary to maintain the desired heating level.

phase crossover frequency See frequency, phase crossover.

phase discriminator A device that detects the phase relationship of a signal to that of a reference.

phase lag Phase shift when the output lags the input [S26].

phase margin The difference between 180° and the absolute value of the open-loop phase angle for a stable feedback system at that frequency where the gain is unity.

phase matching Alignment of a nonlinear crystal with respect to the incident laser beam in the proper way to generate a harmonic of the laser frequency in the material.

phase meter An instrument for measuring electrical phase angles. Also known as phase-angle meter.

phase modulation Modulation of a sinusoidal carrier wave in which the angle of the modulated wave differs from the angle of the carrier wave by an amount proportional to the instantaneous amplitude of the modulating wave.

phase-sequence indicator A device that indicates the sequence in which the fundamental components of a polyphase set of voltages or currents reach some particular value—their maximum positive value, for example.

phase shift 1. Of a transfer function, a change of phase angle with test frequency, as between points on a loop phase characteristic [S51.1]. 2. Of a signal a change of phase angle with transmission [S51.1]. 3. Difference between corresponding points on input and output signal wave shapes, disregarding any difference in magnitude [S26]. 4. The time difference between the input and output signal or between any two synchronized signals, of a control unit, system, or circuit, usually expressed in degrees or radians. 5. A change in phase angle between the sinusoidal input to an element and its resulting output.

phase shift circuit An electronic network whose output voltage is shifted in phase when compared to a specified reference voltage.

phase shifter An electronic device whose output voltage (or current) differs from its input voltage (or current) by some desired phase relationship; in some devices the phase is shifted a fixed amount because of an inherent design fea-

ture, but in others the phase relationship can be adjusted.

phase shift keying (PSK) A form of PCM achieved by shifting the phase of the carrier; e.g., ± 90 degrees to represent "ones" and "zeros."

phase velocity The velocity of an equiphase surface along the normal of a traveling single-frequency plane wave.

Phelps vacuum gage A modified hot-filament ionization gage useful for measuring pressures in the range 10^{-5} to 1 torr.

Philips gage An instrument that measures very low gas pressure (vacuum) indirectly by determining current flow from a glow discharge device.

pH meter An instrument for electronically measuring electrode potential of an aqueous chemical solution and directly converting the reading to pH (a measure of hydrogen ion concentration, or degree of acidity).

phon A unit of loudness level equivalent to a unit pressure level in decibels of a 1000-Hz tone.

phoneme The basic phonological element of speech consisting of a simple sound that, by itself, cannot differentiate one word from another; the American English language, for example, contains 38 to 40 phonemes (14 to 16 vowel sounds and 24 consonant sounds) that are used in conjunction with inflection, volume and emphasis to produce synthetic speech.

phonotelemeter A sophisticated stopwatch for estimating the distance from artillery by measuring the elapsed time from gun flash to arrival of the detonation sound.

phosphatizing Forming an adherent phosphate coating on metal by dipping or spraying with a solution to produce an insoluble, crystalline coating of iron phosphate which resists corrosion and serves for a base for paint.

phosphor A phosphorescent material.

phosphor bronze A hard copper-tin alloy, deoxidized with phosphorus, and sometimes containing lead to enhance its machinability.

phosphorescence Emission of radiant energy —often in the visible-light range—following excitation due to absorption of shorter wavelength radiation; phosphorescent emission may persist for a long time after the exciting radiation stops. Contrast with fluorescence; incandescence.

phot The CGS unit of illuminance, which equals one lumen power cm²; the SI unit, lux, is preferred.

photocell A device that alters its electrical resistance in proportion to the amount of light that impinges on it.

photoconductive Converting a change of measurand into a change in resistance or conductivity of a semiconductor material by a change in the amount of illumination incident upon the material [S37.1].

photoconductive cell A transducer that converts the intensity of EM radiation, usually in the IR or visible bands, into a change of cell resistance.

photoconductor A type of conductor which changes its resistivity when illuminated by light; the changes in resistance can be measured to determine the amount of incident light.

photodarlington A detector in which a phototransistor is fabricated on the same chip with a second transistor which amplifies the signal from the phototransistor. The circuit formed is a Darlington circuit—a simple and inexpensive type of detector with limited performance.

photodiode A diode which detects light. Vacuum photodiodes are tubes in which detection relies on the photoelectric effect producing free electrons which are collected by a positively charged electrode.

photodraft A photographic reproduction of a master layout or design on an emulsion-coated sheet of metal; it is used chiefly as a master in tool- and die-making.

photoelectric cell A device whose electrical properties—electron emittance or conductance, for example—change when a sensitive element within the device is exposed to light.

photoelectric control Modifying a controlled variable in accordance with a control signal whose value is related to the intensity of a light-beam input signal.

photoelectric counter A counting device actuated when a physical object passes through an incident beam of light.

photoelectric effect A physical phenomenon whereby a so-called photoelectric material emits electrons when struck by light—one bound electron being emitted for each photon of light absorbed.

photoelectric hydrometer A device for measuring specific gravity of a continuously glowing liquid, in which a weighted float, similar to a hand hydrometer, rises or falls with changes in liquid density, changing the amount of light that is permitted to fall on a sensitive phototube whose output is calibrated in specific gravity units.

photoelectric photometer A device that uses a photocell, phototransistor or phototube to measure the intensity of light. Also known as electronic photometer.

photoelectric pyrometer An instrument that measures temperature by measuring the photoelectric emission that occurs when a phototube is struck by light radiating from an incandescent object.

photoelectric threshold The amount of energy in a photon of light that is just sufficient to cause photoelectric emission of one bound electron from a given substance.

photoemissive tube photometer A device that uses a tube made of photoemissive material to measure the intensity of light; it is very accurate, but requires electronic amplification of the output current from the tube; it is considered chiefly a laboratory instrument.

photogrammetry 1. The science of making maps or accurately measuring features from aerial photographs. 2. Making surveys by means of aerial photography.

photographic emulsion A light-sensitive coating—usually a silver halide compound in gelatine—used to capture and store the visual image in photography or radiography.

photographic recording Using a signal-controlled light beam or spot to record information—either by recording the position of the spot, its intensity, or both.

photoluminescence Nonthermal emission of electromagnetic radiation that occurs when certain materials are excited by absorption of visible light.

photometer An instrument for measuring the intensity of visible light.

photometry Any of several techniques for determining the properties of a material or for measuring a variable quantity by analyzing the spectrum or intensity, or both, of visible light.

photomultiplier A type of electron tube in which photons incident on a photocathode produce electrons by photoemission. These electrons are then amplified by passing them through an electron multiplier, which increases their numbers. Electrons passing through the multiplier are accelerated by high voltages and hit metal screens, from which they free more electrons.

photomultiplier tube   See multiplier tube.

photon   A quantum of electromagnetic radiation.

photon counting   A measurement technique used for measuring low levels of radiation, in which individual photons generate signals which can be counted.

phototransistor   A transistor in which one of the two junctions is illuminated by light, and electrons are released. The transistor treats this current as an input, which it amplifies, making it a simple detector-amplifier.

phototube   An electron tube containing at least two electrodes, one that functions as a photoelectric emitter.

photovoltaic   Converting a change of measurand into a change in the voltage generated when a junction between certain dissimilar materials is illuminated [S37.1].

photovoltaic cell   A transducer that converts the intensity of EM radiation, usually in the IR or visible bands into a voltage.

physical address space   The set of computer memory locations where information can actually be stored for program execution; virtual memory addresses can be mapped, relocated, or translated to produce a final memory address that is sent to hardware memory units; the final memory address is the physical address.

physical block   A physical record on a mass storage device.

physical input   See measurand.

physical properties   Inherent characteristics of a substance—such as electrical conductivity, magnetic permeability, density or melting point—that can be determined without applying mechanical force.

physical record   The largest unit of data that read/write hardware of an I/O device can transmit or receive in a single I/O operation; the length of a physical record is device-dependent; for example, a punched card can be considered the physical record for a card reader; it is eighty bytes long.

phytometer   A device for determining the transpiration rate of plants; it consists of a soil-filled container in which one or more plants are rooted and sealed so that water can escape only through transpiration.

PI   See proportional-integral derivative (PID) control.

P/I   A pressure to current converter linearly con-verts a signal pressure range into a signal current range (for example, 3-15 psi into 4-20 mA).

piano wire   Carbon steel wire (0.75 to 0.85% C) cold drawn to high tensile strength and uniform diameter.

pica   A unit of measure used in printing; one-sixth of an inch. See point.

pickle liquor   Spent pickling solution.

pickling   Preferential removal of oxide scale from the surface of metal by immersion in a strong alkaline or inhibited acid solution.

pickup   1. A transducer or other device that converts optical, acoustical, mechanical or thermal images or signals into electrical output signals. 2. Electrical noise or interference from a nearby device, system or circuit. 3. The minimum value of an input signal—voltage, current or power, for instance—needed to make a relay function as intended. See also transducer.

PI control   Proportional plus integral control, used in combination to eliminate offset. See both proportional control and integral control. Also called proportional plus reset control.

PI controller   See controller, proportional plus integral.

PID   See proportional integral derivative.

PID action   A mode of controller action in which proportional integral, and derivative action are combined.

PID control   Proportional plus integral plus derivative control, used in processes where the controlled variable is affected by long lag times. See proportional control, integral control, and derivative control.

PID controller   See controller, proportional plus integral plus derivative.

piercing   An operation in which a tool is forced through a metal part in order to cut a hole of a specific shape and size.

piezoelectric   Converting a change in measurand into a change in the electrostatic charge or voltage generated by certain materials when mechanically stressed [S37.1].

piezoelectric accelerometer   A device for measuring variable forces associated with acceleration, such as from an earthquake or from vibration, by means of response of a piezoelectric crystal in physical contact with a mass that reacts to the accelerating forces.

piezoelectric detector   A sensing element for detecting seismic disturbances which consists of a stack of piezoelectric crystals with an inertial mass on top of the stack; metal foil between the crystals collects the charges that develop when the crystals are strained.

piezoelectric effect   The generation of an electric potential when pressure is applied to certain materials or conversely a change in shape when a voltage is applied to such materials. The changes are small, but piezoelectric devices can be used to precisely control small motions of optical components.

piezoelectric gage   A pressure measuring device used to detect and measure blast pressures from explosives and internal pressure transients in guns; it uses a piezoelectric crystal to sense a pressure transient and develop an output voltage pulse in response.

piezoelectric pressure transducer   Any of several sensor designs in which a force acting on the sensing element is converted to an electrical output by a piezoelectric crystal.

piezoid   A piezoelectric crystal adapted for use by attaching electrodes to its surface or by other suitable processing.

piezometer   1. An instrument for measuring fluid pressure. 2. An instrument for measuring compressibility of materials.

piezoresistive accelerometer   A device for measuring variable forces associated with acceleration, such as from an earthquake or from vibration, by means of changes in resistance of two or four semiconductor strain gages connected in a Wheatstone bridge circuit.

pig   1. An in-line scraper for removing scale and deposits from the inside surface of a pipeline; a holder containing brushes, blades, cutters, swabs, or a combination is forced through the pipe by fluid pressure. 2. A crude metal casting, usually of primary refined metal intended for remelting to make alloys.

pigtail   A 270° or 360° loop in pipe or tubing to form a trap for vapor condensate. Used to prevent high temperature vapors from reaching the instrument. Used almost exclusively in static pressure measurement.

pile   1. An assemblage of thermoelectric elements, dissimilar-metal plates or fissile-material components so arranged to produce electrical or thermal power—as in a thermopile, storage battery or atomic reactor. 2. A heap of aggregate or other bulk material stored on a floor or on a flat area outdoors. 3. A long, heavy column made of timber, steel or reinforced concrete which has

been driven or cast in place below grade to support another structure or to hold earth in place.

pilot   1. A mechanical control system, such as may be used to guide an aircraft in flight. 2. A bar extending in front of a reamer to guide the reamer and force it to cut concentric with the original borehole. 3. A flame which is utilized to ignite the fuel at the main burner or burners. See also ignitor.

pilot circuit   That portion of a control circuit or system which carries the control signal from the signal-generating device to the control device.

pilot, constant   A pilot that burns without turndown throughout the entire time the boiler is in service.

pilot, continuous   See pilot, constant.

pilot, expanding   A pilot that normally burns at a low turndown throughout the entire time the burner is in service whether the main burner is firing or not. Upon a call for heat, the pilot is automatically expanded so as to reliably ignite the main burner. This pilot may be turned down at the end of the trial-for-ignition period for the main burner.

pilot flame establishing period   The length of time fuel is permitted to be delivered to a proved pilot before the flame-sensing device is required to detect pilot flame.

pilot light   A light that indicates which of a number of normal conditions of a system or device exists. It is unlike an alarm light, which indicates an abnormal condition. The pilot light is also known as a monitor light [S5.1].

pilot plant   A test facility, built to duplicate or simulate a planned process or full-scale manufacturing plant, used to gain operating experience or evaluate design alternatives before the full-scale plant is built.

pilot, proved   A pilot flame which has been proved by flame failure controls.

pilot stabilization period   A timed interval synonymous on most systems today with timed trial for pilot ignition. Today's programmers prevent main valve operation for a specified number of seconds after commencement of trial for pilot ignition even though pilot is immediately proved.

pin   1. A cylindrical or slightly tapered fastener made of wood, metal or other material which joins two or more members yet allows free angular movement at the joint. 2. In a dot-matrix printer, the tiny cylinders that as a group form a character. Typical computer printers are 9-pin and 24-pin.

pinboard A type of control panel which uses pins rather than wires to control the operation of a computer. On certain small computers which use pinboards, a program is changed by the operator removing one pinboard and inserting another. Related to control panel. See plugboard.

pinch or clamp valve A valve consisting of a flexible elastomeric tubular member connected to two rigid flow path ends whereby modulation and/or shut off of flow is accomplished by squeezing the flexible member into eventual tight sealing contact. The flexible member may or may not be reinforced. The flexible member may or may not be surrounded by a pressure retaining boundary consisting of a metal housing with stem packing box. Squeezing of the flexible member may be accomplished by: a) single stem and leverage acting from both sides so that the total collapse and sealing occurs along the horizontal center line of the flexible member; b) double stem action involving two separate actuator assemblies diametrically opposed, or c) a separate source of fluid pressure applied to an annulus surrounding the flexible member. A clamp valve is a pinch valve but with clamps and shaped inserts used to provide stress relief in the creased area of the tubular member [S75.05]. See one-piece-element clamp, two-piece-element clamp.

pi network A network consisting of three branches connected in series to form a closed mesh; one of the three junctions is an input terminal, one is an output terminal and the third is a common terminal connected to both input and output circuits.

pinhole A fault in a casting or coating resulting from small blisters that have burst or from small voids that formed during plating.

pinhole detector A photoelectric device that can detect small holes or other defects in moving sheets of material.

pinion The smaller of two gear wheels, or the smallest gear in a gear train.

PIN photodiode A semiconductor diode light detector in which a region of intrinsic silicon separates the p and n type materials. It offers particularly fast response and is often used in fiber optic systems.

pipe 1. A tubular structural member used primarily to conduct fluids, gases or finely divided solids; it may be made of metal, clay, ceramic, plastic, concrete or other materials. 2. A general class of tubular mill products made to standard combinations of diameter and wall thickness. 3. A central defect in a metal ingot formed by contraction of the metal as it solidifies and cools. 4. An extrusion defect caused by the oxidized metal surface flowing toward the center of the extrusion at the back end.

pipe elbow meter A variable-head meter used to measure flow around the bend in a pipe.

pipe fitting A piece with an internal cavity for connecting lengths of pipe together or for connecting them to tanks or other process equipment; types include couplings, elbows, nipples, tees and unions.

pipelining The process of increasing data processing speed by simultaneously executing a number of basic instructions.

pipe saddle See leveling saddle.

pipe tap A small hole in the wall of a pipe for sampling its contents or for connecting a control device or pressure measuring instrument.

pipe tee A pipe fitting in the shape of the letter T; it is used to connect a branch line at 90° to the main run of pipe.

pipe thread A type of screw thread used chiefly to connect pipe and fittings; in the usual configuration, it is a 60° thread with flat roots and crests, and with a longitudinal taper of about 3/4 in. per foot (about 6.3%).

piping 1. The term piping includes metal or plastic tube, pipe fittings, valves, and similar components, and the practice of assembling these items into a system [RP60.9]. 2. A system of pipes for carrying a fluid stream or gaseous material.

piping and instrumentation drawing (P&ID) 1. Show the interconnection of process equipment and the instrumentation used to control the process. In the process industry, a standard set of symbols is used to prepare drawings of processes. The instrument symbols used in these drawings are generally based on Instrument Society of America (ISA) Standard S5.1. 2. The primary schematic drawing used for laying out a process control installation.

pipping pressure 1. The pressure at which a safety valve opens. 2. The pressure the pipe cannot withstand without exceeding its design characteristics.

Pirani gage A pressure transducer used to measure very low gas pressure based on measurement of the resistance of a heated wire filament; resistance varies in accordance with thermal conduction of the gas, which in turn is related to gas pressure. Used primarily for pressures less than one atmosphere.

piston A metal cylinder that reciprocates in a tubular housing, either moving against or being moved by fluid pressure.

piston displacement The volume traversed by a piston in a single cycle, or stroke.

piston meter A type of fluid flow meter; it is a variable-area, constant-head device in which the flow rate is indicated by a pointer attached to a piston, which in turn is positioned by the buoyant force of the fluid.

pistonphone A device consisting of a small chamber and reciprocating piston of measurable displacement; it is used to establish a known sound pressure in the chamber.

piston ring seal A seal ring installed in a groove on the piston circumference to minimize the clearance flow between the piston outer diameter and the cylinder bore.

piston type valve A fluid powered device in which the fluid acts upon a movable cylindrical member, the piston, to provide linear motion to the actuator stem [S75.05]. See also actuator, piston type.

piston-type variable-area flowmeter Any of several flowmeter designs in which fluid passing through the meter exerts force on a piston such that the piston moves against a counterbalancing force to expose a portion of an exit orifice, the amount exposed being directly related to volume flow.

pit A small surface cavity in a metal part or coating usually caused by corrosion or formed during electroplating.

pitch 1. An auditory sensation of tone that is directly related to sound-wave frequency. 2. A heavy, black or dark brown liquid or solid residue from distillation of tar or oil; it occurs naturally as asphalt. 3. The distance between similar mechanical elements in an array, such as gear teeth, screw threads or screen wires. 4. The distance between centerlines of tubes, rivets, staybolts, or braces. 5. In computer printers, a measure of the number of characters printed per inch. Typically 10, 12 or 17.

pitot-static tube A combination of a pitot tube and a static tube—the two may be either parallel or concentric.

pitot tube 1. An instrument for measuring stagnation pressure of a flowing liquid; it consists of an open tube pointing upstream, into the flow of fluid, and connected to a pressure indicator or recorder. 2. An instrument which will register total pressure and static pressure in a gas stream, used to determine its velocity.

pitot-venturi tube A combination of a venturi device and a pitot tube.

pitting A concentrated attack by oxygen or other corrosive chemicals in a boiler, producing a localized depression in the metal surface.

pixel 1. The smallest controllable display element on a VDU. Also referred to as picture element (PEL) [S5.5]. 2. In data processing, a portion of a CRT display screen.

PL/1 See programming language/1.

planar network An electronic network which can be drawn or sketched on a plane surface without having any of the branches cross each other.

plane of polarization In a plane polarized electromagnetic wave, the plane that contains both the direction of propagation and the electric field vector.

plane polarized wave An electromagnetic wave in a homogeneous isotropic medium that has been generated, or modified by the use of filters, so that the electric field vector lies in a fixed plane which also contains the direction of propagation.

plan equation An equation for determining horsepower: HP=plan/33,000, where p is mean effective pressure in psi, l is piston stroke in feet, a is net piston area in in.^{2}, and n is number of strokes per minute.

plane wave A wave whose equiphase surfaces form an array of parallel planes.

planimeter A device for measuring area of a plane surface, usually of irregular shape, by tracing its perimeter.

plasma arc welding (PAW) Metals are heated with a constricted arc between an electrode and the workpiece (transferred arc), or the electrode and the constricting nozzle (non-transferred arc). Shielding is obtained from the hot, ionized gas issuing from the orifice which may be supplemented by an auxiliary source of shielding gas.

plastic An imprecise term generally referring to any polymeric material, natural or synthetic. Its plural, plastics, is the preferred term for referring to the industry and its products.

plastic clad silica A step index optical fiber in which a silica core is covered by a transparent plastic cladding of lower refractive index. The plastic cladding is usually a soft material, although hard-clad versions have been introduced.

plastic fibers Optical fibers in which both core and cladding are made of plastic material. Typically their transmission is much poorer than that of glass fibers.

plasticorder A laboratory device for measuring temperature, viscosity and shear-rate in a plastics material which can be used to predict its performance.

plastometer An instrument for determining flow properties of a thermoplastic resin by forcing molten resin through a fixed orifice at specified temperature and pressure.

plate A rolled flat piece of metal; depending on the type of metal, the minimum thickness for the product to be called plate instead of sheet or strip may vary—for instance, plate steel is any hot-finished flat-rolled carbon or alloy steel product more than 8 in. wide and more than 0.230 in. thick, or more than 48 in. wide and more than 0.180 in. thick. See electroplating.

plateau 1. Generally, any portion of a function where the value of the dependent variable is essentially constant over a range of values for the independent variable. 2. Specifically, a portion of the output versus input characteristic of an instrument, electronic component or control device where the output signal level is essentially independent of the input signal level.

plate baffle A metal baffle.

platen A plane surface receiving heat from both sides and constructed with a width of one tube and depth of two or more tubes, bare or with extended surfaces.

plate polarized light Light polarized by means of optical plates set at Brewster's angle to the optic axis. The more plates the greater the purity of the plane polarized exit beam.

PLC See programmable logic controller.

plenum 1. A condition where air pressure within an enclosure is greater than barometric pressure outside the enclosure. 2. An enclosure through which gas or air passes at relatively low velocities.

PL/M A block-structured high-level language for preparing software for Intel microprocessors.

plotter 1. A device for automatically graphing a dependent variable on a visual display or flat board, in which a movable pen or pencil is positioned by one or more instrument control signals. 2. Hardware device which plots on paper the magnitudes of selected data channels, as related to each other or to time.

plotter/printer A plotter that can also print alphanumeric data from the computer.

plug 1. A cylindrical part which moves in the flow stream with linear motion to modify the flow rate and which may or may not have a contoured portion to provide flow characterization. It may also be a cylindrical or conically tapered part, which may have an internal flow path, that modifies the flow rate with rotary motion [S75.05]. 2. A rod or mandrel over which a pierced billet is drawn to form a tube or pipe, or that is inserted into a tube or pipe during cold reduction. 3. A punch or mandrel over which a cup is drawn. 4. A fluid-tight seal made to prevent flow through a leaking pipe or tube. 5. A projecting portion of a die intended to form a recess in a forged part. 6. A term frequently used to refer to the closure component.

plugboard A perforated board that accepts manually inserted plugs to control the operation of equipment, such as a removable panel containing an ordered array of terminals, which may be interconnected by short electrical leads (plugged in by hand) according to a prescribed pattern and thereby designating a specific program or sequence of specified program steps. See pinboard.

plug die See floating plug.

plug fuseholder A receptacle with female threads to accommodate a plug-type fuse.

plug gage A metal member used to check the dimension of a hole. The gaging element may be straight or tapered, plain or threaded, and of any shape cross section.

plugging Physically stopping the flow of fluid, either intentionally or unintentionally, especially by the buildup of material.

plug meter A device for measuring flow rate in which a tapered rod extends through an orifice; when the rod is positioned so that the effective area of the annulus is just sufficient to handle the fluid flow, the rate of flow is read directly from a scale.

plug valve 1. A valve with a closure member that may be cylindrical, conical or a spherical segment in shape. It is positioned, open to closed, with rotary motion [S75.05]. 2. A type of shutoff valve consisting of a tapered rod with a lateral hole through it. As the rod is rotated 90° about its longitudinal axis, the hole is first aligned with the direction of flow through the valve and then aligned crosswise, interrupting the flow.

plumb Indicating a true vertical position with respect to the earth's surface; the condition is usually determined by a plumb bob, which consists of a weight (plummet) suspended on a string and positioned entirely by gravity.

plumb-bob gage 1. A device for determining liquid level in which a weighted plummet is lowered on a calibrated tape or cable until it just touches the liquid surface. 2. A device for detecting solids level in a storage bin or hopper by lowering a plummet until the lowering cable slackens, which is usually detected by an electrical or mechanical triggering device.

plummet gage See plumb-bob gage.

plunger-type instrument A moving-iron instrument in which a pointer is attached to a long, specially shaped piece of iron that moves along the axis of a coil by variable electromagnetic attraction depending on the current flowing in the coil.

PL/Z A high-level language for Zilog microprocessors.

pneumatic 1. A device which converts the energy of a compressible fluid, usually air, into motion [S75.05]. 2. Pertaining to or operated by a gas, especially air. 3. Systems that employ gas, usually air, as the carrier of information and the medium to process and evaluate information.

pneumatic controller 1. A pneumatic controller is a device which compares the value of a variable quantity or condition to a selected reference and operates by pneumatic means to correct or limit the deviation [S7.4]. 2. A device activated by air pressure to mechanically position another device, such as a valve stem. Also known as pneumatic positioner.

pneumatic control system 1. A control system that uses air or gas as the energy source. 2. A system which makes use of air for operating control valves and actuators.

pneumatic control valve A spring-loaded valve that regulates the area of a fluid-flow opening by changing position in response to variable pneumatic pressure opposing the spring force.

pneumatic delivery capability The rate at which a pneumatic device can deliver air (or gas) relative to a specified output pressure change. It is usually determined, at a specified level of input signal, by measuring the output flow rate for a specified change in output pressure. The results are expressed in cubic feet per minute (ft^3/min) or cubic meters per hour (m^3/h), corrected to standard (normal) conditions of pressure and temperature [S51.1].

pneumatic exhaust capability The rate at which a pneumatic device can exhaust air (or gas) relative to a specified output pressure change. It is usually determined, at a specified level of input signal, by measuring the output flow rate for a specified change in output pressure. The results are expressed in cubic feet per minute (ft^3/min) or cubic meters per hour (m^3/h), corrected to standard (normal) conditions of pressure and temperature [S51.1].

pneumatic extensions A system that converts float position to a proportional standard pneumatic signal. A magnetic coupling connects the internal float extension with an external mechanical system linked to a pneumatic transmitter.

pneumatic information transmission system A pneumatic information system is a system for conveying information comprising a) a transmitting mechanism converting input information into a corresponding air pressure, b) interconnecting tubing and c) a receiving element responsive to air pressure which develops an output directly corresponding to the input information [S7.4].

pneumatic signal line 1. An air (pneumatic) signal, usually 3-15 psig is used as the energy medium. 2. Applies to a signal using any gas as the signal medium. If a gas other than air is used, the gas may be identified by a note on the signal symbol or otherwise.

pneumatic supply Air at a nominally constant pressure used to operate pneumatic devices [RP60.9].

pneumatic system A system which makes use of air for operating control valves and actuators (cylinders, motors).

pneumatic telemetering Remote transmission of a signal from a primary sensing element to an indicator or recorder by means of a pneumatic pressure impulse sent through small-bore tubing; may be used to monitor temperature, pressure, flow rate or other variables in a process unit or system. Also known as pneumatic intelligence transmission.

pneumatic to current converter (P/I) A pressure to current converter linearly converts a signal pressure range into a signal current range (for example, 3-15 psi into 4-20 mA).

pneumatic transmission signal A signal used for information transmission which varies in a continuous manner [S7.4].

Pockel's cell A device in which the Pockel's effect is used to modulate light passing through the material. The modulation relies on rotation of beam polarization caused by the application of an electric field to a crystal; the beam then has to pass through a polarizer, which transmits a fraction of the light dependent on its polarization.

pocket chamber A small ionization chamber that can be charged, then carried in a person's pocket and periodically read to determine cumulative radiation dose received since the instrument was last charged. Also known as pocket dosimeter.

pocket meter A pocket-size direct-reading instrument for measuring radiation dose rate.

poidometer An automatic weighing device used in conjunction with a belt conveyer.

point 1. A process variable derived from an input signal or calculated in a process calculation. 2. A unit of measure used in printing; one-seventy second of an inch. 12 points equal one pica.

point drift See drift, point.

pointer 1. A needle-shaped or arrowhead-shaped element whose position over a scale indicates the value of a measured variable. 2. In data processing: A. A data string that tells the computer where to find a specific item. B. Similar to or the same as a cursor on a computer screen.

pointing 1. Reducing the diameter and tapering a short length at one end of a wire, rod or tube; usually done so the pointed end may be inserted through a reducing die and clamped in the moving element of a drawbench. 2. Finishing a mortar joint, or pressing mortar into a raked joint.

point module See alarm module.

points Synonymous with channel [RP55.1].

point-to-point numerical control A simple form of numerical control in which machine elements are moved between programmed positions without particular regard to path or speed control. Also known as positioning control.

poise The CGS unit of dynamic viscosity, which equals one dyne-second per cm^2; the centipoise (cP) is more commonly used.

Poiseuille flow Laminar flow of gases in long tubes at pressures and velocities such that the flow can be described by Poiseuille's equation.

POL See problem-oriented language.

polar diagram 1. A diagram showing the relative effectiveness of an antenna system for either transmitting or receiving. Principally shows directional characteristics. 2. A schematic representation that relates events in the cycle of a piston engine to crankshaft position.

polarimeter An optical device for measuring the degrees and type of polarization of a light source.

polarimetry Chemical analysis in which the amount of substance present in a solution is estimated from the amount of optical rotation (polarization) that occurs when a beam of light passes through the sample.

polarity The relationship between the transducer output and the direction of the applied acceleration; taken as "standard" when a positive charge or voltage appears on the "high" side of the transducer for an acceleration directed from the mounting surface into the body of the accelerometer [S37.2].

polarization maintaining fiber A single-mode optical fiber which maintains the polarization of the light which entered it, normally by including some birefrigence within the fiber itself. Normal single-mode fibers, and all other types, allow polarization to be scrambled in light transmitted through them.

polarized meter A meter with its zero point at the center of the scale so the direction the pointer deflects indicates electrical polarity and the distance it deflects indicates value of a measured voltage or current.

polarizer A filter which transmits light of only a single polarization.

polarizing coating Coatings which influence the polarization of light passing through them, typically by blocking or reflecting light of one polarization and passing light that is orthogonally polarized.

polarographic analysis A method of determining the amount of oxygen present in a gas by measuring the current in an oxygen-depolarized primary cell.

polarography A method of chemical analysis that involves automatically plotting the voltage-current characteristic between a large, non-polarizable electrode and a small polarizable electrode immersed in a dilute test solution; a curve containing a series of steps is produced, the potential identifying the particular cation involved and the step height indicating cation concentration; actual values are determined by comparing each potential and step height with plots generated from test solutions of known concentrations.

pole-dipole array An electrode array for making resistivity or induced-polarization surveys in which one current electrode is placed far away from the area being surveyed while an assembly containing one current electrode and two potential electrodes is moved laterally across the area in a search pattern.

pole face On a magnetized part, the surface through which magnetic lines of flux enter or leave the part.

pole piece A shaped piece of ferromagnetic material, integral with or attached to one end of a magnet, whose function is to control the distribution of magnetic lines of flux.

pole-pole array An electrode array for making resistivity or induced-polarization surveys in which one current electrode and one potential electrode, in close proximity, are moved laterally across the area being surveyed.

polestar recorder An instrument used to determine the amount of cloudiness during the night. It consists of a fixed, long-focus camera positioned so that Polaris is permanently within its field of view; the apparent motion of Polaris is recorded as a circular pattern on the film, with approximate span of cloudiness indicated by interruptions in the arc due to clouds passing between the star and the camera.

poling A stage in fire-refining of copper during which green-wood poles are thrust into the bath of molten metal; the wood decomposes and forms reducing gases that react with oxygen in the bath.

polled access A media access method by which the node that has the right to use the network medium delegates that right to other stations on a per message basis. See master-slave.

polling 1. The act of requesting a station to send data in switching networks [RP55.1]. 2. A method of sequentially observing each channel to determine if it is ready to receive data or requesting computer action. 3. The repetitive search of a LAN system to determine whether a work station is holding data for the main computer.

polymeric material A compound formed by molecular bonding (polymerizing) of two or more simple molecules (monomers). This material is commonly referred to as plastic [S82.01].

polyphase meter An instrument for measuring a quantity such as power factor or electric power in a polyphase electric circuit.

PONA analysis Determination of amounts of paraffins (P), olefins (O), naphthalenes (N) and aromatics (A) in gasoline in ASTM standard tests.

Pope cell A type of relative humidity sensor that employs a bifilar conductive grid on an insulating substrate whose resistance varies with relative humidity over a range of about 15 to 99% RH.

poppet A spring-loaded ball that engages a notch.

poppet valve A mushroom-shaped valve that controls the intake or exhaust of working fluid in a reciprocating engine; it may be cam operated or spring loaded, and its direction of movement is at right angles to the plane of its seat.

popping pressure In compressible fluid systems, the inlet pressure at which a safety relief valve opens.

porcelain enamel See enamel; see vitreous enamel.

porcupine boiler A boiler consisting of a vertical shell from which project a number of dead end tubes.

porosimeter A laboratory device for measuring the porosity of reservoir rock using compressed gas.

port 1. The flow control orifice of a control valve. It is also used to refer to the inlet or outlet openings of a valve [S75.05]. 2. The entry or exit point from a computer for connecting communications or peripheral devices. 3. An aperture for passage of steam or other fluids.

portable 1. Refers to a self-contained, battery-operated instrument that can be carried. 2. Capable of being carried, especially by hand, to any desired location.

portable, continuous-duty A Battery-operated portable or transportable instrument of a type intended to operate continuously for 8 hours (h) or more [S12.13].

portable, continuous-duty—personal A battery-operated, alarm-only portable instrument intended to be operator-worn, and to operate continuously for 8 h or more [S12.13].

portable equipment Equipment specifically designed to be carried by hand from one location to another as determined by the following: A. It is provided with at least one handle and does not exceed 20 kilograms (44 pounds); or B. It has no handle and does not exceed 5 kilograms (11 pounds) [S82.01].

portable, intermittent-duty A battery-operated portable or transportable instrument of a type intended for operation for periods of only a few minutes at irregular intervals [S12.13].

portable standard meter A portable instrument used primarily as a reference standard for testing or calibrating other instruments.

port guide A valve plug with wings or a skirt fitted to the seat ring bore [S75.05].

ports Data channels dedicated to input or output.

position Of a multi-position controller, a discrete value of the output signal [S51.1].

positional notation A numeration system with which a number is represented by means of an ordered set of digits such that the value contributed by each digit depends on its position as well as upon its value.

positioner A position controller, which is mechanically connected to a moving part of a final control element or its actuator, and automatically adjusts its output pressure to the actuator in order to maintain a desired position that bears a predetermined relationship to the input signal. The positioner can be used to modify the action of the valve (reversing positioner), extend the stroke/controller signal (split range positioner), increase the pressure to the valve actuator (amplifying positioner) or modify the control valve flow characteristic (characterized positioner) [S75.05].

positioner, amplifying A pneumatic positioner in which the input control signal is amplified to a proportionately higher pressure, needed to drive the actuator, e.g., 3-15 psig input/6-30 psig output.

positioner, characterized A positioner in which the valve position feedback is modified to produce a nonlinear response.

positioner, double acting A positioner with two outputs, suited to a double acting actuator.

positioner, electro-pneumatic A positioner which converts an electronic control signal input to a pneumatic output.

positioner, reversing A positioner which converts the input control signal into an output which is directionally opposite to the input.

position error See error, position.

positioner, single acting A positioner with one output, suited to a spring opposed actuator.

positioner, split range A positioner which drives an actuator full stroke in proportion to only a part of the input signal range.

positioner types Positioners characterized by their input and output are available as: (a) pneumatic/pneumatic, (b) electric/pneumatic, (c) electric/hydraulic, (d) electric/electric [S75.05].

position independent code (PIC) A code that can execute properly wherever it is loaded in memory, without modification or relinking; generally, this code uses addressing modes that form an effective memory address relative to the central processor's program counter (PC).

position indicator The device, such as a pointer and scale, which indicates the position of the closure member [S75.05]. See also travel indicator.

positioning Manipulating a workpiece in relation to working tools.

positioning action Controller action in which the final position of the control element has a predetermined relation to the value of the controlled variable.

positioning control See point-to-point numerical control.

positioning control system A system of control in which each controlled motion operates in accordance with instructions which only specify the next required position, the movement in the different axes of motion not being coordinated with each other and being executed, simultaneously or consecutively, at velocities which are not controlled by instructions on the input data medium.

position instability Evidenced by uncontrolled fluctuating valve travel, it is caused by the fluid forces interacting with the actuator forces. It is a persistent cyclic motion inconsistent with the control signal to the valve. It is not a static deviation caused by dead band or hysteresis [RP75.18].

position sensor Any device for measuring position and converting the measurement into an electrical, electromechanical or other signal for remote indication or recording.

position switch A pneumatic, hydraulic or electrical device which is linked to the valve stem to detect a single, preset valve stem position [S75.05].

position telemeter A remote-reading instrument for indicating linear or angular position of an object or machine component.

position transmitter A device that is mechanically connected to the valve stem and generates and transmits a pneumatic or electrical signal representing the valve stem position [S75.05].

positive displacement Referring to any device that captures or confines definite volumes of fluid for purposes of measurement, compression or transmission.

positive-displacement flowmeter Any of several flowmeter designs in which volumetric flow through the meter is broken up into discrete elements and the flow rate is determined from the number of discrete elements that pass through the meter per unit time.

positive draft Pressure in a furnace, gas chamber or duct that is greater than ambient atmospheric pressure.

positive feedback 1. A closed loop in which any change is reinforced until a limit is eventually reached. 2. Returning part of an output signal and using it to increase the value of an input signal.

positive-going edge The edge of a pulse going from a low to a high level.

positive meter Any of several devices for measuring fluid flow by alternately filling and emptying a container or chamber of known capacity; in such a device, fluid passes through it in a series of discrete amounts by weight or volume.

positive motion Motion transmitted from one machine part to another without slippage.

positron A positively charged beta particle.

post A vertical support member resembling a pillar or column.

postconversion bandwidth The bandwidth presented to a detector.

post guide Guide bushing or bushings fitted to posts or extensions larger than the valve stem and aligned with the seat [S75.05].

post guiding A design using guide bushing or bushings fitted into the bonnet or body to guide the plug's post.

post processor In numerical control, a computer program which adapts the output of a processor, applicable to a piece part, into a machine program for the production of that part on a particular combination of machine tool and controller.

potential energy Energy related to the position or height above a place to which fluid could possibly flow or a solid could fall or flow.

potential transformer An instrument transformer so connected in the circuit that its primary winding is in parallel with a voltage to be measured or controlled.

potentiometer A device for measuring an electromotive force by comparing it with a known potential difference.

potentiometric Converting a change of measurand into a voltage-ratio change by a change in the position of a movable contact on a resistance element across which excitation is applied [S37.1].

potentiometric element The resistive part of the transduction element upon which the wiper (movable contact) slides and across which excitation is applied. It may be constructed of a continuous resistance or of small diameter wire wound on a form (mandrel) [S37.12] [S37.6].

potentiometric titration A technique of automatic titration where the end point is determined by measuring a change in the electrochemical potential of the sample solution.

potometer A device for measuring transpiration from a leaf, twig or small plant; it consists of a small water-filled container, sealed so that moisture can escape only through the plant.

pound The British or U.S. unit of mass or weight and is equal to .45 Kilograms.

poundal A unit of force in the English system of measurement; it is defined as the force necessary to impart an acceleration of one ft/s/s/ to a body having a mass of one pound.

pour point 1. Temperature at which molten metal is cast. 2. The temperature at which a petroleum-base lubricating oil becomes too viscous to flow, as determined in a standard ASTM test.

pour test Chilling a liquid under specified conditions to determine its ASTM pour point.

powder coating A painting process in which finely ground dry plastic is applied to a part using electrostatic and compressed air transfer mechanisms. The applied powder is heated to its melting point and flows out, forming a smooth film and cures by means of a chemical reaction.

powder pattern The x-ray diffraction pattern consisting of a series of rings on a flat film or a series of lines on a circular strip film which results when a monochromatic beam of x-rays is reflected from a randomly oriented polycrystalline metal or from powdered crystalline material.

power common 1. The reference point for power supplies and return currents from powering equipment. 2. Sometimes referred to simply as common. Power common or common is not to be confused with signal common.

power consumption The maximum amount of electrical power used by a device during normal steady-state operation.

power consumption, electrical The maximum power used by a device within its operating range during steady-state signal condition. NOTE 1: For a power factor other than unity, power consumption shall be stated as maximum volt-amperes used under the above stated condition. 2: For a device operating outside of its operating range, the maximum power might exceed that which is experienced within the operating range [S51.1].

power/energy meter An instrument which measures the amount of optical power (watts) or energy (joules). It can operate in the visible, infrared, or ultraviolet region, and detect pulsed or continuous beams.

power factor 1. The ratio of total watts to the total root-mean-square (rms) volt-amperes. If the voltages have the same waveform as the corresponding currents, power factor becomes the same as phasor power factor. If the voltages and currents are sinusoidal and for polyphase circuits, form symmetrical sets, $F_p = \cos(a\text{-}B)$ [S51.1]. 2. The ratio of actual power to apparent power delivered to an electrical power circuit; in practice this is determined by dividing the resistive load, in watts, by the product of voltage and current, in volt-amperes, and expressing the result in percent.

power-factor meter An instrument for directly indicating power factor in a circuit.

power failure The removal of all power accidentally or intentionally.

power input The energy required to drive a fan, expressed in brake horsepower delivered to fan shaft. See also excitation.

power level At any given point in a system, the amount of power being delivered or used.

power level (dBm) The ratio of the power at a point to some arbitrary amount of power chosen as a reference. This ratio is usually expressed either in decibels based on 1 milliwatt (abbreviated dBm) or in decibels based on 1 watt (abbreviated dBW). See also decibel.

power line protector A device used between a computer and a power outlet to absorb power surges or other interference that could damage the computer. See surge protector.

power loss 1. In a power transmission system or circuit, the difference between input power and output power, often expressed as a percent of input power. 2. In a current or voltage measuring instrument, the active power at its terminals when the pointer is at the upper end of the scale. 3. In any other electrical circuit, the difference between active power and electrical load at a stated value of current or voltage.

power output The energy delivered by the fan, expressed in horsepower based on air or gas pressure and volume.

power spectral density (PSD) A type of frequency analysis on data that can be done by computer using special software, or by an array processor, or by a special-purpose hardware device.

power splitter At the output of a telemetry radio transmitter, the device which splits the transmitter power between two or more antennas.

power supply The device within a computer that transforms external AC power to internal DC voltage.

power supply cord A flexible cord with attachment plug provided to connect equipment to a supply circuit receptacle [S82.01].

power unit The portion of the actuator which converts fluid, electrical or mechanical energy into stem motion to develop thrust or torque [S75.05].

precession The change in orientation of a rapidly spinning body, such as a gyrowheel, that occurs when its axis of spin rotates about a line perpendicular to a plane defined by the original position of the axis of spin and the axis of torque for the moment producing the change in orientation.

precipitate To separate materials from a solution by the formation of insoluble matter by chemical reaction. The material which is removed.

precipitation The removal of solid or liquid particles from a fluid.

precipitator A fly ash separator and collector of the electrostatic type.

precision 1. The degree of reproducibility among several independent measurements of the same true value. 2. The quality of being exactly defined or stated. 3. The value of the smallest incremental difference that can be measured by a given instrument or measurement system. 4. In an approximate number, the decimal position of the rightmost significant digit. See also repeatability and stability. See also repeatability.

precision depth recorder　A machine that plots sonar depth soundings on electrosensitive paper.

predetection　In instrumentation tape recorders, the process of recording a "low" intermediate frequency from the telemetry radio receiver (typically, 900 kHz center frequency) rather than the demodulated output of the receiver.

predictive control　1. A type of automatic control in which the current state of a process is evaluated in terms of a model of the process and controller actions modified to anticipate and avoid undesired excursions. 2. Self-tuning. 3. Artificial intelligence.

predictive maintenance　1. A preventive maintenance program that anticipates failures which can be corrected before total failure. 2. A variation from a normal can indicate a system or equipment is approaching non-conformance. Vibration, eccentricity and noise monitoring are measurements that can predict failure. Also, an increase in diagnostic errors and retries can indicate a failure about to happen.

preface, mag tape　The first few words of each tape record, which identify the record and document the status of the equipment.

preform　1. A cylinder of glass which is made to have a refractive index profile that would be desired for an optical fiber. The cylinder is then heated and drawn out to produce a fiber. 2. Brazing metal foil cut to the exact outline of the mating parts and inserted between the parts prior to placing in a brazing furnace.

preheater air　Air at a temperature exceeding that of the ambient air.

preignition　Spontaneous ignition of the explosive mixture in a cylinder of an internal combustion engine before the spark flashes.

premodulator filter　A lowpass filter at the input to a telemetry transmitter; its purpose is to limit modulation frequencies and thereby limit radiated frequencies outside the desired operating spectrum.

preprocessor　1. A hardware device in front of a computer, capable of making certain decisions or calculations more rapidly than the computer can make them. 2. The first of the two compiler stages. At this stage, the source program is examined for the preprocessor statements which are then executed, resulting in the alteration of the source program text. More generally, a program that performs some operation prior to processing by a main program.

pressed density　The density of a powder-metal compact after pressing and before sintering.

pressure　Measure of applied force compared with the area over which the force is exerted, psia.

pressure, absolute　The pressure measured relative to zero pressure (vacuum) [S37.1].

pressure altimeter　A precision aneroid barometer that measures air pressure at the altitude a plane is flying and converts the reading to indicated height above sea level.

pressure, ambient　The pressure of the medium surrounding a device [S51.1].

pressure connection (pressure port)　The opening and surrounding surface of a transducer used for measured fluid access to transducer sensing element (or isolating element). This can be a standard industrial or military fitting configuration, a tube hose fitting or a hole (orifice) in a base plate. For differential pressure transducers there are two pressure connections: the measurand port and the reference [S37.6].

pressure control　A device or system that can raise, lower or maintain the internal pressure in a vessel or process equipment.

pressure, design　The pressure used in the design of a vessel or device for the purpose of determining the minimum permissible thickness or physical characteristics of the parts for a given maximum working pressure (MWP) at a given temperature [S51.1].

pressure, differential　The difference in pressure between two points of measurement [S37.1].

pressure drop　1. The differential pressure in pascals at a maximum linear flow measured between points four pipe diameters upstream and four pipe diameters downstream from its ends, using a specified liquid, and using pipe size matching the fittings provided [RP31.1]. 2. The difference in pressure between two points in a system, caused by resistance to flow.

pressure elements　The portions of a pressure-measuring gage that move or are temporarily deformed by the system pressure, the amount of movement or deformation being proportional to the pressure.

pressure energized liner A pressure source, either internal flue pressure or an external fluid pressure source, energizes the liner forcing it into tighter contact with the disk [S75.05].

pressure energized seal 1. A seal energized by interference fit between the disk groove and valve liner and also by differential pressure acting across the seal. The seal may be a solid section or have internal pressure ports [S75.05]. 2. A seal ring retained in the body bore with raised flexible lip which contacts an offset disk in the closed position yet is clear of the disk in other positions.

pressure energized stem seal A part and/or packing material deformable by fluid pressure that bears against the stem to make a tight seal [S75.05].

pressure-expanded joint A tube joint in a drum, header or tube sheet expanded by a tool which forces the tube wall outward by driving a tapered pin into the center of a sectional die.

pressure (frequency) response The pressure frequency response (pressure response) of a piezoelectric sound pressure transducer is the ratio, as a function of frequency, of the transducer output to a sound pressure input which is equal in phase and amplitude over the entire sensing surface of the transducer. The pressure frequency response is generally equal to the free-field frequency response at wavelengths long compared to the maximum dimension of the piezoelectric sound-pressure transducer [S37.10].

pressure gage An instrument for measuring pressure by means of a metallic sensing element or piezoelectric crystal.

pressure, gage Pressure measured relative to ambient pressure [S37.1].

pressure, leak The pressure at which some discernible leakage first occurs in a device [S51.1].

pressure level In acoustic measurement, $P = 1 \log(P_s/P_r)$, where P is the pressure level in bels, P_s is the sound pressure, and P_r is a reference pressure, usually taken as 0.002 dyne/cm^2.

pressure, maximum working (MWP) The maximum total pressure permissible in a device under any circumstances during operation, at a specified temperature. It is the highest pressure to which it will be subjected in the process. It is a designed safe limit for regular use. MWP can be arrived at by two methods: a) Designed — by adequate design analysis, with a safety factor; b) Tested — by rupture testing of typical sam-ples. See pressure, design [S51.1].

pressure measurement Any method of determining internal force per unit area in a process vessel, tank or piping system due to fluid or compressed gas; this includes measurement of static or dynamic pressure, absolute (total) or gage (total minus atmospheric), in any system of units.

pressure microphone An acoustic transducer which converts instantaneous sound pressure of impinging sound waves into an electrical signal that directly corresponds in both frequency and amplitude.

pressure, operating The actual pressure at which a device operates under normal conditions. This pressure may be positive or negative with respect to atmosphere pressure [S51.1].

pressure, process The pressure at a specified point in the process medium [S51.1].

pressure rating The maximum allowable internal force per unit area of a pressure vessel, tank or piping system during normal operation.

pressure-regulating valve A valve that can assume any position between fully open and fully closed, or that opens or remains closed against fluid pressure on a spring-loaded valve element, to release internal pressure or hold it and allow it to build up, as desired.

pressure regulator An in-line device that provides controlled venting from a high-pressure region to a lower-pressure region of a closed compressed gas system to maintain a preset pressure value in the lower-pressure region.

pressure relief device A mechanism that vents fluid from an internally pressurized system to counteract system overpressure; the mechanism may release all pressure and shut the system down (as does a rupture disc) or it may merely reduce the pressure in a controlled manner to return the system to a safe operating pressure (as does a spring-loaded safety valve).

pressure, rupture The pressure, determined by test, at which a device will burst. This is an alternate to the design procedure for establishing maximum working pressure (MWP). The rupture pressure test consists of causing the device to burst [S51.1].

pressure, static The steady-state pressure applied to a device; in the case of a differential pressure device, the process pressure applied equally to both connections [S51.1].

pressure, supply The pressure at the supply

291

port of a device [S51.1].

pressure, surge Operating pressure plus the increment above operating pressure to which a device may be subjected for a very short time during pump starts, valve closings, etc. [S51.1]

pressure switch A device that activates or deactivates an electrical circuit when a preselected pressure is exceeded in a process vessel or piping system.

pressure tap A small hole in the wall of an internally pressurized vessel or pipe so that the pressure element of an instrument can be attached to measure static pressure.

pressure transducer An instrument component that senses fluid pressure and produces an electrical output signal that is related to the magnitude of the pressure.

pressure-vacuum gage An instrument for measuring pressure both above and below atmospheric.

pressure vessel A metal container designed to withstand a specified bursting pressure; it is usually cylindrical with hemispherical end closures (but may be of some other shape, such as spherical) and is usually fabricated by welding.

pressurized enclosure Maintained at a pressure higher than the surrounding area and the areas communicated with by conduit runs. The pressurizing medium shall be clean, dry air or an inert gas. In Division 1 locations, the pressure shall be supervised by a suitable pressure switch, to de-energize supply conductors in case of pressure failure [S12.11].

pressurized water reactor (PWR) A nuclear steam supply system in which the pressurized primary coolant fluid is heated by the reactor core, and the process steam is generated in a steam generator by heat transfer from the primary coolant [S67.03].

preventive maintenance Maintenance specifically intended to prevent faults from occurring during subsequent operation. Contrast with corrective maintenance. Corrective maintenance and preventive maintenance are both performed during maintenance time.

primary air Air introduced with the fuel at the burners.

primary-air fan A fan to supply primary air for combustion of fuel.

primary circuit Wiring and components of the equipment supply circuit which are at the voltage of or carry the current of the branch circuit

[S82.01].

primary colors Colors of constant hue and variable brilliance which can be mixed in varying proportions to produce or specify other colors. Also known as primaries.

primary containment The structure that encloses the reactor coolant pressure boundary [S67.03].

primary detector The system element or device that first responds quantitatively to the attribute or characteristic being measured and performs the initial conversion or control of measurement energy.

primary device The part of a flowmeter which generates a signal responding to the flow from which the flow rate may be inferred.

primary element 1. Synonym for sensor. 2. Detector. 3. The first system element that responds quantitatively to the measured variable and performs the initial measurement operation. 4. A primary element performs the initial conversion of measurement energy. 5. Any device placed in a flow line to produce a signal for flow rate measurement. 6. The component of a measurement or control system that first uses or transforms energy from a given medium to produce an effect which is a function of the value of the measured variable. 7. The portion of the measuring means which first either utilizes or transforms energy from the controlled medium to produce an effect in response to change in the value of the controlled variable. 8. The effect produced by the primary element may be a change of pressure, force, position, electrical potential, or resistance. See also element, primary; sensing element; and sensor.

primary element (detector) The first system element that responds quantitatively to the measured variable and performs the initial measurement operation. A primary element performs the initial conversion of measurement energy. For transmitters not used with external primary elements, the sensing portion is the primary element.

primary feedback A signal which is a function of the controlled variable and which is used to modify an input signal to produce an actuating signal.

primary instrument An instrument that can be calibrated without reference to another instrument.

primary loop The outer loop in a cascade system.

primary measuring element A component of a measuring or sensing device that is in direct contact with the substance whose attributes are being measured.

primary/secondary control loop controller The controller which adjusts the set point for the secondary control loop controller in the cascade control action scheme [S77.42].

priming The discharge of steam containing excessive quantities of water in suspension from a boiler, due to violent ebullition.

printed circuit A system of conductors formed or deposited on a nonconducting substrate in a predetermined pattern to allow quick and repetitive construction of electronic devices.

printer In data processing, the device that produces printed paper copy of computer data.

print spooler A computer program that directs the computer to store certain data to be printed so that computer processing is not limited to printer speed.

priority 1. The relative importance attached to different phenomena. 2. Level of importance of a program or device.

priority interrupt The temporary suspension of a program currently being executed in order to execute a program of higher priority. Priority interrupt functions usually include distinguishing the highest priority interrupt active, remembering lower priority interrupts which are active, selectively enabling or disabling priority interrupts, executing a jump instruction to a specific memory location, and storing the program counter register in a specific location. See hardware priority interrupt and software priority interrupt.

privilege A characteristic of a user, or program, that determines what kind of operations a user or program can perform; in general, a privileged user or program is allowed to perform operations that are normally considered the domain of the monitor or executive, or that can affect system operation as a whole.

probe 1. A small, movable capsule or holder that allows the sensing element of a remote-reading instrument, usually an electronic instrument, to be inserted into a system or environment and then withdrawn after a series of instrument readings has been taken. 2. A small tube, movable or fixed, inserted into a process fluid to take physical samples or pressure readings.

probe-type consistency sensor A device in which forces exerted on a cylindrical body in the direction of flow are detected by a strain-gage bridge circuit; if the fluid is water, circuit output is a measure of flow rate, but if a solution or suspension is flowing at a steady flow rate, the output varies with changes in viscosity (or consistency).

problem definition The art of compiling logic in the form of general flow charts and logic diagrams which clearly explain and present the problem to the programmer in such a way that all requirements involved in the program are presented.

problem description In information processing, a statement of a problem. The statement also may include a description of the method of solution, the procedures and algorithms.

problem-oriented language (POL) Programming language designed for ease of problem definition and problem solution of specific classes and problems; for example, a language specifically convenient for expressing a specific problem in mathematical form, such as the ordinary algebraic languages or the symbolic notation of the Boolean algebra applied to a special problem; a special language for machine tool control.

procedural language See procedural-oriented language.

procedure 1. A precise step-by-step method for effecting a solution to a problem. 2. In data processing, a smaller program that is part of a large program. 3. In batch processing, the part of a recipe that defines the generic strategy for producing a batch of material.

procedure-oriented language A programming language designed for convenience in expressing the technique or sequence of steps required to carry out a process or flow. Usually it is a source language and usually is not machine-oriented. Since many classes of problems involve similar procedures for their solution, the procedure-oriented language lends itself more readily to describing how a problem is to be solved. Flow diagrams, process control languages, and many of the common programming languages, such as COBOL, FORTRAN, and ALGOL are considered procedure-oriented languages.

process 1. Physical or chemical change of matter or conversion of energy; e.g., change in pressure, temperature, speed, electrical potential, etc. [S51.1] 2. The collective functions performed in and by industrial equipment, exclusive of computer and/or analog control and monitoring

293

equipment. 3. A series of continuous or regularly recurring steps or actions intended to achieve a predetermined result, as in refining oil, heat treating metal, or manufacturing paper. 4. A general term covering such terms as assemble, compile, generate, interpret, and compute. 5. The functions and operations utilized in the treatment of material. 6. A progressive course or series of actions. 7. Any operation or sequence of operations involving a change of energy, state, composition, dimension, or other properties that may define with respect to a datum. 8. An assembly of equipment and material that relates to some manufacturing sequence.

process alarm Alarm which occurs as a result of a process parameter exceeding some preset limit.

process and instrumentation diagram (P&ID) Show the interconnection of process equipment and the instrumentation used to control the process. In the process industry, a standard set of symbols is used to prepare drawings of processes. The Instrument Society of America (ISA) Standard S5.

process block valve The first valve off the process line or vessel used to isolate the measurement piping. See line class valve.

process calculations Installation-dependent calculations providing derived data to supplement the input signals. e.g., efficiencies, flows by material balance, etc.

process chart A graphical representation of the events in a process.

process computer A computer which, by means of inputs from and outputs to a process, directly controls or monitors the operation of elements in that process. See control computer and industrial computer. See also on-line.

process condition The condition of the monitored variable. The process condition is either normal or abnormal (alarm, alert, or off-normal) [S18.01].

process control 1. The regulation or manipulation of variables influencing the conduct of a process in such a way as to obtain a product of desired quality and quantity in an efficient manner [S51.1]. 2. Descriptive of systems in which computers or controllers are used for automatic regulation of operations or processes. Typical are operations wherein the operation control is applied continuously and adjustments to regulate the operation are directed by the computer

to keep the value of a controlled variable constant. Contrasted with numerical control. 3. An operation that regulates parameters by observation of the parameter, comparison with some desired value, and action to bring the parameter as close as possible to the desired value. 4. Adapting automatic regulatory procedures to the more efficient manufacture of products or the processing of material.

process control chart A table or graph of test results or inspection data for each unit of production, arranged in chronological sequence for the entire assembly or production lot.

process control computer See process computer.

process control engineering A branch of engineering that deals with ways and means of keeping process variables as close as possible to desired values, or keeping them within specified ranges.

process control loop A system of control devices linked together to control one phase of a process.

process database Organized collection of data relating to the operation of a process.

process dynamics A set of dynamic interactions among process variables in a complex system, as in a petroleum refinery or chemical process plant.

process engineering An element of production engineering that involves selecting processes and equipment to be used, establishing the sequence and method of controlling all operations, and acquiring the tools needed to make a product.

process engineer's console A man-machine interface, consisting of various information entry/retrieval devices arranged as a packaged unit, it is used by the person responsible for the performance of a manufacturing process, to adjust the external behavior of the process controller. See also operator's console and programmer's console.

process interrupt See interrupt, process.

process I/O Input and output operations directly associated with a process as contrasted with I/O operations not associated with the process. For example, in a process control system, analog and digital inputs and outputs would be considered process I/O whereas inputs and outputs to bulk storage would not be process I/O. See process.

process I/O bus 1. A circuit over which data or

power is transmitted; often one which acts as a common connection among a number of locations. Synonymous with trunk. 2. A communications path between two switching parts.

process I/O device   An apparatus for performing a prescribed function.

process I/O network   A communication system. A set of OSI subnetworks interconnected by OSI intermediate systems and sharing a common network protocol.

process measurement   The acquisition of information that establishes the magnitude of process quantities [S51.1].

processor   Abbreviated form for central processing unit.

process parameter   A characteristic of a process which can be monitored and measured to provide information on the process.

process pressure   See pressure, process.

process reaction method   A method of determination of optimum controller settings when tuning a process control loop. The method is based on the reaction of the open loop to an imposed disturbance.

process reaction rate   The rate at which a process reacts to a step change.

process steam   Steam used for industrial purposes other than for producing power.

process temperature   See temperature, process.

process time   1. Elapsed time for the portion of the work cycle controlled by machines. 2. Elapsed time for an entire process.

process variable   1. Any variable property of a process. The term process variable is used in this standard to apply to all variables other than instrument signals [S5.1]. 2. In the treatment of material, any characteristic or measurable attribute whose value changes with changes in prevailing conditions. Common variables are flow, level, pressure and temperature.

process variable alarm   Alarm that is set whenever PV exceeds the limits set for a given input.

process visual display   A dynamic display intended for operators and others engaged in process monitoring and control [S5.5].

producer gas   Gaseous fuel obtained by burning solid fuel in a chamber where a mixture of air and steam is passed through the incandescent fuel bed. The process results in a gas, almost oxygen free containing a large percentage of the original heating value of the solid fuel in the form of CO and H_2.

production   Output of a process or manufacturing facility.

production engineering   An element of industrial engineering that deals with planning and control of manufacturing processes, especially for the purpose of improving efficiency and reducing costs associated with mechanical equipment.

productivity   1. Production output per unit of input, such as number of items per labor manhour. 2. Generically, the effectiveness with which labor, materials and equipment are used in a production operation.

products of combustion   The gases, vapors, and solids resulting from the combustion of fuel.

program   1. A repeatable sequence of actions that defines the status of outputs as a fixed relationship to a set of inputs [S5.1]. 2. In data processing, a series of instructions that tell the computer how to operate. 3. Any series of actions proposed in order to achieve a certain result. 4. To design, write, and test a program. 5. A unit of work for the central processing unit from the standpoint of the executive program. See task.

program address counter   Same as location counter.

program control   Descriptive of a system in which a computer is used to direct an operation or process and automatically to hold or to make changes in the operation or process on the basis of a prescribed sequence of events.

program controller   See controller, program.

program counter   A register which contains the address of the next instruction to be executed. At the end of each instruction execution the program counter is incremented by 1 unless a jump is to be carried out in which case the address of the jump label is entered.

program documentation   The complete listing of a program's use, content and installation.

Program Evaluation and Review Technique   A management control tool for managing complex projects; project milestones are defined and interrelated, then using a flowchart or computer progress is measured against the milestones; deviations from the integrated plan are used to trigger decisions or preplanned alternative actions to minimize adverse effects on the overall goal.

program generator   Computer software that translates simple statements into program codes.

program library   A collection of available computer programs and routines.

programmable data distributor (PDD) An optional module for the telemetry frame synchronizer that causes it to send certain predefined words from each telemetry frame out through a separate port (as to a special buffer area).

programmable logic controller (PLC) 1. A controller, usually with mutliple inputs and outputs, that contains an alterable program [S5.1]. 2. A microcomputer-based control device used to replace relay logic. 3. A solid-state control system which has a user-programmable memory for storage of instructions to implement specific functions such as I/O control, logic, timing, counting, three mode (PID) control, communication, arithmetic, data and file manipulation. 4. A PLC consists of a central processor, input/output interface, and memory. 5. A PLC is designed as an industrial control system.

programmable read only memory (PROM) A hardware device that stores digital words; the computer can read the contents but cannot modify them.

programmer's console A man-machine interface, consisting of various information entry/retrieval devices, arranged as a packaged unit. It is used by the programmer of a computer control system for a manufacturing process, to monitor, modify, and control the internal behavior of the digital controller. See also operator's console, process engineer's console.

programming The design, the writing and testing of a program. See convex programming, dynamic programming, linear programming, mathematical programming, nonlinear programming, quadratic programming, macroprogramming, microprogramming, and multiprogramming.

programming language A language used to prepare computer programs.

programming language/1 (PL/1) A high-level programming language for general purpose scientific and commercial applications.

programming module A discrete identifiable set of instructions usually handled as a unit.

programming system A system consisting of a programming language and a computer program (the processor) to convert the language into absolute coding.

program parameter A parameter incorporated into a subroutine during computation. A program parameter frequently comprises a word stored relative to either the subroutine or the entry point and dealt with by the subroutine during each reference. It may be altered by the routine and/or may vary from one point of entry to another. Related to parameter.

program statement A source code instruction which translates into machine code instructions.

program storage A portion of the internal storage reserved for the storage of programs, routines, and subroutines. In many systems, protection devices are used to prevent inadvertent alteration of the contents of the program storage. Contrasted with working storage.

program timer A timing device which actuates a series of switches in programmed sequence.

project engineering 1. Engineering activities associated with designing and constructing a manufacturing or processing facility. 2. Engineering activities related to a specific objective such as solving a problem or developing a product.

projection welding A welding process in which the arc is localized by projections, embossments or intersections.

PROM See programmable read only memory.

PROM programmer A device which allows PROM's to be programmed.

prompt In data processing, instructions that appear on the CRT screen that requests response from the user.

proof pressure The maximum pressure which may be applied to the sensing element of a transducer without changing the transducer performance beyond specified tolerances. NOTE 1: In the case of transducers intended to measure a property of pressurized fluid, proof pressure is applied to the portion subject to the fluid. 2: Differential-pressure transducer specifications should indicate whether the specified differential proof pressure is applicable at ambient or maximum specified reference pressure, or both, and whether a reverse-differential proof pressure, at ambient or maximum specified reference pressure, or both, is additionally applicable [S37.1].

proof transverse acceleration (static) The maximum transverse static acceleration that can be applied without causing permanent degradation in performance beyond specified tolerance [S37.5].

proof transverse acceleration (vibrational)

The maximum transverse dynamic acceleration(s) over a specified frequency range(s) that can be applied without causing permanent degradation in performance beyond specified tolerances [S37.5].

propagation delay The time period between the input of a logic signal to a device and a valid output from that signal at the output of the device.

propagation loss 1. Reduction in amplitude of the radio telemetry signal due to natural laws of attenuation. 2. Reduction in amplitude of optical signal in a fiber optic cable due to losses associated with scattering, absorption and reflection.

propeller meter An instrument for measuring the quantity of fluid flowing past a given point; the flowing stream turns a propeller-like device, and the number of revolutions are related directly to the volume of fluid passed.

proportional band 1. The change in input required to produce a full range change in output due to proportional control action. NOTE 1: It is reciprocally related to proportional gain. 2: It may be stated in input units or as a percent of the input span (usually the indicated or recorded input span). The preferred term is proportional gain. See also gain, proportional [S51.1]. 2. The amount of deviation of the controlled variable from setpoint required to move the final control element through the full range (expressed in % of span). 3. An expression of gain of an instrument (the wider the band, the lower the gain).

proportional bandwidth FM telemetry, where each subcarrier is deviated a fixed percentage of center frequency (and therefore an amount proportional to the center frequency) by data.

proportional control A control mode in which there is a continual linear relationship between the deviation computer in the controller, the signal of the controller, and the position of the final control element.

proportional control action Corrective action which is proportional to the error, that is, the change of the manipulated variable is equal to the gain of the proportional controller multiplied by the error (the activating signal). See also control action, proportional.

proportional controller See controller, proportional.

proportional control mode 1. A controller mode in which the controller output is directly proportional to the controlled variable error. 2.

Produces an output signal proportional to the magnitude of the input signal. 3. In a control system proportional action produces a value correction proportional to the deviation of the controlled variable from set point.

proportional counter An instrument whose primary element is a radiation counter tube or chamber operated in the range where the amplitude of each current pulse in its output is proportional to the energy of the quantum of radiation absorbed.

proportional gain See gain, proportional.

proportional, integral and derivative 1. Three mode controller. 2. Refers to a control method in which the controller output is proportional to the error, its time history, and the rate at which it is changing. The error is the difference between the observed and desired values of the variable that is under control action. 3. Proportional plus integral plus derivative control, used in processes where the controlled variable is affected by long lag times.

proportional integral derivative (PID) control A combination of proportional, integral and derivative control actions. Refers to a control method in which the controller output is proportional to the error, its time history, and the rate at which it is changing. The error is the difference between the observed and desired values of the variable that is under control action. Also called three-mode control.

proportional pitch In computer printers, a typeface in which each character is a different width, such as a W versus an I.

proportional plus derivative control action See control action, proportional plus integral.

proportional plus derivative controller See controller, proportional plus derivative.

proportional plus integral control action See control action, proportional plus integral.

proportional plus integral controller See controller, proportional plus integral.

proportional plus integral plus derivative control action See control action, proportional plus integral plus derivative.

proportional plus integral plus derivative controller See controller, proportional plus integral plus derivative.

proportional plus rate control action See control action, proportional plus derivative.

proportional plus rate controller See controller, proportional plus derivative.

proportional plus reset control A mode of control in which there is a continuous linear relation between the value of the controlled variable and the position of the final control element (proportional) plus an additional change in the position of the final control element based on both the amount and duration of the change in the controlled variable (reset). Same as PI control.

proportional plus reset control action See control action, proportional plus integral. See also control action, proportional plus integral.

proportional plus reset controller See controller, proportional plus integral.

proportional plus reset plus rate control Same as PID control.

proportional plus reset plus rate control action See control action, proportional plus integral plus derivative.

proportional plus reset plus rate controller See controller, proportional plus integral plus derivative.

proportional-position action A type of control-system response where the position of the final control element has a continuous linear relation to the value of the controlled variable.

proportional region The range of operating voltage of a radiation counter tube where the gas amplification factor is greater than 1, usually about 10^3 to 10^5, and the output current pulse is proportional to the number of ions produced by the primary ionizing event, which in turn is proportional to the energy of the radiation quantum absorbed.

proportional speed floating controller See controller, integral.

proportioning probe A probe used in leak testing in which the ratio of air to tracer gas can be changed without changing the amount of flow transmitted to the detector.

PROSPRO IBM process-oriented language.

protected location A computer storage location, reserved for special purposes, in which data cannot be stored without undergoing a screening procedure to establish suitability for storage therein. May be indicated by a set guard bit.

protecting tube 1. A protecting tube is a tube designed to enclose a temperature sensing device and protect it from the deleterious effects of the environment. It may provide for attachment to a connection head but is not primarily designed for pressure-tight attachment to a vessel.

A bushing or flange may be provided for the attachment of a protecting tube to a vessel [ANSI-MC96.1]. 2. A closed-end tube that surrounds the measuring junction of a thermocouple and protects it from physical damage, corrosion or thermochemical interaction with the medium whose temperature is being measured. Also known as thermowell.

protective component A component, or an assembly of components, which is not liable to become defective, in service or in storage, in such a manner as to lower the intrinsic safety of the circuit. Such a component or assembly of components is considered as not subject to fault in that manner when tests of intrinsic safety are made.

protective logic circuits Logic circuits designed to prevent damage to equipment by related system equipment malfunctions, failure, or operator errors [S77.42].

protocol A set of rules and formats which determine the communications behavior of an entity.

protocol data unit (PDU) Each of the seven OSI layers accepts data (SDUs) from the layer above, adds its own header (PCI), and passes the data to the layer below as a PDU. Conversely, each of the layers also accepts data from the layer below, strips off its header, and passes it up to the layer above.

protocols A set of conventions which governs the way in which devices communicate with each other.

proton An elementary atomic particle of mass number 1 and a positive charge equal in magnitude but opposite in sign to the charge on an electron.

prototype A preproduction model suitable for evaluating a product's design, functionality, operability and form, but not necessarily its durability and reliability.

proving Determination of flowmeter performance by establishing the relationship between the volume actually passed through the meter and the volume indicated by the meter.

PROWAY A standard for a process control highway based on IEEE 802.4 Token Bus immediate acknowledged MAC, a physical layer utilizing a phase continuous signaling technique.

proximity detector A sensor that produces an electric signal when the distance from the sensor to another object is less than a predetermined value.

proximity switch A device that senses the presence or absence of an object without physical contact and activates or deactivates an electrical circuit as a result.

PSAP address The fully qualified network address used to access application entities.

PSD See power spectral density.

pseudo code A code that requires translation prior to execution.

pseudo instruction 1. A symbolic representation in a compiler or interpreter. 2. A group of characters having the same general form as a computer instruction, but never executed by the computer as an actual instruction. Synonymous with quasi instruction.

pseudo-operations A group of instructions which, although part of a program, do not perform any application related function. They generally provide information to the assembler.

pseudoplastic A material that exhibits flow (permanent deformation) at all values of shear stress, although in most cases the flow that occurs below some specific value (an apparent yield stress) is low and increases negligibly with increasing stress.

pseudo-random The property of satisfying one or more of the standard criteria for statistical randomness but being produced by a definite calculation process. Related to uniformly distributed random numbers.

pseudo-random number sequence A sequence of numbers, determined by some defined arithmetic process, that is satisfactorily random for a given purpose, such as by satisfying one or more of the standard statistical tests for randomness. Such a sequence may approximate any one of several statistical distributions, such as uniform distribution or normal (gaussian) distribution.

pseudo-variable A variable that requires manipulation prior to calculation or processing.

PSK See phase shift keying.

psophometer An instrument for measuring noise in electric circuits; its output is exactly one-half of the psophometric emf in a circuit when it is connected across a 600-ohm resistance in the circuit.

psychrometer A device consisting of two thermometers, one of which is covered with a water-saturated wick, used for determining relative humidity; for a given set of wet-bulb and dry-bulb temperature readings, relative humidity is read from a chart. Also known as wet-and-dry-bulb thermometer.

PTR See paper tape reader.

puff A minor combustion explosion within the boiler furnace or setting.

pulsating current Unidirectional current whose magnitude alternately rises and falls in a regularly recurring pattern.

pulsating flow 1. Irregular or repeating variations in fluid flow, often due to pressure variations in reciprocating pumps or compressors in the system. 2. A flow rate that varies with time, but for which the mean flow rate is constant when obtained over a sufficiently long period of time.

pulsating pressure Pressure whose magnitude alternately rises and falls in a regularly recurring pattern, and whose variation exceeds 1% per second, or 5% per minute, of the scale on the measuring instrument.

pulsation Rapid fluctuations in furnace pressure.

pulsation dampening A device installed in a gas or liquid piping system to smooth out fluctuations due to pulsating flow and/or pressure.

pulse 1. A variation of a signal whose magnitude is normally constant; this variation is characterized by a rise and a decay, and has a finite duration [S26]. 2. A significant and sudden change of short duration in the level of an electrical variable, usually voltage. 3. A regular or intermittent variation in a normally constant quantity characterized by a relatively rapid rise and subsequent decay within a finite time period.

pulse amplitude modulation (PAM) The process (or the results of the process) in which a series of pulses is generated with amplitudes proportional to the measured signal samples.

pulse-averaging discriminator In an FM system, a subcarrier demodulator which uses the width of each cycle of the subcarrier to derive a data output.

pulse code 1. A code in which sets of pulses have been assigned particular meanings. 2. The binary representations of characters. 3. A series of energy pulses, or a pulse train, modulated in accordance with a data signal. 4. Generally, any data transmission scheme that utilizes pulsed energy to encode the transmitted values.

pulse code modulation (PCM) 1. The process of sampling a signal and encoding the height or amplitude of each sample into a series of uniform pulses. 2. In telemetry, serial data transmission (generally a series of binary-coded words).

pulse count telemetering A method of transmitting information that involves an "off-on" switching signal whose number of signal pulses per unit time represents the transmitted value.

pulse decay time The time between the instant when the amplitude of a pulse begins to drop from a specified upper limit and the instant when it reaches a specified lower limit.

pulse discriminator A device that detects pulses that have defined characteristics.

pulse duration The time between the instant when the amplitude of a pulse reaches some specified fraction of its peak value as it rises and the instant it passes through the same fraction as it falls.

pulse duration modulation (PDM) The process of sampling a signal and encoding each sample into a series of pulses whose duration or widths are proportional to the amplitude of the sample.

pulse duty factor The ratio of average pulse duration to average pulse spacing.

pulse-forming network Electrical circuitry used to generate high voltage pulses of particular shapes, and to modify the shapes of pulses generated by other sources.

pulse-height discriminator An electronic circuit that selects and passes only those voltage pulses that exceed a given minimum amplitude.

pulse-height selector An electronic circuit that selects and passes only those voltage pulses whose peak amplitudes are within a specific range of values.

pulse input In process control systems, a type of input used to measure pulse or tachometer-type signals (speed, rpm, frequency, etc.).

pulse-interval modulation Modulation of a pulsed carrier wave in which the time interval between pulses is varied in accordance with the modulating signal.

pulse mode A type of sequential circuit in which inputs are nonperiodic pulses, as opposed to logic levels used in conjunction with clock pulses (clock mode).

pulse modulation 1. Modulation of a carrier wave by a pulsed modulating wave. 2. Modulation of one or more attributes of a pulsed carrier wave.

pulse motor See stepping motor.

pulse-position modulation A form of pulse-time modulation in which the position in time of a pulse varies in accordance with some attribute of the modulating wave.

pulse repeater An electronic device that receives pulses from one circuit and transmits them into another circuit, with or without altering the frequency and waveform of the pulses.

pulse repetition frequency The rate at which pulses are repeated in a periodic pulse train.

pulse repetition period The reciprocal of pulse repetition frequency.

pulse repetition rate 1. The average rate at which pulses are repeated, whether or not the pulse train is periodic. 2. The number of electric pulses per unit of time experienced by a point in a computer, usually the maximum, normal, or standard pulse rate.

pulse rise time The time between the instant when the amplitude of a pulse reaches some specified fraction of its peak value and the instant when it reaches some specified higher fraction—unless otherwise stated, the two fractions are taken to be 10% and 90%.

pulse spacing The average time interval between corresponding locations on the waveforms of consecutive pulses.

pulse spectrum The frequency distribution of a series of sine waves that can be combined to yield a given periodic pulse train.

pulse switch A switch that provides one pulse of electric current for each cycle of operation.

pulse telemetering Any system for transmitting information in terms of electric pulses that are independent of electrical variations in the transmission channel; they can be classified as pulse duration, pulse count or pulse code systems.

pulse-time modulation Modulation of a pulsed carrier wave in which the time of occurrence of some specific point on each pulse waveform is varied from the unmodulated value in accordance with a modulating signal.

pulse transformer A type of transformer designed to convert an a-c input signal to a pulsed output signal.

pump A machine that draws fluid into itself through an inlet and forces the fluid out through an exit port, often at higher pressure than at the inlet.

pump drive control A control component of the final device that translates a control system demand signal into an electronic, hydraulic, pneumatic, or mechanical signal which affects pump speed [S77.42].

punched card　A durable paper-board card in which patterned punched holes can be read by a computer.

pure code　A computer code that is never modified during execution; it is possible to allow many users to share the same copy of programs that are written as pure code.

purge　1. Increasing the sample flow above normal for the purpose of replacing current sample-line fluid or removing deposited or trapped materials [S67.10]. 2. To cause a liquid or gas to flow from an independent source into the impulse line(s). 3. To introduce air into the furnace and the boiler flue passages in such volume and manner as to completely replace the air or gas-air mixture contained therein.

purged packing box　A packing arrangement consisting of a lantern ring inside the packing rings to permit introduction of a purge fluid to continually flush the space between the stem and body. It is usually used to purge, admit cooling fluid or detect stem seal leakage [S75.05].

purge interlock　A device so arranged that an air flow to the furnace above a minimum must exist for a definite time interval before the interlocking system will permit the automatic ignition torch to be placed in operation. See also purging classifications.

purge meter　A device designed to measure small flow rates of liquids and gases used for purging measurement piping.

purge post　An acceptable method of scavenging the furnace and boiler passes to remove all combustible gases after flame failure controls have sensed pilot and main burner shutdown and safety shut-off valves are closed.

purge, pre-ignition　An acceptable method of scavenging the furnace and boiler passes to remove all combustible gases before the ignition system can be energized.

purging　1. The addition of air or inert gas (such as nitrogen) into the enclosure around the electrical equipment at sufficient flow to remove any hazardous vapors present and sufficient pressure to prevent their re-entry [S12.4]. 2. Elimination of an undesirable gas or material from an enclosure by means of displacing the undesirable material with an acceptable gas or material.

purging classifications　1. Type Z purging. Covers purging requirements adequate to reduce the classification of the area within an enclosure from Division 2 (normally non-hazardous) to non-hazardous [S12.4]. 2. Type Y purging. Covers purging requirements adequate to reduce the classification of the area within an enclosure from Division 1 (hazardous) to Division 2 (normally non-hazardous) [S12.4]. 3. Type X purging. Covers purging requirements adequate to reduce the classification of the area within an enclosure from Division 1 (hazardous) to non-hazardous [S12.4].

purity　The degree to which a substance is free of foreign materials.

pushbutton　A momentary manual switch that causes a change from one sequence state to another. Pushbutton actions include silence, acknowledge, reset, first out reset, and test [S18.1].

push-down list　A list that is constructed and maintained so that the next item to be retrieved is the most recently stored item in the list, that is: last-in, first-out.

push-up list　A list that is constructed and maintained so the next item to be retrieved and removed is the oldest item still in the list, that is: first-in, first-out.

PV tracking　Set point automatically tracks the process variable when the controller is in manual.

pycnometer　A container of precisely known volume that is used to determine density of a liquid by weighting the filled container and dividing the weight by the known volume. Also spelled pyknometer.

pyranometer　An instrument used to measure the combined intensity of direct solar radiation and diffuse sky radiation. Also known as solarimeter.

pyrgeometer　An instrument for measuring radiation from the earth's surface into space.

pyrheliometer　An instrument for measuring the intensity of direct solar radiation only.

pyroelectric detectors　Detectors of visible, infrared, and ultraviolet radiation which rely on the absorption of radiation by pyroelectric materials. Heating of such materials by the absorbed radiation produces electric charges on opposite sides of the crystal, which can be measured to determine changes in the amount of radiation incident on the detector. These detectors usually also exhibit piezoelectric properties and may require isolations from acoustic or acceleration phenomena.

pyrometer Any of a broad class of temperature measuring instruments or devices. The term originally applied only to devices for measuring temperatures well above room temperature, but now it applies to devices for measuring temperatures in almost any range. Some typical pyrometers include thermocouples, radiation pyrometers, resistance pyrometers and thermistors, but usually not thermometers. It is a temperature transducer that measures temperatures by the EM radiation emitted by an object, which is a function of the temperature.

Q

q See volumetric flow rate.

QBC See queue control block.

***Q* factor** 1. A rating factor for electronic components such as coils, capacitors and resonant circuits that equals reactance divided by resistance. 2. In a periodically repeating mechanical, electrical or electromagnetic process, the ratio of energy stored to energy dissipated per cycle.

***Q* meter** A direct-reading instrument that measures the "Q" of an electric circuit at radio frequencies by determining the ratio of inductive reactance to resistance. Also known as quality-factor meter.

Q-switch An optical device which changes the Q (quality factor) of a laser cavity, typically raising it from a value below laser threshold to one well above threshold. This technique produces a short, intense pulse, known as a Q switched pulse. Q-switches can be based on acousto-optic or electro-optic devices, rotating mirrors, frustrated internal reflection, or saturation of absorption in a dye.

quadrant detectors Detectors which are divided up into four angularly symmetric sectors or quadrants. The amounts of radiation incident on each quadrant can be compared to one another for applications such as making sure that a beam is centered on the detector.

quadrant-edged orifice An orifice having a rounded contour at the inlet edge to yield more constant and predictable discharge coefficient at low flow velocity (Reynolds number less than 10,000).

quadratic programming In operations research, a particular case of nonlinear programming in which the function to be maximized or minimized and the constraints are quadratic functions of the controllable variables. Contrast with convex programming, dynamic programming, linear programming, and mathematical programming.

quadrupole mass spectrometer A type of mass spectrometer employing a filter consisting of four conductive rods electrically connected in such a manner that, by varying the absolute potential applied to the rods, all ions except those possessing a specific mass-to-charge ratio are prevented from entering the detector.

quad-slope converter An integrating analog-to-digital converter that goes through two cycles of dual slope conversion, once with zero input and once with the analog input being measured.

qualified 1. Competent, suited, or having met the requirements for a specific position or task. 2. To declare competent or capable.

quality assurance A set of systematic actions intended to provide confidence that a product or service will continually fulfill a defined need.

quality control A set of systematic actions that make it possible to measure significant characteristics of a product or service and to control the characteristics within established limits.

quantity meter A flowmeter in which the flow is separated into known isolated quantities which are separately counted to determine the total volume passed through the meter.

quantization The subdivision of the range of values of a variable into a finite number of non-overlapping, and not necessarily equal, subranges or intervals, each of which is represented by an assigned value within the subrange. For example, a person's age is quantized for most purposes with the quantum of one year.

quantization distortion Inherent distortion introduced when a range of values for a wave attribute is divided into a series of smaller subranges.

quantization level A particular subrange for a quantized wave, or its corresponding symbol.

quantize To convert information from an analog

pulse (as from a multiplexer) into a digital representation of that pulse. See encoder.

quantized pulse modulation Pulse modulation in which either the carrier wave or modulating wave is quantized.

quantizing interval The smallest change in the input signal to a quantizing device which causes a change in the quantized representation of the input signal [RP55.1].

quantum noise Noise due to the discrete nature of light—i.e., its quantization into photons.

quarter amplitude A process-control tuning criteria where the amplitude of the deviation (error) of the controlled variable, following a disturbance, is cyclic so that the amplitude of each peak is one quarter of the previous peak.

quarter-wave plate A polarization retarder which causes light of one linear polarization to be retarded by one quarter wavelength (90°) relative to the orthogonal polarization.

quartz A natural transparent form of silica, which may be marketed in its natural crystalline state, or crushed and remelted to form fused quartz.

quasi instruction Same as pseudo instruction.

query language A means of getting information from a database without the need to write a program.

queue 1. Waiting line resulting from temporary delays in providing service. 2. In data processing, a waiting list of programs to be run next. 3. Any list of items; for example, items waiting to be scheduled or processed according to system- or user-assigned priorities.

queue control block (QCB) A control block that is used to regulate the sequential use of a programmer-defined facility among requesting tasks.

queued access method Any access method that automatically synchronizes the transfer of data between the program using the access method and input/output devices, thereby eliminating delays for input/output operations.

queuing An ordered progression of items into and through a system or process, especially when there is waiting time at the point of entry.

queuing discipline The rules or priorities for queue formation within a system of "customers" and "servers" as well as the rules for arrival time and service time.

queuing theory A form of probability theory useful in studying delays or line-ups at servicing points.

Quevenne scale A specific gravity scale used in determining the density of milk; a difference of 1° Quevenne is equivalent to a difference of 0.001 in specific gravity, and therefore 20° Quevenne expresses a specific gravity of 1.020.

quick-look Essentially, an "instant replay" of data, generally at the same rate at which it was recorded.

quick return A device that makes the return stroke of a reciprocating machine element faster than the power stroke.

R

R See rankine.

rack 1. A standardized steel framework designed to hold 19-in.-wide panels of various heights that have units of electronic equipment mounted on them. Also known as relay rack. 2. A frame for holding or displaying articles. 3. A bar having teeth along one of its long faces for meshing with a gear.

rad See radian.

radar A device or system for detecting distant objects, identifying them, or determining their range, location, speed or direction of travel by means of radio waves reflected from or retransmitted by the object; the term is derived from the original term radio detection and ranging.

radian Metric unit for a plane angle.

radiance Radiant flux per unit solid angle per unit of projected source area.

radiant flux The rate of flow of radiant energy with respect to time.

radiant fluxmeter A device which measures the amount of radiant flux emitted or absorbed. The units typically would be watts per unit area.

radiant intensity The energy emitted per unit time per unit solid angle along a specific linear direction.

radiation Transfer of heat by waves or particles.

radiation damage A general term for the deleterious effects of radiant energy—either electromagnetic radiation or particulate radiation—on physical substances or biological tissues.

radiation loss A comprehensive term used in a boiler-unit heat balance to account for the conduction, radiation, and convection heat losses from the settings to the ambient air.

adiation pyrometer An instrument that uses the radiant power emitted by a hot object in determining its temperature.

adiation thermometer Any of several devices for determining the temperature of a body by measuring its emitted radiant energy, without physical contact between the sensor and the body.

radioactive half life The time it takes for the radioactivity of a specific nuclide to be reduced by one half.

radioactive tracer A radioactive nuclide used at small concentration to follow the progress of some physical, chemical or biological process; typically, a radioactive isotope is substituted for a small proportion of the same element normally present, and the movement and behavior of the elemental substance or of a radioactively "labeled" compound is observed with radiation detection instruments.

radioactivity Any particulate or electro-magnetic radiation emanating from a mass of material due to spontaneous emission from unstable atomic nuclei.

radiobiology A branch of biology that deals with the effect of radiation on biological processes and tissues.

radio channel A band of frequencies wide enough to be used for radio communication.

radio-doppler Determining the radial component of relative velocity between an object and a fixed or moving point of observation by measuring the difference in frequency between transmitted radio-frequency waves and reflected waves returning from the object.

radio engineering A discipline often considered part of electronic engineering that deals with generating, transmitting and receiving radio waves, and with the design, fabrication and testing of equipment for performing these functions.

radio frequency A frequency at which electromagnetic radiation can be used for communication.

radio-frequency heating See electronic heating.

radio frequency interference (RFI)   A type of electrical noise that can affect electronic circuits adversely.

radiography   A form of nondestructive testing that involves the use of ionizing radiation to detect flaws, characterize internal structure or measure thickness of metal parts or the human body; it may involve determining the attenuation of radiation passing through a test object.

radioisotope   1. A radioactive isotope of a chemical element. 2. A nonpreferred synonym for radionuclide.

radioisotope tracer flowmeter   A device for determining flow rate by injecting a radioactive substance into the fluid stream; two types are available—peak timing, in which flow rate is calculated from a measurement of the time it takes a peak concentration of tracer element to pass between two fixed points along the flow path, and dilution method, in which a known concentration of tracer is injected into the fluid stream and its concentration at a point downstream determined by analysis.

radiology   The application of radiation science to the study of medicine, especially the diagnosis and treatment of injury and disease.

radioluminescence   Emission of light due to radioactive decay of a nuclide.

radiometer   An instrument which measures optical power in radiometric units (watts). It measures electromagnetic power linearly over its entire spectral range.

radiometric analysis   A method of quantitative chemical analysis based on measuring the absolute disintegration rate of a radioactive substance of known specific activity.

radionuclide   Any nuclide which undergoes spontaneous radioactive decay.

radiosonde   A balloon-borne instrument for taking meteorological data. It consists of transducers for determining temperature, humidity and barometric pressure; a modulator for converting transducer outputs to radio signals; a selector switch for establishing transmission sequence; and a transmitter for generating a radio-frequency carrier wave.

radix   Also called the base number; the total number of distinct marks or symbols used in a numbering system. For example, since the decimal numbering system uses ten symbols (0, 1, 2, 3, 4, 5, 6, 7, 8, 9), the radix is ten. In the binary numbering system the radix is two, because there are only two marks or symbols (0, 1).

radix-50   A storage format in which three ASCII characters are packed into a sixteen-bit word.

radix complement   A number obtained by subtracting each digit of the given number from one less than the radix, then adding to the least significant digit, executing all required carries; in radix two (binary), the radix complement of a number is used to represent the negative number of the same magnitude.

radix notation   A positional representation in which the significances of any two adjacent digit positions have an integral ratio called the radix of the less significant of the two positions. Permissible values of the digit in any position range from zero to one less than the radix of that position.

radix number   The quantity of characters for use in each of the digital positions of a numbering system. In the more common number systems the characters are some or all of the Arabic numerals. Unless otherwise indicated, the radix of any number is assumed to be 10. For positive identification of a radix 10 number, the radix is written in parentheses as a subscript to the expressed number. Synonymous with base and base number.

radix point   Also called base point, binary point, decimal point, and others, depending on the numbering system; the index that separates the integral and fractional digits of the numbering system in which the quantity is represented.

radome   A weatherproof cover for a primary radar device constructed of a material which is transparent to radio waves.

rails, optical   Long, linear rods which are attached to an optical bench. Optical mounts can be affixed to the rail, creating an optical bench.

RAM   See random access memory.

Raman shifter   A device which alters the wavelength of light by inducing Raman shifts in the light passing through it. Raman shifts are changes in photon energy caused by the transfer of vibrational energy to or from the molecule.

ramp encoder   An analog-to-digital conversion process whereby a binary counter is incremented during the generation of a ramp voltage; when the amplitude of the ramp voltage is equal to the amplitude of the voltage sample, the counter clock is inhibited. The counter contents therefore contain the binary equivalent of the sampled data.

ramp generator An electrical power supply which generates a voltage that increases at a constant rate. A plot of voltage vs. time shows a ramp-like waveform.

ramp response The total (transient plus steady-state) time response resulting from a sudden increase in the rate of change in the input from zero to some finite, constant value. See also response, ramp.

ramp response time See time, ramp response.

random access 1. Pertaining to the process of obtaining data from, or placing data into, storage, where the time required for such access is independent of the location of the data most recently obtained or placed in storage. 2. Pertaining to a storage device in which the access time is effectively independent of the location of the data. 3. Synonymous with direct access. See also access, random.

random access device A device in which the access time is effectively independent of the location of the data. Synonymous with direct access device.

random access memory (RAM) This is memory that can be written into or read from and allows access to any address within the memory. RAM is volatile in that contents are lost when the power is switched off. A type of computer memory used for temporary storage of data; allows the CPU to have fast access to read or change any of its memory locations.

random access storage A storage technique in which the time required to obtain information is independent of the location most recently obtained.

random error Precision or repeatability data that deviate from a mean value in accordance with the laws of chance.

random file A collection of records, stored on random access devices such as discs. Algorithms are used to define the relationship between the record key and the physical location of the record.

randomized non-return-to-zero A coding scheme which provides the greatest packing density on instrumentation tape with low possibility of DC components.

randomizer A hardware device to inject a pseudo-random bit sequence into an NRZ wavetrain, thereby guaranteeing frequent data transitions so that the low-frequency component is not too low for transmission or recording. A "deran-

domizer" removes the sequence and restores data to the original form; a form of data enhancement.

random number generator A special routine or hardware designed to produce a random number or series of random numbers according to specified limitations.

random numbers 1. A series of numbers obtained by chance. 2. A series of numbers considered appropriate for satisfying certain statistical tests. 3. A series of numbers believed to be free from conditions which might bias the result of a calculation. 4. See pseudo-random number sequence.

random sampling Selecting a small number of items for inspection or testing (the sample) from a much larger number of items (the lot or population) in a manner that gives each item in the population an equal probability of being included in the sample.

random sequence In welding, the technique of depositing a longitudinal weld bead in increments of random length and location.

random walk The path followed by a particle which makes random scattering collisions with other particles in a gaseous or liquid medium.

random walk method In operations research, a variance-reducing method of problem analysis in which experimentation with probabilistic variables is traced to determine results of a significant nature. Uninteresting walks add only to the variance of the process and thus contribute nothing. An interesting walk tends to lead toward a predictive solution.

range 1. The region between the limits within which a quantity is measured, received, or transmitted, expressed by stating the lower and upper range-values. NOTE A: Unless otherwise modified, input range is implied. B: The following compound terms are used with suitable modifications in the units: measured variable range, measured signal range, indicating scale range, chart scale range, etc. C: For multi-range devices, this definition applies to the particular range that the device is set to measure [S51.1]. 2. The measurand values, over which a transducer is intended to measure, specified by their upper and lower limits [S37.1]. 3. The range of an instrument is the region covered by the span and is expressed by stating the two end-scale values [S26]. 4. The series of outputs corresponding to values of concentrations of hydrogen sulfide over which accuracy is ensured by calibration

[S12.15]. 5. For instrumentation, the set of values over which measurements can be made without changing the instrument's sensitivity. 6. The extent of a measuring, indicating or recording scale. 7. An area within defined boundaries or landmarks used for testing vehicles, artillery or missiles, or for other test purposes. 8. The maximum distance a vehicle, aircraft or ship can travel without refueling. 9. The distance between a weapon and a target. 10. The maximum distance a radio, radar, sonar or television transmitter can send a signal without excessive attenuation. 11. The set of values that a quantity or function may assume. 12. The difference between the highest and lowest value that a quantity or function may assume. 13. The difference between the maximum and minimum values of physical output over which an instrument is designed to operate normally. See also error range.

rangeability 1. Describes the relationship between the range and minimum quantity that can be measured. 2. The ratio of the maximum flow rate to the minimum flow rate of a meter. 3. Installed rangeability may be defined as the ratio of maximum to minimum flow within which limits the deviation from a desired installed flow characteristic does not exceed some stated limits. 4. Inherent rangeability, a property of the valve alone, may be defined as the ratio of maximum to minimum flow coefficients between which the gain of the valve does not deviate from a specified gain by some stated tolerance.

rangeability, inherent The ratio of the largest flow coefficient (Cv) to the smallest flow coefficient (Cv) within which the deviation from the specified inherent flow characteristic does not exceed the stated limits [S75.05].

range check In data processing, a validation that data is within certain limits.

range, elevated-zero A range in which the zero value of the measured variable, measured signal, etc., is greater than the lower range-value. NOTE 1: The zero may be between the lower and upper range-values, at the upper range-value, or above the upper range-value. 2: Terms suppression, suppressed range or suppressed span are frequently used to express the condition in which the zero of the measured variable is greater than the lower range-value. The term "elevated-zero range" is preferred [S51.1].

range-limit, lower The lowest value of the measured variable that a device can be adjusted to measure. NOTE: The following compound terms are used with suitable modifications to the units: measured variable lower range-limit, measured signal lower range limit, etc. [S51.1].

range-limit, upper The highest value of the measured variable that a device can be adjusted to measure. NOTE: The following compound terms are used with suitable modifications to the units: measured variable upper range-limit, measured signal upper range-limit, etc. [S51.1].

range of an analog dc current signal The range of an analog dc current signal is determined by stating the lower and the upper limit of the signal current [S50.1].

range of a pneumatic transmission signal The range determined by the lower limit and the upper limit of the signal pressure [S7.4].

range, suppressed-zero A range in which the zero value of the measured variable is less than the lower range value. (Zero does not appear on the scale.) NOTE: Terms elevation, elevated range or elevated span are frequently used to express the condition in which the zero of the measured variable is less than the lower range-value. The term "suppressed-zero range" is preferred [S51.1].

range-value, lower The lowest value of the measured variable that a device is adjusted to measure. NOTE: The following compound terms are used with suitable modifications to the units: measured variable lower range-value, measured signal lower range-value, etc. [S51.1].

range-value, upper The highest value of the measured variable that a device is adjusted to measure. NOTE: The following compound terms are used with suitable modifications to the units: measured variable upper range-value, measured signal upper range-value, etc. [S51.1].

ranging The measurement of distance by timing how long it takes a light, radio frequency sound or ultrasound pulse to make a round trip from the source to a distant object.

rank 1. To arrange in an ascending or descending series according to importance. 2. Position in some ascending or descending series.

Rankine An absolute temperature scale where the zero point is defined as absolute zero (the point where all spontaneous molecular motion ceases) and the scale divisions are equal to the scale divisions in the Fahrenheit system; 0 °F equals approximately 459.69 °R.

rapid traverse A machine tool mechanism that quickly moves the workpiece to a new position while the cutting tool is retracted.

raster A set of lines that provide essentially uniform coverage of a given area—for instance, the set of parallel lines on a television picture tube that are most easily seen when there is no picture.

raster scan A form of display similar to television in which the signal is scanned backwards and forwards across the screen.

rate See control action, derivative.

rate action 1. Another name for the derivative control mode. 2. A control action which produces a corrective signal proportional to the rate at which the controlled variable is changing. 3. Rate action produces a faster corrective action than proportional action alone.

rate control See derivative control.

rate control action See control action, derivative.

rated capacity The manufacturers stated capacity rating for mechanical equipment, for instance, the maximum continuous capacity in pounds of steam per hour for which a boiler is designed.

rated flow 1. Design flow rate for a piping system or process vessel. 2. Normal operating flow rate for a fluid passing through a piping system.

rated horsepower The maximum or allowable power output of an engine, turbine or other prime mover under normal, continuous operating conditions.

rated load The maximum design load for a machine, structure or vehicle.

rated supply voltage The supply voltage, or range of voltages, for which the manufacturer has designed the equipment [S82.01].

rated travel The amount of movement of the valve closure member from the closed position to the rated full open position [S75.05].

rated value The value or one of the values assigned by the equipment manufacturer or component manufacturer [S82.01].

rate gain See gain, derivative action.

rate gyroscope A gyro wheel mounted in a single gimbal ring in such a manner that rotation about an axis perpendicular to both the gimbal axis and gyro axis produces a precessional torque proportional to the rotation rate.

rate of blowdown A rate normally expressed as a percentage of the water fed.

rate-of-climb indicator A navigation instrument that indicates the rate at which an aircraft gains or loses altitude.

rate response A relationship describing the output of a control system as a function of its input signal.

rate time In the action of a proportional-plus-rate or proportional-plus-reset-plus-rate controller, the time by which the rate action advances the proportional action on the controlled device.

rating See load.

ratio controller 1. A controller that maintains a predetermined ratio between two or more variables. 2. Maintains the magnitude of a controlled variable at a fixed ratio to another variable. See also controller, ratio.

ratio meter A measuring instrument whose pointer deflection is proportional to the ratio of the currents passing through two coils; it measures the quotient of two electrical qualities.

ratio of specific heats Specific heat at constant pressure divided by specific heat at constant volume.

ratio spectrofluorometer A type of instrument used in chemical assaying to determine proportions when two compounds are similar in bioassay and spectrophotometry but differ markedly in fluorescence; quantitative assays can be made either by proportionality or by use of a linearity curve.

ratio thermometer A device consisting essentially of two radiation thermometers in the same housing, the output of each thermometer having a separate wavelength response; the output is a ratio signal that is a function of temperature, but that is relatively insensitive to target size and therefore is as accurate for small radiating bodies as it is for larger ones.

ratio-type telemeter A telemeter that translates data in terms of the relative phase relation between two electrical quantities, or their relative magnitudes.

raw data In data processing, information that has not been processed or analyzed by the computer.

raw water Water supplied to the plant before any treatment.

Rayleigh disc A special form of acoustic radiometer used to measure particle velocities.

RC **oscillator** Any oscillator whose frequency is determined by the interaction between resistors and capacitors in an electronic circuit.

RCTL See resistor-capacitor-transistor logic.

Re See Reynolds number.

reach rod A valve extension mechanism used to provide for manual operation of valves which are inaccessible [S67.10].

reactance A component in an electrical circuit which is due to the presence of capacitive or inductive elements and not resistive elements, and which opposes the flow of electric current.

reactance drop The voltage drop 90° out of phase with the current.

reaction A chemical transformation or change brought about by the interaction of two substances.

reactive volt-ampere meter See varmeter.

reactor 1. A circuit element that introduces capacitative or inductive reactance. 2. A vessel in which a chemical reaction takes place. 3. An enclosed vessel in which a nuclear chain reaction takes place.

reactor coolant pressure boundary (RCPB) All those pressure-containing components of boiling and pressurized water-cooled nuclear power reactors, such as pressure vessels, piping, pumps, and valves, which are: (a) part of the reactor coolant system, or (b) connected to the reactor coolant system, up to and including any and all of the following: (i) The outermost containment isolation valve in system piping which penetrates primary reactor containment. (ii) The second of two valves normally closed during normal reactor operation in system piping which does not penetrate primary reactor containment. (iii) The reactor coolant system safety and relief valves. For nuclear power reactors of the direct cycle boiling water type, the reactor coolant system extends to and includes the outermost containment isolation valve in the main steam and feedwater piping [S67.03].

reactor, endothermic A reactor which absorbs heat from the surroundings.

reactor, exothermic A reactor which generates heat.

read 1. In data processing, to retrieve information from any memory medium. 2. To playback data from a tape, disk, and the like, or to obtain the contents of computer memory.

readability The smallest fraction of the scale on an instrument which can be easily read—either by estimation or by use of a vernier.

reading The indicated value determined from the scale of an indicating instrument, or from the position of the index on a recording instrument with respect to an appropriate indicating scale.

read-only memory (ROM) Storage containing data that cannot be changed by computer instruction, but requires alteration of construction circuits; therefore, data that is nonerasable and reusable, or fixed.

readout In data processing, the display of information on the CRT screen or other display unit.

read time See access time.

read write memory Memory whose contents can be written and read from. Read write memory is not necessarily random access memory.

real number Any number that can be represented by a point on a number line.

real time 1. In data processing, the actual time that is required to solve a problem. 2. Pertaining to the actual time during which a physical process transpires. 3. Pertaining to the performance of a computation during the actual time that the related physical process transpires in order that results of the computation can be used in guiding the physical process.

real-time clock A clock which indicates the passage of actual time, in contrast to a fictitious time set up by the computer program; such as, elapsed time in the flight of a missile, wherein a 60-second trajectory is computed in 200 actual milliseconds, or a 0.1 second interval is integrated in 100 actual microseconds. See clock, real time.

real-time input Input data inserted into a system at the time of generation by another system.

real-time interrupt process (RIP) Software within TELEVENT that responds to "events" (interrupts).

real-time language (RTL) A computer language designed to work on problems of a time-critical nature.

real-time operation See real-time processing

real-time output Output data removed from a system at time of need by another system.

real-time processing 1. The processing of information or data in a sufficiently rapid manner so that the results of the processing are available in time to influence the process being monitored or controlled. 2. Computation that is performed while a related or controlled physical activity is occurring so that the results of the computation can be used to guide the process.

real-time program A program which operates

concurrently with an external process which it is monitoring or controlling, meeting the needs of that process with respect to time.

real-time system A system which responds within the time scale of the process being controlled, i.e. whose response time depends upon the process dynamics or time constants.

reasonableness test A test providing a means for detecting a gross error in calculation by comparing results against upper and lower limits representing an allowable reasonable range.

reassociation The recombination of the products of dissociation.

Réaumur scale A temperature scale having 0° as the ice point and 80° as the steam point; the scale is little used outside the brewing, wine-making and distilling industries.

reboiler 1. The heat exchanger at the bottom of a distillation column. The reboiler generates the vapors that ascend through the column from liquid which comes down the column. The term is derived from "re-boil". 2. A closed heat exchanger which uses one medium to heat a second. The purpose is to maintain a separation between the two mediums due to noncompatibility or to prevent contamination between systems while transferring thermal energy.

receiver 1. An apparatus near the outlet of a compressor which collects excess oil or moisture in the compressed air, and which reduces or eliminates pressure pulsations and stores compressed air for later use. 2. An electronic device which detects radio-wave signals and amplifies them into electrical signals of varying frequency and amplitude to drive an output device such as a loudspeaker. Also known as radio receiver. 3. A device which detects an optical signal, converts it into electronic form, then processes it further so it can be used by electronic equipment.

receiver gage A gage, calibrated in engineering units, which receives the output of a pneumatic transmitter.

receiving gage A fixed gage designed to inspect several dimensions on a part, and also inspect the relationships between dimensions.

receptacle A connector device permanently connected to the supply circuit (usually a branch circuit) into which an attachment plug is inserted [S82.01].

recipe The complete set of data and procedure that defines the control requirements of a particular product manufactured by a batch process. A recipe consists of a header, equipment requirements, procedure and formula.

reciprocal transducer A transducer in which the output signal is proportional to the reciprocal of the level of the stimulus.

recirculation The reintroduction of part of the flowing fluid to repeat the cycle of circulation.

recombination The reaction between an ion and one or more electrons that returns the ionized element or molecule to the neutral state.

record 1. A group of related facts or fields of information treated as a unit, thus a listing of information, usually in printed or printable form [RP55.1]. 2. To put data into a storage device [RP55.1]. 3. A segment of a file consisting of an arbitrary number of words or characters. 4. In data processing, a group of data that contains all the information about a single item.

record (noun) On a computer tape, the smallest group of words which can be located and input by the computer. In telemetry systems, this is usually 1,000 to 4,000 data words.

record (verb) To store on some permanent (generally magnetic) medium, as on a magnetic tape or disk.

recorded value The value of a measured variable as determined from the position of a trace or mark on chart paper or as determined from a permanent or semipermanent effect on an alternative recording medium.

recorder 1. An instrument that makes and displays a continuous graphic, acoustic or magnetic record of a measured variable. 2. A measuring instrument in which the values of the measured variable are recorded.

record format The content and organization of a single record.

record gap An area between two consecutive records.

recording channel In a recording system that incorporates any means of producing multiple traces on a single chart or multiple tracks on an alternative recording medium, the circuit and associated apparatus needed to produce a single trace.

recording extensions A recorder is attached directly to the meter body with the recorder pen positioned by the metering float through a magnetic coupling.

recording instrument See instrument, recording.

recording stylus See stylus.

recovery time 1. The time interval, after a specified event (e.g., overload, excitation transients, output shortcircuiting) after which a transducer again performs within its specified tolerances [S37.1]. 2. In a radiation counter tube, the time that must elapse after detection of a photon of radiation until the instrument can deliver another pulse of substantially full response level upon interaction with another photon of ionizing radiation.

recrystallization A nucleation and growth process in which new, strain-free grains in a metal or alloy form from a distorted structure that has undergone at least a threshold amount of cold work; it occurs in a few metals at room temperature but requires annealing at elevated temperature for detectable amounts to occur in most.

recrystallization temperature The minimum temperature at which complete recrystallization of a cold worked metal occurs in a specified time, usually one hour.

rectifier Any of several devices for converting alternating current to a relatively steady or pulsating direct current signal.

rectifier instrument An instrument for measuring certain quantities of an alternating current circuit in which the a-c input is converted to direct current by a rectifier and the d-c signal is actually measured.

rectifier-type voltmeter An instrument that incorporates four semiconductor rectifier elements arranged in a square (full wave bridge configuration), with the a-c input connected across one diagonal and a permanent magnet moving-coil detector connected across the other diagonal; the d-c output is proportional to the average current or voltage over any given half cycle of the a-c input, which makes the instrument useful for measuring nominal voltage or low-range (milliampere) current.

rectifying section The section of trays in a distillation column above the feed plate. In this section the vapor is enriched in the light components that are taken overhead.

recursion The property which allows a callable program to call itself.

recursive Pertaining to a process which is inherently repetitive. The result of each repetition is usually dependent upon the result of the previous repetition.

recycle synchronizer A method of telemetry subframe recognition in which a specific subcommutator word (or words) contains a unique signal that marks the recycling of the subcommutator.

redox potential The electrochemical potential prevailing in a chemical reaction involving an exchange of electrons (reduction-oxidation potential).

reducer A pipe fitting used to couple a pipe of one size to a pipe of a different size. When the flow is from the smaller pipe to the larger, increaser may be used to differentiate.

reducing atmosphere An atmosphere which tends to: a) promote the removal of oxygen from a chemical compound; b) promote the reduction of immersed materials.

reducing coupling A pipe fitting for connecting two pipes of different sizes.

reduction 1. Gain of electrons by a constituent of a chemical reaction [S71.04]. 2. Removal of oxygen from a chemical compound.

redundancy 1. In the transmission of information, that fraction of the gross information content of a message which can be eliminated without loss of essential information. Any circuitry or program instructions present solely to handle faults or errors and not necessary for normal system operation. 2. A parallel or secondary system that takes over when the primary system fails so control can continue uninterrupted.

redundancy check An automatic or programmed check based on the systematic insertion of components or characters used especially for checking purposes.

redundant character A character specifically added to a group of characters to insure conformity with certain rules which can be used to detect computer malfunction.

redundant check See redundancy check.

redundant (redundancy) Duplication or repetition of elements in electronic or mechanical equipment to provide alternative functional channels in case of failure of the primary device [S77.42].

Redwood scale A time-based viscosity scale used predominantly in Great Britain; it is similar in concept to the Saybolt scale used in the United States.

reed A thin bar of metal, wood or cane that is clamped at one end and set into transverse elastic vibration by a stimulating force such as wind pressure; it is used to create sound in musical instruments and to act as a frequency standard in some meters.

re-enterable load module   A load module that can be used concurrently by two or more tasks.

re-entrancy   The property which allows a callable program to be called and executed before it has completed the execution from a previous call. The results of the previous call are not affected.

reentrant   The property of a program that enables it to be interrupted at any point by another program, and then resume from the point where it was interrupted.

re-entrant program   A program which can be used for various tasks.

re-entry point   The instruction at which a program is re-entered from a subroutine.

refereed test   A predetermined destructive or nondestructive test made by a regulatory body or a disinterested organization, often to fulfill a regulatory requirement; in some cases, the test may be done by the regulated organization and merely witnessed by an agent of the regulatory body.

reference accuracy   In process instrumentation, a number or quantity that defines a limit that error will not exceed when a device is used under specified operating conditions. Error represents the difference between the measured value and the standard or ideal value [S67.04]. See accuracy rating.

reference dimension   A dimension without tolerance on a mechanical drawing which is given for information only and is not to be used in making or inspecting the part or assembly.

reference flowmeter   Flowmeter used as a transfer standard for in-system and comparison calibrations of turbine flowmeters [RP31.1].

reference ground   A datum level from which input-output signals are measured [RP55.1].

reference input   A signal from an independent reference source that is used as one of the inputs to an automatic controller.

reference-input element   See element, reference-input.

reference-input signal   See signal, reference-input.

reference junction   1. That thermocouple junction which is at a known or reference temperature. The reference junction is physically that point at which the thermocouple or thermocou-

ple extension wires are connected to a device or where the thermocouple is connected to a pair of lead wires, usually copper [S51.1]. 2. A device used with a thermocouple transducer which couples it to copper wires without introducing an error.

reference junction compensation   A means of counteracting the effect of temperature variations of the reference junction, when allowed to vary within specified limits.

reference level   The basis for comparison of an audio-frequency signal level given in decibels or volume units. See datum plane.

reference operating conditions   See operating conditions, reference.

reference performance   Performance attained under reference operating conditions. Performance includes such things as accuracy, dead band, hysteresis, linearity, repeatability, etc. [S51.1].

reference pressure   The pressure relative to which a differential-pressure transducer measures pressure [S37.1].

reference pressure error   1. The maximum change in output at specified measurand values due to a specified change in the reference pressure applied at both ports simultaneously [S37.6]. 2. The error resulting from variations of a differential-pressure transducer's reference pressure within the applicable reference pressure range. NOTE: It is usually specified as the maximum change in output, at any measurand value within the specified range, when the reference pressure is changed from ambient pressure to the upper limit of the specified reference pressure range [S37.1].

reference pressure range   The range of reference pressures which can be applied without changing the differential-pressure transducer's performance beyond specified tolerances for reference pressure error. When no such error is specified, none is allowed [S37.1].

reference pressure sensitivity shift   The sensitivity shift resulting from variations of a differential-pressure transducer's reference pressure within specified limits [S37.1].

reference pressure zero shift   The change in the zero-measurand output of a differential-pressure transducer resulting from variations of reference pressure (applied simultaneously to both pressure ports) within its specific limits [S37.1].

reference sensitivity (charge or voltage) The ratio of the change in charge or voltage generated by a transducer to the change in value of the acceleration that is measured under a set of defined conditions. (Amplitude, frequency, temperature, total capacitance, amplifier, frequency, temperature, total capacitance, amplifier input resistance, mounting torque). Deviations in sensitivity should be reported as deviations from the reference sensitivity [RP37.2].

refinery gas The commercially non-condensible gas resulting from fractional distillation of crude oil, or the cracking of crude oil or petroleum distillates. Refinery gas is either burned at the refineries or supplied for mixing with city gas.

reflash (multiple input) 1. An auxillary logic circuit that allows two or more abnormal process conditions to initiate or initiate the alarm state of one alarm point at any time. The alarm point cannot return to normal until all related process conditions return to normal [S18.1]. 2. One type of auxiliary output [S18.1].

reflectance The fraction of incident light which is reflected by the surface.

reflectometer A photoelectric instrument for measuring the proportion of light reflected from a given surface.

reflux The recycle stream that is returned to the top of the column. This stream supplies a liquid flow for the rectifying section that enriches the vapor stream moving up the column. Material in the stream is condensate from the overhead condenser. Reflux closes the energy balance by removing heat introduced at the reboiler.

reflux ratio A quantity usually expressed as the ratio of the reflux flow to the distillate flow. The ratio is used primarily in column design.

refracted wave The resultant wave train produced when an incident wave crosses the boundary between its original medium and a second medium; in many cases, only a portion of the wave crosses the boundary, with the remainder being reflected from the boundary.

refraction loss Reduction in amplitude or some other wave characteristic due to the refraction occurring in a nonuniform medium.

refractive index The ratio of phase velocity of a wave in free space to phase velocity of the same wave in the specific medium.

refractometer An instrument for measuring the index of refraction of a transparent substance; measurement can be accomplished in any of several ways, including measuring the critical angle, measuring refraction produced by a prism, observing interference patterns in transmitted light, and measuring the substance's dielectric constant.

refractory baffle A baffle of refractory material.

refractory lined fire-box boiler A horizontal fire tube boiler, the front portion of which sets over a refractory or water cooled refractory furnace, the rear of the boiler shell having an integral or separately connected section containing the first pass tubes through which the products of combustion leave the furnace, then returning through the second-pass upper bank of tubes.

refractory wall A wall made of refractory material.

refuse The solid portions of the products of combustion.

regain moisture content Same as dry basis moisture content.

regenerator A repeater, that is, a device which detects a weak signal in a fiber optic communication system, amplifies it, cleans it up, and retransmits it in optical form.

register 1. Accurate matching or superimposition of two or more images. 2. Alignment with respect to a reference position or set of coordinates. 3. A subassembly of the burner on a furnace or oven that directs airflow into the combustion chamber. 4. The component of a meter which counts the revolutions of a rotor or individual pulses of energy and indicates the number of counts detected. 5. In data processing, the specific location of data in memory.

registration The accurate positioning relative to a reference [RP55.1].

regulate The act of maintaining a controlled variable at or near its setpoint in the face of load disturbances.

regulating transformer A transformer used to adjust the voltage, phase relation, or both, in steps, without interrupting the load; it generally consists of one or more windings connected in series with the load circuit, and one or more windings excited from the load circuit or from a separate source.

regulation 1. Control of flow or of some other process variable. 2. The difference between maximum and minimum anode voltage drop over a range of anode current for a cold-cathode glow-discharge tube.

regulator A device for controlling pressure or flow in a process. See also controller, self-operated (regulator).

regulator, gas pressure A spring loaded, dead weighted or pressure balanced device which will maintain the gas pressure to the burner supply line.

regulatory control Maintaining the outputs of a process as close as possible to their respective setpoint values despite the influences of setpoint changes and disturbances.

reheated steam Superheated steam which derived its superheat from a reheater.

reheating The process of adding heat to steam to raise its temperature after it has done part of its intended work. This is usually done between the high pressure and low pressure sections of a compound turbine or engine.

reinforced-concrete stack A stack constructed of concrete reinforced by steel.

REJ (reject) algorithm See compressor.

relational data base In data processing, an information base that can draw data from another information base outside the original information base.

relational operator In data processing, a symbol used to determine a relationship to be tested, such as "greater than".

relative accuracy The maximum deviation from a straight line that passes through the end point of an ADC or a DAC transfer junction. Expressed as percent, ppm of the full-scale range or in LSBs.

relative address An address to which the base address must be added in order to find the machine address.

relative addressing Technique for addressing data in memory locations. A number contained in the address part of an instruction, when added to a base address, gives the actual address.

relative code A code in which all addresses are specified or written with respect to an arbitrarily selected position, or in which all addresses are represented symbolically in a computable form.

relative damping See damping, relative.

relative density bottle See specific gravity bottle.

relative flow coefficient The ratio of the flow coefficient (Cv) at a stated travel to the flow coefficient (Cv) at rated travel [S75.11].

relative gain An open-loop gain determined with all other manipulated variables constant, divided by the same gain determined with all other controlled variables constant.

relative humidity 1. The ratio of the amount of water vapor contained in the air at a given temperature and pressure to the maximum amount it could contain at the same temperature and pressure under saturated conditions [S7.3]. 2. The ratio of the weight of water vapor present in a unit volume of gas to the maximum possible weight of water vapor in unit volume of the same gas at the same temperature and pressure.

relative luminosity The ratio of measured luminosity at a particular wavelength to measured luminosity at the wavelength of maximum luminosity.

relative response The ratio of the response of a device or system under some specific condition to its response under stated reference conditions.

relative travel The ratio of the travel at a given opening to the rated travel [S75.05] [S75.11].

relative wind The velocity of airflow with respect to a specific stationary or moving body, neglecting any localized disturbances due to the presence of the body in the airstream.

relaxation Decrease of stress with time at constant strain.

relaxation oscillator A device which generates a periodic nonsinusoidal electrical signal by gradually storing electrical energy and then rapidly releasing it.

relay 1. A device whose function is to pass on information in an unchanged form or in some modified form. Relay is often used to mean computing device. The latter term is preferred. The term "relay" also is applied specifically to an electric, pneumatic, or hydraulic switch that is actuated by a signal. The term also is applied to functions performed by a relay [S5.1]. 2. An electromechanical device which is operated by a change in a relatively low power electric signal to control the flow of electric current in one or more electric circuits generally not interconnected with the relay-control circuit.

relay-operated controller A control system or device in which the signal that operates the final control element or device is produced by supplementing the energy from the primary control element with energy from another source.

relay rack See rack.

reliability 1. The probability that a device will

perform its objective adequately, for the period of time specified, under the operating conditions specified [S51.1]. 2. The probability that a component, piece of equipment or system will perform its intended function for a specified period of time, usually operating hours, without requiring corrective maintenance. 3. The closeness of agreement among a number of consecutive measurements of the output for the same value of the input under the same operating conditions, approaching from the same direction, for full range traverses.

relief Clearance around the cutting edge of a tool, provided by tapering or contouring the adjacent surfaces.

relief valve A device used to protect piping and components from overpressure.

relief valve (safety) An automatic pressure relieving device actuated by the pressure upstream of the valve and characterized by opening pop action with further increase in lift with an increase in pressure over popping pressure.

relocatable A program that can be moved about and located in any part of a system memory without affecting its execution.

relocatable coding Absolute coding containing relative addresses, which when derelativized, may be loaded into any portion of a computer's programmable memory and will execute the given action properly. The loader program normally performs the derelativization.

relocate In programming, to move a routine from one portion of storage to another and to adjust the necessary address references so that that routine, in its new location, can be executed.

relocation dictionary The part of an object or load module that identifies all relocatable address constants in the module.

reluctance Resistance of a substance to the passage of magnetic lines of force; it is the reciprocal of magnetic permeability. Also known as magnetic resistance.

reluctive Converting a change of measurand into an a-c voltage change by a change in the reluctance path between two or more coils or separated portions or one coil when a-c excitation is applied to the coil(s). Included among reluctive transducers are those employing differential-transformer, inductance-bridge, and synchro elements [S37.1].

reluctive pressure transducer A type of pressure sensor in which a moving armature attached to a pressure-sensitive element varies the reluctance of a magnetic circuit—either a permanent magnet or an electromagnet—thus producing an output current in a measuring coil.

REM See roentgen equivalent man.

remedial maintenance The maintenance performed following equipment failure, as required, on an unscheduled basis. Contrasted with preventive maintenance.

remote In data processing, a term used to refer to any devices that are not located near the main computer.

remote access Pertaining to communication with a data processing facility by one or more stations that are distant from that facility.

remote control Operating a mechanism from a point some distance away by means of electronic or electrical signals transmitted by a radio, cable or other means to servo units mounted on the mechanism.

remote logic annunciator An annunciator that locates visual displays and sequence logic circuits in separate assemblies [S18.1].

remote manipulation Using electromechanical or hydromechanical equipment to enable a person to perform manual operations while remaining some distance from the work location; usually used for handling radioactive or otherwise hazardous materials.

remote processing unit (RPU) Field station with input/output circuitry and the main processor. These devices measure analog and discrete inputs, convert these inputs to engineering units, perform analog and logical calculations (including control calculations) on these inputs and provide both analog and discrete (digital) outputs.

remote sensing Detecting, measuring, indicating or recording information without actual contact between an instrument and the point of observation—for example, as in optical pyrometry.

REP See roentgen equivalent physical.

repair 1. The act of restoring an item to serviceable condition following a failure or malfunction. 2. Narrowly, restoring an item to serviceable condition, but not exactly to original design specifications.

repeatability 1. The ability of a transducer to reproduce output readings when the same measurand value is applied to it consecutively under the same conditions, and in the same direction.

Repeatability is expressed as the maximum difference between output readings; it is expressed as "within ± _____ percent of full scale output." Two calibration cycles are used to determine repeatability unless otherwise specified [S37.1]. 2. The closeness of agreement among a number of consecutive measurements of the output for the same value of the input under the same operating conditions, approaching from the same direction, for full-range traverses [S67.04]. It is usually measured as a non-repeatability and expressed as repeatability in percent of span. It does not include hysteresis [S51.1]. 3. In data processing, normally considered to be a measurement of performance made over a very short period of time; a period short enough so that gain and offset instabilities of system components are of no significance. In terms of the distribution of readings, it is a measure of the maximum deviation of the readings from the mean value of the distribution [RP55.1].

repeater A device that amplifies or regenerates data signals in order to extend the distance between data stations. Also called a regenerator.

replicated optics Optical components formed by transferring a master pattern to a roughly machined substrate, using an epoxy layer to form a final optical surface. The epoxy layer is then coated with a reflective layer to form the final component. The process allows mass production of complex surfaces much less expensively than conventional polishing techniques.

report generator A computer program that gives a less experienced user the ability to create reports from various files.

reproducibility 1. In process instrumentation, the closeness of agreement among repeated measurements of the output for the same value of input made under the same operating conditions over a period of time, approaching from both directions. NOTE A: It is usually measured as a nonreproducibility and expressed as reproducibility in percent of span for a specified time period. Normally, this implies a long period of time, but under certain conditions the period may be a short time during which drift may not be included. B: Reproducibility includes hysteresis, dead band, drift and repeatability. C: Between repeated measurements the input may vary over the range and operating conditions may vary within normal operating conditions [S51.1]. 2. The ability of an instrument to dupli-

cate, with exactness, measurements of a given value. Usually expressed as a percent of span of the instrument. See also repeatability.

rerun In data processing, executing a program again.

resealing pressure The inlet pressure at which fluid no longer leaks past a relief valve after it is closed.

reserved variable Any variable available only to specific programs in the system. Contrast with global variable.

reserved words Certain words in a language which may only appear in the context reserved for them.

reservoir A holding tank, cistern or pond for storing reserves of potable or make-up water.

reset 1. The sequence action that returns the sequence to the normal state [S18.1]. 2. To restore a storage device to a prescribed initial state, not necessarily that denoting zero. 3. To place a binary cell into the state denoting zero. 4. See integral control action (reset).

reset action 1. A control action which produces a corrective signal proportional to the length of time the controlled variable has been away from the set point. 2. Takes care of load changes. 3. Another name for the integral control mode.

reset, automatic Reset occurs after acknowledge when the process condition returns to normal [S18.1].

reset control See integral control.

reset control action See control action, integral (reset).

reset cycle To return a cycle index to its initial value.

reset, first out Reset of the first out indication occurs when the acknowledge or first out reset pushbutton is operated, whether the process condition has returned to normal or not, depending on the sequence [S18.1].

reset, manual Reset occurs after acknowledge when the process condition has returned to normal and the reset pushbutton is operated [S18.1].

reset rate See integral action rate.

reset windup Saturation of the integral mode of a controller developing during times when control cannot be achieved; this condition often causes the controlled variable to overshoot its setpoint when the obstacle to control is removed.

resident In data processing, a program that is permanently stored in the memory of the computer.

residual error The error remaining after attempts at correction.

residual fuels Products remaining from crude petroleum by removal of some of the water and an appreciable percentage of the more volatile hydrocarbons.

residue check 1. Any modulo n check. 2. A check of numerical data or arithmetic operations in which the number A is divided by n and the remainder B accompanies A as a check digit.

resistance 1. Impediment to gas flow, such as pressure drop or draft loss through a dust collector. Usually measured in inches water gage (w.g.). 2. The opposition to the flow of electricity in an electric circuit measured in ohms.

resistance drop The voltage drop in phase with the current.

resistance, input The resistance appearing at the input terminal of a device [RP55.1].

resistance magnetometer A device for measuring magnetic field strength by means of a change in electrical resistance of a material immersed in the magnetic field.

resistance meter Any instrument for measuring electrical resistance. Also called an ohmmeter or megger.

resistance strain gage A fine wire or similar device whose electrical resistance changes in direct proportion to the amount of elastic strain it is subjected to.

resistance temperature detector 1. A component of a resistance thermometer consisting of a material whose electrical resistance is a known function of temperature. 2. A temperature transducer that provides temperature information as the change in resistance of a metal wire element, often platinum, as a function of temperature.

resistance thermometer A temperature measuring device in which the sensing element is a resistor of a known variation in electrical resistance with temperature. See RTD.

resistance-thermometer element The temperature-sensitive unit in a resistance thermometer which consists of the resistance temperature detector, its supporting structure, and terminations or other means of attaching it to the detection circuit of the instrument.

resistance welding A group of welding processes in which heat is obtained from resistance of the work to electrical current in a circuit of which the work is one part. Spot welding is an example of resistance welding.

resistive Converting the change of measurand into a change of resistance [S37.1].

resistive flowmeter A device for measuring liquid flow rates in which an electrical output signal proportional to flow rate is determined from the rise and fall of a conductive differential-pressure manometer fluid in contact with a resistance-rod assembly.

resistivity Electrical resistance per unit length and unit cross section.

resistor An electrically conductive material shaped and constructed so that it offers a known resistance to the flow of electricity.

resistor-capacitor-transistor logic (RCTL) A type of computer circuit used to perform the "not or" logic function at speeds higher than can be achieved with RTL circuits. Similar to RTL except that capacitors reduce switching time.

resistor-protected shunt diode barrier A network identical to a fuse-protected shunt diode barrier, except that the fuse is replaced by a resistor [RP12.6].

resistor-transistor logic (RTL) A form of logic circuit using resistors and transistors that performs not or nor logic.

resolution 1. The least interval between two adjacent discrete details which can be distinguished one from the other [S51.1] [S75.05]. 2. The magnitude of output step changes as the measurand is continuously varied over the range. NOTE A: This term relates primarily to potentiometric transducers. B: Resolution is best specified as average and maximum resolution; it is usually expressed in percent of full scale output. C: In the sense of the smallest detectable change in measurand use threshold [S37.1]. 3. A measure of the smallest input change which can be detected (not necessarily measured) at the output of the system. 4. Ability of a telescope or microscope to separate detailed features of the subject in the resulting image. 5. The minimum distance by which two objects or features must be separated to be observed as two images in the output of a telescope, microscope, radar set, television camera or similar imaging device. 6. The error associated with the ability to resolve a flowmeter output signal to the smallest measurable unit. For example, only + or - 1 pulse is measurable in any pulse output device. 7. The maximum capability of a system used to convert an analog signal to a proportional digital value; generally, resolution is expressed in bits, from which the actual resolution may be determined (bits-per-word). 8. The minimum detectable change of some variable in a measurement system.

resolution, average The reciprocal of the total number of output steps over the range, multiplied by 100 and expressed in percent voltage-ratio (for a potentiometric transducer) or in percent of full-scale output [S37.1].

resolution, maximum The magnitude of the largest of all output steps over the range, expressed as percent voltage-ratio (for a potentiometric transducer) or in percent of full-scale output [S37.1].

resolution sensitivity The smallest change in an input that produces a discernible response.

resolver Any means for determining the mutually perpendicular components of a vector quantity.

resolving power 1.A measure of the ability to respond to small changes in input. 2.The ability of an optical device to separate the images of two objects very close together. 3.The ability of a monochromator to separate two lines in a multiline spectrum.

resolving time The minimum separation time between events that will enable a counting device to detect and respond to both events.

resonance 1. Of a system or element, a condition evidenced by large oscillatory amplitude, which results when a small amplitude of periodic input has a frequency approaching one of the natural frequencies of the driven system [S51.1]. 2. A condition existing between an externally excited system and the external sinusoidal excitation when any small increase or decrease in the frequency of the excitation signal causes the peak-to-peak amplitude of a specified response to decrease.

resonance bridge An electrical network used in measuring inductance, capacitance or frequency; it normally consists of four arms—one containing both inductance and capacitance and the other three containing only nonreactive resistances—and an adjustment device which balances the network by establishing resonance.

resonances Amplified vibrations of transducer components, within narrow frequency bands, observable in the output, as vibration is applied along specified transducer axes [S37.1].

resonant frequency The wave frequency at which mechanical or electronic resonance is achieved. See also frequency, resonant.

resonant frequency amplification factor The ratio of the maximum sensitivity of a transducer at its lowest resonant frequency to its nominal sensitivity [S37.10].

resonator Generally a pair of mirrors located at either end of a laser medium, which cause light to bounce back and forth between them while passing through the laser medium.

resource Any facility of the computing system or operating system required by a job or task, and including main storage, input/output devices, the central processing unit, data sets, and control processing programs.

resource manager A general term for any control program function responsible for the allocation of a resource.

response 1. The change in output of a device in relation to a change of input [S26]. 2. Defined output for a given input under explicitly stated conditions.

response-critical That aspect of controlling a process which implies the need to react to random disturbances in time to prevent impairment of yield, or dangerous conditions. Real time is often used synonymously.

response, dynamic The behavior of the output of a device as a function of the input, both with respect to time [S51.1].

response, ramp The total (transient plus steady-state) time response resulting from a sudden increase in the rate of change from zero to some finite value of the input stimulus [S51.1].

response, step The total (transient plus steady-state) time response resulting from a sudden change from one constant level of input to another [S51.1].

response time 1. The length of time required for the output of a transducer to rise to a specified percentage of its final value as a result of a step change of measurand. NOTE A: To indicate this percentage it can be worded so as to precede the main term, e.g., "98%-response time: — milliseconds, max." B: Also see time constant and rise time [S37.1]. 2. The time period between the process condition becoming abnormal and initiation of the alarm state. The minimum momentary alarm duration required for annunciator operation [S18.1]. 3. The time required for the absolute value of the difference between the output and its final value to become and remain less than a specified amount, following the application of a step input or disturbance. 4. The time between the initiation of an operation from a computer terminal and the receipt of results at the terminal. Response time includes transmission of data to the computer, processing, file, access and transmission of results to the terminal. 5. The time required for the output to first reach a definite value after the application of a step input or disturbance. 6. The time it takes for a controlled variable to react to a change in input.

response, time An output expressed as a function of time, resulting from the application of a specified input under specified operating conditions [S51.1].

response time characteristics Those properties (e.g., transfer function, time constant, delay time, power spectral density) of the equipment from which its response time can be determined [S67.10].

response time, fluid transport The response time associated with fluid transport from the location at which a property is to be measured to the sensor location. This delay may include contributions from both the transport time associated with fluid velocity and mixing times determined by mass flow rate and system configuration [S67.10].

responsiveness The ability of an instrument or control device to follow wide or rapid changes in the value of a dynamic measured variable.

restart In electronic computing, the process of recommencing a computing function from a known point in a program following computer system failure or other unusual event during execution of a task.

restart address The address at which a program can be restarted; normally, the address of the code required to initialize variables, counters, and the like.

restoring torque gradient The rate of change, with respect to deflection, of the resultant of electric and mechanical torques that tend to restore an instrument's moving element to any position of equilibrium.

retarded elastic-chamber gage A pressure gage whose sensitive element is an elastic chamber that moves freely only through the lower portion of its indicating range.

retarder A straight or helical strip inserted in a

fire tube primarily to increase the turbulence.

retarding magnet A magnet used in a motor-type meter to limit rotor speed to a value proportional to the quantity being measured.

reticle A glass window on which is etched or printed a pattern, typically for use in measurement or alignment. The simplest type of reticle is the crosshairs of an alignment telescope.

retreat idler The first idler reached after the material on the belt leaves the weigh carriage. Also called departure idler [RP74.01].

retrieve In data processing, to search for and extract data that is contained in a computer file.

retrofit 1. A word derived from 'retroactive retrofit', which describes modification of a piece of equipment to incorporate design changes made in later models of the same equipment. 2. Modification and upgrading of older control systems. 3. Parts, assembly, or kit that will replace similar components originally installed on equipment. 4. Retrofits are generally performed on a machine to correct a deficiency or to improve performance.

retroreflector An optical device which reflects an incident beam of light back to the source. The corner-cube prism is an example.

return alert See ringback.

return flow oil burner A mechanical atomizing oil burner in which part of the oil supplied to the atomizer is withdrawn and returned to storage or to the oil line supplying the atomizer.

return key In data processing, a frequently used key on a keyboard that activates a variety of instructions.

return signal See signal, return.

reverberation Persistence of sound at a particular location after sound waves from the source are no longer being received.

reverberation time The time in seconds it takes for average sound-energy density to decrease to one-millionth of its original steady-state value after sound from the source has stopped.

reverberation time meter An instrument for determining reverberation time of an acoustic enclosure.

reverse-acting controller A controller in which the value of the output signal decreases as the value of the input (measured variable or controlled variable) increases. See controller, reverse acting.

reverse drawing Drawing, especially a deep-drawn part, a second time in a direction opposite to the original draw direction.

reverse polarity 1. An electrical circuit in which the positive and negative electrodes have been interchanged. 2. An arc welding circuit in which the electrode is electrically positive and the workpiece electrically negative.

reverse video 1. The interchange of foreground and background attributes, such as intensity, color, etc. [S5.5]. 2. A CRT screen display of dark characters on a light background—the opposite of the usual CRT screen display.

reversible seat Refers to the seat ring with seating surfaces on both sides such that when one surface has worn, the ring may be reversed to present a new surface to contact the closure member [S75.05].

reversible transducer A transducer in which the transducer loss is independent of the direction of energy transmission through the transducer.

reversing switch An electrical switch whose function is to reverse connections, on demand, of one part of the circuit.

revolutions per minute (rpm) A standard unit of measure for rotational speed.

rework Restoring an item to a condition exactly conforming to original design specifications; usually applied to corrective action taken when an item has failed an inspection but requires a relatively simple operation such as replacing a part to enable the item to pass an identical inspection.

Reynolds number A dimensionless criterion of the nature of flow in pipes. It is proportional to the ratio of dynamic forces to viscous forces: the product of diameter, velocity and density, divided by absolute viscosity.

RF See radio frequency.

RFI See radio frequency interference.

RFI protector A device that protects a computer from strong radio or television transmissions.

rheopectic substance A fluid whose apparent viscosity increases with time at any constant shear rate.

rheostat An adjustable variable resistor.

rifled tube A tube which is helically grooved on the inner wall.

rigidity Resistance of a body to instantaneous change of shape.

ring A sequential network topology where each node is connected to exactly two nodes, and serves as a repeater when it is not sourcing data

onto the network.

ring back (return alert) A sequence feature that provides a distinct visual or audible indication or both when the process condition returns to normal [S18.1].

Ringelmann chart A series of four rectangular grids of black lines of varying widths printed on a white background, and used as a criterion of blackness for determining smoke density.

ringing An oscillating transient in an output signal that occurs following a sudden rise or fall in the input signal.

ringing period 1. The period of time during which the amplitude of measurand step-function-excited oscillations exceed 10% of the step amplitude [S37.6]. 2. The period of time during which the amplitude of output oscillations, excited by a step change in measurand, exceed the steady-state output value. NOTE: Unless otherwise specified, the ringing period is considered terminated when the output oscillations no longer exceed ten percent of the subsequent steady-state output value [S37.1].

ringing time In ultrasonic testing, the length of time that a piezoelectric crystal continues to vibrate after the ultrasonic pulse has been generated.

RIP See real-time interrupt process.

ripple A small alternating-current signal superimposed on a larger direct-current signal; it usually results from imperfect filtering of the d.c. output from the rectifier.

ripple content The ripple content is the ratio between the peak to peak value of the ac part and the range of the dc current signal [S50.1].

riser See feedhead.

rise time 1. The length of time for the output of a transducer to rise from a small specified percentage of its final value to a large specified percentage of its final value as a result of a step-change of measurand. NOTE A: Unless otherwise specified, these percentages are assumed to be 10 and 90 percent of the final value. B: Also see time constant [S37.1]. 2. The time required for the leading edge of a pulse to rise from one tenth of its final value to nine-tenths of its final value. Rise time is proportional to the principal time constant of the circuit. 3. The time required for the output voltage of a digital circuit to change from a logical low level (0) to a logical high level (1). See also time, rise.

rms value See also root mean square value and

value, rms.

RNRZ See randomized non-return-to-zero.

robot An intelligent multi-purpose device, usually programmable, which carries out pick and place, assembly or other manipulative operations.

robotics The area of artificial computer intelligence as applied to the use of industrial robots.

rod-out The act of pushing a specially designed rod through a valve or opening to loosen deposits.

Roentgen A quantity of x-ray or gamma-ray radiation that produces, in air, ions carrying one electrostatic unit of electrical charge of either sign per 0.001293 gram of air.

roentgen equivalent man (REM) The unit of dose in radiation dosimetry; it equals the amount of radiation of any type that produces the same amount of biological damage in human beings as a dose of 1 roentgen of 200-kV x-rays.

roentgen equivalent physical (REP) A unit of radiation equal to the amount of radiation of any type that results in energy absorption of 93 ergs/g in soft tissue.

Roentgen rays An alternative term for x-rays.

rolled joint A joint made by expanding a tube into a hole by a roller expander.

roll-over error For an analog-to-digital converter with bipolar input range, the output difference for inputs of equal magnitude but opposite polarity. Specified in counts or LSBs.

ROM See read-only memory.

room conditions Ambient environmental conditions, under which transducers must commonly operate, which have been established as follows: (a) temperature: $25 + 10$ °C ($77 + 18$ °F). (b) relative humidity: 90 percent or less. (c) barometric pressure: 26 to 32 inches Hg. NOTE: Tolerances closer than shown above are frequently specified for transducer calibration and test environments [S37.1].

root 1. The bottom of a screw thread. 2. The region of closest approach between two members of a weld joint. 3. The points at which the fusion zone of a weld intersect the base metal.

root directory In MS-DOS, the primary directory of a floppy or hard disk that contains subdirectories.

root mean square value (RMS) The square root of the mean of the squares of sample values.

root valve The first valve located in a sample line after it taps off the process [S67.02]. It is typically located in close proximity to the sample

tap [S67.10].

rosette strain gage See strain rosette.

rosette-type strain gage A type of resistance strain gage having three individual gage elements arranged to measure strain in three different directions simultaneously; typical arrangement has two elements oriented 90° to each other and the third at 45° to the first two.

rotameter A variable-area, constant-head, indicating type rate-of-flow volume meter in which fluid flows upward through a tapered tube, lifting a shaped plummet to a position where upward fluid force just balances the weight of the plummet.

rotary actuator A device that converts electric energy into controlled rotary force; it usually consists of an electric motor, a gear box, a control relay and one or more limit switches.

rotary oil burner A burner in which atomization is accomplished by feeding oil to the inside of a rapidly rotating cup.

rotary valve A valve for the admission or exhaust of working fluid, where the valve is a ported piston or disk that turns on its axis.

rotating-cup viscometer A laboratory device for measuring viscosity in terms of the drag torque on a stationary element, such as a paddle or cylinder, immersed in a liquid contained in a cup that rotates at constant speed.

rotating meter See velocity-type flowmeter.

rotational transition A change in the rotational state of a molecule. Rotational transitions involve less energy than either electronic or vibrational transitions, and typically correspond to wavelengths in the far infrared, longer than about 20 micrometers.

rotational viscometer A device for measuring the apparent viscosity of non-Newtonian fluids by determining the torque required to rotate a spindle in a container filled with the substance; in some instruments, the container may rotate while the spindle does not.

rotor 1. The rotating member of a turbine, electric motor, compressor, pump or similar machine. 2. Any rotating assembly of vanes or airfoils.

rotor-type vacuum gage A device for measuring low pressures, down to 10-7 torr, by sensing the deceleration of a rotor (usually a steel ball) levitated in a rotating magnetic field, the rotor being exposed directly to the evacuated space.

round-chart instrument A recording instrument whose output trace is written on a circular paper chart.

rounded orifice   An orifice whose inlet side is rounded rather than sharp edged.

rounding error   The error resulting from rounding off a quantity by deleting the less significant digits and applying some rule of correction to the part retained; e.g., 0.2751 can be rounded to 0.275 with a rounding error of .0001. Synonymous with round-off error and contrasted with truncation error.

round-off   Synonymous with round. See also rounding error and half-adjust.

round-off error   Same as rounding error.

router   A network device that interconnects two computer networks that have the same network architecture. A router requires OSI Level 1, 2 and 3 protocols. See bridge and gateway.

routine   1. In data processing, a set of instructions arranged in proper sequence to cause a computer to perform a desired task [RP55.1]. 2. A subdivision of a program consisting of two or more instructions that are functionally related; therefore, a program. Clarified by subroutine and related to program.

routine, executive   A routine that controls the execution of other routines. Synonymous with supervisory routine [RP55.1].

rpm   See revolutions per minute.

RS232   A logic level and connector specification for serial ASCII data transmission; sometimes called the "EIA interface".

RS-232C   1. An EIA standard, originally introduced by the Bell System, for the transmission of data over a cable less than 50 feet in length; it defines pin assignments, signal levels, etc. for receiving and transmitting devices. 2. A communications interface between a modem and other computer devices that complies with EIA Standard RS-232C.

RS422   Standard for serial data transmission.

RS511   A messaging standard under development in EIA for communication between factory floor devices. It uses ASN.1 for data encoding. RS511 is being considered for inclusion in MAP/EPA and MINI-MAP.

RSX-11   Real-time resource sharing executive. A real-time, multiprogramming operating system that controls the sharing of system resources among any number of user-prepared tasks.

RT-11   A single-user foreground/background real-time disk-operating system; real-time does not mean that it is capable of accepting real-time data directly, but rather is an indication that it can support a real-time system such as TELE-VENT.

RTD   See resistance temperature detector.

RTL   See real-time language; also see resistor-transistor logic.

ruled diffraction gratings   Diffraction gratings in which the lines are mechanically ruled by a precision ruling engine. These gratings are normally replicated and the replicas are sold for most applications.

run   1. In data processing, to start a program on the computer. 2. The TELEVENT executive command to initiate a background program.

runback   An action by the boiler control system initiated by the loss of any auxiliary equipment that limits the capabilities of the unit to sustain the existing load. Upon runback initiation, the boiler demand signal is reduced at a preset rate to the capability of the remaining auxiliaries [S77.42].

rundown   An action by the boiler control system initiated by an unsafe operating condition—i.e., fuel/air limit (cross-limiting), temperature limits, etc. Upon rundown initiation, the boiler demand signal is reduced in a controlled manner to the load point where the unsafe operating condition is eliminated [S77.42].

rung   Group of program elements in a ladder diagram. The group controls a single output element (coil or function).

running fit   Any of a class of clearance fits that allow assembled parts to run freely, especially shafts within their bearings or pinned joints in linkages. See also sliding fit.

run time   The length of time between the beginning and the end of a program execution.

run-time error   In data processing, an error that occurs during a program operation that may or may not cause the program to stop.

rupture disc   A diaphragm designed to burst at a predetermined pressure differential.

rupture disc device   A nonreclosing pressure relief device that relieves excessive static inlet pressure via a rupture disc.

rupture pressure   See pressure, rupture.

Rutherford   1. 10^{-6} radioactive disintegration per second. 2. A quantity of a nuclide having an activity equal to one Rutherford.

S

s   See second; has largely replaced sec as the preferred abbreviation.

S   See siemens.

S100 bus   A hobbyist and small business user standard board and bus system which has become a de facto standard for microcomputers. The S100 bus uses 100 pins.

Sabin   A unit of measure for sound absorption equivalent to one square foot of a perfectly absorptive surface.

saddle   A casting, fabricated chair, or member used for the purpose of support.

safe area   1. Non-hazardous area. 2. An area in which explosive gas/air mixture are not expected to be present so that special precautions for the construction and use of electrical apparatus are not required. See non-hazardous area.

safety can   A metal can with a special closure used for storing, handling and transporting flammable liquids.

safety control   See control, safety.

safety ground   1. A connection between metal structures, cabinets, cases, etc. which is required to prevent electrical shock hazard to personnel. 2. Safety ground is not a signal reference point.

safety hoist   A hoisting device that stops automatically when tension is released.

safety hook   A lifting hook with a spring-loaded latch that prevents the lifting sling from accidentally slipping off.

safety limit   A limit on an important process variable that is necessary to reasonably protect the integrity of physical barriers that guard against uncontrolled release of radioactivity [S67.04].

safety plug   A nonreclosable pressure-relief device containing a fusible element that melts at a predetermined temperature.

safety relief valve   An automatic pressure relieving device actuated by the pressure upstream of the valve and characterized by opening pop action with further increase in lift with an increase in pressure over popping pressure.

safety rod   See control rod.

safety shut down   The action of shutting off all fuel and ignition energy to the burner by means of safety control or controls such that restart cannot be accomplished without operator action.

safety stop   1. A device on a hoisting apparatus to prevent the load from falling. 2. A device on a hoisting engine that automatically prevents it from overwinding. 3. A device that prevents mechanical over-travel on a piece of equipment.

safety switch   A switch employed for the purpose of providing protection according to this standard. Such a switch may be known as an interlock [S82.01].

safety valve   A spring loaded valve that automatically opens when pressure attains the valve setting. Used to prevent excessive pressure from building up in a boiler.

safe working pressure   See design pressure.

salinometer   An instrument for measuring water salinity; may utilize electrical conductivity measurement or a hydrometer calibrated to read percent salt content directly.

SAMA   See Scientific Apparatus Makers Association.

sample and hold   A device which senses and stores the value of an analog signal units subsequently updated [RP55.1].

sample-and-hold per channel   A method of sampling and holding analog data channels in advance of a normal multiplex sample-and-hold A/D encoding process.

sampled-data control   That branch of automatic control theory concerned with the control of variables whose current values are not continuously available for comparison with the setpoint but instead are sampled only at given intervals.

sample draw 1. Refers to a method used to cause deliberate flow of the atmosphere being monitored to a gas-sensing element [S12.15]. 2. A method to cause flow of the atmosphere being monitored to be directed to a gas-sensing element [S12.13].

sampled signal A signal which is updated only at given intervals by a new observation of the variable.

sample/hold amplifier An amplifier which samples the input signal and holds the value for sufficient time before it is input to an analog to digital (A/D) converter.

sample interval The time interval between measurements or observations of a variable.

sample line A piping and/or tubing system which removes fluid from a process either continuously or periodically for the purpose of determining the constituents or the physical properties of the process fluid. The sample line begins at the process tap or nozzle used for sampling, and terminates where the flow of sample fluid ends as a discrete and controlled entity [S67.10].

sample plan The plan designed by a telemetry engineer to sample and encode data incrementally so that it may be accurately decoded and re-created.

sample sink An installed device with controlled drainage and/or ventilation at which a grab sample may be obtained [S67.10].

sample tap The point where the sample line taps into the process line (pipe, duct, container) and the point where sample flow begins. It may also be referred to as "sample connections," "sample nozzle," or "process tap" [S67.10].

sample vessel An integrally valved, portable sample container designed to obtain pressurized samples at process pressure [S67.10].

sampling 1. Obtaining the values of a function for discrete, regularly or irregularly spaced values of the independent variable. 2. Selecting only part of a production lot or population for inspection, measurement or testing. 3. The removal of a portion of a material for examination or analysis. 4. In statistics, obtaining a sample from a population.

sampling action A type of controller action in which the value of the controlled variable is measured at intermittent intervals rather than continuously.

sampling controller A controller using intermittently observed values of a signal such as the setpoint signal, the actuating error signal, or the signal representing the controlled variable, to effect control action. See controller, sampling.

sampling period The time interval between observations in a periodic sampling control system [S51.1].

sampling rate For a given measurement, the number of times that it is sampled per second in a time-division-multiplexed system. Typically, it is at least five times the highest data frequency of the measurement.

sampling theorem Nyquist's result that equispaced data, with two or more points per cycle of highest frequency, allow reconstruction of band-limited functions; the theorem states: "If the rms spectrum/G(t), is identically zero at all frequencies above W cycles-per-second, then g(t) is uniquely determined by giving its ordinates at a series of points spaced 1/2W seconds apart, the series extending through the time domain."

sandblasting Grit blasting, especially when the abrasive is ordinary sand. See grit blasting.

sanding Smoothing a surface with abrasive cloth or paper, usually implies use of paper covered with adhesive-bonded flint or quartz fragments.

sandwich braze A joining technique for reducing thermal stress in a brazed joint, in which a shim is placed between the opposing surfaces to act as a transition layer.

sandwich construction A technique of producing composite materials that consists of gluing hard outer sheets onto a center layer, usually a foamed or honeycomb material.

sanitary engineering A field of civil engineering that deals with construction and operation of facilities that protect public health.

sans-serif In typesetting, a type style with no straight or curved decorative additions. See serif.

SAP See service access point.

satin finish A type of metal finish produced by scratch brushing a polished metal surface to produce a soft sheen.

saturable-core magnetometer A magnetometer whose output is derived from changes in magnetic permeability of a ferromagnetic core as a function of the strength of the magnetic field being measured.

saturable-core reactor A device for introducing inductive reactance into a circuit, where the effective reactance can be varied by varying auxiliary direct-current excitation of a ferromagnetic core.

saturated air Air which contains the maximum amount of water vapor that it can hold at its temperature and pressure.

saturated steam Steam at the temperature corresponding to its pressure.

saturated temperature The temperature at which evaporation occurs at a particular pressure.

saturated water Water at its boiling point.

saturation 1. A device characteristic exhibited when a further change in the input signal produces no significant additional change in the output [S67.04]. 2. A characteristic curve exhibits saturation when further change of the input variable beyond a certain value results in a negligible additional change of the output variable.

saturation current 1. In ionic conduction, the current obtained when the applied voltage is sufficient to collect all of the ions present. 2. In an electromagnet, the excitation current required to produce magnetic saturation.

saturation voltage The minimum applied voltage that produces saturation current.

saturator A device, equipment or person that saturates one material with another.

sawtooth wave A type of cyclic direct-current waveform in which voltage and current rise gradually from zero to a peak value, then drop instantaneously to zero and repeat the cycle.

Saybolt color scale A standardized color scale used primarily in the petroleum and pharmaceutical industries to grade the yellowness of pale products; it is based on matching the color of a column of the sample liquid with one of a set of color-controlled glass disks, as described in ASTM standard D156.

Saybolt Furol viscosimeter An instrument similar to a Saybolt Universal viscosimeter, but with a larger diameter tube for measuring the viscosity of very thick oils.

Saybolt Universal viscosimeter An instrument for determining viscosity by measuring the time it takes an oil or other fluid to flow through a calibrated tube.

s-band In telemetry, the portion of the radio frequency spectrum between 2200 and 2300 MHz.

scab A surface defect on a casting or rolled metal product consisting of a thin, flat piece of metal partly detached from the substrate.

SCADA See supervisory control and data acquisition.

scaffold A movable or temporary platform that allows workers to perform tasks at considerable heights above the ground; it may be either supported from ground level on a framework or suspended from above on ropes or cables.

scalar quantity Any quantity that can be described by magnitude alone, as opposed to a vector quantity which can only be described by both magnitude and direction.

scale 1. A graduated series of markings, usually used in conjunction with a pointer to indicate a measured value. 2. A graduated measuring stick, such as a ruler. 3. A device for weighing objects. 4. A thick metallic oxide, usually formed by heating metals in air. 5. A hard coating or layer of materials on surfaces of boiler pressure parts.

scale factor 1. The factor by which the number of scale divisions indicated or recorded by an instrument should be manipulated to compute the value of the measured variable. NOTE: Deflection factor is a more general term than scale factor in that the instrument response may be expressed alternatively in units other than scale divisions [S51.1]. 2. The coefficients used to multiply or divide quantities in a problem in order to convert them so as to have them lie in a given range of magnitude, e.g., plus one to minus one. 3. A constant multiplier which converts an instrument reading in scale divisions to a measured value in standard units. 4. In analog computing, a proportionality factor that relates the value of a specified variable to the circuit characteristic which represents it in the computer. 5. In digital computing, an arbitrary factor applied to some of the numerical quantities in the computer to adjust the position of the radix point so that the significant digits occupy specific positions.

scale length The distance that the pointer of an indicating instrument, or the marking device of a recording instrument, travels in moving from one end of the instrument scale to the other, measured along the baseline of the scale divisions.

scale-of-ten circuit A decade scaler.

scale-of-two circuit A binary scaler.

327

scaler A measuring-circuit or control-circuit component that produces one output pulse each time a specific number of input pulses have been received.

scale span The algebraic difference, measured in scale units, between the highest value that can be read from the scale and the lowest value.

scale units The units of measure stated on an instrument scale.

scale-up Using data from an experimental model or pilot plant to design a larger (scaled-up) facility or device, usually of commercial size.

scaling 1. A misnomer for descaling. 2. Forming a thick layer of oxide on a metal, especially at high temperatures. 3. Depositing solid, adherent inorganic layers on the internal surface of a boiler tube, process pipe or vessel, usually from a very dilute water solution.

scaling circuit An electronic circuit that produces an output pulse whenever a predetermined number of input pulses has been received.

scaling factor A factor used in heat-exchange calculations to allow for reduced thermal conductivity across a tube or pipe wall due to scaling.

scalp To remove the surface layer of a billet, slab or ingot, thereby removing surface defects that might persist through later operations.

scan 1. To sample, in a predetermined manner, each of a number of variables intermittently. The function of a scanning device is often to ascertain the state or value of a variable. The device may be associated with other functions such as recording and alarming [S5.1]. 2. Collection of data from process sensors by a computer for use in calculations, usually obtained through a multiplexer. 3. Sequential interrogation of devices or lists of information under program control. 4. A single sweep of PC applications program operation. The scan operates the program logic based on I/O status, and then updates outputs and input status. The time required for this is called the scan time. 5. To examine an area, volume or portion of the electromagnetic spectrum, point by point, in an ordered manner.

scanner 1. An instrument which automatically samples or interrogates the state of various processes, files, conditions, or physical states and initiates action in accordance with the information obtained. 2. Any device that examines a region or quantity point by point in a continuous systematic manner. 3. A device that repeatedly and systematically samples or measures several quantities in a predetermined sequence.

scanning rate (or speed) The speed at which a computer can select and convert an analog input variable.

scan rate 1. A single sweep of PC applications program operation. The scan operates the program logic based on I/O status, and then updates output and inputs status. The time required for this is called scan time. 2. Sample rate, in a predetermined manner, each of a number of variables intermittently.

scattering A collision or other interaction that causes a moving particle or photon of electromagnetic energy to change direction.

scattering loss A reduction in the intensity of transmitted radiation due to internal scattering in the transmission medium or to roughness of a reflecting or transmitting surface.

scavenger 1. A reactive metal added to molten metal to combine with and remove dissolved gases or other impurities. A chemical added to boiler water to remove oxygen.

scavenging 1. Removing spent gases from the cylinder of an internal combustion engine and replacing them with a fresh charge or with air. 2. Removing dissolved gases or other impurities from molten metal by reaction with an additive.

schedule 160 A term used to define the wall thickness of pipe (schedule 40, 80, 160 and others).

scheduled maintenance 1. Maintenance carried out in accordance with an established plan. 2. Related to preventive maintenance. 3. Scheduled maintenance may be scheduled on hours uses, sequences or calendar.

Schering bridge A type of a-c bridge circuit particularly useful for measuring the combined capacitive and resistive qualities of insulating materials and high-quality capacitors.

Schmitt trigger A bi-stable pulse generator in which an output pulse or constant amplitude exists only as the input voltage exceeds a certain DC value; the circuit can convert a slowly changing input waveform to an output waveform with sharp transitions.

Scientific Apparatus Makers Association (SAMA) An industrial association in the United States.

scintillation A flash of light produced as a result of an ionizing event in a phosphor.

scintillation counter A radiation measuring

instrument consisting of a phosphor, a photomultiplier tube, and associated electronic circuitry; the amount of radiation is measured by counting the number of scintillations it produces in the phosphor per unit time interval.

scintillation spectrometer A scintillation system so designed that it can separate and determine the energy distribution in heterogeneous radiation.

scissor jack A lifting jack whose operating mechanism consists of parallelogram linkages driven by a horizontal screw.

sclerometer An instrument that determines hardness of a material by measuring the force needed to scratch or indent the surface with a diamond point.

scleroscope An instrument that determines hardness of a material by measuring the height to which a standard steel ball rebounds when dropped from a standard height.

scoring Deep scratches on the surface of a metal.

scotch boiler A cylindrical steel shell with one or more cylindrical internal steel furnaces located (generally) in the lower portion and with a bank or banks (passes) of tubes attached to both end closures.

Scotch yoke A type of four-bar linkage used to convert uniform rotation into simple harmonic motion. Also converts linear motion to rotary motion.

scouring 1. Physical or chemical attack on internal surfaces of process equipment. 2. Mechanical finishing or cleaning using a mild abrasive and low pressure.

SCR See silicon-controlled rectifier.

scrap 1. Solid material or inspection rejects suitable for recycling as feedstock in a primary operation such as plastic molding, alloy production or glass remelting. 2. Narrowly, any unusable reject at final inspection.

scraper ring A piston ring that scrapes oil from the cylinder wall to prevent it from being burnt.

scratch hardness 1. A measure of the resistance of minerals or metals to scratching; for minerals it is defined by comparison with 10 selected minerals comprising the Mohs scale. 2. A method of measuring metal hardness in which a cutting point is drawn across a metal surface under a specified pressure, and hardness is determined by the width of the resulting scratch.

scratch pad An intermediate work file which stores the location of an interrupted program, and retrieves the program when the interruption is complete.

scratchpad memory A high-speed, limited-capacity computer information store that interfaces directly with the central processor unit; it is used to supply the central processor with data for immediate computation, thus avoiding delays that would be encountered by interfacing with main memory. (The function of the scratchpad memory is analogous to that of a pad of paper used for jotting down notes.)

scratch register Addresses of scratch pad storage locations which can be referenced by the use of only one character.

screen In data processing, the plane surface of a CRT that is visible to the user.

screen analysis A method of finding the particle size distribution of any loose, flowing aggregate by sifting it through a series of standard screens with holes of various sizes and determining the proportion that passes each screen.

screen grid An internal electron-tube element positioned between a control grid and an anode; it is usually maintained at a fixed positive potential so that electrostatic influence of the anode is reduced in the region between the screen grid and the cathode.

screening 1. Separation of an aggregate mixture into two or more portions according to particle size, by passing the mixture through one or more standard screens. 2. Removing solids from a liquid-solid mixture by means of a screen. 3. Graded material that has passed through a screen. 4. Elimination of defective items from a lot by inspecting for specific defects and setting aside any items having the defects.

screw 1. A cylindrical machine element with a helical groove cut into its surface. 2. A type of fastener with threads cut into a cylindrical or conical shank and with a slitted, recessed, flat or rounded head that is usually larger in diameter than the shank.

screw conveyor A device for moving bulk material which consists of a helical blade or auger rotating within a stationary trough or casing; a screw elevator moves material in a vertical direction; a screw feeder moves material into a process unit. Also known as auger conveyor; spiral conveyor; worm conveyor.

screwed ends See end connections, threaded.

screw thread Any of several forms of helical ridges formed on or cut into the surface of a cylindrical or conical member; standard thread designs are used to connect pipes to fittings or to construct threaded fasteners.

scriber A sharp pointed tool for drawing lines on metal or plastic workpieces.

scroll In data processing, to move what is visible on the CRT screen up and down.

scrubber 1. A device for removing entrained dust or moisture from a process gas stream. 2. A device for washing out or otherwise removing an undesirable gaseous component from a process gas stream. 3. An apparatus for the removal of solids from gases by entrainment in water.

scuffing 1. A dull mark or blemish, sometimes due to abrasion, on a smooth or polished surface. 2. A form of mild adhesive wear generally exhibited as a dulling of the worn surface.

scum 1. A film of impurities floating on the surface of a liquid. 2. A slimy film on a solid surface.

S/D See synchro-to-digital converter.

SDLC See synchronous data link control.

seal 1. Any device or system that creates a non-leaking union between two mechanical components. 2. A perfectly tight closure or joint. 3. A device to close openings between structures to prevent leakage.

seal chambers Enlarged pipe sections in measurement impulse lines to provide a) a high area-to-volume displacement ratio to minimize error from hydrostatic head difference when using large volume displacement measuring elements, and b) to prevent loss of seal fluid by displacement into the process.

seal coat 1. A layer of bituminous material flowed onto macadam or concrete to prevent moisture from penetrating the surface. 2. A preliminary coating to seal the pores in a material such as wood or unglazed ceramic.

sealed device A device which is so constructed that it cannot be opened during normal operational conditions or operational maintenance; it has a free internal volume less than 100 cm^3 (6.1 in^3) and is sealed to restrict entry of an external atmosphere. It may contain normally arcing parts or internal hot surfaces [S12.12].

sealed reference differential pressure transducer A transducer which measures the pressure difference between an unknown pressure and the pressure of a gas in an integral sealed reference chamber [S37.6] [S37.3].

sealing 1. Impregnating castings with resins to fill regions of porosity. 2. Immersing anodized aluminum parts in boiling water to reduce porosity in the anodic oxide film.

sealing voltage The voltage required to move the armature of a magnetic relay from the position where contacts first touch to its fully closed position; a similar term applies to current.

seal leg The piping from the instrument to the top elevation of the seal fluid in the impulse line.

seal on disk A seal ring located in a groove in the disk circumference. The body is unlined in this case [S75.05].

seal pot See seal chambers.

seal weld A weld used primarily to obtain tightness and prevent leakage.

seal welded bonnet A bonnet welded to a body, at assembly, to provide a zero leakage joint. This construction consists of a low-strength weld with the bonnet retained to the body by other means to withstand the body pressure load acting on the bonnet area [S75.05].

seam 1. An extended length weld. 2. A mechanical joint, especially one made by folding edges of sheet metal together so that they interlock. 3. A mark on ceramic or glass parts corresponding to the mold parting line. 4. An unwelded fold extending inward from the surface of a metal casting or wrought product; to nondestructive testing it appears as a crack, but when examined metallographically it exhibits an oxide layer along the free surface.

seamless tubing Tubular products made by piercing and drawing a billet, or by extrusion.

seam welding A process for making a weld between metal sheets that consists of a series of overlapping spot welds; it is usually done by resistance welding but may be done by arc welding.

search In data processing, to seek out data meeting specific criteria.

season cracking A term usually reserved for describing stress-corrosion cracking of copper or copper alloys in an environment that contains ammonium ions.

seat 1. The area of contact between the closure component and its mating surface which establishes valve shutoff [S75.05]. 2. The fixed area of a valve into which the moving part of a valve rests when the valve is closed to retain pressure and prevent flow.

seat angle The angle between the axis of the seat orifice and the seating surface. A flat seated valve has a seat angle of 90°. The seat angle of the closure member and seat may differ slightly to provide line contact [S75.05].

seating, downstream Seating assisted by pressure differential across the closure component in the closed position, moving the closure component slightly downstream into tighter contact with the seat ring seal that is supported by the body.

seating, upstream A seat on the upstream side of a ball, designed so that the pressure of the controlled fluid causes the seat to move toward the ball.

seat joint The area of contact between the closure member and the valve seat which establishes the sealing action [S75.05].

seat load The total net contact force between the closure member and seat with stated static conditions [S75.05].

seat ring A part that is assembled in the valve body and may provide part of the flow control orifice. The seat ring may have special material properties and may provide the contact surface for the closure member [S75.05].

seat, spring loaded A seat utilizing a design that exerts a greater force at the point of closure component contact to improve the sealing characteristics, particularly at low pressure differential.

second A unit of time, metric and English systems.

secondary air Combustion air introduced into a combustion chamber over the burner flame to provide excess air and ensure complete combustion.

secondary circuit The part of an electrical circuit that conducts current output from a transformer to perform the circuit's function.

secondary combustion Combustion which occurs as a result of ignition at a point beyond the furnace. See also delayed combustion.

secondary device Part of a flowmeter which receives a signal from the primary device and displays, records, and/or transmits it as a measure of the flow rate.

secondary electron An energetic electron set in motion by the transfer of momentum from primary electromagnetic or particulate radiation.

secondary emission The emission of electrons from a substance as a direct result of being bombarded by a stream of electrons or ions.

secondary hardening Hardening of certain alloy steels by precipitation hardening during tempering; the hardening occurring during this stage supplements hardening achieved by controlled cooling from above the critical temperature in a step that precedes tempering.

secondary loop The inner loop of a cascade system.

secondary storage The storage facilities not an integral part of the computer but directly connected to and controlled by the computer, e.g., magnetic drum and magnetic tapes.

secondary treatment Treatment of boiler feed water or internal treatment of boiler-water after primary treatment.

secondary winding The output winding of a transformer or similar electrical device.

sectional conveyor A belt conveyor that can be made shorter or longer by removing or adding interchangeable sections.

sectional header boiler A horizontal boiler of the longitudinal or cross-drum type, with the tube bank comprised of multiple parallel sections, each section made up of a front and rear header connected by one or more vertical rows of generating tubes and with the sections or groups of sections having a common steam drum.

sector 1. In magnetic information storage medium, a defined area of a track or band. 2. The smallest addressable portion of storage on some disk and drum storage units.

sector gear A toothed machine component that looks like part of a gear wheel containing the bearing and part of the rim and its teeth.

sector link In an elastic-chamber pressure gage, the connecting link between the elastic chamber and the sector gear that positions the pointer.

sediment 1. Matter in water which can be removed from suspension by gravity or mechanical means. 2. A non-combustible solid matter which settles out at the bottom of a liquid; a small percentage is present in residual fuel oils.

sedimentation 1. Classification of metal powders according to the rate at which they settle out of a fluid suspension. 2. Removal of suspended matter either by quiescent settling or by continuous flow at high velocity and extended retention time to allow the matter to deposit out.

sediment trap 1. A device for measuring the rate at which sediment accumulates on the floor of a body of water. 2. A device used for removing sediment from an instrument sensing line on a boiler.

seed A small, single crystal of semiconductor material used to start the growth of a single large crystal from which semiconductor wafers are cut.

seek To position the access mechanism of a direct-access storage device at a specified location.

seek time The time taken to execute operation.

segment In computer software programming, the division of a routine.

segmented ball A closure member that is a segment of a spherical surface which may have one edge contoured to yield a desired flow characteristic [S75.05].

segregation 1. Keeping process streams apart. 2. Nonuniform distribution of alloying elements and impurities in a cast metal microstructure. 3. A series of close, parallel, narrow and sharply defined wavy lines of color on the surface of a molded plastics part that differ in shade from surrounding areas and make it appear as if the components have separated. 4. The tendency of refuse of varying compositions to deposit selectively in different parts of the unit.

seismic detector An instrument that registers seismic impulses.

seismic profiler A continuous seismic reflection system used to study geologic structure beneath the oceans' floor to depths of 10,000 ft or more; the reflections are recorded on a rotating drum.

seismochronograph A device for precisely determining the time at which an earthquake shock arrives at the instrument's location.

seismograph An instrument that detects and records vibrations in the earth, such as an earthquake.

seismoscope A device that records the occurrence or time of occurrence of an earthquake, but does not record the frequency or amplitude of earthquake shocks.

seizure See freeze-up.

selection Addressing a terminal and/or a component on a selective calling circuit. See also lockout and polling.

selective-ion electrode A type of pH electrode that involves use of a metal-metal-salt combination as the measuring electrode, which makes the electrode particularly sensitive to solution activities of the anion in the metal salt.

selective plating Any of several methods of electrochemically depositing a metallic surface layer at only localized areas of a base metal, the remaining unplated areas being masked with a nonconductive material during the plating step.

selectivity The characteristic of an electronic receiver which determines the extent to which it can differentiate between a desired signal and electronic noise or undesired signals of other frequencies.

selector 1. A device for choosing objects or materials according to predetermined attributes. 2. A device for starting or stopping a mechanism at predetermined positions or locations. 3. A gearshift lever for operating an automatic transmission in a motor vehicle. 4. The part of a gearshift in a motor vehicle transmission that selects the required gearshift bar. 5. A converter that separates purified copper from residue in a single operation. 6. A device which selects one of a plurality of signals.

self absorption Attenuation of radiation due to absorption within the substance that emits the radiation.

self-adapting Pertaining to the ability of a system to change its performance characteristics in response to its environment.

self-checking code Same as error detecting code.

self-cleaning A descriptor for any device fitted with a mechanism that removes accumulated deposits from its interior without disassembly.

self-contained apparatus Apparatus that is not necessarily connected to equipment in the non-hazardous area. (Usually therefore is self powered). Self-contained apparatus is normally portable e.g. walkie-talkie radios—but the term does not imply that it must be.

self-contained instrument An instrument that contains all of its component parts within a single case or enclosure, or has them incorporated into a single assembly.

self-documenting language Pertains to languages which permit comments to be interspersed with the commands. These comments become documentation. Also, the language is generally very readable.

self-extinguishing The quality of a material to stop burning once the source of a flame is removed.

self-generating Providing an output signal without applied excitation. Examples are piezoelectric, electromagnetic, and thermoelectric transducers [S37.1].

self-heating Internal heating resulting from

electric energy dissipated within a device [S51.1] [S37.1].

self-operated controller See controller, self-operated (regulator) [S51.1].

self-organizing environment A class of equipment which may be characterized loosely as containing a variable network in which the elements are organized by the equipment itself, without external intervention, to meet criteria of successful operation.

self-quenched counter tube A type of radiation counter tube in which internal interactions inhibit the reignition of electron discharge.

self-regulation (inherent regulation) The property of a process or machine which permits attainment of equilibrium, after a disturbance, without the intervention of a controller [S51.1].

self-supporting steel stack A steel stack of sufficient strength to require no lateral support.

self-tapping screw A threaded fastener with specially designed and hardened threads that form internal threads in a hole in sheet metal or soft materials as the screw is driven.

self-test A circuit used by a computer used to check its operation when power is first turned on.

self-tuning See adaptive control.

Selsyn A trade name for a synchro device.

semantics The relationships between symbols and their meanings.

semaphore A task synchronization mechanism. See synchronization of parallel computational processes.

semiautomatic controller A control device in which some of the basic functions are performed automatically.

semi-conductor 1. Materials, used for sensing elements or transduction elements, whose resistivity falls between that of conductors and insulators (e.g. germanium, silicon, etc.). Examples of useful phenomena associated with these materials are: Hall effect, temperature coefficient of resistance, photo-resistivity, photovoltaic effect, piezoresistance, etc. [S37.1]. 2. An electronic device such as an integrated circuit or transistor.

semiconductors, II-VI Semiconductors composed of elements from group II and VI of the periodic table—sometimes extended to cover elements with valances of 2 and 6. Typical II-VI compounds are cadmium telluride and cadmium selenide.

semiconductors, III-V Semiconductors composed of atoms group II and V of the periodic table, such as gallium (III) and arsenic (V), which form gallium arsenide.

semiconductor strain gage A type of strain measuring device particularly well suited to use in miniature transducer elements; it consists of a piezoresistive element that is either bonded to a force-collecting diaphragm or beam or diffused into its surface.

semiconductor temperature sensor See thermistor.

semi-graded index An optical fiber with refractive index profile intermediate between step-index and graded-index. Strictly speaking, this might be considered a type of graded index fiber with refractive index profile somewhat steeper than normal.

semikilled steel Steel that is partly deoxidized during teeming so that only a small amount of dissolved gas is evolved as the metal solidifies.

semirigid plastic Any plastics material having an apparent modulus of elasticity of 10,000 to 100,000 psi.

sense 1. To examine, particularly relative to a criterion. 2. To determine the present arrangement of some element of hardware, especially a manually set switch. 3. To read punched holes or other marks.

sense switch See alteration switch.

sensibility reciprocal A balance characteristic equal to the change in load required to vary the equilibrium position by one scale division at any load.

sensible heat Heat that causes a temperature change.

sensing element 1. That part of the transducer which responds directly to the measurand. NOTE: This term is preferred to primary element, primary detector, primary detecting element [S37.1]. 2. The portion of a device directly responsive to the value of the measured quantity. It may include the case protecting the sensitive portion. See element, sensing.

sensing element elevation The difference in elevation between the sensing element and the instrument. The elevation is considered positive when the sensing element is above the instrument [S51.1].

sensing line A pipe or tube of relatively static fluid that connects the process being sensed to the sensor (transducer) and is filled with the process fluid, or is a fluid filled capillary [S67.02].

sensing, mark A technique for detecting special pencil marks entered in special places on a punch card and automatically translating the marks into punched hole [RP55.1].

sensing mode of transduction element The method used to stress the transduction element such as compression, bending or shear [RP37.2].

sensitive time A characteristic of a cloud chamber equal to the amount of time after expansion when the degree of supersaturation is sufficient to allow a track to form and be detected.

sensitivity 1. The ratio of the change in transducer output to a change in the value of the acceleration. Where one sensitivity under defined conditions is the basis for determining deviations in performance, use "reference sensitivity" [RP37.2]. 2. The ratio of the change in output magnitude to the change of the input which causes it after the steady-state has been reached [S75.05]. NOTE A: It is expressed as a ratio with the units of measurement of the two quantities stated. (The ratio is constant over the range of a linear device. For a nonlinear device the applicable input level must be stated.) B: Sensitivity has frequently been used to denote the dead band. However, its usage in this sense is deprecated since it is not in accord with accepted standard definitions of the term [S51.1]. 3. The ratio of the change in transducer output to a change in the value of the measurand. NOTE: In the sense of the smallest detectable change in measurand use threshold [S37.1]. 4. Ratio of change of output to change of input. 5. Also defined as the least signal input capable of causing an output signal having desired characteristics. 6. The smallest change in actual value of a measured quantity that will produce an observable change in an instrument's indicated or recorded output. 7. The minimum value of an observed quantity that can be detected by a specific instrument. 8. The degree to which a process characteristic can be influenced or changed by a small change in some physical or chemical stimulus.

sensitivity shift A change in the slope of the calibration curve due to a change in sensitivity [S37.1].

sensitivity stability A measure of the irreversible change in transducer sensitivity level after exposure to temperature and/or pressure extremes, or with time [S37.10].

sensitized stainless steel Any austenitic stain-less steel having chromium carbide deposited at the grain boundaries. This deprives the base alloy of chromium resulting in more rapid corrosion in aggressive media.

sensitometer An instrument for determining the sensitivity of light-sensitive materials.

sensitometry Technology related to the measurement of the way photographic film responds to light under specified conditions of exposure and development.

sensor 1. That part of a loop or instrument that first senses the value of a process variable, and that assumes a corresponding, predetermined, and intelligible state or output. The sensor may be separate from or integral with another functional element of a loop. The sensor is also known as a detector or primary element [S5.1]. 2. That portion of a channel which responds to changes in a plant variable or condition, and which converts the measured process variable into an instrument signal [S67.06] [S67.04]. 3. That portion of a device directly responsive to the value of the measured quantity. It may include the case protection and the sensitive portion. 4. The portion of a channel which responds to changes in a plant variable or condition, and converts the measured process variable into an instrument signal. 5. A device that produces a voltage or current output representative of some physical property being measured (speed, temperature, flow, etc.). 6. Generally, the output of a sensor requires further processing before it can be used elsewhere. 7. A generic name for a device that detects either the absolute value of a physical quantity or a change in value of the quantity and converts the measurement into a useful input signal for an indicating or recording instrument. Also known as primary detector; sensing element. 8. A device which reacts to changes in process parameters. See transducer.

separation 1. An action that disunites a mixture of two phases into the individual phases. 2. Partition of aggregates into two or more portions of different particle size, as by screening. 3. The degree, in decibels, to which right and left channels of a stereophonic radio or sound system are isolated from each other. 4. The parting of two connected members of a structure or system, as occurs at preplanned times after launching a multistage rocket. 5. The removal of dust from a gas stream.

separator 1. Any machine for dividing a mixture of materials according to some attribute such as size, density or magnetic properties. 2. A device for separating materials of different specific gravity using water or air. 3. A cage in a ball-bearing or roller-bearing assembly. See cage.

separator-filter A piece of process equipment that removes solids and entrained liquid from a fluid stream by passing the fluid both through a set of baffles or a coalescer and through a screen.

sequence The chronological series of actions and states of an annunciator after an abnormal process condition or manual test initiation occurs [S18.1].

sequence action A signal that causes the sequence to change from one sequence state to another. Sequence actions include process condition changes and manual operation of pushbuttons [S18.1].

sequence checking routine A routine which checks every instruction executed, and prints out certain data, e.g., to print out the coded instructions with addresses, and the contents of each of several registers, or it may be designed to print out only selected data, such as transfer instructions and the quantity actually transferred.

sequence control A system of control in which a series of machine movements occurs in a desired order, the completion of one movement initiating the next, and in which the extent of the movements is not specified by numerical input data.

sequence diagram A graphic presentation that describes sequence actions and sequence states [S18.1].

sequence module See alarm module.

sequence monitor Computer monitoring of the step-by-step actions that should be taken by the operator during a startup and/or shutdown of a power unit. As a minimum, the computer would check that certain milestones had been reached in the operation of the unit. The maximum coverage would have the computer check that each required step is performed, that the correct sequence is followed, and that every checked point falls within its prescribed limits. Should an incorrect action or result occur, the computer would record the fault and notify the operator.

sequencer A mechanical or electronic control device that not only initiates a series of events but also makes them follow each other in an or-dered progression.

sequence state The condition of the visual display and audible device provided by an annunciator to indicate the process condition or pushbutton actions or both. Sequence states include normal, alarm (alert), silenced, acknowledged, and ringback [S18.1].

sequence table A presentation that describes sequence actions and sequence states by lines of statements arranged in columns [S18.1].

sequencing Planning a series of operations or tasks to optimize the use of available production facilities and resources.

sequential access A data access method in which records or files are read one at a time in the order in which they appear in the file or volume.

sequential control 1. A mode of computer operation in which instructions are executed in consecutive order by ascending or descending addresses of storage locations, unless otherwise specified by a jump. 2. A class of industrial process control functions in which the objective of the control system is to sequence the process units through a series of distinct states (as distinct from continuous control). See sequence monitor.

sequential files Collection of related records stored on secondary storage devices such as magnetic tapes and discs. The records are physically stored in the same order as the key number and it is not necessary to use any index or algorithms to locate a particular record.

sequential logic A logic circuit in which the output depends on the inputs to the circuit and the internal states of the circuit.

sequential sampling A method of inspection that involves testing an undetermined number of samples, one by one, until enough test results have been accumulated to allow an accept/reject decision to be made.

serial 1. Pertaining to the time-sequencing of two or more processes [RP55.1]. 2. Pertaining to the time-sequencing of two or more similar or identical processes, using the same facilities for the successive processes [RP55.1]. 3. Pertaining to the time-sequential processing of the individual parts of a whole, such as the bits of a character or the characters of a word, using the same facilities for successive parts [RP55.1]. 4. In reference to digital data, the presentation of data as a time-sequential bit stream, one bit after another. 5. In PCM telemetry, the transfer of information on a bit by bit basis. 6. In data transfer operations, a procedure that handles the data one bit at a time in contrast to parallel operations.

serial access See access, serial.

serial computer 1. A computer having a single arithmetic and logic unit. 2. A computer, some specified characteristic of which is serial, for example, a computer that manipulates all bits of a word serially. Contrast with parallel processing.

serial I/O Method of data transmission in sequential mode, one bit at a time. Only one line is needed for the transmission. However, it takes longer to send/receive the data than parallel I/O.

serial operation A mode of computer operation in which information flows sequentially in time using only one digit, word, line or channel at any given time.

serial output In data processing, programming that instructs the computer to send only one bit at a time between interconnecting devices.

serial-to-parallel converter In PCM telemetry, the circuitry which converts a serial bit stream into bit-parallel data outputs, each transfer representing one measurement.

serial transmission In telemetering, sending bits of information from different sensors or devices over a single channel in sequence, with each bit of information coded to identify its source as well as its value.

series cascade action A type of control-system interaction whereby the output of each controller in a series (except the last one) serves as an input signal to the next controller.

series element Any of a number of two-terminal electronic elements that form a path from one node of a network to another in such a way that only elements of the path terminate at intermediate nodes along the path; alternatively, any of a number of two-terminal elements connected in such a way that any mesh containing one of the elements also contains the others.

series resistor A resistive element of the voltage circuit of an instrument that adapts the instrument to operate on some designated voltage.

serif In typesetting, a type style that has decorative additions at the top and bottom of letters. Commonly called a roman typeface. See sans-serif.

service access point (SAP) The connection point between a protocol in one OSI layer and a protocol in the layer above. SAPs provide a mechanism by which a message can be routed through the appropriate protocols as it is passed up through the OSI layers.

service factor 1. For a facility such as a chemical processing plant or electric generating station, the proportion of time the facility is operating—actual operating time in hours divided by total elapsed time in hours, expressed as a percent. 2. In electric motors, a factor in which a motor can be operated above rated current without damage. For example, an electric motor with a service factor of 1.15 can be operated up to 115% of rated current without damage.

service idlers Those idlers in the weighing area, including scale-borne idlers and several idlers on either side of the scale-borne idlers. They must be of the same type and grade and receive maintenance as weigh idlers [RP74.01].

service life The length of time a mechanism or piece of equipment can be used before it becomes either unreliable or economically impractical to maintain in good working order.

service water General purpose water which may or may not have been treated for a special purpose.

servicing time Same as engineering time.

servo brake 1. A motor vehicle brake in which vehicle motion is used to increase the pressure on one of the brake shoes. 2. A power-assisted braking device.

servomechanism 1. A transducer type in which the output of the transduction element is amplified and fed back so as to balance the forces applied to the sensing element or its displacements. The output is a function of the feedback signal [S37.1]. 2. An automatic feedback control device in which the controlled variable is mechanical position or any of its time derivatives [S51.1]. 3. A feedback control system in which at least one of the system signals represents mechanical motion. 4. An automatic control system incorporating feedback that governs the physical position of an element by adjusting either the values of the coordinates or the values of their time derivatives. 5. Any feedback control system.

session Layer 5 of the OSI model.

set 1. A collection. 2. To place a storage device into a specified state, usually other than that denoting zero or blank. 3. To place a binary cell into the state denoting one. 4. In simulation theory, sets consist of entities with at least one common attribute. Additionally, entities may own any number of sets. Sets may be arranged (topologically ordered) on a "first-in, first-out," "last-in, last-out," or ranked basis. 5. A combination of units, assemblies, or parts connected together or used together to perform a single function, as in a television or radar set. 6. A group of tools, often with at least some of the individual tools differing from others only in size. 7. In plastics processing, conversion of a liquid resin or adhesive into a solid material. 8. Hardening of cement, plaster or concrete. 9. Permanent strain in a metal or plastics material.

setpoint 1. An input variable which sets the desired value of the controlled variable. NOTE A: The input variable may be manually set, automatically set or programmed. B: It is expressed in the same units as the controlled variable [S51.1] [S5.1]. 2. A predetermined level at which a bistable device changes state to indicate that the quantity under surveillance has reached the selected value [S67.06]. 3. The position at which the control point setting mechanism is set. This is the same as the desired value of the controlled variable.

setpoint control A control technique in which the computer supplies a calculated setpoint to a conventional analog instrumentation control loop.

set pressure The inlet pressure at which a safety relief valve opens; usually a pressure established by specification or code.

set screw A small, headless machine screw used for holding a knob, gear or collar on a shaft; it usually has a sharp or cupped point on one end and a slot or recessed socket on the other end.

settling Partial or complete separation of heavy materials from lighter ones by gravity.

settling time The time interval between the step change of an input signal and the instant when the resulting variation of the output signal does not deviate more than a specified tolerance from its steady-state value. See also time, settling.

setup 1. An arrangement of data or devices to solve a particular problem. 2. In a computer which consists of an assembly of individual computing units, the arrangement of interconnections between the units, and the adjustments needed for the computer to solve a particular problem. 3. Preliminary operations—such as control adjustments, installation of tooling or filling of process fluid reservoirs—that prepare

a manufacturing facility or piece of equipment to perform specific work.

set-up driver A routine capable of accepting raw set-up and control information, converting this information to static stores, dynamic stores, or control words, and loading or transmitting the converted data to the associated module in order to achieve the desired effect.

sexadecimal number A number, usually of more than one figure representing a sum in which the quantity represented by each figure is based on a radix of sixteen. Synonymous with hexadecimal number.

shackle An open or closed link having extended arms, each with a hole to accommodate a single pin that spans the gap between the arms.

shading Controlling the phase distribution and amplitude distribution of transducer action at the active face in order to control its directionality.

shaft 1. The mechanical input element of the transducer [S37.12]. 2. The mechanical member used to support a rotary closure component [S75.05]. 3. A cylindrical metal rod used to position, and sometimes to drive or be driven by, rotating parts such as gears, pulleys or impellers, which transmit power and motion.

shaft balancing A method of reducing vibrations in rotating equipment by redistributing the mass to eliminate asymmetrical centrifugal forces.

shaft bearings Devices used in rotary valves to support the shaft and guide the closure component through its travel [S75.05].

shaft encoder A device for indicating the angular position of a cylindrical member. See gray code and cyclic code.

shaft horsepower 1. The power output of an engine, turbine or motor. 2. The power input to a pump or compressor.

shaft position An indication of the position of the wiper relative to a reference point [S37.12].

shakedown test An equipment test made during installation or prior to its initial production operation.

shakeout Removing sand castings from their molds.

shake table See vibration machine.

shake-table test A durability test in which a component or assembly is clamped to a table or platen and subjected to vibrations of predetermined frequencies and amplitudes.

shall, should, and may The word "SHALL," is to be understood as a REQUIREMENT, the word "SHOULD" as a RECOMMENDATION, and the word "MAY" as PERMISSIVE, neither mandatory nor recommended [S77.42].

sample-hold A device that takes a "snapshot" of an analog signal so that it is held stationary for an A/D conversion.

shank The end of a tool that fits into a collet, chuck or other holding device.

shaping A machining process in which a reciprocating, single-point tool cuts a flat or simply contoured surface.

shareable program A (reentrant) program that can be used by several users at the same time.

shared controller 1. A controller, containing preprogrammed algorithms that are usually accessible, configurable, and assignable. It permits a number of process variables to be controlled by a single device [S5.1]. 2. A control device that contains a plurality of pre-programmed algorithms which are user retrievable, configurable, and connectable, and allows user defined control strategies or functions to be implemented. Control of multiple process variables can be implemented by sharing the capabilities of a single device of this kind [S5.3].

shared display 1. The operator interface device (usually a video screen) used to display process control information from a number of sources at the command of the operator [S5.1]. 2. The operator interface device used to display signals and/or data on a time shared basis. The signals and/or data, i.e. alphanumeric and/or graphic, reside in a data base from where selective accessibility for display is at the command of a user [S5.3]. 3. VDU, visual display unit.

shared time control See control, shared time.

shareware Computer software that can be freely copied. Additional payment to the program author is expected from those who find frequent use for the program.

sharpen To impart a keen edge or acute point to a cutting or piercing tool.

shaving 1. Cutting a thin layer of material off the surface of a workpiece, such as to bring gear teeth to final shape. 2. Trimming thin layers or burrs from forgings, stampings or tubing to smooth parting lines, uneven edges or flash.

shear 1. A tool that cuts plate or sheet material by the action of two opposing blades that move along a plane approximately at right angles to the surface of the material being cut. 2. A type of stress tending to separate solid material by moving the portions on opposite sides of a plane through the material in opposite directions.

shearing Separation of material by the cutting action of shears, or by similar action in a punch-and-die set.

shear lip A characteristic of ductile fractures in which the final portion of the fracture separation occurs along the direction of principal shear stress, as exhibited in the cup and cone fracture of a tensile-test specimen made of relatively ductile material.

shear pin 1. A pin or wire designed to hold parts in a fixed relative position until sufficient force to cut through the pin is applied to the assembly. 2. A pin through the hub and shaft of a powertrain member which is designed to fail in shear at a predetermined force, thereby protecting the mechanism from being overloaded.

shear spinning A metal forming process in which sheet metal or light plate is formed into a part having rotational symmetry by pressing a tool against a rotating blank and deforming the metal in shear until it comes in contact with a shaped mandrel. The resulting part has a wall thinner than the original blank thickness.

shear test Any of various tests intended to measure the shear strength of a solid.

shear wave A wave in an elastic medium in which any element of the medium along the wave changes its shape without changing its volume.

sheathed thermocouple A sheathed thermocouple is a thermocouple having its thermoelements, and sometimes its measuring junction, embedded in mineral oxide insulation compacted within a metal protecting tube [ANSI-MC96.1].

sheave A pulley or wheel with a grooved rim to guide a rope, cable or belt.

sheet Any flat material intermediate in thickness between film or foil and plate; specific thickness limits for sheet depend on the type of material involved, and sometimes also on other dimensions such as width.

sheet metal A flat-rolled metal product generally thinner than about 0.25 in.

shell 1. A thin metal cylinder. 2. The outer wall of a tank or pressure vessel. 3. A mold wall made of sand and a thermosetting plastics material used in certain casting processes. 4. A cast tube used as starting stock for certain types of drawn seamless tubing. 5. The metal tube that remains when a billet is extruded using a dummy block of smaller diameter than the billet. 6. A hollow, pierced forging. 7. The outer member of a pulley block surrounding the sheave.

shellac A flammable resinous material, produced by a species of insect found in India, that is used to make a water-resistant coating for wood by dissolving the resin in alcohol.

shell-and-tube heat exchanger A device for transferring heat from a hot fluid to a cooler one in which one fluid passes through the inside of a bundle of parallel tubes while the other fluid passes over the outside of the tubes but inside the vessel shell; heat is transferred by conduction across the walls of the tubes.

sheltered area See area, sheltered.

sheltered locations Locations where neither air temperature nor humidity is controlled and equipment is protected against direct exposure to such climatic elements as direct sunlight, fall of rain and other precipitation, and full wind pressure. Indoor locations which are neither heated nor cooled are sheltered locations [S82.03].

shield 1. Any barrier to the passage of interference-causing electrostatic or electromagnetic fields. An electrostatic shield is formed by a conductive layer, like a foil, surrounding a cable core. 2. An electromagnetic shield is a ferrous metal cabinet or wireway. 3. An attenuating body that blocks radiation from reaching a specific location in space, or that allows only radiation of significantly reduced intensity to reach the specific location.

shielded conductor An insulated conductor encased in one or more conducting envelopes, usually made of woven wire mesh or metal foil; similar products containing one or more insulated conductors are known as shielded cable, shielded conductor cable, or shielded wire.

shielded metal arc welding (SMAW) Metals are heated with an arc between a covered metal electrode and the work. Shielding is obtained from decomposition of the electrode covering. Pressure is not used and filler metal is obtained from the electrode.

shielding 1. Surrounding an electronic circuit or signal-transmission cable with a ground plane, such as a foil or woven-metal sheath, so that capacitive coupling between the circuit and ground plane remains stable. 2. Interposing a radiation absorbing material between a source of ionizing radiation and personnel or equipment to reduce or eliminate radiation damage.

shift register A data storage location in which data is stored and moved (shifted) from one position to another in the register. This shifting is usually to perform some logic or arithmetic operation on the stored data.

shim A thin piece of material, usually metal, that is placed between two surfaces to compensate for slight variations in dimensions between two mating parts, and to bring about a proper alignment or fit.

ship auger A wood boring tool consisting of a spiral body having a single cutting edge instead of two and without a spur at the outer end of the cutting edge; it may or may not have a central feed screw.

shock A substantial disturbance characterized by a rise and decay of acceleration from a constant value in a short period of time [RP37.2].

shock absorber A component connected between a piece of equipment and its frame or support to damp out relative motion between them and reduce the effect of acceleration forces; it normally consists of a dashpot or a combination of a dashpot and a spring.

shock motion A sudden transient motion of large relative displacement.

shock mount A supporting structure that isolates sensitive equipment from the effects of mechanical shock or relatively high amplitude vibrations.

shock resistance The ability to absorb mechanical shock without cracking, breaking or excessively deforming.

shock wave An extremely thin wave in an elastic or compressible medium characterized by a sharp wave front and high intensity; it is typically generated by supersonic flow in a fluid, an explosion, a sudden intense pressure transient, or a sharp intense blow on the surface of an elastic solid.

shoe 1. A renewable friction element whose contour fits that of a drum and stops it from turning when lateral pressure is applied. Also known as brake shoe. 2. A metal block used as a form or support during bending of tubing, wire, rod or sheet metal. 3. A generic term for machine elements that provide support, or separate two members, while allowing relative sliding motion.

shop fabrication Making components and assemblies in a workshop for later transportation to the jobsite for installation.

shop weld A weld made during shop fabrication.

short An electrical short circuit; an electrical circuit with nearly zero resistance.

shortcoming A characteristic or operational deviation which does not prevent an item from functioning, but which should be corrected to achieve optimum efficiency and serviceability.

shortness A form of brittleness in alloys, usually brought about by grain boundary segregation; it may be referred to as hot, red or cold, depending on the temperature range in which it occurs.

short run Failure of molten metal to completely fill the mold cavity.

shorts Large particles remaining on a sieve after the finer portion of an aggregate has passed through the screen.

shot 1. Small spherical particles of a metal. 2. Small, roughly spherical steel particles used in a blasting operation to remove scale from a metal surface. 3. An explosive charge.

shot effect In an electron tube, random variation in electron emission from the cathode, or random variation in electron distribution between electrodes.

shoulder A portion of a cylindrical machine element such as a shaft, screw or flange that is larger in diameter than the remainder.

shrinkage 1. A decrease (shrinkage) in drum level due to a decrease in steam bubble volume. This condition is due to a decrease in load (steam flow), with a resulting increase in drum pressure and decrease in heat input [S77.42]. 2. Contraction of a metal or plastics material upon cooling, or in the case of plastics, upon curing (polymerization).

shrink fit A tight interference fit between mating parts where the amount of interference varies almost directly with diameter; parts are assembled by heating the outer member so that it expands, assembling the parts, and then allowing the outer member to cool and shrink onto the inner member. See also force fit.

shrink forming A process for forming metal parts that uses a combination of mechanical force and shrinkage of a heated blank to achieve final shape.

shrink ring A heated ring or collar that is placed over an assembly and allowed to contract to hold the parts firmly in position.

shrink wrapping A method of packaging that involves heating a plastics film, releasing internal strain and causing the film to shrink tightly over the object being packaged.

shroud A machine element used chiefly as a protective covering over other elements, especially in a rotating assembly.

shunt 1. In an electric circuit, a low-resistance element connected in parallel with another portion of the circuit and used chiefly to carry most of the current flowing in the circuit. 2. To divert all or part of a process flow away from the main stream and into a secondary operation, holding area, or bypass.

shunt calibration resistor A shunt resistor which, when placed across specified points of the electrical circuit of the transducer, will electrically simulate a specified percentage of the full scale output of the transducer at room conditions [S37.5] [S37.3].

shunt diode barrier 1. A fuse- or resistor-protected diode barrier [RP12.6]. 2. A network consisting of a fuse or resistor provided to protect voltage limiting shunt diodes and a current limiting resistor or other limiting components provided to limit current and voltage to intrinsically safe circuits.

shunting resistance The electrical resistance observed between the two terminals of a piezoelectric transducer or its integral cable [S37.10].

shunt valve A valve that allows a fluid under pressure to escape into a passage that is of lower pressure or can accommodate higher flow rates than the normal passage.

shut-down circuit An electronic, electrical, hydraulic or pneumatic circuit that provides controlled steps for turning off or closing down process equipment; it is usually designed to automatically sequence shutdown actions and prevent equipment damage due to performing them out of sequence; it may be used for normal or emergency situations.

shutoff head The pressure developed by a centrifugal or axial-flow pump at its discharge when the discharge flow is zero.

shuttle 1. A back and forth motion in a machine, where the moving element continues to face one direction only. 2. The machine element that undergoes shuttle motion.

SI See Systeme Internationale d'Unites.

sidebands The frequency components on either side of center frequency (fo) which are generated when a carrier wave is modulated.

side-draw product A product stream removed from the column midway between the top and bottom trays. If the side draw is above the feed tray, a vapor is generally removed; if it is below the feed tray, it is usually a liquid.

side milling Using a milling machine and a cutter with teeth on one or both sides to machine a vertical surface.

side rake The angle between a reference plane and the tool face of a single-point turning tool.

side relief angle (SRF) In a cutting tool, the angle between a plane normal to the base and the flanks of the tool below the cutting edge.

side rod 1. In a side-lever engine, one of the members linking the piston-rod crossheads to the side levers. 2. In a railroad locomotive, a large link joining the crankpins of adjacent drive wheels on one side of the engine.

siderograph An instrument combining a clock and a navigation instrument which keeps a reference time equivalent to the time at 0° longitude (Greenwich meridian).

siemens Metric unit of conductance.

sieve 1. A meshed or perforated sheet, usually of metal, used for straining liquids, classifying particulate matter or breaking up masses of loosely adherent or softly compacted solids. 2. A meshed sheet with apertures of uniform standard size used as an element of a set of screens for determining particle size distribution of a loose aggregate.

sieve fraction The portion of a loose aggregate mass that passes through a standard sieve of given size number but does not pass through the next finer standard sieve; usually expressed in weight percent.

sight glass A glass tube, or a glass-faced section of a process line, used for sighting liquid levels or taking manometer readings.

sighting tube A tube, usually made of a ceramic material, that is used primarily for directing the line of sight for an optical pyrometer into a hot chamber.

sigma phase A brittle, nonmagnetic, intermetallic compound generally formed between iron and chromium during long periods of exposure at 1050 to 1800 °F.

sign 1. In arithmetic, a symbol which distinguishes negative quantities from positive ones. 2. An indication of whether a quantity is greater than zero or less than zero. The signs often are the marks, + and -, respectively, but other arbitrarily selected symbols may be used, such as 0 and 1, or 0 and 9; when used as codes at a predetermined location, they can be interpreted by a person or machine.

signal 1. In process instrumentation, a physical variable, one or more parameters of which carry information about another variable (which the signal represents) [S51.1]. 2. Information conveyed from one point in a transmission or control system to another. Signal changes usually call for action or movement. 3. The event or phenomenon that conveys data from one point to another. 4. A time-dependent value attached to a physical phenomenon and conveying data.

signal, actuating error In process instrumentation, the reference-input signal minus the feedback signal. See also deviation, system [S51.1].

signal air Air at varying pressure used to represent process or control information [RP60.9].

signal amplitude sequencing (split ranging) Action in which two or more signals are generated or two or more final control elements are actuated by an input signal, each one responding consecutively, with or without overlap, to the magnitude of that input signal [S51.1].

signal, analog A signal representing a variable which may be continuously observed and continuously represented [S51.1].

signal attenuation The reduction in the strength of electrical signals.

signal booster relay A pneumatic relay that is used to reduce the time lag in pneumatic circuits by reproducing pneumatic signals with high volume and/or high pressure output. These relays may be either volume boosters, amplifying or a combination of both [S75.05].

signal common 1. The signal common shall refer to a point in the signal loop which may be connected to the corresponding points of other signal loops. It may or may not be connected to earth ground [S50.1]. 2. The reference point for all voltage signals in a system. Current flow into signal common is minimized to prevent IR drops which induce inaccuracy in the signal common reference.

signal conditioner (analog data) Hardware device which accepts data from some type of transducer and conditions it to a common scale for multiplexer input.

signal conditioner (PCM data) One of the functions of the bit synchronizer whereby serial data are accepted in the presence of perturbations (noise, jitter) and are reconstructed into coherent data.

signal conditioning To process the form or mode of a signal so as to make it intelligible to or compatible with a given device, including such manipulation as pulse shaping, pulse clipping, digitizing, and linearizing.

signal contact See field contact.

signal contact voltage See field contact voltage.

signal converter See signal transducer.

signal, digital Representation of information by a set of discrete values in accordance with a prescribed law. These values are represented by numbers [S51.1].

signal distance The path length which a signal is required to traverse.

signal, error In a closed loop, the signal resulting from subtracting a particular return signal from its corresponding input signal. See also signal, actuating error [S51.1].

signal, feedback In process instrumentation, the return signal which results from a measurement of the directly controlled variable [S51.1].

signal, feedforward See control, feedforward.

signal generator An instrument used in testing and calibrating other electronic instruments that delivers an output wave-form at an accurately calibrated frequency any where in the audio to microwave range.

signal, input A signal applied to a device, element or system [S51.1] [RP55.1].

signal isolation Signal isolation refers to the absence of a connection between the signal loop and all other terminals and earth ground [S50.1].

signal, measured The electrical, mechanical, pneumatic or other variable applied to the input of a device. It is the analog of the measured variable produced by a transducer (when such is used). Example 1: In a thermocouple thermometer, the measured signal is an emf which is the electrical analog of the temperature applied to the thermocouple. 2: In a flowmeter, the measured signal may be a differential pressure which is the analog of the rate of flow through the orifice. 3: In an electric tachometer system, the measured signal may be a voltage which is the electrical analog of the speed of rotation of the part coupled to the tachometer generator. See also variable, measured [S51.1].

signal, output A signal delivered by a device, element or system [S51.1] [RP55.1].

signal piping That piping interconnecting instruments, instrument devices or bulkhead fittings [RP60.9].

signal, reference-input One external to a control loop, serving as the standard of comparison for the directly controlled variable [S51.1].

signal, return In a closed loop, the signal resulting from a particular input signal, and transmitted by the loop and to be subtracted from the input signal. See signal, feedback [S51.1].

signal selector A device which automatically selects either the highest or the lowest input signal from among two or more input signals. This device is sometimes referred to as a signal auctioneer [S51.1].

signal simulator A hardware device that generates a signal similar in most respects to actual data from a test vehicle.

signal-to-noise ratio 1. Ratio of signal amplitude to noise amplitude. For sinusoidal and non-sinusoidal signals, the amplitude may be peak or rms and should be so specified [S51.1]. 2. The ratio of the power of the signal conveying information to the power of the signal not conveying information.

signal transducer (signal converter) A transducer which converts one standardized transmission signal to another [S51.1].

signal validation Equipment "error" indications and program limit checks that detect faulty signals.

signature analysis A process can be identified as having a particular signature when operating correctly. This can be noise spectrum or vibration spectrum. Signature analysis involves identifying departures from the reference signature and recognizing the source of the departure.

sign bit A single bit, usually the most significant bit in a word, which is used to designate the algebraic sign of the information contained in the remainder of the word.

sign-check indicator An error-checking device, indicating no sign or improper signing of a field used for arithmetic processes. The machine can, upon interrogation, be made to stop or enter into a correction routine.

sign digit In coded data, a digit incorporating 1 to 4 binary bits which is associated with an item of data to indicate its algebraic sign.

significance In positional representation, the factor, dependent on the digit position, by which a digit is multiplied to obtain its additive contribution in the representation of a number. Synonymous with weight.

significant digits A set of digits, usually from consecutive columns beginning with the most significant digit different from zero and ending with the least significant digit whose value is known and assumed relevant, e.g., 2300.0 has five significant digits; whereas, 2300 probably has two significant digits; however, 2301 has four significant digits and 0.0023 has two significant digits.

significant event An event or condition that indicates a change in system status in an event-driven system; a significant event is declared, for example, when an I/O operation completes. A declaration of a significant event indicates that the executive should review the eligibility of task execution since the event might unblock the execution of a higher-priority task. The following are considered to be significant events: I/O queuing, I/O request completion, a task request, a scheduled task execution, a mark time expiration, a task exit.

sign position A position, normally located at one end of a numeral, that contains an indication of the algebraic sign of the number represented by the numeral.

silence The sequence action that stops the sound of the audible device [S18.1].

silica glass A transparent or translucent material consisting almost entirely of fused silica (silicon dioxide). Also known as fused silica; vitreous silica.

silicon A basic material used to make semiconductors that has limited capacity for conductivity.

silicon bronze A corrosion-resistant alloy of copper and 1 to 5% silicon that has good mechanical properties.

silicon-controlled rectifier (SCR) A semiconductor device used to provide stepless control of an electric power circuit without the necessity of load matching; usually, two rectifiers are used in the circuit to provide full-wave control of the heater element, but in some instances two SCRs are used in a single package, known as a triac.

silicone A generic name for semiorganic polymers of certain organic radicals; they can exist as fluids, resins or elastomers, and are used in diverse materials such as greases, rubbers, cosmetics and adhesives.

siliconizing Producing a surface layer alloyed by diffusing silicon into the base metal at elevated temperature.

silky fracture A type of fracture surface appearance characterized by a fine texture, usually dull and nonreflective, typical of ductile fractures.

silver solder A brazing alloy composed of silver, copper and zinc that melts at a temperature below that of silver but above that of lead-tin solder.

similitude Resembling something else—for example, a process that has been scaled up from a laboratory or pilot plant operation to commercial size.

simmer Detectable leakage from a safety relief valve at a pressure below the popping pressure.

simple apparatus A device which will neither generate nor store more than 1.2 V, 0.1 A, 25 mW, or 20 μJ. Examples are : switches, thermocouples, light-emitting diodes, and resistance temperature devices (RTDs) [RP12.6].

simple balance A weighting device consisting of a bar resting on a knife edge and two pans, one suspended from each end of the bar; to determine precise weight, an unknown weight on one of the pans is approximately balanced by known weights placed in the other pan, and a precise balance is obtained by sliding a very small weight along the bar until a pointer attached to the bar at the balance point indicates a null position.

simple buffering A technique for controlling buffers in such a way that the buffers are assigned to a single data control block and remain so assigned until the data control block is closed.

simple electrical apparatus and components Those items (e.g. thermocouples, photocells, junction boxes) which do not generate or store more than 1.2V, 0.1A, 20μJ and 25 mW in the intrinsically safe system.

simple engine A machine for converting thermal energy to mechanical power by expanding a working fluid in a single stage, after which the fluid passes out of the engine through an exhaust port.

simple machine Any of several elementary mechanical devices that form the basis for creating more complex devices—usually the lever, wheel, pulley, inclined plane and screw are the only devices considered simple machines.

simple sound source A sound source that radiates uniformly in all directions in an unrestrictive airspace.

simple systems Simple systems, in which all the electrical apparatus is certified intrinsically safe or certified associated electrical apparatus, do not require to be certified provided it is completely clear from the information given on the electrical apparatus certification documents that the system is intrinsically safe.

simplex mode Method of data transmission whereby the data is transmitted in one direction only, i.e. send or receive.

simplex pump A reciprocating pump with only one power cylinder and one pumping cylinder.

SIMSCRIPT A generic class of discrete, event-oriented simulation languages.

simulate 1. To artificially create behavior, environmental conditions or operating conditions pertaining to one system by using another, different system; usually done to accomplish testing, experimentation or training that would be difficult or hazardous to accomplish with the real system. 2. The representation of physical phenomena by use of mathematical formulas.

simulation 1. The representation of certain features of the behavior of a physical or abstract system by the behavior of another system. For example, the representation of physical phenomena by means of operations performed by a computer or the representation of operations of a computer by those of another computer. 2. Using computers, electronic circuitry, models or other imitative devices to gain knowledge about operations and interactions that take place in real physical systems.

simulation framework A simulation system capable of integrating and running two or more simulation algorithms in a single simulation environment.

simulation tests Tests on the weighing unit in which either the movement of the belt, the effect of the material thereon, or both are simulated by using known weights and forces [RP74.01].

simulator 1. A program which simulates the operation of another device or system. In the case of microprocessors a simulator allows the execution of a microprocessor object program on a computer which is different from the microprocessor for which the program has been written. The simulator provides a range of debugging tools which allows the programmer to correct errors in the program. 2. A device, system, or computer program that represents certain features of the behavior of a physical or abstract system.

sine bar An accurately constructed layout aid consisting essentially of a straight bar with cylindrical rests at each end; one end of the sine bar is placed on a surface plate or gage block and the other end on a stack of gage blocks equal to the sine of a desired angle to the surface plate or another reference plane.

sine galvanometer A magnetometer whose measuring element consists of a small magnet suspended in the center of a pair of Helmholtz coils; the magnitude of a magnetic field is determined from the position of the magnet when various known currents are passed through the coils.

sine wave 1. A signal varying with time which can be obtained through projection of a rotating vector of constant magnitude with constant angular velocity on a linear scale [S26]. 2. A waveform in which the value of wave parameters—such as voltage and current in certain alternating-current circuits—vary directly as the sine of another variable—such as time.

sine-wave response See frequency response.

single acting 1. An actuator in which the power supply acts in only one direction. In a spring and diaphragm actuator, for example, the spring acts in a direction opposite to the diaphragm thrust. Single acting spring and diaphragm actuators may be further classified as to direction of stem movement on increasing fluid pressure: a) air to extend actuator stem, b) air to retract actuator stem [S75.05]. 2. Producing power or motion in one direction only.

single acting positioner A positioner is single acting if it has a single output [S75.05].

single-axis tracking antenna A receiving antenna which tracks the transmitting station automatically in azimuth, but not in elevation.

single board computer (SBC) A complete computer, including memory, clock and input/output ports assembled on a single board.

single cascade action A type of control-system action whereby the input to the second of two automatic controllers is supplied by the first.

single density A computer diskette that can store approximately 3,400 bits per inch.

single-element feedwater control A control system whereby one process variable, drum level, is used as the input to the control loop that regulates feedwater flow to the drum to maintain the drum level at set point [S77.42].

single-ended amplifier An electronic amplifier in which each stage operates asymmetrically with respect to ground; each stage contains one tube or amplifying transistor; alternatively, each contains two or more, connected in parallel.

single flange (lugged) A thin annular section body whose end surfaces mount between the pipeline flanges, or may be attached to the end of a pipeline without any additional flange or retaining parts, using either thru bolting and/or tapped holes [S75.05].

single-mode An optical fiber which can carry only a single waveguide mode of light. Components such as connectors used with such fiber are also labeled single-mode.

single-phase meter An instrument for determining power factor in a single-phase alternating-current circuit; it contains a fixed coil that carries the load current and crossed coils connected to the load voltage; the moving system is not restrained by a spring and therefore takes a position related directly to the phase angle between voltage and current.

single pole A type of device such as a switch, relay or circuit breaker that is capable of either opening or closing one electrical path.

single sampling A type of inspection where an entire lot or production run (population) is accepted or rejected based on results of inspecting a single group of items (sample) selected from the population.

single-sideband modulation A type of modulation whereby the modulating wave's spectrum is translated in frequency by a specific amount, either with or without inversion.

single-sided A computer diskette that can record data on only one side.

single-speed floating control A type of controller action in which the final control element moves toward one extreme of its travel or the other extreme, depending on the direction of the deviation; the rate of movement is at a predetermined speed regardless of the amount of deviation; in most practical systems, there is a neutral zone so that controller action is suspended when the controlled variable is within the allowable range. See also controller, single speed floating.

single-stage compressor A machine that raises pressure in a compressible fluid in a sin-

gle pass through a single set of machine elements.

single-stage pump A machine that develops pressure to drive a relatively incompressible fluid through a system by passing the fluid through a single set of machine elements.

single-stream batch A method of batch processing in which only one stream of batch commands is processed.

sink A reservoir into which material or energy is rejected.

sinkhead See feedhead.

sintering Heating a powder metal compact at a temperature below the melting point to form diffusion bonds between the particles.

siphon A tube, hose or pipe for moving liquid from a higher to a lower elevation by a combination of gravity acting on liquid in the longer leg and atmospheric pressure acting to keep the shorter leg filled.

site license In data processing, a software agreement that allows unlimited use of a program to a single organization at one location.

size 1. A specified value for some dimension that establishes an object's comparative bulk or magnitude. 2. One of a set of standard dimensions used to select an object from among a group of similar objects to obtain a correct fit. 3. In welding, the joint penetration of a groove weld or the nugget diameter of a spot weld or the length of the nominal legs of a fillet weld. 4. A material such as casein, gum, starch or wax used to treat the surface of leather, paper or textiles.

skew 1. Having an oblique position in relation to a specific reference plane, reference direction, or physical object. 2. A tape motion characterized by an angular velocity between the gap center line and a line perpendicular to the tape center line.

skid 1. A metal bar or runner on an object that provides support or wear resistance when the object contacts a floor, runway, apron or other flat areaway. 2. A wood or metal platform on legs, runners or wheels used to support parts or materials a short distance off the ground during storage or material handling. 3. A brake for a power machine. 4. A device placed under a wheel to prevent it from turning while a heavy wheeled object descends a steep hill.

skin A general term for a thin exterior covering-may be applied to the exterior walls of a building, the exterior covering of an airplane, a protective covering made of wood or plastics sheeting, or a thin layer on a mass of metal that differs in composition or some other attribute from the main mass of metal.

skirting Stationary sideboards or sections of the belt conveyor attached to the conveyor support frame or other stationary support to prevent the bulk material from falling off the side of the belt [RP74.01].

slab 1. A flat piece of concrete that spans beams, piers, columns or walls to make a floor, roof or platform. 2. A relatively thick piece of metal whose width is at least twice its thickness— generally used to describe a mill product intermediate between ingot and a flat rolled product such as sheet or plate.

slack Looseness or play in a mechanism, sometimes due to normal tolerances in the assembly and sometimes due to wear.

slag Molten or fused refuse.

slave A mechanical or electronic device that is under the control of a another device.

sleeve 1. A tubular member through a wall to permit passage of pipe or other connections. 2. A cylindrical part that fits over another part.

sleeve bearing A cylindrical machine element that fits around a shaft and supports the shaft while it turns.

sleeve coupling A hollow cylinder that fits over the ends of two adjacent shafts or pipes to hold them together.

slewing Rapidly moving a device to a new rotational position or a new elevation direction, or both.

slew rate 1. The maximum rate of change of an output signal from a device. 2. The limitation in a device or circuit in the rate of change of output voltage, usually imposed by some basic circuit considerations.

slide 1. Any mechanism that moves with predominantly sliding motion. 2. The main reciprocating member of a mechanical press, which moves up and down in the press frame and carries the punch or upper die. 3. A flat-bottomed chute.

sliding-block linkage A mechanism for converting rotary motion into linear motion, or vice versa, which consists of a crank, a block that slides back and forth in a slot or on ways, and a link bar attached to the crank and block with pin joints.

sliding disk A flat or wedge-shaped closure component that modifies the flow rate with linear motion across the flow path [S75.05].

sliding fit A type of clearance fit used to accurately locate parts that must assemble together without perceptible play (close-sliding fit) or used to allow assembled parts to move or turn easily but not run freely (sliding fit). See also running fit.

sliding gear A gear set whose speed can be changed by sliding gears along their axes to put them in or out of mesh with other gears of different sizes.

sliding-vane rotary flowmeter A type of positive-displacement flowmeter in which radial vanes slide in or out to trap and release discrete volumes of the metered fluid as a rotor containing the vanes revolves about a central cam surface which controls vane position.

slime 1. A soft, viscous or semisolid surface layer —often resulting from corrosion or bacterial action, and often having a foul appearance or odor. 2. A mudlike deposit in the bottom of a chemical process or electroplating tank. 3. A thick slurry of very fine solids. Also known as mud; pulp; sludge.

sling A length of rope, wire rope or chain used to support a load hanging from a crane hook.

sling psychrometer A device for determining relative humidity that consists of a wet-and-dry bulb thermometer mounted in a frame that can be whirled about, usually by means of a handle and short piece of chain or wire rope attached to the upper end of the frame.

slip 1. A term commonly used to express leakage in positive-displacement flowmeters. 2. A suspension of ground flint or fine clay in water that is used in making porcelain or in decorating ceramic ware. Sometimes called slurry.

slip-in liner In a butterfly valve body, an annular shaped liner which makes a slight interference fit with the body bore and which may be readily forced into position through the body end. May be plain or reinforced [S75.05].

slip joint 1. A telescoping joint between two parts in an assembly. 2. A mechanical union that allows limited axial movement of one member, such as a pipe or duct, with respect to a mating member. 3. In civil engineering, a type of contraction joint consisting of a tongue and groove that allows independent movement between two members such as wall sections, slabs or precast structural units. 4. A type of scarf joint used in flexible-bag molding in which plastics veneers are laid up so that their beveled edges overlap.

slippage 1. Fluid leakage along the clearance between a reciprocating-pump piston and its bore. Also known as slippage loss. 2. Movement that unintentionally displaces two solid surfaces in contact with each other. 3. Movement of a gas phase through or past a gas-liquid interface instead of driving the interface forward; especially applicable to certain phenomena in petroleum engineering.

slip plane A crystallographic plane along which dislocations move under local shear stresses to produce permanent plastic strain in ductile metals.

slip ratio The quotient of actual advance of a screw propeller divided by the theoretical advance determined by blade pitch angle and number of revolutions in a specific period of time.

slip seal A seal between members designed to permit movement of either member by slipping or sliding.

slit 1. A long, narrow opening—often used for directing and shaping streams of radiation, fluids or suspended particulates. 2. To cut sheet metal, rubber, plastics or fabric into sheet or strip stock of precise width using rotary cutters, knives or shears.

sliver 1. Any thin, elongated, often sharp edged fragment of solid material. 2. A thin fragment attached at one end to the surface of flat-rolled metal and rolled into the surface during reduction.

slope control Electronically producing specific changes in a parameter with time, especially applied to a method of varying welding current.

slot 1. Any of certain apertures in an airfoil to improve aerodynamic behavior. 2. Any elongated opening in a machine part or structural member. 3. A special socket in a PC designed to accept an additional circuit board.

slotted nut A hexagon nut with slots cut across the flats of the hexagon so that a lockwire or cotter pin can be used to prevent the nut from turning.

slotted ring An access procedure that uses a pattern that circulates around the ring to determine access. The pattern is divided into sections called slots. When a node has data to transmit, it waits for an unused slot, writes its data into the slot, and marks the slot used. When the slot rotates to the destination node, it reads the data from the slot. When the slot rotates back to the original node, it marks the slot unused again.

slot washer 1. A washer with a slot extending to one edge so that the washer can be removed from a bolt without completely disassembling the fastened joint. 2. A type of lockwasher with an indentation on its rim so that it can be held in place with a nail or screw.

slow neutron A free (uncombined) neutron having a kinetic energy of about 100 eV or less.

slow (time) code A spread-out time code, modulated in amplitude and width, suitable for display on a relatively low-speed chart recorder.

sludge 1. A soft water-formed sedimentary deposit which normally can be removed by blowing down. 2. Fine sediment such as may be found in the bottom of an oil crankcase or boiler drum. See also slime.

slug 1. A large "dose" of chemical treatment applied internally to a steam boiler intermittently. Also used sometimes instead of "priming" to denote a discharge of water out through a boiler steam outlet in relatively large intermittent amounts. 2. A small, simply shaped piece of metal used as starting stock for forging, upsetting or extrusion. 3. The offal resulting from piercing a hole in sheet metal. 4. Liquid that completely fills the internal passage of a tube for a short distance.

slugging Producing a substandard weld joint by adding a separate piece of material which is not completely fused into the joint.

sluice 1. A waterway fitted with a vertical sliding gate for controlling the flow of water. 2. A channel for draining away excess water.

slump test A quality control test for determining the consistency of concrete; the amount of slump is expressed as the decrease in height that occurs when a conical mold filled with wet concrete is inverted over a flat plate and then removed, leaving the concrete behind.

slurry 1. A suspension of fine solids in a liquid which can be pumped or can flow freely in a channel. 2. See slip. 3. A semiliquid refractory material used to repair furnace linings. 4. An emulsion of soluble oil and water used as a cutting fluid in certain machining operations.

small scale integration Low density of integrated circuits per unit area.

SMAW   See shielded metal arc welding.

SME   See CASA/SME.

smoke   1. A dispersion of fine solid or liquid particles in a gas. 2. Small gas borne particles of carbon or soot, less than 1 micron in size, resulting from incomplete combustion of carbonaceous materials and of sufficient number to be observable.

smoke detector   A device that produces an alarm signal when the density of smoke in an area exceeds a preset value; usually, smoke density is detected photoelectrically.

smoke point   The maximum flame height in a standard test that kerosene or jet fuel will burn without smoking.

SNAP   See sub-network access protocol.

snap fastener   A type of fastening device used primarily to hold the edges of fabric articles together; it consists of a flange with a protruding ball attached to one edge of the fabric, and a mating flange with a socket attached to the opposing edge.

snap gage   A device with two flat, parallel surfaces that are precisely spaced apart for checking one limit of tolerance on a diameter or length dimension. Sometimes, go-no-go tolerance limits are built into a single snap gage frame to permit checking both high and low limits of tolerance at the same time.

snap ring   A type of retaining fastener in the shape of the letter C which is expanded across its diameter and allowed to snap back into a groove to hold parts in position, and especially keep them from sliding axially along a shaft.

snapshot dump   A selective dynamic dump performed at various points in a machine run.

snatch block   A pulley whose side can be opened to allow a loop or rope to be inserted.

snorkel   A tube that supplies air for an underwater operation.

snow   A speckled background on an intensity-modulated CRT display that is produced by electronic noise; the appearance is similar to the display on a television screen when the station is not broadcasting; snow may or may not make a transmitted image or data display unsuitable for its intended purpose.

snubber   1. A device which is used to damp the motion of the valve stem. This is usually accomplished by an oil filled cylinder/piston assembly. The valve stem is attached to the piston and the flow of hydraulic fluid from one side of the piston to the other is restricted [S75.05]. 2. A mechanical or hydraulic device for restraining motion.

snub pulley   Any pulley used to increase the arc of contact between the belt and the drive pulley [RP74.01].

soap bubble test   A leak test consisting of applying soap solution to the external surface or joints of a system under internal pressure and observing the location, if any, where bubbles form indicating the existence of a gas leak.

socket weld   An external weld joining the plain-ended male portion and the corresponding socket. Used here as a male valve inlet in a process line or vessel socket.

softening   The act of reducing scale forming calcium and magnesium impurities from water.

softening agent   A substance—often an organic chemical—that is added to another substance to soften it.

soft hammer   A hammer with a head made of annealed copper, leather or plastic which keeps it from damaging finished surfaces.

soft seated trim   Globe valve trim with an elastomeric, plastic or other readily formable material used either in the valve plug or seat ring to provide tight shutoff with minimal actuator forces [S75.05]. See also trim, soft seated.

soft-sector disk   In data processing, a disk that accepts magnetic patterns to define the boundaries of each sector.

software   1. Digital programs, procedures, rules, and associated documentation required for the operation and/or maintenance of a digital system [S5.3]. 2. A set of programs, procedures, rules, and possibly associated documentation concerned with the operation of a computer system, for example, compilers, library routines, manuals, circuit diagrams. 3. Contrast with hardware.

software license   An agreement between the seller and buyer of a computer program that usually limits the use of the program to one person at one time at one location. See site license.

software link   The interconnection of system components or functions via software or keyboard instruction [S5.3].

software priority interrupt   The programmed implementation of priority interrupt functions. See priority interrupt.

soft water   Water which contains little or no calcium or magnesium salts, or water from which scale forming impurities have been removed or reduced.

soil mechanics A branch of civil engineering that deals with the application of principles of solid and fluid mechanics to the design, construction and maintenance of earthworks and stable foundations.

soil pipe A vertical drain for carrying sewage from a building into a sewer or septic system.

solar blind A detector which contains filters to block sunlight, so the detector becomes essentially "blind" to the sun. In most cases, this involves blocking wavelengths longer than approximately 300 nm, simulating the absorptive effects of upper-atmosphere ozone.

solar engine A device for converting thermal energy from the sun into electrical or mechanical energy or for using thermal energy from the sun to run a refrigeration system.

solar furnace A device for producing high temperatures by focusing solar radiation.

solar heating The use of solar radiation to produce enough heat for cooking, industrial operations, or heating buildings.

solarimeter See pyranometer.

solar power Any of several methods of using energy from the sun to perform useful work.

solder A joining alloy with a melting point below about 450 °F, such as certain lead-base or tin-base alloys.

solder glass A special glass that softens below about 900 °F, and that is used to join two pieces of higher melting glass without deforming them.

soldering embrittlement Penetration by molten solder along grain boundaries of a metal with resulting loss of mechanical properties.

solenoid A type of electromechanical operator in which reciprocal axial motion of a ferromagnetic core within an electromagnetic coil performs some mechanical function; common applications include opening or closing valves or electrical contacts; normally a solenoid armature is spring-loaded so that the core moves against the action of the spring when the coil is energized, and the spring returns the core to its original position when the energizing electric current is turned off.

solenoid valve A shutoff valve whose position is determined by whether or not electric current is flowing through a coil surrounding a moving-iron valve stem; the valve may be normally open, in which case gas or liquid flows through the valve when electricity to the coil is turned off; normally closed, in which case gas or liquid flows only when electricity is turned on; or three-way, in which gas or liquid flows in one path through the valve when electricity is off and in a different path when electricity is on.

soleplate 1. A flat member used as the supporting base of a machine. 2. A flat member in a machine's frame on which a bearing can be mounted and, if necessary, adjusted slightly.

solid coupling A device used to rigidly connect two shafts together and usually capable of transmitting full torque from one shaft to the other.

solid die A one-piece tool with internal threads used for cutting screw threads on rod stock or small-diameter pipe.

solidification The change in state from liquid to solid in a material as its temperature passes through its melting temperature or melting range on cooling.

solid state Pertaining to an electronic device or circuit whose operation is controlled by some combination of electrical, magnetic and optical phenomena within a circuit element consisting largely of a single piece of solid material, usually a crystalline semiconductor.

solid-state welding Any welding process that produces a permanent bond without exceeding the melting point of the base materials and without using a filler metal.

soluble oil An oil-based fluid that can form a stable emulsion or colloidal suspension with water; used principally as a cutting fluid or coolant.

solution A liquid, such as boiler water, containing dissolved substances.

solution heat treatment Heating an alloy into a temperature range where the principal alloying element(s) become dissolved in a single solid phase, then cooling the material rapidly enough to prevent precipitation of secondary phases.

sonar 1. An assembly of electronic equipment for generating or detecting underwater sound, and using that sound to locate objects or to merely detect their presence, or to provide a communications link between two underwater locations. Also known as sonar set. 2. Any method or system that depends on underwater sound to passively listen for noises made by ships or sea creatures, to search for underwater objects by sending out sound pulses and listening for returning echoes, or to communicate by sending sound signals from one undersea location to another. An acronym derived from "Sound Navigation And Ranging."

sonic barrier A popular term for the large increase in drag encountered when the speed of an aircraft or missile approaches the speed of sound in air; the speed at which this occurs is somewhat indefinite, and depends on altitude and general atmospheric conditions.

sonic opacity A characteristic of a medium such as one containing a large quantity of particles or small bubbles that results in sound or ultrasound being reflected randomly from the discontinuities rather than being transmitted through the medium.

sonic speed The speed of sound in the specific medium of concern.

sonobuoy A combination of a passive sonar set and a radio transmitter mounted in a buoy that can be dropped by parachute from an airplane; underwater sounds, such as those from a submarine, are picked up by the sonar set and transmitted by radio to a receiver on the aircraft or a ship; the source of underwater sound can be determined by triangulation using several sonobuoys dropped in a known pattern, and comparing time-delay data from their signals by computer analysis.

sonograph 1. An instrument for recording sound or seismic vibration patterns. 2. An instrument for converting sound into percussive (seismic) vibrations.

soot A black deposit containing impure carbon and oily compounds resulting from the incomplete combustion of resinous materials, oils, wood and coal.

soot blower A mechanical device for discharging steam or air to clean heat absorbing surfaces.

sort In data processing, a routine that puts data in a specific order based on established criteria.

sorting table Any horizontal conveyor where operators stationed along one or both sides manually sort bulk material, packages or individual items by selectively removing them from the conveyor.

sound 1. A pressure wave in an elastic or compressible material which exists in the form of alterations in pressure, stress, particle displacement or particle velocity; the term is usually restricted to such waves whose frequency is in the range of human hearing. 2. The human auditory sensation produced by such pressure waves.

sound analyzer A device for measuring the band pressure level at various sound frequencies.

sounding 1. Determining the depth of a body of water, either by echo ranging or by taking depth readings with a weight attached to a line having knots or telltales regularly spaced along its length. 2. Measuring the depth of soil above bedrock by driving a steel rod into the soil. 3. Generically, any penetration of the natural environment for the purpose of taking scientific measurements.

sound-level meter An electronic instrument for measuring noise or sound levels in either decibels or volume units.

sound-powered telephone A type of telephone usually used for emergency communications over short distances; electric current for transmitting the signal is generated by the speaker's voice in a specially designed microphone, and no external source of power is required.

sound pressure The total instantaneous pressure at a given point in the presence of a sound wave, minus the static pressure of that point [S37.1].

sound pressure level (SPL) The intensity of a sound wave which, in decibels, equals 20 log (P_s/P_r), where P_s is the pressure produced by the sound and Pr is a stated reference pressure.

sound reproduction Any process for detecting sound at one location and time and regenerating the same sound at the same location and time, or at another location, at another time, or both, and with any desired intensity.

source A reservoir from which material or energy is drawn.

source code 1. The program instructions written in high-level languages or assembly languages. The program must be translated into object code before it can be executed by the computer. 2. Software generated by a programmer in assembly language, generally with comments, headings, and other annotation.

source impedance The impedance of the excitation supply presented to the excitation terminals of the transducer [S37.1]. See impedance, source.

source language The system of symbols and syntax, easily understood by people, used to describe a procedure a computer can execute.

source module A series of statements in the symbolic language of an assembler or compiler, which constitutes the entire input to a single execution of the assembler or compiler.

source program A program written in a source language.

sour crude Crude oil containing excessive amounts of sulfur, which liberate corrosive sulfur compounds during refining. Contrast with sweet crude.

sour gas Natural gas that contains corrosive sulfur-bearing compounds such as H_2S or mercaptans.

space In data processing, a unit of storage that is empty.

space charge The electric charge carried by a stream of electrons or ions in a region of low gas pressure (vacuum); the charge must be sufficient to cause local variations in the distribution of electric potential within the affected space.

spacer A simple mechanical member designed to keep two other members in an assembly a specific distance apart.

space technology The systematic application of science and engineering to the exploration and exploitation of outer space.

spade bolt A bolt having a flattened head shaped like a spade with a transverse hole; it is used predominantly for fastening components such as shielded coils and capacitors to the chassis of heavy-duty electronic equipment.

spade drill A drill made from round or square stock by hammering one end, tapered to a thin edge, then sharpening a point on the thin edge with the finished piece looking like a pointed spade.

spade lug A device consisting of a body, that can be clamped or crimped onto the end of an electrical wire or cable, and a flat two-pronged projection, that can be slipped under a screw or nut on a terminal block to complete an electrical connection yet allow the connection to be disassembled without completely removing the screw or nut.

spall To detach material from a surface in the form of thin chips whose major dimensions are in a plane approximately parallel to the surface.

spalling The breaking off of the surface of refractory material as a result of internal stresses.

span 1. A structural dimension measured in a straight line between two specific extremities, such as the ends of a beam or two columnar supports. 2. The dimension of an airfoil, such as the wings of an aircraft, from tip to tip, measured in a straight line. 3. The algebraic difference between the upper and lower range-values [S51.1] [S37.1] [S26] [S12.15]. Note A: The following compound terms are used with suitable modifications to the units; measured variable span, measured signal span, etc. B: For multi-range devices, this definition applies to the particular range that the device is set to measure [S51.1]. Thus, a temperature in the range of 20 °C to 250 °C has a span of 230 °C. See range. 4. The difference between maximum and minimum calibrated measurement values. Example: an instrument having a calibrated range of 20-120 has a span of 100.

span adjustment See adjustment, span.

span error The difference between actual span and ideal span, usually expressed as a per cent of ideal span. See also error, span.

spanner 1. A wrench with a semicircular head having a projection or hole at one end. 2. A horizontal structural brace. 3. An attachment for a sextant that establishes an artificial horizon.

span of pneumatic transmission signal The difference between the stated high and low pneumatic pressure values of a transmission range [S7.4].

span shift Any change in slope of the input-output curve [S51.1].

spare alarm point See alarm point.

spark arrester 1. A device that reduces or prevents electric sparks at a point where a power circuit is opened or closed, such as at a circuit breaker or knife switch. 2. A device that prevents airborne embers from escaping from a chimney.

spark recorder A type of recorder where sparks passing between a metal pointer and an electrically grounded plate periodically burn small holes in recording paper as it moves slowly across the face of the plate; sparks are produced at regular intervals by a circuit powering an induction coil, and the varying lateral position of the moving pointer creates the trace.

spatial coherence The coherence of light over an area of the wavefront of a beam; where the beam hits the surface.

spatter Particles of molten metal expelled during a welding operation and becoming adhered to an adjacent surface.

specific acoustic impedance The complex ratio of sound pressure to particle velocity at a given point within the medium.

specific acoustic reactance The imaginary component of specific acoustic impedance.

specific acoustic resistance The real component of specific acoustic impedance.

specific address See absolute address.

specification 1. A list of requirements that must be met when making a material, part, component or assembly; installing it in a system; or testing its attributes or functions. 2. A set of standard requirements applicable to any product or process within the jurisdiction of a given standards-making organization; an industry consensus standard.

specific code See absolute code.

specific fuel consumption The amount of fuel required to produce a given unit of power-expressed, for example, in pounds per horsepower-hour.

specific gravity (sp gr) The ratio of the density of a material to the density of the water at the same conditions. Specific gravity: G_f = liquid at flowing condition referred to water at 60 °F; G_g = gas referred to air, both at STP.

specific gravity bottle A small flask used to determine density; its precise weight is determined when empty, when filled with a reference liquid such as water, and when filled with a liquid of unknown density. Also known as density bottle; relative density bottle.

specific gravity, gas The density of a gas compared to the density of air.

specific gravity, liquid The density of a liquid compared to the density of water.

specific heat (sp ht) 1. The quantity of heat, expressed in Btu, required to raise the temperature of 1 lb of a substance 1 °F. 2. The ratio of the thermal capacity of a substance to that of water. The specific heat at constant pressure of a gas is designated cp. The specific heat at constant volume of a gas is designated c_v. The ratio of the two (c_p/c_v), is called the ratio of specific heats, k.

specific humidity The weight of water vapor in a gas water vapor mixture per unit weight of dry gas.

spectral analysis A frequency decomposition of the analog input signals. Identification of the frequency spectrum.

spectral density (As in PCM-coded data) the amount of a signal level at each frequency or portion of the spectrum.

spectral emissivity The ratio, at a specified wavelength, of thermal radiation emitted from a non-blackbody to that emitted from a blackbody at the same temperature.

spectrofluorometer An instrument for determining chemical concentration by fluorometric analysis using two monochromators—one to analyze the wavelength of strongest emission and the other to select the wavelength of best excitation in the sample.

spectrometer A spectroscope which includes an angular scale for measurement of the angular deviation and wavelengths of the components of the spectrum.

spectrophotometric titration Instrumented titration in which the end point is determined by measuring a change in absorbed radiation with a spectrophotometer.

spectroradiometer An instrument which measures power as a function of wavelength.

spectroreflectometer A device which measures the reflectance of a surface as a function of wavelength.

spectroscope A device which spreads out the spectrum for analysis. The simplest type is a prism or diffraction grating which spreads out the spectrum on a piece of paper or ground glass.

spectroscopic analysis Identification of chemical elements by characteristic emission and absorption of light rays.

spectrum Frequency band.

spectrum analyzer 1. An instrument for measuring the distribution of energy among the frequencies emitted by a pulse magnetron. 2. An electronic instrument for analyzing the output, amplitude and frequency of audio or radio frequency generators or amplifiers under normal or abnormal operating conditions.

spectrum display unit 1. An adjunct to a radio receiver that displays the radio spectrum in and on each side of the carrier being received. 2. On a telemetry receiver, a device which displays the spectrum at and on both sides of the frequency to which the receiver is tuned.

specular transmission density The value of photographic density obtained when only the normal component of transmitted flux is measured for source illumination whose rays are perpendicular to the plane of the film.

speed of response See response time, time constant.

speed reducer A gear train for transmitting power from a motor to the machinery it drives at a rotational speed less than that of the motor.

spent liquor The liquid effluent from the pulping stage of papermaking; it consists of wood chemicals such as lignin and partly reacted digestion chemicals (caustic, sulfite or sulfate, depending on which pulping process was used).

sp gr See specific gravity.

spherical aberration A lens defect that makes rays from the peripheral part of the lens focus at a different point than do rays from the central portion of the lens, which produces an image lacking in contrast.

spherical wave A wave whose equiphase surfaces form a series of concentric spheres.

spherometer A device for measuring the spherical curvature of a surface.

spherulitic-graphite cast iron See ductile iron.

sp ht See specific heat.

spill The accidental release of a hazardous chemical or radioactive liquid from a process system or a container.

spinning 1. Production of plastics filament by extrusion through a spinneret. 2. Forming sheet metal into rotationally symmetrical shapes such as bowls or cones by pressing a round-ended tool against the flat stock and forcing it to conform to the shape of a rotating mandrel.

spiral bevel gear A bevel gear with curved oblique teeth, which provide for gradual engagement and bring more teeth into contact with each other at any given time than for an equivalent straight bevel gear.

spiral flow test Determining the flow characteristics of thermoplastic resins by measuring the length and weight of resin that flow along a spiral cavity.

spiral gear A helical gear that transmits power from a driving shaft to a nonparallel driven shaft.

spiral welded pipe Pipe made by forming steel plate into long helical strips, fitting the strips together, and welding the spiral seams.

SPL See sound pressure level.

splash lubrication A method of lubricating a piston engine where the connecting-rod bearings dip into troughs filled with oil, splashing it onto other engine parts.

splice 1. To connect two pieces, forming a single longer piece, as in connecting the ends of wire, rope or tubing; the connection may be made by any of several methods including weaving and welding, and may be made with or without a connector. 2. A permanent junction between two optical fiber ends. It can be a mechanical splice, formed by gluing or otherwise attaching the ends together mechanically, or a fusion splice, formed by melting the ends together.

splice housing A housing designed to protect a splice in an optical fiber from damage such as from the application of stress on the fiber.

splice plate A piece of flat-rolled stock used to connect the webs or flanges of two girders together.

spline One of a set of axial keyways or gearlike ridges on the end of a shaft or the interior of a hub; in use, the splined shaft fits into a mating splined hub to transmit rotational power and motion, while permitting limited axial play between the two members.

split-beam colorimeter An instrument for determining the difference in radiation absorption by the sample at two wavelengths in the visible or ultraviolet region.

split-beam ultraviolet analyzer An instrument for monitoring the concentration of a specific chemical substance in a process stream or coating by measuring the amount of ultra-violet light absorbed at one wavelength and comparing it to the amount at a reference wavelength that is only weakly absorbed by the sample. Also known as a dual beam analyzer.

split bearing A journal bearing consisting of two semicylindrical pieces bolted together.

split body A body divided in half by a plane containing the longitudinal flow path axis [S75.05].

split clamp ends Valve end connections of various proprietary designs using split clamps to apply gasket loading [S75.05].

splitnut A nut that is cut in half lengthwise and hinged so that it can be rapidly engaged, on closing, and rapidly disengaged, on opening. Found on all thread cutting lathes.

split ranging See signal amplitude sequencing (split ranging).

splitter Plates spaced in an elbow of a duct so disposed as to guide the flow of fluid through the elbow with uniform distribution and to minimize pressure drop.

splitter vanes A set of curved, parallel strips of metal placed along the flow direction in a gas conduit to guide gas flow around a sharp bend in the conduit.

spoke A bar, rod or wire connecting the hub of a wheel to its rim.

spokeshave   A small tool for planning concave or convex surfaces.

sponge metal   Any metal mass produced by decomposition or chemical reduction of a compound at a temperature below the metal's melting temperature.

spontaneous combustion   Ignition of combustible materials following slow oxidation without the application of high temperature from an external source.

spool   1. The drum of a hoist. 2. The movable member of a slide-type hydraulic valve. 3. A reel or drum for winding up thread or wire. 4. A relatively short transition member (also known as a spool piece) for making a welded connection between two lengths of pipe.

spooling   The technique by which output to low-speed devices is placed into queues on faster devices to await transmission to the slower devices.

spot check   A type of random inspection in which only a very small percentage of total production is checked to verify that a process remains within its control limits.

spot drilling   Drilling a small, shallow hole in a surface to act as a centering guide in a subsequent machining operation.

spot face   A machined annular surface around a bolt hole on the side of a through bolted flange, opposite the gasket face, that is provided for nut seating [S75.05].

spot facing   Producing a flat, machined surface concentric with a drilled hole to serve as a seat for a washer or bolthead, or to allow for flush mounting of mating parts.

spot welding   A form of resistance welding where a weld nugget is produced along the interface between two pieces of metal, usually sheet metal, by passing electric current across the joint which is clamped between two small-diameter electrodes or between an electrode and an anvil or plate.

spray   A mechanically produced dispersion of liquid drops in a gas stream; the larger the drops, the greater must be the gas velocity to keep the drops from separating out by gravity.

spray angle   The angle included between the sides of the cone formed by liquid fuel discharged from mechanical, rotary atomizers and by some forms of steam or air atomizers.

sprayer plate   A metal plate used to atomize the fuel in the atomizer of an oil burner.

spray nozzle   A nozzle from which a liquid fuel is discharged in the form of a spray.

spray painting   A process in which compressed air atomizes paint and carries the resulting spray to the surface to be painted.

spray tower   A duct through which liquid particles descend countercurrent to a column of gas; a fine spray is used when the object is to concentrate the liquid, a coarse spray when the object is to clean the gas by entrainment of the solid particles in the liquid droplets.

spread (data processing)   In some performance measures, the total dispersion or spread of readings required. The spread S may be defined as, S R1 + R2 where R1 and R2 refer to the maximum deviation of the readings from either side of the mean value of the distribution. See repeatability [RP55.1].

spreadsheet program   In data processing, a program that will do a variety of calculations frequently needed by accountants and other businessmen.

spring   A machine element whose chief purpose is to store mechanical energy or to induce mechanical force through elastic deformation of the element's material; the element may be shaped in the form of a plate, leaf, flat-wound helix, coil or washer; it may be made of almost any relatively hard metal or alloy; it may be stressed in tension, compression, bending or torsion; and in most spring designs the amount of deflection is directly proportional to applied load—if the load is released, the element returns to its normal, unstressed shape or position.

springback   1. Movement of a part in the direction of recovering original size or shape upon release of elastic stress. 2. The amount of elastic deflection that occurs in cold-formed material upon release of the forming force; movement is in a direction opposite to the direction of plastic flow. 3. In flash, upset or pressure welding, the amount of deflection in the welding machine due to the upsetting pressure.

spring clip   1. A U-shaped fastener that attaches a leaf spring to an axle. 2. A fastener used chiefly in electrical connections that grips a part by elastic force.

spring coupling   A flexible coupling with resilient parts.

spring hook   A hook-shaped device with a spring-loaded member spanning the gap to form an eye; the spring-loaded member allows a bight of rope or cable to be quickly inserted into the eye and prevents the rope from slipping off the hook unless the member is deliberately depressed toward the center of the eye.

spring loaded seat A seat design that utilizes a mechanical means, such as a spring, to exert a greater force at the point of ball contact to improve the sealing characteristics, particularly at low pressure differential. The spring action may be accomplished by a metal spring arrangement or a compressed elastomer [S75.05].

spring rate The force change per unit change in length of a spring. This is usually expressed as pounds per inch or Newtons per millimeter [S75.05].

spring steel Carbon or low-alloy steel that is cold-worked or heat treated to give it the high yield strength normally required in springs; if it is a heat treatable composition, the springs may be formed prior to heat treatment (hardening).

spring temper A level of hardness and strength for nonferrous alloys and some ferrous alloys corresponding approximately to a cold worked state two-thirds of the way from full hard to extra spring temper.

sprocket A tooth on the periphery of a wheel or spool for engaging the links of a chain or the perforations in computer paper or motion picture film, or some other similar device, so that the chain or paper or film can be driven without slippage or will traverse the wheel without lateral movement, or both.

sprocket chain A flat chain, usually with pinned links, that meshes with the teeth of a sprocket for transmitting motion and mechanical power from one sprocket (the driving sprocket) to another (the driven sprocket).

sprocket hole Any of a series of perforations along the edge of motion picture film, paper tape, computer paper or continuous stationery which engage the teeth of a sprocket wheel or spool so the material can be driven through a mechanical device such as a camera, projector, printer, or recording instrument.

sprung arch An arch in the form of a segment of a circle supported by skew blocks at the two ends.

spur gear A toothed wheel whose teeth run parallel to the axis of the hub.

spurious error Errors due to instrument malfunction or to human goof-ups.

sputter-ion pump See getter-ion pump.

square mesh A weave in wire cloth or textile fabric where the number of wires or threads per inch is the same both with the weave and in the cross-weave direction.

square thread A machine thread with a square cross section; the widths of land and groove are each equal to one-half the pitch.

square wave A wave in which the dependent variable assumes one fixed value for one-half of the wave period, then assumes a second fixed value for the other half, with negligible time of transition between the two fixed values at each transition point.

squeegee A tool for spreading liquids onto a surface or scraping them off; it consists of a simple handle and a transverse blade with a flexible scraping edge usually made of rubber.

squeeze roll One of two opposing rollers designed to exert pressure on a material passing between them.

squeeze time In resistance welding, the time from initial application of pressure until welding current begins to flow.

SRF See side relief angle.

stability 1. The ability of a transducer to retain its performance characteristics for a relatively long period of time. NOTE: Unless otherwise stated, stability is the ability of a transducer to reproduce output readings obtained during its original calibration, at room conditions, for a specified period of time; it is then typically expressed as "within ____ percent of full scale output for a period ____ months." [S37.1]. 2. In data processing, a measure of the ability of a device to maintain constant volumes for one or more parameters describing its operation [RP55.1]. 3. Freedom from undesirable deviation. 4. A measure of the controllability of a process. 5. The relative ability of a substance to retain its mechanical, physical and chemical properties during service. 6. The relative ability of a chemical to resist decomposition during storage. 7. The ability of an electronic device or circuit to maintain specified operating characteristics over extended periods of service. 8. The ability of a machine element to retain its original dimensions when exposed to heat, humidity or other environmental conditions. 9. The relative ability of a body such as an aircraft or missile to maintain an attitude or to resist displacement from its flight path, and to restore its attitude or flight path if displaced. 10. The relative ability of a waterborne vessel to remain upright in a moving sea. 11. The state of a system if the magnitude of the response produced by an input variable, either constant or varied in time, is limited and related to the magnitude of the input variable.

stability of a linear system A linear system is stable if, having been displaced from its steady state by an external disturbance, it comes back to that steady state when the disturbance has ceased.

stabilizer 1. An airfoil or combination of airfoils, considered a single unit, with the principal function of maintaining stable flight for an aircraft or missile. 2. Any chemical added to a formulation for the chief purpose of maintaining mechanical or chemical stability throughout the useful life of the substance.

stabilizing treatment Any of various treatments—mechanical or thermal—intended to promote dimensional or microstructural stability in a metal or alloy.

stable element Any device, such as a gyroscope, used to maintain a stable spatial position for devices such as instrumentation or ordnance mounted in a ship or aircraft.

stack 1. The portion of a chimney above roof level. 2. Any structure that contains flues for discharging waste gases to the atmosphere. 3. A vertical conduit, which due to the difference in density between internal and external gases, creates a draft at its base. 4. An area of memory set aside for temporary storage, or for procedure and interrupt linkages. A stack uses the last-in, first-out (LIFO) concept. As items are added to ("pushed on") the stack, the stack pointer decrements; as items are retrieved from ("popped off") the stack, the stack pointer increments.

stack draft The magnitude of the draft measured at inlet to the stack.

stack effect That portion of a pressure differential resulting from difference in elevation of the points of measurement.

stack effluent Gas and solid products discharged from stacks.

stacker A machine for lifting goods on a platform or fork and placing them in tiered storage, such as in a warehouse.

stacker, card An output device that accumulates punched cards in a deck [RP55.1].

stack pointer (SP) The SP contains the address of the top (lowest) address of the processor-defined stack.

stage In electronics, that portion of a circuit between the control tap of one tube or transistor and the control tap of another.

staggered-intermittent fillet welding Welding a T joint on both sides of the tee in such a manner that the weld bead is segmented, with the segments on either side being opposite gaps between segments on the opposing side.

stagnation The condition of being free from movement or lacking circulation.

stagnation pressure A theoretical pressure that could be developed if a flowing fluid could be brought to rest without loss of energy (isentropically).

stagnation temperature The temperature that would be attained if all of the kinetic energy of a moving stream of fluid were converted to heat.

stain 1. A nonprotective liquid coloring agent used to bring out the grain in decorative woods. 2. A permanent or semipermanent discoloration on wood, metal, fabric or plastic caused by a foreign substance. 3. Any colored organic compound used to prepare biological specimens for microscopic examination.

stainless alloy Any member of a large and complex group of alloys containing iron, at least 5% chromium, and often other alloying elements, and whose principal characteristic is resistance to atmospheric corrosion or rusting. Also known as stainless steel.

stamping Virtually any metal forming operation carried out in a press.

standard air Dry air weighing 0.075 lb per cu ft at sea level (29.92″ barometric pressure) and 70 °F.

standard atmospheric pressure A reference pressure approximately equal to the mean atmospheric pressure at sea level; because atmospheric pressure varies with elevation and is not constant with time, standard atmospheric pressure is defined arbitrarily as an absolute pressure of 14.695 psi, 30.0 in. of mercury or 760 mm Hg (using mercury of density 13.595 g/cm^3).

standard cell A reference cell for electromotive force.

standard fit Any fit between mating parts whose allowance and tolerance have been standardized.

standard flue gas Gas weighing 0.078 lb per cu ft at sea level (29.92″ barometric pressure) and 70 °F.

standard gage 1. A highly accurate gage used

only as a reference standard for checking or calibrating working gages. 2. A set span across tracks of a railroad that measures 4 ft 8-1/2 in. (1.44 m).

standard leak A controlled finite amount of tracer gas allowed to enter a leak detector during adjustment and calibration.

standard sphere gap The maximum distance between the surfaces of two metal spheres, measured along a line connecting their centers, at which spark-over occurs when a dynamically variable voltage is applied across the spheres under standard atmospheric conditions; this value is a measure of the crest value of an alternating-current voltage.

standard wire rope Wire rope made of six wire strands laid around a sisal core.

standing wave A wave in which, for any of the dependent wave functions, the ratio of its instantaneous value at one point on the wave to its instantaneous value at any other point does not vary with time.

standing wave meter An instrument for measuring the standing-wave ratio in a radio frequency transmission line.

standpipe A vertical tube filled with a liquid such as water.

staple A fastener consisting of a U-shaped piece of wire with pointed ends; the fastener may be driven into a solid material such as wood as if it were a double-pointed nail, or it may be driven through thin sheets of paper or fabric and the ends folded over to hold the sheets together.

star A wiring technique where devices are interconnected via a central hub or wiring closet.

star coupler A coupler in which many fibers are brought together to a single optical element in which their signals are mixed. The mixed signals are then transmitted back through all the fibers.

starlan StarLAN is a proposed standard for 1 Mbit CSMA/CD on twisted pair medium. It is wired as a star with active hubs and is designed to use existing wiring. See TTP.

star network A set of three or more branches in an electronic network where one terminal of each branch is connected at a common node.

start bit The first bit in any asynchronous serial data transmission. Used to wake up the system; it carries none of the message information.

starter 1. An electric motor and gear used to turn the crankshaft of an internal combustion engine until its operation becomes self-sustaining. 2. In some chemical processes, a reactive mixture used to initiate a reaction between less reactive chemicals. Also known as starting mix.

starting resistance The force needed to produce an oil film in a set of journal bearings supporting a shaft when the shaft first begins to turn.

state 1. Condition of a circuit, system, etc., such as the condition at the output of a circuit that represents logic 0 or logic 1. 2. A description of the process in terms of its measured variables, or a description of the condition of a circuit or device as in "logic state 1".

statement A software instruction to a computer telling it to perform some sequence of operations.

state variables The output(s) of the memory element(s) of a sequential circuit.

static calibration 1. A calibration performed under room conditions and in the absence of any vibration, shock, or acceleration (unless one of these is the measurand) [S37.1]. 2. A calibration procedure during which the quantity of liquid is measured while the liquid is not flowing into or out of the measuring vessel [S37.1].

static connection A pipe tap on a manifold used to connect process pressure to an instrument.

static efficiency The mechanical efficiency multiplied by the ratio of static pressure differential to the total pressure differential, from fan inlet to fan outlet.

static friction See stiction.

static gain See gain, static.

static-head liquid-level meter A pressure-sensing device, such as a gage, so connected in the piping system that any dynamic pressures in the system cancel each other and only the pressure difference due to liquid head above the gage position is registered.

static model See steady-state model.

static pressure 1. The pressure of a fluid that is independent of the kinetic energy of the fluid. 2. Pressure exerted by a gas at rest, or pressure measured when the relative velocity between a moving stream and a pressure-measuring device is zero. See also pressure, static.

static pressure gage An indicating instrument for measuring pressure.

static pressure tube See static tube.

static RAM Random access memory which requires continuous power but does not need to be refreshed as with dynamic RAM. Memory density is not as high as for dynamic RAM.

static register A computer register which retains information in static form.

static seal See gasket.

static stability The property of a physical system which maintains constancy in its static and dynamic responses despite changes in its internal conditions and variations in its environment. Compare with dynamic stability.

static stores Digital registers in telemetry devices that hold set-up instructions from the computer.

static subroutine A subroutine which involves no parameters other than the addresses of the operands. Contrasted with dynamic subroutine.

static temperature The temperature of a fluid as measured under conditions of zero relative velocity between the fluid and the temperature-sensitive element, or as measured under conditions that compensate for any relative motion.

static test 1. Any measurement taken in a normally dynamic system under static conditions—for instance, a pressure test of a hydraulic system under no-flow conditions. 2. Specifically, a test to verify structural characteristics of a rocket, or to determine rocket-engine thrust, while a rocket is in a stationary or hold-down position.

static tube A device used to measure static pressure in a stream of fluid. Normally, a static tube consists of a perforated, tapered tube with a branch tube for connecting it to a manometer; a related device called a static pressure tube consists of a smooth tube with a rounded nose that has radial holes in the tube behind the nose.

static weighing A method in which the net mass of liquid collected is deduced from tare (empty tank) and gross (full tank) weighings respectively made before the flow is diverted into the weighing tank and after it is diverted to the by-pass.

stationary A gas detection instrument intended for permanent installation in a fixed location [S12.13] [S12.15].

stationary wave A standing wave in which the energy flux is zero at all points on the wave.

station management The portion of network management that applies to the lowest two OSI layers.

statistical error 1. Generally, any error in measurement resulting from statistically predictable variations in measurement system response. 2. Specifically, an error in radiation-counter response resulting from the random time distribution of photon-detection events.

statistically significant number of readings A statistically significant number of readings is a sample whose statistics closely approximate the true statistics of the parameter under consideration. That is, increasing the number of readings or repeatedly performing the data collection procedure will not result in substantially different calculated statistical parameters such as the mean and deviation [RP55.1].

statistical quality control Any method for controlling the attributes of a product or controlling the characteristics of a process that is based on statistical methods of inspection.

stator The stationary portion of a machine that interacts with a rotor to produce power or motion.

statoscope 1. A barometer for recording small changes in atmospheric pressure. 2. An instrument for indicating small changes in altitude of an aircraft.

statuary bronze Any of several copper alloys used chiefly for casting ornamental objects such as statues; a typical composition is 90% Cu-6% Sn-3% Zn-1% Pb.

status words Sixteen-bit words, available for computer input, that tell the status of telemetry or magnetic tape equipment.

stay A tensile stress member to hold material or other members rigidly in position.

staybolt A bolt threaded through or welded at each end, into two spaced sheets of a firebox or box header to support flat surfaces against internal pressure.

steady flow A flow in which the flow rate in a measuring section does not vary significantly with time.

steady state A characteristic of a condition, such as value, rate, periodicity, or amplitude, exhibiting only negligible change over an arbitrary long period of time. It may describe a condition in which some characteristics are static, others dynamic [S51.1] [S77.42].

steady-state deviation The system deviation after transients have expired. See offset; see also deviation, steady-state.

steady-state model A mathematical model that represents the process at equilibrium (infinite time) conditions.

steady-state optimization A method of optimizing some criterion function of a process usually using a steady-state model of the process.

Linear programming is frequently the optimization method used and a function approximating the profit of the process is a typical optimizing criterion. Contrast with dynamic optimization.

steady-state vibration A condition within a vibrating system where the velocity of each moving particle can be described by a periodic function.

steam The vapor phase of water substantially unmixed with other gases.

steam atomizing oil burner A burner for firing oil which is atomized by steam. It may be of the inside or outside mixing type.

steam attemperation Reducing the temperature of superheated steam by injecting water into the flow or passing the steam through a submerged pipe.

steam binding A restriction in circulation due to a steam pocket or a rapid steam formation.

steam cock A valve for admitting or releasing steam.

steam cure To hasten the curing cycle of concrete or mortar by the use of heated water vapor, at either atmospheric or higher pressure.

steam dryer A device for removing water droplets from steam. See steam scrubber.

steam-free water Water containing no steam bubbles.

steam gage A device for measuring pressure in a steam system.

steam generating unit A unit to which water, fuel, and air are supplied and in which steam is generated. It consists of a boiler furnace, and fuel burning equipment, and may include as component parts water walls, superheater, reheater, economizer, air heater, or any combination thereof.

steam jacket A casing around the cylinders and heads of a steam engine, or around some other mechanism or space, to keep the surfaces hot and dry.

steam jacketed valve See jacketed valve.

steam-jet blower A device which utilizes the energy of steam flowing through a nozzle or nozzles to induce a flow of air to be supplied for combustion.

steam purity The degree of contamination. Contamination usually expressed in ppm.

steam quality The percent by weight of vapor in a steam and water mixture.

steam scrubber A series of screens, wires, or plates through which steam is passed to remove entrained moisture.

steam separator A device for removing the entrained water from steam.

steam trace The technique of preventing freezing in a pipe or tubing line with an adjacent steam line; usually 1/4″ to 1/2″ copper tubing. See heat tracing.

steam tracing An arrangement for heating a process line or instrument-air line to keep liquids from freezing or condensing—often, a piece of pipe or tubing carrying live steam is simply run alongside or coiled around the line to be heated.

steam trap A device that automatically collects condensate in a steam line and drains it away.

steel Any alloy of iron with up to 2% carbon that may or may not contain other alloying elements to enhance strength or other properties.

stellite Any of a family of cobalt-containing alloys known for their wear resistance, corrosion resistance, and resistance to softening at high temperature.

stem 1. The rod, shaft or spindle which connects the valve actuator with the closure member [S75.05]. 2. The part, usually a rod or shaft, which connects to the valve stem and transmits motion (force) from the actuator to the valve. The actuator stem delivering an output thrust may or may not be the same stem as that on the power unit stem [S75.05]. 3. A rod connecting a knob or handwheel to the moving part it operates.

stem, actuator The port which connects the actuator to the valve stem or shaft and transmits motion (force) from the actuator to the valve [S75.05].

stem anti-rotation device A mechanical means of preventing rotation of the linear actuator stem and/or valve stem [S75.05].

stem bearings Butterfly stem bearings are referred to as either the outboard or the inboard type, depending on their location, outside or inside of the stem seals [S75.05].

stem boot A protective device similar to a flexible bellows, used outside the bonnet to protect the valve stem from the surrounding atmosphere [S75.05].

stem connector The device which connects the actuator stem to the valve stem [S75.05].

stem guide A guide bushing closely fitted to the valve stem and aligned with the seat [S75.05].

stem rotation A phenomenon which occurs in

linear motion valves when the hydraulic forces from the process fluid cause the closure component to rotate about the stem axis.

stem seals The part or parts needed to effect a pressure-tight seal around the stem while allowing movement of the stem [S75.05].

stem, valve In a linear motion valve, the part which connects the actuator stem with the closure component [S75.05].

step 1. One operation in a computer routine. 2. To cause a computer to execute one operation.

step bearing A bearing that supports the lower end of a vertical shaft. Also known as pivot bearing.

step brazing Making a series of brazed joints in a single assembly by sequentially making up individual joints and heating each one at a lower temperature than the previous joint to maintain joint integrity of earlier joints; the process requires a lower melting brazing alloy for each successive joint in the assembly.

step change The change from one value to another in a single increment in negligible time.

step gage 1. A plug gage consisting of a series of cylindrical gages of increasing diameter mounted on the same axis. 2. A gage for measuring the height of a step or shoulder; it consists of a gage body and a sliding blade.

step-index fiber An optical fiber in which there is a discontinuous change in refractive index at the boundary between fiber core and cladding. Such fibers have a large numerical aperture (light accepting angle), and are simple to connect; but have lower bandwidth than other types of optical fibers.

stepping motor A motor useful for low torque applications and suitable for computer interfacing. Pulse input results in a precise rotary step, typically $0.8°$ per pulse or $1.6°$ per pulse. It is often operated in open loop mode.

step response 1. Of a system or a component, the time required for an output to go through a specified percentage of the total excursion either before, or (in the absence of overshoot) as a result of a step change to the input. NOTE: This is usually stated for 90, 95, or 99 percent change [S67.06]. 2. The time response of a device or process when subjected to an instantaneous change in input from one steady-state value to another. See also response, step.

step response time See time, step response.

step soldering Making a series of joints by sol-

dering them sequentially at successively lower temperatures.

steradian The solid angle subtended at the center of a sphere by an area on the surface equal to a square with sides of length equal to the radius of the sphere.

stereophonics Reproducing or reinforcing sound by using two or more audio channels so that the sound gives three-dimensional sensations similar to those of the sound sources.

stick gage A vertical rod or stick with a graduated scale or markings that is fixed in an open tank or vessel so that liquid level changes can be observed directly.

stiction (static friction) Resistance to the start of motion, usually measured as the difference between the driving values required to overcome static friction upscale and downscale [S51.1].

stiffener A plate, angle, channel or similar structural element attached to a slender beam or column to prevent it from buckling by increasing its stiffness.

stiffness In process instrumentation, the ratio of change of force (or torque) to the resulting change in deflection of a spring-like element. Stiffness is the opposite of compliance [S51.1].

Stilb A unit of luminescence equal to one candela per cm^2; it is rarely used, as the candela per m^2 is preferred.

stilling basin An area ahead of the weir plate large enough to pond the liquid so that it approaches the weir plate at low velocity, also called weir pond.

stimulate To cause an occurrence or action artificially, rather than waiting for it to occur naturally, as to stimulate an event.

stimulus See measurand.

stitch bonding A method of making wire connections on an integrated circuit board using impulse welding or heat and pressure to bond a connecting wire at two or more points while feeding the wire through a hole in the welding electrode.

stitching 1. Making a seam in fabric using a sewing machine. 2. Progressive welding of thermoplastics by successively pressing two small induction-heated electrodes against the material along a seam in a manner resembling the action of a sewing machine.

stitch welding Making a welded seam using a series of spot welds that do not overlap.

stochastic Pertaining to direct solution by trial-

and-error, usually without a step-by-step approach, and involving analysis and evaluation of progress made, as in a heuristic approach to trial-and-error methods. In a stochastic approach to a problem solution, intuitive conjecture or speculation is used to select a possible solution, which is then tested against known evidence, observations or measurements. Intervening or intermediate steps toward a solution are omitted. Contrast with algorithmic and heuristic.

stock Material, parts or components kept in storage until needed.

stockpile A reserve stock of supplies in excess of normal usage.

Stoddard solvent A specific type of petroleum naphtha used chiefly in dry cleaning, but also used in small quantities for cleaning soiled surfaces by hand.

stoichiometric conditions In chemical reactions, the point at which equilibrium is reached, as calculated from the atomic weights of the elements taking part in the reaction; stoichiometric equilibrium is rarely achieved in real chemical systems but, rather, empirically reproducible equivalence points are used to closely approximate stoichiometric conditions.

stoker A mechanized means of feeding coal or other solid combustibles into a furnace, burning them under controlled conditions, and carrying away solid combustion products.

Stokes A unit of kinematic viscosity (dynamic viscosity divided by sample density); the centistoke is more commonly used.

stoneware Glazed ceramic ware used in certain laboratory and industrial applications involving corrosive chemicals.

stop bit The last bit in an asynchronous serial transmission. Like the start bit, it is used for timing control and carries none of the message information.

stop cock A small valve for roughly controlling or shutting off the flow of fluid in a pipe.

stop nut 1. A nut positioned on an adjusting screw to restrict its travel. 2. A nut with an insert made of a compressible material that keeps the nut tight without requiring a lock washer.

storage 1. Pertaining to a device in which data can be stored and from which it can be obtained at a later time. The means of storing data may be chemical, electrical or mechanical. 2. A device consisting of electronic electrostatic, electrical hardware, or other elements into which

data may be entered and from which data may be obtained as desired. 3. The erasable storage in any given computer. Synonymous with memory.

storage address register A portion of core memory in a computer that contains the address of a storage location to be activated, either for reading the contents of the location or for storing information at the location.

storage allocation The process of reserving storage for specified information.

storage block A contiguous area of main or secondary storage.

storage buffer 1. A synchronizing element between two different forms of storage, usually between internal and external. 2. An input device in which information is assembled from external or secondary storage and stored, ready for transfer to internal storage. 3. An output device into which information is copied from internal storage and held for transfer to secondary or external storage. Computation continues while transfers between buffer storage and secondary or internal storage or vice versa take place. 4. Any device which stores information temporarily during data transfer. Clarified by buffer.

storage calorifier See cylinder.

storage capacity The amount of data that can be contained in a storage device.

storage cell An elementary unit of storage, for example, a binary cell, a decimal cell.

storage cycle A periodic sequence of events occurring when information is transferred to or from the storage device of a computer. Storing, sensing, and regeneration form parts of the storage sequence.

storage device A device into which data can be inserted, in which it can be retained, and from which it can be retrieved.

storage dump A listing of the contents of a storage device, or selected parts of it. Synonymous with memory dump. See also core dump.

storage key An indicator associated with a storage block or blocks, which requires that tasks have a matching protection key to use the blocks.

storage location A storage position holding one machine word and usually have a specific address.

storage, main See main storage.

storage protection An arrangement of preventing access to storage for either reading or writing, or both. See memory protect.

storage register A register in the storage of the computer, in contrast with a register in one of the other units of the computer.

store 1. To enter data into a storage device. 2. To retain data in a storage device.

stored program See stored routine.

stored program computer A computer controlled by internally stored instructions that can synthesize, store, and in some cases alter instructions as though they were data, and that can subsequently execute these instructions.

stored routine A series of instructions in storage to direct the step-by-step operation of the machine. See stored program.

straightening vanes Horizontal vanes inside a fluid conduit or pipe to reduce turbulent flow ahead of an orifice or venturi meter.

straight polarity Arc welding in which the electrode is connected to the negative terminal of the power supply.

straight-tube boiler See boilers.

strain The deformation per unit length produced in a solid as a result of stress [S37.1].

strain aging A change in properties of a metal or alloy that occurs at room or slightly elevated temperature following cold working.

strainer A screen or porous medium positioned in a flowing stream of fluid (such as a water intake) to separate out harmful objects or particles before the fluid enters process equipment.

strain error The error resulting from a strain imposed on a surface to which the transducer is mounted. NOTE 1: This term is not intended to relate to strain transducers (strain gages). 2: Also see mounting error [S37.1].

strain foil A type of strain gage made by photoetching a resistance element out of thin foil.

strain gage 1. Converting a change of measurand into a change or resistance due to strain [S37.1]. 2. A device that can be attached to a surface, usually with an adhesive, and that indicates strain magnitude in a given direction by changes in electrical resistance of fine wire; it may be used to measure strain due to static or dynamic applied loading, in tension or compression, or both, depending on design of the gage, bonding technique, and type of instrumentation used to determine resistance changes in the strain element. 3. A high-resistance, fine-wire or thin-foil grid for use in a measuring bridge circuit. When the grid is securely bonded to a

specimen, it will change its resistance as the specimen is stressed. These devices are used in many forms of transducers. 4. A transducer that converts information about the deformation of solid objects, called the strain, into a change of resistance.

strain hardening The increase in tensile and yield strengths, and the corresponding reduction in ductility, associated with plastic deformation of a metal at temperatures below its recrystallization range.

strain rosette An assembly of two or more strain gages used for determining biaxial stress patterns. Also known as rosette strain gage.

strain sensitivity 1. The sensitivity to strains applied to the base by bending, in the absence of any rigid body motion of the transducer. It is expressed as 10^{-6} times the equivalent acceleration level in g's for a strain in the plane of the base [RP37.2]. 2. A characteristic of a conductor that describes its resistance change in relation to a corresponding length change; it can be calculated as $\Delta R/R$ divided by $\Delta L/L$; when referring to a specific strain gage material, strain sensitivity is commonly known as the gage factor.

strand 1. One of several wires that are twisted together to form wire rope, cable or electrical conductors. 2. One of the fibers or filaments used to produce yarn, thread, rope or cordage. 3. A piece of cable, rope, string, thread, wire or yarn of specified length. 4. A bar, billet, bloom or slab produced by continuous casting.

strand casting See continuous casting.

strap bolt 1. A bolt with a hook or flattened extension instead of a head. 2. A double-ended bolt with a flattened, nonthreaded center section that can be bent around an object to form a U-bolt.

strategy, frame synchronizer The procedure defined by an operator to emphasize rapid acquisition, or to emphasize accuracy of acquisition, or any point between those extremes.

stratification Non-homogeneity existing transversely in a gas stream.

stray current corrosion Galvanic corrosion of a metal or alloy induced by electrical leakage currents passing between a structure and its service environment.

stream An input data path to the computer from a single telemetry source, as PCM, PAM, and so on.

streamline flow A type of fluid flow in which

flow lines within the bulk of the fluid remain relatively constant with time.

streamlining   Contouring the exterior shape of a body to reduce drag due to relative motion between the body and a surrounding fluid.

stream tube   In the characterization of fluid flow, an imaginary tube whose wall is generated by streamlines passing through a closed curve.

street elbow   A pipe elbow with an external thread at one end and an internal thread at the other end.

strength weld   A weld capable of withstanding a design stress.

stress amplitude   One-half the algebraic difference between the maximum stress and minimum stress in one cycle of repeated variable loading.

stress-corrosion cracking   Deep cracking in a metal part due to the synergistic action of tensile stress and a corrosive environment, causing failure in less time than could be predicted by simply adding the effects of stress and the corrosive environment together. The tensile stress may be a residual or applied stress, and the corrosive environment need not be severe but only must contain a specific ion that the material is sensitive to.

stress raiser   A discontinuity or change in contour that induces a local increase in stress in a structural member.

stress relieving   Heating to a suitable temperature, holding long enough to reduce residual stress, and then cooling slowly enough to avoid inducing new residual stresses.

stretcher leveling   Removing warp and distortion in a piece of metal by gripping it at both ends and subjecting it to tension loading at stresses higher than the yield strength.

stretch forming   Shaping a piece of sheet metal or plastics sheet by applying tension and then wrapping the sheet around a die form; the process may be performed cold or the sheet may be heated first. Also known as wrap forming.

striation technique   A method of making sound waves in air visible by using their individual ability to refract light.

strike   1. A thin electroplated film to be followed by other plated coatings. 2. A plating solution of high covering power and low efficiency used for electroplating very thin metallic films. 3. A local crater or remelted zone caused by accidental contact between a welding electrode and the surface of a metal object; also known as arc strike.

string   1. In data processing, a group of consecutive characters. 2. A linear sequence of entities, such as characters of physical elements.

string lines   Wires, piano wire, or monofilament line of suitable tensile strength and visibility, strung over each of the three rolls of the weigh idlers to confirm idler alignment and elevation (three-wire line alignment) [RP74.01].

string manipulation   The handling of string data by various methods generally in terms of bits, characters, and sub-strings.

string-shadow instrument   An indicating instrument in which the measured value is indicated by means of the shadow of a filamentary conductor whose position in an electric or magnetic field depends on the magnitude of the quantity being measured.

strip   1. A flat-rolled metal product of approximately the same thickness range as sheet but having a width range narrower than sheet. 2. To remove insulation from the end of a wire or cable. 3. To mine stone, coal or ore without tunneling, but rather by removing broad areas of the earth's surface to relatively shallow depths.

strip chart   A hardware device that records analog data (generally, six or eight channels) on a continuous chart.

strip-chart recorder   Any instrument that produces a trace or series of data points, using one or more pens or a print wheel, on a grid printed on a continuous roll of paper that is moved at a uniform rate of travel in a direction perpendicular to the motion of the instrument's indicating mechanism. The resulting trace is a graph of the measured variable as a function of time.

stripper   A distillation column that has no rectifying section. In such a column, the feed enters at the top, and there is no other reflux.

stripping section   That section of a distillation column below the feed. This section strips the light components from the liquid moving down the column.

strip printer   A device that prints the output from a computer, telegraph or recording instrument on a very narrow, continuous length of paper tape.

strobe pulse   A pulse of light whose duration is less than the period of a recurring event or periodic function, and which can be used to render a specific event or characteristic visible so it can be closely observed.

stroboscope A device for intermittently viewing or illuminating moving bodies so that they appear to be motionless, either by placing an intermittent shutter between the object and an observer or by repeatedly flashing a brilliant light on the object. In this manner, a vibrating or rotating object can be made to appear stationary by adjusting the stroboscope's frequency; the indicated frequency of the stroboscope is equal to the object's vibrational or rotational frequency.

stroboscopic tachometer A stroboscopic lamp and variable-flashing-rate control circuit that enables the frequency to be adjusted until a rotating object appears to stand still; the frequency is read from a calibrated dial, and represents either the fundamental rotational speed in cycles per unit time or one of its harmonics; sometimes, a patterned disk centered on the axis of rotation is used to make it easier to determine fundamental frequency.

stroke The linear extent of movement of a reciprocating mechanical part. See also travel.

stroke cycle Travel of the closure member from its closed position to the rated travel opening and return to the closed position [S75.05]. See also travel cycle.

stroke time The time required for one-half a stroke cycle at specified conditions [S75.05]. See also travel time.

Strouhal number A nondimensional parameter defined as: $S=fh/V$, where f is frequency, V is velocity and h is reference length.

structural analysis Determination of the stresses and strains in a structural member due to combined gravitational and applied service loading.

structural steel Hot-rolled steel produced in standard sizes and shapes for use in constructing load-bearing structures, supports and frameworks; some of the standard shapes are angles, channels, I-beams, H-beams and Z-sections.

stud 1. A headless bolt threaded at both ends. 2. A threaded fastener with one end intended for welding to a metal surface. 3. A rivet, boss or nail with a large ornamental head. 4. A projecting pin serving as a support or means of attachment.

stud arc welding (SW) Metals are heated with an arc between a metal stud, or similar part, and the work. Once the surfaces to be joined are properly heated, they are brought together under pressure.

stud welding Producing a joint between the end of a rod-shaped fastener and a metal surface, usually by drawing an arc briefly between the two members then forcing the end of the fastener into a small weld puddle produced on a metal surface.

stuffing See packing.

stuffing box A cavity around a rod or shaft that penetrates a pump casing, valve body or other portion of a pressure boundary which can be filled with packing material and compressed to form a leak-tight seal while still permitting axial or rotary motion of the shaft.

stylus 1. Generically, any device that produces a recorded trace by direct contact with a chart or similar recording medium. 2. A needle-shaped device that follows the grooves in a phonograph record and converts the resulting mechanical vibrations into an audio-frequency signal.

subassembly An assembled group of parts intended for incorporation into a device or mechanism as a unit; often a subassembly performs a specific function independently or in conjunction with other subassemblies, and can be removed from the device for maintenance or repair without completely disassembling the device itself.

subcarrier A carrier applied as a modulating wave to another carrier or an intermediate subcarrier.

subcarrier band A band (of frequencies) associated with a given subcarrier and specified in terms of maximum subcarrier deviation.

subcarrier channel The channel required to convey telemetry information involving a subcarrier band.

subcarrier discriminator In FM telemetry, the device which is tuned to select a specific subcarrier and demodulate it to recover the data.

subcarrier oscillator The basic subcarrier frequency generator whose output frequency is used as the transmission or carrier medium of desired signal information; in telemetry, the desired signal information is most often used to frequency modulate the subcarrier for transmission.

subcommutation Commutation of a number of channels with the output applied to an individual channel of the primary commutator; subcommutation is synchronous if its rate is a submultiple of that of the primary commutator. Unique identification must be provided for the subcommutation frame pulse.

subcommutation frame In PCM systems, a recurring integral number of subcommutator words, which includes a single subcommutation frame synchronization word. The number of words in subcommutation frame is equal to an integral number of primary commutator frames. The length of a subcommutation frame is equal to the total number of words or bits generated as a direct output of the subcommutator.

subdirectory In MS-DOS, a file that is stored in another directory. See root directory.

subframe A multiplex generated at a slower rate than a frame, and input to the frame through one of the channels.

subharmonic A sinusoidal function whose frequency is a submultiple of some other periodic function to which it is related.

sublayer A subdivision of an OSI layer; e.g. the IEEE 802 Standard divides the link layer into the LLC and MAC sublayers.

submerged-arc welding An electric-arc welding process in which the arc between a bare-wire welding electrode and workpiece is completely covered by granular flux during welding.

submergence The distance measured from the crest level to the downstream water surface when the flow is submerged, i.e., no air is contained beneath the nappe.

submersible pump A pump and electric motor housed together in a water-tight enclosure so that the unit may operate when submerged.

submultiplexer boundary See submiltiplexer group.

submultiplexer group In analog signal multiplexer construction, it is common to separate the analog input channels into groups of, typically, 4 to 64 channels. The outputs of the channels included in any single group are bussed together and provide an input to another (second level) multiplexer, the output of which is commonly connected to the subsystem amplifier or analog to digital converter. The group of input channels connected together to form a single group in such a multi-level multiplexer is called a submultiplexer group in this standard. The submultiplexer boundary for such a group is the electrical boundary defined by the second level multiplexer switching device connected to the common output of the submultiplexer group

[RP55.1].

submultiplexing See block switching.

sub-network access protocol (SNAP) Provides a mechanism to uniquely identify private protocols above LLC.

sub-optimization The process of fulfilling or optimizing some chosen objective which is an integral part of a broader objective. Usually the broad objective and lower-level objective are different.

subprogram A part of a larger program which can be converted into machine language independently.

subroutine 1. The set of instructions necessary to direct the computer to carry out a well-defined mathematical or logical operation. 2. A sub-unit of a routine. A subroutine is often written in relative or symbolic coding even when the routine to which it belongs is not. 3. A portion of a routine that causes a computer to carry out a well-defined mathematical or logical operation. 4. A routine which is arranged so that control may be transferred to it from a master routine and so that, at the conclusion of the subroutine, control reverts to the master routine. Such a subroutine is usually called a closed subroutine. 5. A single routine may simultaneously be both a subroutine with respect to another routine and a master routine with respect to a third. Usually control is transferred to a single subroutine from more than one place in the master routine and the reason for using the subroutine is to avoid having to repeat the same sequence of instructions in different places in the master routine. Clarified by routine. 6. Any of several branches in a computer program or repetitive task which are used to perform a specific function when certain defined conditions are encountered during execution of the main process routine.

subroutine call The subroutine, in object coding, that performs the call functions.

subroutine library A set of standard and proven subroutines which is kept on file for use at any time.

subscale Subsurface oxides formed by reaction of a metal with oxygen that diffuses into the interior of the section rather than combining with metal in the surface layer.

subset 1. In data processing, any set of items that relate to a larger set. 2. A set within a set. 3. A subscriber apparatus in a communications network.

subsidence See damping, also subsidence ratio.

subsidence ratio In process instrumentation, the ratio of the peak amplitudes of two successive oscillations of the same sign measured from an ultimate steady-state value, the numerator representing the first oscillation in time [S51.1].

subsieve analysis Determination of particle-size distribution in a powdered material, none of which is retained on a standard 44-micrometer sieve.

subsieve fraction The portion of a powdered material that passes through a standard 44-micrometer sieve.

subsonic 1. A generic term roughly designating a speed less than the speed of sound in a given fluid medium. 2. For an aircraft, any speed from hovering (zero) up to about 85% of the speed of sound in the atmosphere at ambient temperature.

substitute power supply Supply equipment that may be used instead of a battery supply for the equipment [S82.01].

substrate A surface underlying a coating such as paint, porcelain enamel or electroplate.

sub-subcommutation Commutation of a number of channels with the output applied to an individual channel of a subcommutator; unique identification must be provided for sub-subcommutation frame synchronization.

subsystem 1. Any assemblage of components which operates as a part of a system, and which is collectively capable of performing a task within, and as defined by that system [RP55.1]. 2. A portion of a larger system consisting of several components or process units which, together, have the characteristics of a system by themselves.

subsystems Interconnected elements provided by a single supplier [S50.1].

successive approximation A type of analog-to-digital conversion that compares the unknown input with sums of accurately known binary fractions of full scale, starting with the largest, and rejecting any that changes the comparator's state. At the end of conversion, the output of the converter is a digital representation of the ratio of the input to full scale by a fractional binary code.

suction lift The pressure, in feet of fluid, that a pump must induce on the suction side to raise the fluid from the level in the supply well to the level of the pump. Also known as suction head.

suction line A tube, pipe or conduit that leads fluid from a reservoir or intake system to the intake port of a pump or compressor.

sulfonated oil Mineral or vegetable oil treated with sulfuric acid to make an emulsifiable form of oil.

sulfurized oil Any of various oils containing active sulfur to increase film strength and load carrying ability.

sulphate-carbonate ratio The proportion of sulphates to carbonates, or alkalinity expressed as carbonates, in boiler water. The proper maintenance of this ratio has been advocated as a means of inhibiting caustic embrittlement.

sum The quantity resulting from the addition of an addend to an augend.

summation action A type of control-system action where the actuating signal is the algebraic sum of two or more controller output signals, or where it depends on a feedback signal which is the algebraic sum of two or more controller output signals.

summing point Any point at which signals are added algebraically [S51.1].

sump A tank or pit for temporarily storing drainage.

sump pump A small, single-stage vertical pump used to remove drainage from a shallow well or pit.

supercalendered finish A shiny, smooth finish on paper obtained by subjecting the material to steam and pressure while passing it between alternating fiber-filled and steel rolls.

supercharger A device such as an air pump or blower fitted into the intake of an internal combustion engine to raise the pressure of combustion air above the pressure that can be developed by natural aspiration.

supercommutation Commutation at a rate higher than once per commutator cycle, accomplished by connecting a single data input source to equally spaced contacts of the commutator (cross-patching); corresponding cross-patching is required at the decommutator.

supercompressibility The extent to which behavior of a gas departs from Boyle's law.

superconductor A compound capable of exhibiting superconductivity—that is, an abrupt and large increase in electrical conductivity as the material's temperature approaches absolute zero.

superfines The portion of a metal powder whose particle size is less than 10 micrometers.

superfinishing Producing a finely honed surface by rubbing a metal with abrasive stones.

superheat To raise the temperature of steam above its saturation temperature. The temperature in excess of its saturation temperature.

superheated steam Steam at a higher temperature than its saturation temperature.

superheater A nest of tubes in the upper part of a steam boiler whose function is to raise the steam temperature above saturation temperature.

superluminescent diode A compromise between a diode laser and LED, which is operated at the high drive currents characteristic of diode lasers, but lacks the cavity-mirror feedback mechanisms that produce stimulated emission. It is used when high power output is desired, but coherent emission is not wanted.

supernatant liquor The liquid above settled solids, as in a gravity separator.

superplasticity The unusual ability of some metals and alloys to elongate uniformly by several thousand percent at elevated temperatures without separating.

supersonic 1. A generic term roughly designating a speed that exceeds the speed of sound in a given fluid medium. 2. For an aircraft, any speed that exceeds Mach 1, which is about 650 to 750 mph depending on atmospheric conditions and altitude.

supervisory control 1. A term used to imply that a controller output or computer program output is used as an input to other controllers, e.g., generation of setpoints in cascaded control systems. Used to distinguish from direct digital control. 2. An analog system of control in which controller setpoints can be adjusted remotely, usually by a supervisory computer. 3. Also known as a digitally-directed analog (DDA) control system. See also control, supervisory.

supervisory control and data acquisition (SCADA) Technique in industrial control/monitor work.

supervisory control system (SCS) Remote setpoint information to single loop analog controllers provided by a digital computer.

supervisory program 1. A program used in supervisory control. 2. Same as executive program.

supervisory set point control system The generation of set point and/or other control information by a computer control system for use by shared control, shared display or other regulatory control devices [S5.3].

supply circuit The circuit supplying electrical energy to the equipment from a branch circuit, a battery, or a power supply [S82.01].

supply equipment Equipment that takes energy from an electrical supply, generally a branch circuit, and supplies it in a modified form to energize other equipment [S82.01].

supply pressure 1. The pneumatic pressure supply which enables the system element to generate the pneumatic transmission signals specified in this standard or to provide the valve or final element with required operational force [S7.4]. 2. In a hydraulic or pneumatic system, the output pressure from the primary source of pressure, which is subsequently regulated or controlled to provide desired system functions. See also pressure, supply.

supporting electrode An electrode in a spectroscopic apparatus, other than a self electrode, that is designed to hold the analytical sample on or inside it.

support system A programming system used to support the normal translating functions of machine-oriented, procedural-oriented, and problem-oriented language processors.

suppressed range A suppressed range is an instrument range which does not include zero. The degree of suppression is expressed by the ratio of the value at the lower end of the scale to the span. See range, elevated-zero.

suppressed span See range, elevated-zero.

suppressed weir A rectangular weir in which the width of the approach channel is equal to the crest width, i.e., there are no end contractions.

suppressed-zero instrument Any indicating or recording instrument whose zero (no load) indicator position is offscale, below the lower limit of travel for the pointer or marking device.

suppressed-zero range See range, suppressed-zero.

suppression See range, elevated-zero.

suppression ratio (of a suppressed-zero range) The ratio of the lower range-value to the span [S51.1].

surface The exterior skin of a solid body, considered to have zero thickness.

surface analyzer An instrument that measures irregularities in the surface of a body by moving a stylus across the surface in a predetermined pattern and producing a trace showing minute differences in height above a reference plane magnified as much as 50,000 times.

surface area 1. The total amount of exterior area on a solid body. 2. The sum of the individual surface areas of all the particles in a mass of particulate matter.

surface blowoff Removal of water, foam, etc. from the surface at the water level in a boiler. The equipment for such removal.

surface combustion The non-luminous burning of a combustible gaseous mixture close to the surface of a hot porous refractory material through which it has passed.

surface condenser Any of several designs for inducing a change of state from gas to liquid by allowing the gas phase to come in contact with a surface such as a plate or tube which is cooled on the opposite side, usually by being in direct contact with flowing cooled water.

surface density Any amount distributed over a surface, expressed as amount per unit area of surface.

surface finish The roughness of a surface after finishing, measured either by comparing its appearance with a set of standards of different patterns and lusters or by measuring the height of surface irregularities with a profilometer or surface analyzer.

surface gage 1. A scribing tool in an adjustable stand that is used to check or lay out heights above a reference plane. 2. A gage for measuring height above a reference plane.

surface grinder A machine for grinding a plane surface; usually consists of a motor-driven wheel made of bonded abrasive and mounted on an arbor above a reciprocating table that holds the workpiece.

surface hardening Any of several processes for producing a surface layer on steel that is harder and more wear resistant than the softer, tougher core; the process usually involves some kind of heat treatment, and may or may not involve changing the chemical composition of the surface layer.

surface plate A table, usually made of granite or steel at least 2 ft square, that has a very accurate flat plane surface; it is used primarily in inspection and layout work as a reference plane for determining heights.

surface roughness Minute pits, projections, scratches, grooves and the like which represent deviations from a true planar or contoured surface on solid material.

surface treating Any of several processes for altering properties of a metal surface, making it more receptive to ink, paint, electroplating, adhesives or other coatings, or making it more resistant to weathering or chemical attack.

surfacing Depositing filler metal on the surface of a part by welding or thermal spraying.

surge 1. A transient variation in the current and/or potential at some point in the circuit. 2. An upheaval of liquid in a process system, which may result in carryover of liquid into vapor lines. 3. An unstable pressure buildup in a process system. 4. The peak system pressure. 5. The sudden displacement or movement of water in a closed vessel or drum.

surge, in compressors An unstable operating regime in which internal oscillations persist.

surge pressure See pressure, surge.

surge protector A device positioned between a computer and a power outlet designed to absorb power bursts that could damage the computer.

surge tank 1. A vessel used to absorb fluctuations in flow so that they are not passed on to other units. 2. A standpipe or storage reservoir in a downstream channel or conduit to absorb sudden rises in pressure and to prevent starving the conduit during sudden drops in pressure. 3. An open tank connected to the top of a surge line which maintains steady loading on a pump.

surveillance 1. Systematic observation of an area—usually by visual, electronic or photographic means—to gather intelligence for military or law-enforcement purposes. 2. Systematic observation of a process or operation while it is being performed to verify the use of proper equipment, materials, procedures and methods.

survivor curve A type of reliability curve that shows the average percent of total production of a given model or type of machine still in service after various lengths of service life in hours.

susceptibility meter An instrument for determining magnetic susceptibility at low magnitudes.

susceptometer A device for measuring the magnetic susceptibility of ferromagnetic, paramagnetic or diamagnetic materials.

suspended arch An arch in which the refractory blocks or shapes are suspended by metallic hangers.

suspended solids Undissolved solids in boiler water.

suspension 1. A fine wire or coil spring that

supports the moving element of a meter or other instrument. 2. A system of springs, shock absorbers and other devices that support the chassis of a motor vehicle on its running gear. 3. A mixture of finely divided insoluble particles of solid or liquid in a carrier fluid (liquid or gas). 4. A system of springs or other devices that support an instrument or sensitive electronic equipment on a frame and reduce the intensity of mechanical shock or vibration transmitted through the frame to the instrument.

SW See stud arc welding.

swaging Any of several methods of tapering or reducing the diameter of a rod or tube, most commonly involving hammering, forging, or squeezing between simple concave dies.

swapping The process of copying areas of memory to mass storage, and back, in order to use the memory for two or more purposes. Data are swapped out when a copy of the data in memory is placed on a mass storage device; data are swapped in when a copy on a mass storage device is loaded in memory.

swapping device A mass storage device that is especially suited for swapping because of its fast transfer rate.

sweat The condensation of moisture from a warm saturated atmosphere on a cooler surface. A slight weep in a boiler joint but not in sufficient amount to form drops.

sweet crude Crude petroleum containing very little sulfur.

sweet gas Natural gas containing no hydrogen sulfide or mercaptans.

swell 1. An increase (swell) in drum level due to an increase in steam bubble volume. This condition is due to an increase in load (steam flow), with a resulting decrease in drum pressure and an increase in heat input. Swelling also occurs during a cold start-up as the specific volume of the water increases [S77.42]. 2. The sudden increase in the volume of steam in the water steam mixture below the water level.

swell plug Consists of a piston actuator coaxial with the flow-path axis and suspended from the body wall by radial fins. The piston compresses an annular elastomer member which forces it to close the annulus between the piston outside diameter and the body internal diameter [S75.05].

swinging load A load that changes at relatively short intervals.

swing joint A connection between two pipes that allows them to be repositioned with respect to each other.

swing pipe A discharge pipe whose inlet end can be raised or lowered within a tank.

swirl A qualitative term, describing tangential motions of liquid flow in a pipe or tube [RP31.1].

switch 1. A device that connects, disconnects, selects, or transfers one or more circuits and is not designated as a controller, a relay, or a control valve. As a verb, the term is also applied to the functions performed by switches [S5.1]. 2. A device for controlling whether an electrical circuit is 'on' or 'off', usually by making or breaking an electrical connection between the circuit and its power supply. 3. Any electrical or mechanical device for placing another device or circuit into an operating or nonoperating condition. Also known as switching device; switching mechanism. 4. A device for re-routing signals from one optical fiber into another. 5. Any device for controlling which of two fixed paths a railway train, tram, subway car or conveyor line will follow.

switch, air flow proving A device installed in an air stream which senses air flow or loss thereof and electrically transmits the resulting impulses to the flame failure circuit.

switch, high pressure A device to monitor liquid, steam or gas pressure and arranges to open and/or close contacts when the pressure value is exceeded.

switching point A point in the input span of a multiposition controller at which the output signal changes from one position to another [S51.1].

switching time 1. The time interval in a switching device between the reference time, or time at which the leading edge of a switching or driving pulse occurs, and the last instant at which the instantaneous output response reaches a stated fraction of its peak value. 2. The time interval between the reference time and the first instant at which the instantaneous output response reaches a stated fraction of its peak value.

switch, low pressure A device to monitor liquid steam or gas pressure and arranged to open and/or close contacts when the pressure drops below the set value.

switch, oil temperature limit A device to monitor the temperature of oil between preset limits and arranged to open and/or close contacts should improper oil temperature be detected.

swivel A mechanical device that can move freely about a pin joint.

symbol 1. Symbols are used on drawings to indicate devices, instruments, types of communication lines, connection points, valves, actuators, primary elements, and other graphic representations. 2. A name which represents a quantity of operation.

symbolic address A label, alphabetic or alphanumeric, that is used to specify a storage location in the context of a particular program; programs are often first written using symbolic addresses in some convenient code and are then translated into absolute addresses by an assembly program.

symbolic code A code which expresses programs in source language, i.e., by referring to storage locations and machine operations by symbolic names and addresses which are independent of their hardware determined names and addresses. Synonymous with pseudo code and contrasted with machine language code.

symbolic instruction An instruction in an assembly language directly translatable into a machine code.

symbolic logic The discipline that treats formal logic by means of a formalized artificial language or symbolic calculus whose purpose is to avoid the ambiguities and logical inadequacies of natural languages.

symbolic notation A method of representing a storage location by one or more figures, or labels.

symbolic number A numeral, used in writing routines, for referring to a specific storage location; such numerals are converted to actual storage addresses in the final assembling of the program.

symbolic programming The use of arbitrary symbols to represent addresses in order to facilitate programming.

symmetrical transducer A transducer in which all possible termination pairs may be interchanged without affecting transducer function.

synchronization Keeping one part of a process in a fixed relationship to another part of a process—for example, keeping the advance of strip stock through a stamping press timed to move ahead only when the press has raised the die free of the work.

synchronization of parallel computational processes Controlling the execution of parallel computational processes so as to maintain some desired sequential relationship between programmed actions within the processes. See semaphore.

synchronization pattern A sequence of ones and zeros that signals the start of each frame of PCM data; the pattern generally chosen is a pseudo-random sequence, one that is unlikely to occur randomly in data.

synchronize To lock one element of a system into step with another. The term usually refers to locking a receiver to a transmitter, but it can refer to locking the data terminal equipment bit rate to the data set frequency.

synchronizer A hardware device that can recognize a predetermined pattern in a telemetry format and generate clock pulses that coincide with those occurrences.

synchronous 1. Electrically or mechanically in phase or in step—as applied to two or more circuits, motors, machines or other devices. 2. Describes data transmission or other logic events that use a timing signal or clock pulse to control the rate at which events occur. 3. The performance of a sequence of operations controlled by an external clocking device; implies that no operation can take place until the previous operation has been completed.

synchronous data link control (SDLC) A type of data link protocol.

synchronous system trap (SST) A system condition that occurs as a result of an error or fault within the executing tasks.

synchroscope An instrument for indicating whether two periodic quantities are synchronous by means of a rotating pointer or cathode-ray oscilloscope; the position of the pointer or pattern on the oscilloscope tube indicates the instantaneous phase difference between the two quantities.

synchro-to-digital converter (S/D) An electronic device for converting the analog output signal of a rotary transformer (synchro) into a digital word for further processing.

synergism An action where the total effect of two components or agents is greater than the individual effects of the components when simply added together; for instance, in stress-corrosion cracking, cracks form and propagate deep into a material in a much shorter time and at a much lower stress than could be predicted from known effects of stress and the corrosive environment.

syntax 1. The structure of expressions in a language. 2. The rules governing the structure of a language. 3. In data processing, grammatical rules for software programming that specify how instructions can be written.

synthetic lubricant Any of a group of lubricating substances that can perform better than straight petroleum products in the presence of heat, chemicals or other severe environmental conditions.

synthetic relationship A relation existing between concepts which pertains to empirical observation. Such relationships are involved not in defining concepts or terms, but in reporting the results of observations and experiments.

SYSGEN See system generation.

system 1. An organized collection of parts united by regulated interaction [RP55.1]. 2. An organized collection of men, machines, and methods required to accomplish a specific objective [RP55.1]. 3. An assembly of procedures, processes, methods, routines, or techniques united by some form of regulated interaction to form an organized whole. 4. The complex or hardware and software utilized to affect the control of a process. 5. An assemblage of equipment, machines, control devices, or a combination thereof, interconnected mechanically, hydraulically, pneumatically or electrically, and intended to act together to perform a predetermined function. 6. In data processing, any group of software and hardware that is connected to operate as a unit.

systematic error 1. An error in a set of measurements or control that can be predicted from scientific principles; individual errors from the same cause bias the value of the mean because they all act in the same direction (sense); the amount of error in each individual value may or may not have a direct mathematical relationship to true value of the quantity. 2. That which cannot be reduced by increasing the number of measurements if the equipment and conditions remain unchanged. 3. Any constant or reproducible error introduced into a measured or controlled value due to failure to control or compensate for a specific side effect. See error, systematic.

system board The control center of a computer.

system check A check on the overall performance of the system, usually not made by built-in computer check circuits, e.g., control totals, hash totals, and record counts.

system control See control system.

system, controlled The collective functions performed in and by the equipment in which the variable(s) is (are) to be controlled. NOTE: Equipment as embodied in this definition should be understood not to include any automatic control equipment [S51.1].

system, controlling 1. Of a feedback control system, that portion which compares functions of a directly controlled variable and a set point, and adjusts a manipulated variable as a function of the difference. It includes the reference-input elements; summing point; forward and final controlling elements, and feedback elements (including sensing element) [S51.1]. 2. Of a control system without feedback, that portion which manipulates the controlled system [S51.1].

system crash The sudden and complete failure of a computer system.

system deviation See deviation, system.

system device The device on which the operating system is stored.

system, directly controlled The body, process, or machine directly guided or restrained by the final controlling element to achieve a prescribed value of the directly controlled variable [S51.1].

Systeme Internationale d'Unites (SI) The current International System of Units.

system error In a control system, the difference between the value of the ultimately controlled variable and its ideal value.

system generation (SYSGEN) The process of building an operating system on or for a particular hardware configuration with software configuration modifications.

system generator A program that performs system-level functions; any program that is part of the basic operating system; a system utility program.

system, idealized An imaginary system whose ultimately controlled variable has a stipulated relationship to a specified set point. It is the basis for performance standards [S51.1].

system, indirectly controlled The portion of the controlled system in which the indirectly controlled variable is changed in response to changes in the directly controlled variable [S51.1].

system, linear One of which the time response

to several simultaneous inputs is the sum of their independent time responses. It is represented by a linear differential equation and has a transfer function which is constant for any value of input within a specified range. A system not meeting these conditions is described as nonlinear [S51.1].

systems analysis The examination of an activity, procedure, method, technique, or a business to determine: a) behavioral relationships or, b) what must be accomplished and how.

systems engineering 1. Designing, installing and operating a system in a manner intended to achieve optimum output while conserving manpower, materials and other resources. 2. Designing and operating a system to achieve a predetermined level of performance, taking into consideration all of the factors that contribute to system performance, including manpower utilization and human factors.

system test Determining performance characteristics of an integrated, interconnected assemblage of equipment under conditions that evaluate its ability to perform as intended and that verify suitability of its interconnections.

T

t See temperature.

T See tesla; also see temperature.

table 1. A flat plate, with or without legs, used primarily to support workpieces or other items at a given vertical height. 2. The flat portion of a machine tool such as a grinder that directly or indirectly supports and positions the work. 3. A collection of data in a form suitable for ready reference, frequently as stored in sequenced machine locations or written in the form of an array of rows and columns for ready entry and in which an intersection of labeled rows and columns serves to locate a specific piece of data or information. 4. In data processing, any group of data organized as an array.

table look up A procedure for obtaining the function value corresponding to an argument from a table of function values.

tableting A method of compacting powdered or granular solids using a punch and die; used to make certain food products, dyes, and pharmaceuticals.

tachometer An instrument for measuring speed of rotation, usually in revolutions per minute.

tack 1. A small, sharp nail with a broad head. 2. The quality of an adhesive, paint, varnish or lacquer to remain sticky to the touch for a prolonged period of time.

tackiness agent An additive that imparts adhesive qualities to a nonadhesive material.

tackle Any arrangement of ropes and pulleys used to produce a mechanical advantage.

tack weld 1. Any small, isolated arc weld especially one that does not bear load but rather merely holds two pieces in a fixed relationship. 2. A weld joint made by arc welding at small, isolated points along a seam.

tag A unit of information whose composition differs from that of other members of the set so that it can be used as a marker or label; also called a flat or sentinel.

tag (from data compressor) A unique sixteen-bit word, preselected by the operator, that precedes each data output word and identifies it.

tag number Instrument loop identification number.

Tag-Robinson colorimeter A laboratory device used to compare shades of color in oil products by varying the thickness of a column of the oil until its color matches that of a standard.

tail pulley The pulley at the opposite end of the conveyor from the head pulley [RP74.01].

takeup (gravity) A device plus a calculated quantity of dead weight to provide sufficient tension in a conveyor belt to ensure that the belt will be positively driven by the drive pulley. A counter-weighted takeup consists of a horizontal pulley free to move in either the vertical or horizontal direction, with dead weights applied to the pulley shaft to provide the tension required [RP74.01].

tandem networks An arrangement of two-terminal-pair networks such that the output terminals of one network are directly connected to the input terminals of the other network.

tang 1. The slim, tapered end of a hand file that fits into a handle. 2. A tonguelike projection on the shank end of a drill that fits into the spindle of a drill press and ensures transmission of torque to the drill body.

tangent galvanometer A galvanometer consisting of a small compass mounted horizontally in the center of a large vertical coil of wire; the current through the coil is proportional to the tangent of the angle the compass needle makes with its rest (no current) position.

tank A large container, covered or open, for holding, storing or transporting liquids.

tank circuit A resonant electronic circuit that

consists of a capacitor and an inductor connected in parallel.

tap 1. A connection to a potentiometric element along its length, frequently at the element's center for use in providing bidirectional output [S37.12] [S37.6]. 2. A threaded plug, where the threads are of accurate form and dimensions, and have cutting edges that form internal threads in a hole as the plug is screwed into the hole. 3. A small hole in the wall of a pipe or process vessel, usually threaded, where an instrument, control device or sampling device is attached. 4. A coupler in which part of the light carried by one fiber is split off and inserted into another fiber, essentially the same as a tee coupler. 5. To withdraw a quantity of molten metal from a refining or remelting furnace.

tap drill A drill used to make a small, precise hole for tapping.

tape 1. A graduated steel ribbon used to measure lengths, as in surveying. 2. A ribbon made of plastic, metal, paper or other flexible material suitable for data recording by means of electromagnetic imprinting; punching or embossing patterns; or printing. 3. An adhesive-backed ribbon used for sealing packages, attaching labels, and various other purposes.

tape-and-plumb-bob liquid-level gage See plumb-bob gage.

tape-controlled machine A machine tool operated automatically by means of control signals read off a length of magnetic or punched paper tape.

tape drive A device that moves magnetic tape past a head that can "read" the tape.

tape formatter A device, including buffers and controls, for recording ordered data on magnetic tape in gapped form, in a format recognized by a computer.

tape header data Several recorded characters at the beginning of a magnetic tape, used to identify the content of the tape.

taper A dimensional feature where thickness, height, diameter or some other measurement varies linearly with distance along a given axis.

tapered-roller bearing A roller bearing having tapered rollers that run in conical races; it can support both radial and thrust loads.

tapered-tube rotameter A type of variable-area flowmeter in which a float that has greater density than the fluid rides inside a tapered tube in such a manner that fluid flowing upward through the tapered section carries the float with it until the upward force exerted by the flowing fluid just balances the downward force due to float weight; as the float rides upward, the annular area around it becomes larger and force on the float decreases; if the tube is made of glass, it can be graduated so that flow rate is read directly by observing float position; otherwise flow rate must be determined from an indirect indication of float position.

tapered waveguide A waveguide section having a continuous change in cross section.

tape search A hardware process by which an operator or computer can cause instrumentation tape to be searched automatically for specific start and stop times for data reduction.

tape speed compensation signal A signal recorded on instrumentation tape along with the data (preferably on the same track as the data) to correct electrically for tape speed errors during playback.

tape-type liquid-level gage A liquid-level gage consisting of a tape wound around a drum which is attached to a pointer or other level indicator, with one end of the tape attached to a float and the other counterweighted to keep the tape taut.

taphole A hole in the side or bottom of a furnace or ladle for draining off molten metal.

tappet An oscillating part such as a lever, operated by a cam or push rod, and used to tap or push another machine element such as a valve.

tappet rod A pivot rod carrying one or more tappets and acting as a fulcrum for their motion.

tapping See dither.

tare weight In any weighing operation, the residual weight of any containers, scale components or residue that is included in total indicated weight and must be subtracted to determine weight of the live load.

target 1. A goal or standard against which some quantity such as productivity is compared. 2. A point of aim or object to be observed by visual means, electromagnetic imaging, radar, sonar or similar noncontact method. 3. In MS-DOS, the location where data is to be copied and stored.

target computer 1. The computer in which the target program is used. 2. A computer which has its programs prepared by a host processor. Same as object machine.

target flowmeter A device for measuring fluid

flow rates by means of the drag force exerted on a sharp-edged disk centered in a circular flow-path due to differential pressure created by fluid flowing through the annulus; usually, the disk is mounted on a bar whose axis coincides with the tube axis, and drag force is measured by a secondary device attached to the bar.

target language The language into which some other language is to be translated.

target program An object program which has been assembled or compiled by a host processor for a target computer.

target system The microcomputer system to be used in the final product.

target-type flowmeter An instrument for measuring fluid flow in which the fluid exerts force on a small circular disc suspended in the center of the flow conduit by means of a pivoted bar; the force exerted on the bar by a force-balance transmitter to counteract the fluid force on the target is an indication of the flow rate.

tarnish Discoloration of a finished surface by a thin film of corrosion products.

task 1. A unit of work for the computer central processing unit from the stand-point of the executive program. 2. A specific "run time" execution of a program and its subprograms. 3. In the MULTICS sense, a virtual processor. (A single processor may be concurrently simulating many virtual processors.) 4. The execution of a segment on a virtual processor. See virtual processor. 5. In RSX-11 terminology, a load module with special characteristics; in general, any discrete operation performed by a program.

task control block (TCB) The consolidation of control information related to a task.

task dispatcher The control program that selects from the task queue the task that is to have control of the central processing unit and gives control to the task.

task management Those functions of the control program that regulate the use by tasks of the central processing unit and other resources, except for input/output devices.

task queue A queue of all the task control blocks present in the system at any one time.

task/surround lumination ratio The luminance ratio between the keyboard and screen (TASK) and workplace (SURROUND) within the operator's field of view [S5.5].

taut-band ammeter An instrument for measuring electric current in which a moving coil is mounted on a taut metal band held rigidly at the ends; when current flows through the coil, it deflects within the gap of a permanent magnet, twisting the metal band; the magnitude of current is indicated by a pointer attached to the coil when the torque exerted by magnetic-field interaction is balanced by restoring torque in the twisted band.

T-bolt A bolt shaped like the letter T, used primarily in conjunction with a dog or other hold-down device to secure workpieces against a machine bed or table containing a number of T-shaped slots that the bolt head fits into.

T/C See thermocouple.

TDM See time division multiplex.

TDMA See time division multiplex access.

tear-down time The amount of time required to disassemble a machine set-up following a production run and prior to setting up the jigs and fixtures for the next order.

Technical and Office Protocol (TOP) A specification for a suite of communication standards for use in office automation developed under the auspices of Boeing Computer Services. The development of this specification is being taken over by the MAP/TOP Users Group under the auspices of CASA/SME.

technical characteristics Those attributes of equipment that pertain to the engineering principles governing its functions.

technical evaluation An investigation to determine the suitability of materials, equipment or systems to perform a specific function.

technical specifications A detailed description of the technical characteristics of an item or system in sufficient detail to form the basis for design, development, production and, in some cases, operation.

technician 1. An expert in a technical process. 2. A person whose occupation requires training in a specific technical process.

tee coupler A fiber optic coupler in which the three fiber ends are joined together, and a signal transmitted from one fiber is split between the other two.

tee joint A junction, such as in piping or a weldment, where a branch member is connected at one end to a cross member running at right angles to the branch.

teeming Pouring molten metal into an ingot mold; most often used with reference to steel production.

teldata In these systems, the common parallel interface for telemetry data, as from a frame synchronizer to a buffered data channel.

telecommunications Pertaining to the transmission of signals over long distances, such as by telegraph, radio, or television.

telemetering 1. The transmission of a measurement over long distances, usually by electromagnetic means. 2. Using radio waves, wires or other means to transmit instrument readings to a remote location. Also known as remote metering; telemetry.

telemetry The science of measuring quantities, transmitting the results to a distant station, and interpreting, indicating, and/or recording the quantities measured.

telemetry front end (TFE) Hardware devices that accept multiplexed data and time, establish synchronization, convert to parallel data, and provide timing pulses, status, and the like for computer entry.

telemetry input channel A device which prepares telemetry data for input to a real-time computer.

telephone twisted pair (TTP) A network medium that uses existing telephone wiring. Standard work is in progress on a TTP standard for IEEE 802.3 StarLAN and IEEE 802.5 Token Ring.

teleran An aircraft navigation system that combines radar position information with a television image; ground plan position indicator, map and weather information are displayed together in the aircraft.

telescoping gage An adjustable gage for measuring inside dimensions such as hole diameters; it consists of a spring-loaded member that extends until it touches both sides of a hole, then is locked in place to prevent further extension when it is withdrawn from the hole; an outside micrometer or vernier calipers is used to measure the length of the locked gage member.

teletypewriter A popular hard copy device attached to the computer, which prints one character at a time.

telltale A marker on the outside of a tank that indicates water level on the inside of the tank.

TELSET In these systems, the common parallel interface for telemetry set-up, as from a buffered data channel.

temper 1. The relative hardness and strength of flat-rolled steel or stainless steel that cannot be further hardened by heat treatment. 2. The relative hardness and strength of nonferrous alloys, produced by mechanical or thermal treatment (or both) and characterized by a specific structure, range of mechanical properties or reduction of area during cold working. 3. In the production of casting molds, to moisten mold sand with water. 4. In the heat treatment of ferrous alloys, to reheat after hardening for the purpose of decreasing hardness and increasing toughness without undergoing a eutectoid phase change. 5. In tool steels, an imprecise shop term sometimes used to denote carbon content. 6. In glass manufacture, to anneal or toughen by heating below the softening temperature. 7. To moisten and mix clay, mortar or plaster to a consistency suitable for use. 8. A master alloy added to tin to make the finest pewter.

temperature Indication of how hot or cold a substance is.

temperature, ambient The temperature of the medium surrounding a device. Note 1: For devices which do not generate heat this temperature is the same as the temperature of the medium at the point of device location when the device is not present. 2: For devices which do generate heat this temperature is the temperature of the medium surrounding the device when it is present and dissipating heat. 3: Allowable ambient temperature limits are based on the assumption that the device in question is not exposed to significant radiant energy sources [S51.1].

temperature coefficient The rate of change of some physical property—electrical resistivity, for instance—with temperature; the coefficient may be constant or nearly constant, or it may vary itself with temperature.

temperature compensation Any construction or arrangement that makes a measurement device or system substantially unaffected by changes in ambient temperature.

temperature error 1. The maximum change in output, at any measurand value within the specified range, when the transducer temperature is changed from room temperature to specified temperature extremes [S37.1]. 2. An error in an instrument reading caused by a difference between the ambient temperature of the instrument and some desired standard temperature.

temperature error band The error band applicable over stated environmental temperature limits [S37.1].

temperature gradient error The transient deviation in output of a transducer at a given measurand value when the ambient temperature or the measured fluid temperature changes at a specified rate between specified magnitudes [S37.1].

temperature, process The temperature of the process medium at the sensing element [S51.1].

temperature range, compensated See temperature range, operating.

temperature range, fluid The range of temperature of the measured fluid, when it is not the ambient fluid, within which operation of the transducer is intended. NOTE 1: Within this range of fluid temperature all tolerances specified for temperature error, temperature error band, temperature gradient error, thermal zero shift and thermal sensitivity shift are applicable. 2: When a fluid temperature range is not separately specified, it is intended to be the same as the operating temperature range [S37.1].

temperature range, operating 1. The interval of temperatures in which the transducer is intended to be used, specified by the limits of this interval [RP37.2]. 2. The range of ambient temperatures, given by their extremes, within which the transducer is intended to operate; (S) within this range of ambient temperature all tolerances specified for temperature error, temperature error band, temperature gradient error, thermal zero shift and thermal sensitivity shift are applicable [S37.1].

temperature sensitivity error The change in sensitivity of a transducer from its reference sensitivity as a result of changes in its ambient temperature over a specified operating temperature range. Note: If changes in voltage sensitivity are specified, the total associated capacitance must be stated [RP37.2].

tempering 1. Heating hardened ferrous alloys below the transformation temperature to reduce hardness and improve toughness. 2. Adding moisture to molding sand, clay, mortar or plaster. 3. Heating glass below its softening temperature.

tempering air Air at a lower temperature added to a stream of pre-heated air to modify its temperature.

tempilstick A crayon made of a material having a sharp reaction at a specific temperature; in use, a crayon sensitive to a specific temperature is used to mark the surface of a metal to be heated; confirmation that the intended temperature was reached or exceeded is indicated by a change in color of the mark.

template 1. A guide or pattern used in laying out parts to be manufactured. 2. A guide used in drawing standard shapes on an engineering or architectural drawing.

temporal coherence The coherence of light over time. Light is temporally coherent when the phase change during an interval T remains constant regardless of when the interval is measured.

tensile specimen A bar, rod or wire of specified dimensions used in a tensile test. Also known as tensile bar; test specimen.

tensile strength The maximum load per unit area that a material can withstand before fracture, usually computed as maximum load divided by original cross sectional area of a standard specimen pulled to fracture in uniaxial tension.

tensile test A method of determining mechanical properties of a material by loading a machined, cast or molded specimen of specified cross-sectional dimensions in uniaxial tension until it breaks; the test is used principally to determine tensile strength, yield strength, ductility and modulus of elasticity. Also known as pull test.

terminal 1. An I/O device that includes a keyboard and a display mechanism; a terminal is used as the primary communication device between a computer system and a person. A terminal can be dumb with no processing capability or intelligent when some processing capability is included within the terminal. 2. The connection points where the field wiring is brought to the I/O modules.

terminal-based conformity See conformity, terminal-based.

terminal-based linearity See linearity, terminal-based.

terminal board A structural component that provides one or more electrical terminals which are electrically insulated from the chassis or mounting, and almost always from each other.

terminal device A part used to facilitate the making of external connections [S82.01].

terminal line A theoretical slope for which the theoretical end points are 0 and 100% of both measurand and output [S37.1].

terminal pair A set of two associated terminals, so arranged as to be accessible for connecting a pair of associated leads.

terrain clearance indicator An instrument for measuring absolute altitude; also known as absolute altimeter.

tertiary air Air for combustion supplied to the furnace to supplement the primary and secondary air.

tesla Metric unit for magnetic flux density.

test 1. An annunciator sequence initiated by operation of the test pushbutton to reveal lamp or circuit failure [S18.1]. 2. A standard procedure for determining an attribute or performance characteristic of a material, part, component, assembly or system; a test may be used to determine basic properties, verify a function or condition, establish a response characteristic or calibration, or provide information about operating behavior.

test-block fan requirements The operating conditions for which a fan is designed which are to be proven by test, following the procedure outline by the Test Code of the National Association of Fan Manufacturers.

test chain A calibrating device consisting of a series of rollers or wheels linked together to ensure that their weight is uniform and they move freely (so that chain weight loss due to wear is minimized) [RP74.01].

test fixture A test fixture is a device to close off the pipe connections and/or moving stem seal areas of the control valve to allow pressurization for hydrostatic shell testing [S75.19].

test gage A pressure gage specially built for test service or other types of work requiring a high degree of accuracy and repeatability.

test gas Hydrogen sulfide diluted with clean air to a known concentration [S12.15].

test interval 1. The elapsed time between the performance of tests [S67.06]. 2. The elapsed time between the initiation (or successful completion) of tests on the same sensor, channel, load group, safety group, safety system, or other specified system or device [S67.04].

test, lamp Test of the visual display lamps [S18.1].

test, operational (functional) Test of the sequence, visual display lamps, audible devices, and pushbuttons [S18.1].

test point A process connection to which no instrument is permanently connected, but which is intended for the temporary or intermittent connection of an instrument [S5.1].

test ports Calibration connection points on the manifold between the manifold block valves and the instrument.

test stand A framework, rig or table equipped with instrumentation, power sources and auxiliary equipment necessary to perform an operating test on a machine, electronic device, engine or instrument.

text In data processing, any information that has a specific meaning.

text editor In data processing, a function used to process textual instructions rather than other program language.

TFE See telemetry front end.

theodolite An optical instrument for accurately measuring horizontal and vertical angles.

theoretical air The amount of air required to completely burn a given amount of a combustible material.

theoretical curve The specified relationship (table, graph, or equation) of the transducer output to the applied measurand over the range [S37.1].

theoretical cutoff frequency Disregarding any dissipation effects, the characteristic frequency at which the image attenuation constant of a transducer changes from zero to a positive value, or vice versa.

theoretical draft The draft which would be available at the base of a stack if there were no friction or acceleration losses in the stack.

theoretical end points The specified points between which the theoretical curve is established and to which no end point tolerances apply. The points can be other than 0 and 100% of both measurand and output [S37.1].

theoretical flame temperature See adiabatic temperature.

theoretical plate A hypothetical device for bringing two streams of material into such perfect contact that they leave the device in equilibrium with each other.

theoretical slope The straight line between the theoretical end points [S37.1].

therm A unit of heat applied especially to gas. One therm equals 100,000 Btu.

thermal-agitation voltage The electrical potential difference induced in circuits by the agitated motion of electrons in the circuit conductors.

thermal analysis Determining transformation temperatures and other characteristics of materials or physical systems by making detailed observations of time-temperature curves obtained during controlled heating and cooling.

thermal-arrest calorimeter A device for measuring heats of fusion in which a sample is frozen under vacuum at subzero temperatures and thermal measurements are taken as the calorimeter warms to room temperature.

thermal bulb A device for measuring temperature in which the liquid in a bulb expands and contracts with changes in temperature, causing a Bourdon-tube element to elastically deform, thereby moving a pointer in direct relation to the temperature at the bulb.

thermal coefficient of resistance The relative change in resistance of a conductor or semiconductor per unit change in temperature over a stated range of temperature. Expressed in ohms per ohm per degree F or C [S37.1].

thermal compensation See compensation.

thermal conductivity Heat flow per unit cross section per unit temperature gradient.

thermal conductivity gage A device for measuring pressure in a high-vacuum system by observing changes in thermal conductivity of an electrically heated wire that is exposed to the low-pressure gas in the system.

thermal converter A device consisting of one or more thermoelectric junctions in contact with or integral with an electric heater; the output of the thermoelectric component is directly related to the current flowing in the electric heater.

thermal cutout A device for protecting a circuit or electrical device from excessive current; it consists of a heater element and a replaceable fusible link which melts and opens the circuit when too much current flows through the heater element.

thermal delay timer A timing device relying on the movement of a heated bimetal to actuate a set of contacts.

thermal detector See bolometer.

thermal diffusion Spontaneous movement of solvent atoms or molecules to establish a concentration gradient as a direct result of the influence of a temperature gradient.

thermal electromotive force The electromotive force developed across the free ends of a bimetallic couple when heat is applied to a physical junction between the opposite ends of the couple. Also known as thermal emf.

thermal emf The electrical potential generated in a conductor or circuit due to thermal effects, usually differences in temperature between one part of the circuit and another.

thermal energy Energy that flows between bodies because of a difference in temperature.

thermal expansion 1. The increase in a volume of liquid caused by an increase in temperature [RP31.1]. 2. A physical phenomenon whereby raising the temperature of a body causes it to change dimensions (usually increasing) in a manner characteristic of the material of construction.

thermal instrument Any instrument that measures a physical quantity by relating it to the heating effect of an electric current, such as in a hot-wire instrument.

thermal neutron A free (uncombined) neutron having a kinetic energy approximately equivalent to the kinetic energy of its surroundings.

thermal power plant A facility or system for converting thermal energy into electric power.

thermal printer Prints characters on paper using a high speed heating element activating chemicals in the paper to form an image.

thermal radiation 1. Electromagnetic radiation that transfers heat out of a heated mass. 2. Electromagnetic radiation resulting from thermal agitation. 3. Electromagnetic radiation that is absorbed by a grey or black body from a source at a higher temperature than the absorbing body.

thermal sensitivity shift (S) The sensitivity shift due to changes of the ambient temperature from room temperature to the specified limits of the operating temperature range [S37.1].

thermal shock An abrupt temperature change applied to a device [S51.1].

thermal spraying A method of coating a substrate by introducing finely divided refractory powder or droplets of atomized metal wire into a high temperature plasma stream from a special torch, which propels the coating material against the substrate.

thermal transducer Any device which converts thermal energy into electric power or other useful measuring medium. An example is a thermocouple.

thermal-type flowmeter An apparatus in which heat is injected into a flowing fluid stream and flow rate is determined from the rate of heat dissipation; either the rise in temperature at some point downstream of the heater or the amount of thermal or electrical energy required to maintain the heater at a constant temperature is measured.

thermal-type liquid-level meter Any of several devices which indicate the position of liquid level in a vessel by means of a thermally activated property such as an abrupt change in temperature, evaporation or condensation effects, or thermal expansion effects.

thermal variable A characteristic of a material or system that depends on its thermal energy— temperature, thermal expansion, calorific value, specific heat or enthalpy, for instance.

thermal zero shift (S) The zero shift due to changes of the ambient temperature from room temperature to the specified limits of the operating temperature range [S37.1].

thermionic emission Spontaneous ejection of electrons from an emitter as a result of a temperature effect.

thermionic tube An electron tube in which at least one of its electrodes is heated to induce electron or ion emission.

thermistor A temperature transducer constructed from semiconductor material and for which the temperature is converted into a resistance, usually with negative slope and highly nonlinear. Its usual applications are as a nonlinear circuit element (either alone or in combination with a heater), as a temperature compensator in a measurement circuit, or as a temperature-measurement element.

thermit welding A fusion welding process in which a mixture of finely divided iron oxide and aluminum particles is ignited, reducing the iron oxide and producing a molten ferrous alloy that is then cast in a mold built up around the joint to be welded.

thermoammeter A device used chiefly to measure radio-frequency currents where the current is run through a wire of appropriate size; on this wire is mounted a thermocouple whose output is proportional to the temperature of the wire, which is a function of the R.F. current passing through the wire. Also called a hot wire ammeter.

thermocouple A temperature measuring instrument that develops an electric voltage when heated because of the combined thermoelectric effect due to dissimilar composition between two electrically connected conductors (usually wires) and to temperature difference between the connection (hot junction) and the other end of the conductors (cold junction).

thermocouple extension wire A matched pair of wires having specific temperature-emf properties that make the pair suitable for use with a thermocouple to extend the location of its reference junction (cold junction) to some remote location; alloys for such wires are specially designed and processed to make the pair suitable for use with only one type of thermocouple.

thermocouple instrument An electrothermic instrument having a direct-current mechanism, such as a permanent-magnet moving coil, which is driven by the output of one or more thermojunctions heated directly or indirectly by electric current.

thermocouple vacuum gage A device for measuring pressures in the range of about 0.005 to 0.02 torr by means of current generated by a thermocouple welded to the midpoint of a small heating element exposed to the vacuum chamber; alternatively, the instrument current may be generated in a specially constructed thermopile that serves as both heater and thermoelectric element.

thermoelectric Converting a change of measurand into a change in the emf generated by a temperature difference between the junctions of two selected dissimilar materials [S37.1].

thermoelectric cooling A method of cooling a chamber based on the Peltier effect, in which an electric current is circulated in a thermocouple whose cold junction is coupled to the chamber; the hot junction dissipates heat to the environment. Also known as thermoelectric refrigeration.

thermoelectric heating A method of heating involving a device similar to one used for thermoelectric cooling, except that the direction of current is reversed in the circuit.

thermoelectric hygrometer A condensation-type hygrometer in which the mirror element is chilled thermoelectrically.

thermoelectric series A tabulation of metals and alloys, arranged in order according to the magnitude and sign of their characteristic thermal emf.

thermoelectric thermometer A thermometer which uses a thermocouple or thermocouple array in direct contact with the body whose temperature is to be measured, and whose reading is given in relation to the reference junction whose temperature is known or automatically compensated for.

thermoforming A method of forming sheet plastic by heating it then pulling it over a contoured mold surface.

thermograd probe An instrument for recording temperature versus depth as it is lowered to the ocean floor; it records the flow of heat through the ocean floor.

thermograph An instrument for recording air temperature. Also known as recording thermometer.

thermography Either of two methods—contact thermography or projection thermography—for measuring surface temperature using thermoluminescent materials.

thermojunction Either of the two locations where the conductors of a thermocouple are in electrical contact; one, the measuring junction, is in thermal contact with the body whose temperature is being determined, and the other, the reference junction, is generally held at some known or controlled temperature.

thermometer An instrument for measuring temperature—usually involving a change in a physical property such as density or electrical resistance of a temperature-sensitive material.

thermophone An electroacoustic transducer that produces sound waves when a conductor whose temperature varies in response to a varying electric current causes air adjacent to the conductor to expand and contract.

thermopile An array of thermocouples used for measuring temperature or radiant energy, or for converting radiant energy into power; the thermocouples may be connected in series to give a higher-voltage output, or in parallel to give a higher-current output.

thermoplastic resin An organic solid that will repeatedly soften when heated and harden when cooled; examples include styrene, acrylics, polyethylene, vinyl and nylon.

thermoregulator A highly accurate or highly sensitive thermostat—for instance, a mercury-in-glass thermometer with sealed-in electrodes, which turns an electric circuit on and off as the level of mercury rises and falls past the position of the electrodes.

thermosetting resin An organic solid that sets up (solidifies) under heat and pressure, and cannot be softened and remolded readily; examples include phenolic, epoxy, melamine and urea.

thermostatic switch An electric switch whose contacts open and close in response to the amount of heat received by conduction or convection from the device whose power is regulated by the switch.

thermowell A thermowell is a pressure-tight receptacle adapted to receive a temperature sensing element and provided with external threads or other means for pressure-tight attachment to a vessel [ANSI-MC96.1]. See protecting tube.

thickener Equipment for removing free liquid from a slurry or other liquid-solid mixture to give a solid or semisolid mass without using filtration or evaporation; usually, the process involves centrifuging or gravity settling.

thickness 1. A physical dimension usually considered to represent the shortest of the three principal measurement axes—in flat products such as sheet or plate, for instance, it is the distance between top and bottom surfaces measured along an axis mutually perpendicular to them. 2. The distance from the external surface of a coating to the substrate, measured along a direction perpendicular to the coating-substrate interface. 3. The short transverse dimension of rolled, drawn or extruded stock.

thickness gage A device for measuring thickness of sheet material; it may involve physical gaging, but more often involves methods such as radiation absorption or ultrasonics.

thin-film potentiometer A potentiometer in which the conductive element is a thin film of a cermet (metal mix), conductive plastic or deposited metal; usually thin-film potentiometers are useful when a stepless output is desired.

thin-film strain gage A strain gage in which the gage is produced by depositing an insulating layer (usually a ceramic) onto the structural element then depositing a metal gage element onto the insulation layer by sputtering or vacuum deposition through a mask which defines the strain gage configuration; thin-film techniques are used almost exclusively for transducer applications; for greatest sensitivity, four bridge elements, wiring between the elements, balance components and temperature-compensation components are deposited simultaneously.

thin-layer chromatography (TLC) A form of chromatography in which a small amount of sample is placed at one end of thin sorbent layer deposited on a metal, glass or plastic plate; after washing a solvent through the sorbent bed by capillary action, the individual components of the sample may be detected by visual means, ultraviolet analysis, radiochemistry or other suitable technique, or by a combination of techniques; thin-layer chromatography is a rapid and inexpensive method for screening and selecting solvent–stationary-phase systems for liquid chromatographic analysis.

thinner An organic liquid added to a mixture such as paint to reduce its viscosity and make it more free-flowing.

thixotropic substance A substance whose flow properties depend on both shear stress and agitation; at a given shear stress, flow increases with increasing time of agitation; when agitation stops, internal shear stress exhibits hysteresis.

thread 1. A continuous helical rib used to provide interconnection by twisting the ribbed member into a mating ribbed member; used extensively in pipe-and-fitting connections and in threaded fasteners. 2. A thin single-strand or twisted filament of natural or synthetic fibers, plastics, metal, glass or ceramic.

threaded ends Valve end connections incorporating threads, either male or female [S75.05]. See also end connections, threaded.

three-element feedwater control A control system whereby three process variables (steam flow, feedwater flow and drum level) are used as inputs to the control loop that regulates feedwater flow to the drum to maintain the drum level at set point. This is a cascaded feedforward loop with drum level as the primary variable steam flow as the feedforward input, and feedwater flow (feedback) as the secondary variable [S77.42].

three-mode controller Another name for a PID controller.

three-position controller See controller, three-position.

three-quarters hard A temper of nonferrous alloys and some ferrous alloys that corresponds approximately to a hardness and tensile strength midway between those of half hard and full hard tempers.

three term control Proportional integral derivative control.

three-way ball A closure member that is a spherical surface with one or more flow passages through it. The passages may be round, contoured or otherwise modified to yield a desired flow characteristic [S75.05].

three-way valve A control valve with three end connections used for mixing or diverting flow [S75.05].

threshold 1. The smallest change in the measurand that will result in a measurable change in transducer output. When the threshold is influenced by the measurand values, these values must be specified [S37.1]. 2. Generally, an energy level, power level or other minimum value that must be reached before some specific phenomenon can occur or some specific action can take place. 3. Specifically, the least change in a measured quantity that a given instrument can detect with certainty under specified conditions within a specified time.

threshold limit value (TLV) The maximum value that nearly all workers may be repeatedly exposed to day after day without any adverse effects [S12.15].

threshold limit value-short term exposure limit (TLV-STEL) A 15-minute, time-weighted average exposure that should not be exceeded at any time during a work day, even if the 8-hour time-weighted average is within the TLV. Exposures at the STEL should not be longer than 15 minutes and should not be repeated more than four times per day. There should be at least 60 minutes between successive exposures at the STEL [S12.15].

threshold limit value-time weighted average (TLV-TWA) The time-weighted average concentration for a normal 8-hour work day and a 40-hour work week, to which nearly all workers may be repeatedly exposed, day after day [S12.15].

threshold sensitivity The lowest value of a measured quantity that a given instrument or controller responds to effectively.

throat 1. The narrowest point along a constricted duct or passage, as in a venturi or nozzle. 2. The shortest distance from the root of a fillet weld to its face. 3. The distance in a machine such as a resistance spot welder or arbor press from the centerline of the electrodes or punch to the nearest point of interference with the frame or other machine component; it establishes the maximum distance from the edge of flat work that the machine can perform an operation. 4. In a C-clamp, gage or micrometer, the lateral distance from the anvil to the inner edge of the C-shaped frame. 5. The smaller end of an acoustical horn or tapered waveguide. 6. The neck portion of a passageway.

throttle valve A device for regulating flow of a fluid by alternatively opening up or closing down a restriction in a passage or inlet.

throttling The actions to regulate fluid flow through a valve by restricting its orifice opening. See also modulating [S75.05].

throttling calorimeter An instrument that determines the moisture content of steam by admitting steam to a well-insulated expansion chamber through an orifice and then measuring steam temperature; moisture content is found by referring to steam tables.

throughput rate The net rate at which data can be received by a device, manipulated as specified, and output to some other specified device.

through stay A brace used in fire-tube boilers between the heads or tube sheets.

throwing power The relative ability of an electroplating solution or electrophoretic paint to cover irregularly shaped parts with a uniform coating. Contrast with covering power.

throwout The device for disengaging the driving and driven plates of a motor vehicle clutch.

thrust 1. Weight or pressure applied to a drill bit or other tool to make it cut. 2. The pushing force developed by a rocket or jet engine. 3. Generically, the force any body exerts on another body —both can be stationary, both can be in motion, or one can be stationary and the other in motion.

thrust bearing A bearing that supports axial load on a shaft and prevents the shaft from moving in an axial direction.

thryatron An electronic switching tube filled with gas. Application of a voltage to the control grid turns on the current. Normally used for high voltage switches.

thumbscrew A threaded fastener with a head that is flattened along the axis of the threads so that it can be gripped between the thumb and forefinger for tightening or loosening.

thumbwheel A multiple position switch driven by a notched disc that can be rotated by thumb action. Frequently used to enter numerical information into a PC from a machine location.

thyratron A hot-cathode gas tube having one or more control electrodes which initiate anode current, but do not limit it except under specific operating conditions.

tide gage Any of several types of devices, ranging from a simple graduated rod observed visually to an elaborate recording instrument, used for determining the rise and fall of tides.

tide indicator The part of the tide gage—at water level or remote—that indicates the current height of a tide.

tie plate A plate, through which a bolt or tie rod is passed, to hold brick in place.

tie rod 1. A structural brace designed to bear tensile loads. 2. A part of the linkage in an automotive steering gear that connects the steering mechanism to the wheel supports and transmits steering forces to the wheels. 3. A rodlike member that connects mechanical or structural members of a machine together. 4. A tension member between buckstays or tie plates.

tight 1. Inadequate clearance, or barest minimum clearance between moving parts. 2. In a pressurized or vacuum system, freedom from leaks. 3. A class of fit having slight negative allowance, which requires light to moderate levels of force to assemble mating parts together.

TIG welding See gas tungsten-arc welding.

tile A preformed refractory, usually applied to shapes other than standard brick.

tile baffle A baffle formed of preformed refractory shapes.

tilt-switch level detector A relatively simple device for detecting high level in a bulk solids container by means of a free-hanging sensor that produces a switch action when the rising level of bulk material tilts the sensor from its normal vertical position.

TIMDATA In these systems, the common parallel interface for time data, as from a time code generator/translator to a buffered data channel.

time A fundamental measurement whose value indicates the magnitude of an interval between successive events.

time base In PC timer programming, refers to the basic increment of time used in the timer operation.

time base error In instrumentation tape recording and playback, the data error which results from a difference between tape recording speed and tape playback speed.

time code A serial BCD code, superimposed on a carrier so that it can be recorded on instrumentation tape, to annotate the time of day at which all data were recorded.

time code translator A hardware device to accept the serial time code (as from a separate track of an instrumentation tape), recognize synchronization, and prepare the time of day in parallel format for computer entry.

time constant   1. The value τ in an exponential response term. For the output of a first-order system forced by a step or an impulse. τ is the time required to complete 63.2% of the total rise or decay; at any instant during the process, τ is the quotient of the instantaneous ratio of change divided into the change still to be completed. In higher-order systems, there is a time constant for each of the first-order components of the process. 2. In process instrumentation, the value T in an exponential response term A exp (-t/T) or in one of the transform factors $1+sT$, $1+j\omega T$, $1/(1+sT)$, $1/(1+j\omega T)$, where: s=complex variable; t=time, seconds; T=time constant; j=square root of -1; ω=angular velocity, radians per second. Note: For the output of a first-order system forced by a step or an impulse, T is the time required to complete 63.2% of the total rise or decay; at any instant during the process, T is the quotient of the instantaneous rate of change divided into the change still to be completed. In higher order systems, there is a time constant for each of the first-order components of the process. In a bode diagram, break points occur at w1/T [S51.1] [S67.06]. 3. The length of time required for the output of a transducer to rise to 63% of its final value as a result of a step change of measurand [S37.1].

time constant, derivative action   Of proportional plus derivative control action, a parameter the value of which is equal to $1/2\pi f_d$ where f_d is the frequency (in hertz) on a Bode diagram of the lowest frequency gain corner resulting from derivative control action [S51.1].

time constant, integral action   1. Of proportional plus integral control action, a parameter whose value is equal to $1/2\pi f_i$ where f_i is the frequency (in hertz) on a Bode diagram of the highest frequency gain corner resulting from integral control action. Note: The use of integral action rate is preferred [S51.1]. 2. It is the reciprocal of integral action rate.

time, correction   See time, settling.

time, dead   The interval of time between initiation of an input change or stimulus and the start of the resulting observable response [S51.1].

time, derivative action   In proportional plus derivative control action, for a unit ramp signal input, the advance in time of the output signal (after transients have subsided) caused by derivative control action, as compared to the output signal due to proportional control action only [S51.1].

time division multiplex (TDM)   A system for the transmission of information about two or more quantities (measurands) over a common channel, by dividing available time intervals among the measurands to form a composite pulse train; information may be transmitted by variation of pulse duration, pulse amplitude, pulse position, or by a pulse code. Abbreviations of the codes used are PDM, PAM, PPM, and PCM.

time division multiplex access (TDMA)   A type of protocol for access to communication system network.

time-division multiplexing   A digital technique for combining two or more signals into a single stream of data by interleaving bits from each signal. Bit one might be from signal one, bit two from signal two, etc.

time gate   A transducer that gives an output signal only during specific time intervals.

time, proportioning control   See control, time proportioning.

timer   1. A device that automatically starts or stops a machine function, or series of functions, depending on either time of the day or elapsed time from an arbitrary starting point. 2. An instrument that measures elapsed time from some arbitrary starting point. 3. A device that fires the ignition spark in an internal combustion engine at a preset point in the engine cycle. 4. A device that opens or closes a set of contacts, and automatically returns them to their original position after a preset time interval has elapsed. Also known as interval timer. 5. A device providing the system with the ability to read elapsed time in splitsecond increments and to inform the system when a specified period of time has passed.

time, ramp response   The time interval by which an output lags an input, when both are varying at a constant rate [S51.1].

time response   The variation of an output variable of an element or a system, produced by a specified variation of one of the input variables. See also response, time.

time-rise   The time required for the output of a system (other than first-order) to change from a small specified percentage (often 5 to 10) of the steady-state increment to a large specified percentage (often 90 to 95), either before or in the absence of overshoot. Note: If the term is unqualified, response to a unit step stimulus is understood; otherwise the pattern and magnitude of the stimulus should be specified [S51.1].

timers Hardware devices or PC program elements that can produce control actions based on elapsed time.

timer, watchdog An electronic interval timer which will generate an interrupt unless periodically recycled by a computer. It is used to detect program stall or hardware failure conditions [RP55.1].

time-schedule controller A controller in which the setpoint (or reference input signal) automatically adheres to a predetermined time schedule. See also, controller, time schedule.

time series The discrete or continuous sequence of quantitative data assigned to specific moments in time, usually studied with respect to their distribution in time.

time, settling The time required, following the initiation of a specified stimulus to a system, for the output to enter and remain within a specified narrow band centered on its steady-state value [S75.05]. Note: The stimulus may be a step impulse, ramp, parabola, or sinusoid. For a step or impulse, the hand is often specified as + 2%. For nonlinear behavior both magnitude and pattern of the stimulus should be specified [S51.1].

time sharing 1. A computer system in which CPU time and system resources are shared with a number of tasks or jobs under the direction of a scheduling formula or plan. 2. The use of a device for two or more purposes during the same overall time interval, accomplished by interspersing component actions in time. 3. Participation in available computer time by multiple users, via terminals. Characteristically, the response time is such that the computer seems dedicated to each user.

time slice The time allocated by the operating system for processing a particular program.

time slicing A method of scheduling programs in a multiprogramming environment in which specific fixed time periods or "time slices" are assigned to each task in a cyclic fashion.

time, step response Of a system or an element, the time required for an output to change from an initial value to a large specified percentage of the final steady-state value either before or in the absence of overshoot, as a result of a step change to the input. Note: usually stated for 90, 95 or 99 percent change. See "time constant" for use of 63.2% value [S51.1].

tinning 1. Coating with a thin layer of molten solder or tin to prevent corrosion or prepare a connection for soldering. 2. Covering or preserving a metal surface with tin. 3. A protective surface layer of tin or solder.

titration curve A plot with pH as the ordinate and units of reagent added per unit of sample as the abscissa.

TLC See thin-layer chromatography.

T network A network consisting of three branches, one terminal of each branch being connected at a common node and the remaining terminals being connected, respectively, to an input junction, an output junction and a common input and output junction.

toe 1. The junction between the face of a weld and the adjacent base metal. 2. The portion of the base of a dam, earthwork or retaining wall opposite to the retained material.

toe crack A crack in the weldment that runs into the base metal from the toe of a weld.

toggle 1. A flip-flop. 2. Pertaining to a manually operated on-off switch, i.e., a two-position switch. 3. Pertaining to flip-flop, see-saw, or bistable action. 4. A pinned lever that can be used to amplify forces.

toggle switch A manually operated electric switch with a small projecting knob or arm that may be placed in either of two positions, "on" or "off," and will remain in that position until changed. 2. An electronically operated circuit that holds either of two states until changed.

token A token represents the right to use the network medium.

token bus An access procedure where the right to transmit is passed from device to device via a logical ring on a physical bus.

token passing Networking procedure in which access to the bus for transmission is conditional on possession of a circulating token signal.

token ring An access procedure where the right to transmit is passed from device to device around the physical ring.

tol See tolerance.

tolerance (tol) 1. Permissible variation in the dimension of a part. 2. Permissible deviation from a speciified value; may be expressed in measurement units or percent.

tolerance limits The extreme upper and lower boundaries of a specified range; it is computed from the nominal value and its tolerance.

ton 1. A weight measurement equal to 2,000 lb (avoirdupois), short ton; 2,240 lb (avoirdupois), long ton; or 1,000 kg, metric ton. 2. A unit volume of sea freight equal to 40 cu ft. 3. A unit of measurement for refrigerating capacity equal to 200 Btu/min, or about 3517 W; derived from the capacity equal to the rate of heat extraction needed to produce a short ton of ice having a latent heat of fusion of 144 Btu/lb from water at the same temperature in 24 hr.

tong hold The end of a forging billet where the forger grips it with his tongs; the end is cropped off and scrapped after the operation is completed.

tongs Any of various tools designed to grip, hold or manipulate hot metal or other solid materials; they consist generally of two bars connected by a pivot joint near one end, and are used much like a large pliers.

tool 1. Any device used to assist man in his work. 2. In manufacturing, any hand-held or machine-operated device for shaping material into a finished product—includes cutting, piercing or hammering devices; jigs and fixtures; dies and molds; shaped or flat rolls; and abrasives. 3. To equip a factory for production, or to design and build special devices needed to manufacture a specific product or model.

tool bit A piece of hardened metal with a sharpened edge or point used in a metal-cutting operation.

tool post A device attached to the tool slide on a lathe or similar machine tool clamping and positioning a tool holder.

tool steel Any of various steel compositions containing sufficient carbon and alloying elements to permit hardening to a level suitable for use in cutting tools, dies, molds, shear blades, metalforming rolls and other tooling applications.

tooth 1. One of the shaped projections on the rim or face of a gear. 2. A projection on a tool such as a comb, rake or saw.

TOP See Technical and Office Protocol.

top dead center The position of a piston and its connecting rod when at the extreme outer end of its stroke.

topology The logical interconnection between devices. Local area networks typically use either a broadcast topology (bus) in which all stations receive all messages, or a sequential topology (ring) where each station receives messages from the station before them and transmits (repeats) messages to the station after them. Wide area networks typically use a mesh topology where each station is connected to one or more other stations and acts as a bridge to pass messages through the network.

topworks A nonstandard term for actuator.

torch A device used to control and direct a gas flame, such as in welding, brazing, flame cutting or surface heat treatment.

torching The rapid burning of combustible material deposited on or near boiler-unit heating surfaces.

torque A rotary force, such as that applied by a rotating shaft at any point on its axis of rotation.

torque amplifier A two-shaft device that supplies power to rotate the output shaft, maintaining corresponding position between output and input shafts, but without imposing any additional torque on the input shaft.

torque-coil magnetometer An instrument for measuring properties of a magnetic field whose output is related to the torque developed in a coil that can turn within the field being measured.

torque converter Any of several mechanisms designed to change or vary the torque, speed or mechanical advantage between an input shaft and an output shaft.

torque error See mounting error.

torque-tube flowmeter A device for measuring liquid flow through a pipe in which differential pressure due to the flow operates a bellows, whose motion is transmitted to a recorder arm by means of a flexible torque tube.

torque-type viscometer An instrument that can measure viscosity of Newtonian fluids, non-Newtonian fluids, and suspensions by determining the torque needed to rotate a vertical paddle or cylinder submerged in the fluid.

torque wrench 1. A hand or power tool that can be adjusted to deliver a preset rotary force to a nut or bolt. 2. A wrench that can measure the torque required to start rotary motion when tightening or loosening a bolt.

torr Also spelled tor. A unit of pressure equal to the pressure exerted by a column of mercury 1 mm high at 0 °C.

torsiometer An instrument consisting of angular scales mounted around a rotating shaft to determine the amount of twist in the loaded shaft, and thereby determine the power transmitted. Also known as torsionmeter.

torsion balance An instrument for measuring minute magnetic, electrostatic or gravitational forces by means of the rotational deflection of a horizontal bar suspended on a torsion wire whose other end is fixed.

torsion bar A type of spring that flexes by twisting about its axis rather than by bending.

torsion galvanometer A galvanometer whose reading is determined by the angle through which the moving system must be rotated to bring it to its zero position while under the influence of a specific force between the fixed and moving systems.

torsion hygrometer An instrument for measuring humidity in which a substance sensitive to humidity is twisted or spiraled under tension in such a manner that changes in length of the sensitive element will rotate a pointer in direct relation to atmospheric humidity.

total absorption spectrometer An instrument that measures the total amount of x-rays absorbed by a sample and compares it to the amount absorbed by a reference sample; the sample may be solid, liquid or gas.

total accuracy See accuracy, total.

total adjusted error The maximum output deviation from the ideal expected values. Expressed as LSBs or percent of full-scale range at a fixed reference voltage.

total air The total quantity of air supplied to the fuel and products of combustion. Percent total air is the ratio of total air to theoretical air, expressed as percent.

total emissivity The ratio of the total amount of thermal radiation emitted by a non-blackbody to the total amount emitted by a blackbody at the same temperature.

total error band See error band.

totalizer A device used with a belt-conveyor scale to indicate the total weight of the material that has been conveyed over the scale. The master weight totalizer is the primary indicating element of the belt-conveyor scale. An auxiliary vernier counter used for scale calibration should not be part of the master weight totalizer. Totalizers can be remote auxiliary totalizers as well as local masters. The totalizer shows the accumulated weight; a totalizer may be nonresettable or resettable to zero to measure a definite amount of conveyed material [RP74.01].

total pressure The sum of the static and velocity pressures. See stagnation pressure.

total radiation pyrometer See wideband radiation thermometer.

total range The portion of a system of units that is between an instrument's upper and lower scale limits, and therefore defines the values of the measured quantity that can be indicated or recorded.

total solids concentration The weight of dissolved and suspended impurities in a unit weight of boiler water, usually expressed in ppm.

touch feedback A type of interaction in a manipulator in which servos provide force feedback to the manipulator fingers, providing a sense of resistance so the operator does not crush the object.

touch panel See membrane switch.

towbar A detachable bar or rigid linkage use to tow a vehicle.

towing tank See model basin.

trace 1. A graphical output from a recorder, usually in the form of an ink line on paper. 2. An interpretive diagnostic technique which provides an analysis of each executed instruction and writes it on an output device as each instruction is executed. See also sequence checking routine.

tracer A colored thread or filament visible in the insulation on an electrical wire so that the wire can be easily identified or traced between connections.

tracer gas A gas used in connection with a leak detecting instrument to find minute openings in a sealed vacuum system.

tracer milling Cutting a duplicate of a three-dimensional form by using the position of a stylus that traces across the form to operate the quill and table controls on a milling machine.

track 1. The portion of a moving storage medium, such as a drum, tape, or disc, that is accessible to a given reading head position [RP55.1]. 2. The line on the surface of the earth corresponding to the projection of the flight path of an aircraft or rocket. 3. A pair of parallel metal rails for a railroad, tram, or similar wheeled vehicle. 4. A crawler mechanism for earth-moving equipment or military vehicles. 5. A band of data on recording tape or the spiral groove in a phonograph record. 6. An overhead rail for repositioning hoisting gear. 7. To follow the movement of an object—for instance, by continually repositioning a telescope or radar set so its line of slight is always on the object. 8. In data processing, a specific area on any storage medium that can be read by drive heads.

track, disk   The path on one disk plotter traversed by a head during one revolution.

tracking error   In lateral mechanical recording equipment, the angle between the vibration axis of the pickup and a plane that is both perpendicular to the record surface and tangent to the unmodulated recording groove at the point where the needle rides in the groove.

tracking system   Any device that continually repositions a mechanism or instrument to follow the movement of a target object.

track, tape   The path traversed by one head during the record or playback process.

tractor feed   On a computer printer, a mechanical feed mechanism that uses gears with teeth that mesh with holes on the side of computer paper to pull the paper through the printer.

trailing edge   The second transition of a pulse.

train   A collection of one or more associated units and equipment modules, arranged in serial and/or parallel paths, used to make a complete batch of product.

training aid   Any object or device designed, constructed or adapted chiefly for the purpose of instructing personnel.

training idlers   Idlers of special design or mounting that are intended to counteract any tendency of the belt to shift sideways [RP74.01].

transceiver   A device that provides the electrical interface to the physical medium. Typically used to connect a node to a baseband network. Also see MODEM.

transducer   1. An element or device which receives information in the form of one quantity and converts it to information in the form of the same or another quantity [S51.1]. 2. A device which provides a usable output in response to a specified measurand. NOTE: The term transducer is usually preferred to "sensor" and "detector" and to such terms as "flowmeter", "accelerometer" and "tachometer"; it is always preferred to "pickup", "gage" (when not equipped with a dial-indicator), "transmitter" (which has an entirely different meaning in telemetry technology), "cell", and "end instrument" [S37.1]. 3. A device to convert one form of signal to another [S75.05]. 4. A general term for a device that receives information in the form of one or more physical

quantities, modifies the information and/or its form, if required, and produces a resultant output signal. Depending on the application, the transducer can be a primary element transmitter, relay, converter or other device. Because the term transducer is not specific, its use for specific applications is not recommended [S5.1]. 5. Any device or component that converts an input signal of one form to an output signal of another form—for instance, a piezoelectric transducer converts pressure waves into electrical signals, or vice versa. See also primary element, signal transducer, and transmitter.

transducer dissipation loss   The ratio of the power delivered by a specified source to a transducer connected to a specified load, to the power available from the transducer connected to the same source.

transducer gain   The ratio of the power that a transducer delivers to a specified load under specific operating conditions to the power available from a specified source.

transducer loss   The reciprocal of transducer gain.

transduction element   The electrical portion of a transducer in which the output originates [S37.1].

transfer   1. The conveyance of control from one mode to another by means of instructions or signals. 2. The conveyance of data from one place to another. 3. An instruction for transfer. 4. To copy, exchange, read, record, store, transmit, transport, or write data. 5. An instruction which provides the ability to break the normal sequential flow of control. Synonymous with jump.

transfer admittance   The complex ratio of current at a second pair of terminals of an electrical transducer to the emf across a given pair of terminals, at a specified frequency, both pairs being terminated in a specified manner.

transfer chamber   In plastics molding, an intermediate chamber or vessel for softening a thermosetting resin with heat and pressure before admitting it to the mold for final curing.

transfer characteristic   The current at one electrode expressed as a function of voltage at another electrode, with all other voltages held constant; the relationship is usually shown graphically.

transfer constant   A transducer rating consisting of a complex number equal to 1/2 the natural logarithm of the complex ratio of the product of voltage and current entering a transducer to that leaving the transducer when the transducer is connected to its image impedance—the real part of the transfer coefficient is the image attenuation constant and the imaginary part is the image phase constant; transfer constants also can be determined for pressure and volume flow rate or force and velocity, instead of voltage and current. Also known as transfer factor.

transfer fluid   A degassed liquid used between an isolating element and a sensing element to provide hydraulic coupling of the pressure between both elements [S37.6].

transfer function   1. A mathematical, graphical, or tabular statement of the influence which a system or element has on a signal or action compared at input and at output terminals [S51.1]. 2. A mathematical expression frequently used by control engineers, which expresses the relationship between the outgoing and the incoming signals of a process, or control element. The transfer function is useful in studies of control problems. Transfer functions are generally presented in terms of the Laplace transform. 3. A mathematical expression which describes the relationship between physical conditions at two different points in time or space in a given system, and perhaps, also describes the role played by the intervening time or space. 4. The response of an element of a process-control loop that specifies how the output of the device is determined by the innput.

transfer impedance   The complex ratio of applied a-c voltage, force or pressure at one point in a transducer to a-c current, velocity or volume velocity at another point in the same transducer, all inputs and outputs being connected to the system in some specified manner.

transfer instruction   See branch instruction.

transfer lag   See capacity lag.

transfer of control   See branch and jump.

transfer operation   An operation which moves information from one storage location or one storage medium to another, e.g., read, record, copy, transmit, or exchange. Transfer is sometimes taken to refer specifically to movement between different storage media.

transfer ratio   The ratio of the number of turns in the secondary winding to the number of turns in the primary winding.

transfer switch   A switch that controls whether a given conductor is connected to one circuit or to another.

transfer time   The time interval between the instant the transfer of data to or from storage commences and the instant it is completed.

transfer vector   A transfer table used to communicate between two or more programs. The table is fixed in relationship with the program for which it is the transfer vector. The transfer vector provides communication linkage between that program and any remaining subprograms.

transform   To change the form of data according to specific rules.

transformation temperature   The temperature at which a phase change occurs in a crystalline solid; sometimes, the term is applied to the upper or lower limit of a transformation range.

transformer   An electrical device that uses electromagnetic induction to transfer power from one electric circuit to another at the same frequency, usually increasing voltage and decreasing current (or vice versa) in the process.

transformer voltage divider   An inductive-type voltage divider used in some a-c bridge circuits to provide high accuracy, much as a KVVD is used in some d-c bridge circuits.

transient   1. In process instrumentation, the behavior of a variable during transition between two steady states [S51.1]. 2. The behavior variable during the transition between two steady states [S77.42]. 3. A dynamic condition or characteristic—power level, voltage, magnetic field strength, force or pressure, for example—that is not periodically repeated; often, it implies an anomalous, temporary departure from a steady-state condition, the latter being either constant or cyclic. 4. Pertaining to rapid change. 5. In process instrumentation, the behavior of a variable during transition between two steady-states.

transient analyzer   An electronic device used to capture a record of a transient event for later analysis.

transient deviation   See deviation, transient.

transient digitizer   A device which records a transient analog waveform and converts the information it has collected into digital form.

transient overshoot   1. The maximum excursion beyond the final steady-state value of output as the result of an input change [S51.1]. 2. An excursion beyond the final steady-state value of the output as the result of a step-input change.

It is usually referred to as the first excursion; expressed as a percentage of the steady-state output step.

transient overvoltage A momentary excursion in voltage occurring in a signal or supply line of a device which exceeds the maximum rated conditions specified for that device.

transient response The response of a transducer to a step-change in measurand. NOTE: Transient response, as such, is not shown in a specification except as a general heading, but is defined by such characteristics as time constant, response time, ringing period, etc. [S37.1]

transient temperature error The output of a transducer as a result of a specified transient temperature change within a specified operating temperature range. Note: The associated capacitance and load resistance, as well as the time, after the applied transient, at which the amplitude peak occurs must be specified [RP37.2].

transistor A three terminal solid state semiconductor device which can be used as an amplifier, switch, detector or wherever a three terminal device with gain or switching action is required.

transistor/transistor logic (TTL) A type of digital circuitry.

transit A surveying instrument having a telescope mounted for measuring both horizontal and vertical angles. Also known as transit theodolite; theodolite.

transition The switching from one state (for example, positive voltage) to another (negative) in a serial transmission.

transitional flow Flow between laminar and turbulent flow; generally between a pipe Reynolds number 2000 and 7000.

transition frequency See crossover frequency.

transition loss The ratio of signal power delivered to the portion of a transmission system following a discontinuity, after insertion of an ideal transducer, to the signal power delivered to the same portion prior to insertion of the ideal transducer.

transit time The time it takes for a particle, such as an electron or atom, to move from one point to another in a system or enclosure.

translate To convert from one language to another language.

translator 1. A program whose input is a sequence of statements in some language and whose output is an equivalent sequence of state-ments in another language. 2. A translating device.

transmission line A continuous conductor or other pathway capable of transmitting electromagnetic power from one location to another while maintaining the power within a system of material boundaries.

transmissometer An instrument for measuring the extinction coefficient of the atmosphere, and for determining visual range. Also known as hazemeter; transmittance meter.

transmittance The ratio of transmitted electromagnetic energy to incident electromagnetic energy impinging on a body that is wholly or partly transparent to the particular wavelength(s) involved.

transmitter 1. A transducer which responds to a measured variable by means of a sensing element and converts it to a standardized transmission signal which is a function only of the measured variable [S51.1]. 2. A device that senses a process variable through the medium of a sensor and has an output whose steady-state value varies only as a predetermined function of the process variable. The sensor may or may not be integral with the transmitter [S5.1]. 3. A device that translates the low-level output of a transmission to a site where it can be further processed. 4. In process control, a transmitter mounted together with a sensor or transducer in a single package designed to be used at or near the point of measurement. 5. A light source (LED or diode laser) which is combined with electronic circuitry to drive it. A transmitter operates directly from the signal generated by the other electronic equipment to produce the drive current needed for an LED or diode laser. 6. In process control, a device that converts a variable into a form suitable for transmission of information to another location (for example, resistance changed to current that is propagated on wires to a control installation).

transmitting extensions, electric or electronic A system that converts float position to a proportional electric signal (either AC or DC), or to a proportional shift or unbalance in impedance which is balanced by a corresponding shift in impedance in the receiving instrument [RP16.4].

transmitting extensions, pneumatic A system that converts float position to a proportional standard pneumatic signal. A magnetic coupling connects the internal float extension with an external mechanical system linked to a pneumatic transmitter [RP16.4].

transmutation A nuclear reaction that changes a nuclide into a nuclide of a different element.

transparent 1. The quality of a substance that permits light, some other form of electromagnetic radiation, or particulate radiation to pass through it. 2. In data processing, a programming routine that allows other programs to operate identically regardless of whether the transparent instructions are installed or not installed.

transponder A type of transmitter-receiver designed to automatically transmit a signal upon receipt of a signal having a predetermined frequency, pulse pattern or other unique characteristic.

transport Layer 4 of the OSI.

transportability A measure of the ability to reuse computer programs on an industrial computer.

transportation and storage conditions The conditions to which a device may be subjected between the time of construction and the time of installation. Also included are the conditions that may exist during shutdown [S51.1].

transport, tape A hardware device that moves magnetic tape past heads for recording or playback.

transverse acceleration An acceleration perpendicular to the sensitive axis of the transducer [S37.1].

transverse electric wave A type of electromagnetic wave having its electric field vector everywhere perpendicular to the direction of propagation in a homogeneous isotropic medium.

transverse electromagnetic wave A type of electromagnetic wave having both its electric field vector and its magnetic field vector everywhere perpendicular to the direction of propagation in a homogeneous isotropic medium.

transverse interference See interference, normal mode.

transverse response See transverse sensitivity.

transverse sensitivity 1. The maximum sensitivity of a uni-axial transducer to a transverse acceleration, within a specified frequency range, usually expressed in percent of the reference sensitivity in the intended measuring direction [RP37.2]. 2. The sensitivity of a transducer to transverse acceleration or other transverse measurand. NOTE: It is specified as maximum transverse sensitivity when a specified value of measurand is applied along the transverse plane in any direction, and is usually expressed in percent of the sensitivity of the transducer in its sensitive axis [S37.1].

trap 1. Conditional jump to a known location, automatically activated by hardware or software, with the location from which the jump occurred recorded. Often a temporary measure taken to determine the source of a computer bug. 2. A vertical S-, U- or J-bend in a soil pipe that always contains water to prevent sewer odors from backing up into the building. 3. A device on the intake, or high-vacuum side, of a diffusion pump to reduce backflow of oil or mercury vapors from the pumping medium into the evacuated chamber. 4. A receptacle for the collection of undesirable material.

trapped-air process A method of forming closed blow-molded plastics objects, in which sliding machine elements pinch off the top of the object after blowing to form a sealed, inflated product.

trapped fuel Any fuel in a fuel-delivery system, such as the fuel system of an internal combustion engine, that is not contained in the tanks.

trapping A feature of some computers whereby an unscheduled jump is made to a predetermined location in response to a machine condition, e.g., a tagged instruction, or an anomalous arithmetic situation. Such a feature is commonly used by monitor routines to provide automatic checking or for communication between input-output routines and the programs using them.

trapping mode A scheme used mainly in program diagnostic procedures for certain computers. If the trapping mode flip-flop is set and the program includes any one of certain instructions, the instruction is not performed and the next instruction is taken from location 0. Program counter contents are saved in order to resume the program after executing the diagnostic procedure.

travel The amount of movement of the closure member from the closed position to an intermediate or the rated full open position [S75.05].

travel characteristic The relationship between signal input and travel [S75.05].

travel cycle Travel of the closure component from its closed position to the rated travel opening and its return to the closed position.

travel indicator A means of externally showing position of the closure member; typically in terms of percent of or degrees of opening. Can be a visual indicator at or on the valve or a remote indicating device by means of transmitter or appropriate linkage [S75.05].

travel indicator scale A scale or plate fastened to a valve and marked with graduations to indicate the valve opening position [S75.05].

traveling block In a block-and-tackle system, the portion of the hoisting apparatus—excluding any slings or special rigging—that is raised and lowered with the load; it usually consists of the sheaves, pulley frame, clevis and hook.

traveling wave A wave in which the ratio of the instantaneous value for any component of the wave field at one point to the instantaneous value at any other point varies with time; it also has the property of transmitting energy from one point to another along its direction of propagation.

travel time The time required for one-half a travel cycle at specified conditions.

traverse 1. To swivel a gun, antenna, tracking device or similar mechanism in a horizontal plane. 2. A survey consisting of a set of connecting lines of known length which meet each other at specific angles.

tray A horizontal plate in a distillation column that temporarily holds a pool of descending liquid until it flows into a vertical "downcomer" and onto the next tray. Each tray has openings to permit passage of ascending vapors.

tread 1. The outer surface of a wheel or tire that contacts the roadway or rails. 2. The horizontal portion of a stairstep. 3. The horizontal distance between successive risers in a stairway.

treadle A bar or machine element that is pivoted at one end and connected to one or more other machine elements so that when it is stepped on, power or motion, or both, are transmitted to the other elements.

treated water Water which has been chemically treated to make it suitable for boiler feed.

tree 1. A decoder, the diagrammatic representation of which resembles the branching of a tree. 2. In microcomputing, the arrangement of DOS directories and subdirectories.

triac Semiconductor switching element. Commonly used in a-c output modules for PC's.

trial-for-ignition That period of time during which the programming flame failure controls permit the burner fuel valves to be open before

the flame sensing device is required to detect the flame.

trial for main flame ignition A timed interval when with the ignition means proved, the main valve is permitted to remain open. If the main burner is not ignited during this period, the main valve and ignition means are cut off. A safety switch lockout follows.

trial for pilot ignition A timed interval when the pilot valve is held open and an attempt made to ignite and prove it. If the presence of the pilot is proved at the termination of the interval, the main valve is energized; if not, the pilot and ignition are cut off followed by a safety lock-out.

triangulation 1. In navigation, determining position by laying out lines of sight to three celestial bodies or landmarks, widely spaced around the horizon; if properly corrected for time of observation, for the ship's or aircraft's speed and heading, and for current or wind, the lines will meet at a point or will form a small triangle on a map that indicates position at the time of observation. 2. In surveying, a method of measuring a large land area by establishing a baseline, then building up a network of triangles each having at least one side common with an adjacent triangle.

trifilter hydrophotometer An instrument for measuring the transparency of water at three wavelengths using red, green and blue optical filters.

trim The internal parts of a valve which are in flowing contact with the controlled fluid [S75.05].

trim, anti-cavitation A combination of control valve trim that by its geometry reduces the tendency of the controlled liquid to cavitate.

trim, anti-noise A combination of control valve trim that by its geometry reduces the noise generated by fluid flowing through the valve.

trim, balanced Control valve trim designed to minimize the net static and dynamic fluid flow forces acting on the trim.

trimming 1. Removing irregular edges from a stamped or deep drawn part. 2. Removing gates, risers and fins from a casting. 3. Removing parting-line flash from a forging. 4. Adding or removing small amounts of R, L or C from electronic circuits to cause minor changes in the circuit performance or to bring into specification.

trim, reduced Control valve trim which has a flow area smaller than the full flow area for that valve.

trim, restricted Control valve trim which has a flow area less than the full flow area for that valve [S75.05].

trim, soft seated Valve trim with an elastomeric, plastic or other readily deformable material used either in the closure component or seat ring to provide shutoff with minimal actuator forces.

triode An electron tube containing three electrodes—an anode, a cathode and a control electrode, or grid.

trip 1. To release a catch or free a mechanism. 2. An apparatus for automatically dumping minecars.

trip hammer A large power hammer that falls by gravity when released from its raised position by a cam or lever.

triple point A temperature at which all three phases of a pure substance—solid, liquid and gas—are in mutual equilibrium.

tripod A three-legged support for a transit, camera or other instrument which can be readily set up, collapsed and adjusted.

tripper 1. A device for unloading a belt conveyor at a point between the loading point and the head pulley [RP74.01]. 2. A device that discharges a load from a conveyor by snubbing the conveyor belt.

trip setpoint A predetermined value at which a bistable device changes state to indicate that the quantity under surveillance has reached the selected value [S67.04].

tri-state A type of logic device that has a high impedance state in addition to a high- and low-level output state. The high impedance state effectively disconnects the output of the device from the circuit; useful in the design of bus-oriented systems.

tri states A three-way switch, 1, 0 and a neutral state (effectively disconnected).

tristimulus values The amounts of each primary color that must be mixed together to obtain a specific hue or to match the color of a specific sample.

tritium An isotope of hydrogen having atomic weight of 3 (one proton and two neutrons in the nucleus).

trochotron A multiple-electrode electron tube which generates an output signal proportional to an input signal by charging elements in sequence with an electron beam manipulated by a magnetic field.

trolley A wheeled car running on an overhead track or rail.

tropical finish A coating applied to electronic equipment to protect it from insects, fungi and high humidity characteristic of tropical climates.

trouble contact See field contact.

trouble contact voltage See field contact voltage [S18.1].

trouble-shoot To search for the cause of a malfunction or erroneous problem behavior, in order to remove the malfunction. See debug.

trouble signal A signal (contact transfer and/or visible or audible signal) advising an instrument user of conditions such as input power failure, an open circuit breaker, a blown fuse, loss of continuity to the detector head, defective gas-sensing element, or significant downscale indication [S12.15].

true complement See complement.

true mass flow A measurement that is a direct measurement of mass and independent of the properties and the state of the fluid.

true ratio A characteristic of an instrument transformer equal to root-mean-square primary current (or voltage) divided by root-mean-square secondary current (or voltage) determined under specified conditions.

true value (data processing) The true value of a quantity is the value which would be measured with a perfect (i.e., error-free) measurement instrument. From a practical standpoint, the true value is often considered to be the value of a quantity measured by comparison to a primary standard such as those maintained by the National Bureau of Standards. For the purposes of this standard, the recommended true value is the value of a quantity obtained by comparison with a defined reference quantity. This reference quantity to which measurements are to be referred in the tests described in this standard shall be specified by the vendor or agreed upon by the vendor and user [RP55.1].

truncate 1. To terminate a computational process in accordance with some rule, e.g., to end the evaluation of a power series at a specified term. 2. To drop digits of a number of terms of a series thus lessening precision, e.g., the number 3.14159265 is truncated to five figures in 3.1415, whereas one may round off to 3.1416.

truncated address An operand address whose address field is shorter than the programmable memory's memory address register. The remaining fields in the instruction determine the algorithm for computing an effective address from the truncated address.

truncation error The error resulting from the use of only a finite number of terms of an infinite series, or from the approximation of operations in the infinitesimal calculus for operations in the calculus of finite differences. It is frequently convenient to define truncation error, by exclusion, as an error generated in a computation not due to rounding, initial conditions, or mistakes. Contrasted with rounding error.

trunk See bus.

trunk communication system A communication link joining two telephone central offices or other large switching facilities. It is distinguished by its large capacity, and by the fact that all signals go from point to point, without branching off to many separate points except at the end points.

trunnion Extensions of the ball used to locate, support and turn the ball within the valve body. May be integral or attached to the ball [S75.05].

truth table A table that describes a logic function by listing all possible combinations of input values and indicating, for each combination, the true output values.

TTL See transistor/transistor logic.

TTP See telephone twisted pair.

tube 1. A long hollow cylinder used for conveying fluids or transmitting pressure. Also known as tubing. 2. An evacuated glass-enveloped device used in electronic equipment to modify operating characteristics of a signal. Also known as electron tube.

tube cleaner A device for cleaning tubes by brushing, hammering, or by rotating cutters.

tube hole A hole in a drum, header, or tube sheet to accommodate a tube.

tube plug A solid plug driven into the end of a tube.

tubercle A localized scab of corrosion products covering an area of corrosive attack.

tube seat That part of a tube hole with which a tube makes contact.

tube sheet A perforated plate for mounting an array of tubes so that fluid on one side of the plate is admitted to the interior of the tubes and is kept separate from fluid on the outside of the tubes, such as in a shell-and-tube heat exchanger.

tube turbining Passing a power-driven rotary

device through a length of tubing to clean its interior surface.

tubular-type collector A collector utilizing a number of essentially straight-walled cyclone tubes in parallel.

tumbling 1. A process for smoothing and polishing small parts by placing them in a barrel with wooden pegs, sawdust and abrasives, or with metal slugs, and rotating the barrel about its axis until the desired surface smoothness and lustre is obtained. 2. Loss of control in a two-frame free gyroscope due to a slowing of the wheel.

tundish A pouring basin for molten metal.

tuner In a telemetry receiver, the input circuitry which selects and amplifies the desired frequency band.

tungsten inert-gas welding A nonpreferred term for gas tungsten-arc welding. Also known as TIG welding.

tuning The adjustment of control constants in algorithms or analog controllers to produce the desired control effect.

turbidity The optical obstruction to the passing of a ray of light through a body of water, caused by finely divided suspended matter.

turbine 1. A bladed rotor which turns at a speed nominally proportional to the volume rate of flow [S37.1]. 2. A machine for converting thermal energy in a flowing stream of fluid into rotary mechanical power by expanding the working fluid through one or more sets of vanes on the periphery of a rotor. 3. A machine for converting fluid flow into mechanical rotary motion, such as steam, water or gasturbines of single or multiple stages.

turbine flowmeter with an electrical output A flow measuring device in which the action of the entire liquid stream turns a bladed turbine at a speed nominally proportional to the volume flow, and which generates or modulates an output signal at a frequency proportional to the turbine speed [RP31.1].

turbine meter A volumetric flow measuring device using the rotation of a turbine type element to determine flow rate.

turbining See tube turbining.

turboblower An axial-flow or centrifugal compressor.

turbofan An air-breathing jet engine in which additional thrust is gained by extending a portion of the compressor or turbine blades outside the inner engine casing.

turbojet A type of jet engine in which part of the energy of the exhaust gases is used to power a turbine mounted along the axis of the exhaust jet; the turbine drives a compressor mounted in the engine intake, which enables the engine to develop more power for its size.

turbosupercharger A gas-turbine-driven air compressor used to increase air-intake pressure of a reciprocating internal-combustion engine.

turbulent burner A burner in which fuel and air are mixed and discharged into the furnace in such a manner as to produce turbulent flow from the burner.

turbulent flow A flow regime characterized by random motion of the fluid particles in the transverse direction as well as motion in the axial direction. This occurs at high Reynolds numbers and is the type of flow most common in industrial fluid systems.

turnaround time The particular amount of time required for a computation task to get from the requester, to the computer and back to the requester with desired results.

turnbuckle A device for tightening stays or tension rods in which a sleeve with a thread in one end and a swivel at the other (or with threads at both ends) is turned about its axis, drawing the ends of the device together.

turndown The ratio of the maximum plant design flow rate to the minimum plant design flow rate. See rangeability, inherent.

turner fluorometer A type of uv fluorometer in which primary filters pass only uv radiation to excite the sample and secondary filters pass only visible light to the photomultiplier tube; the intensity of emitted light is proportional to sample concentration even when exciting and measured light are not at optimum wavelengths.

turn-key system A system that includes all computer hardware and software, ready to operate.

turnover frequency See crossover frequency.

Twaddle scale A specific gravity scale that attempts to simplify measurement of liquid densities heavier than water, such as industrial liquors; the range of density from 1.000 to 2.000 is divided into 200 equal parts, so that one degree Twaddle equals a difference in specific gravity of 0.005; on this scale, 40° Twaddle indicates a specific gravity of 1.200.

twist The number of turns per unit length in the lay of fiber, rope, thread, yarn or cord.

twist drill A sharpened cylindrical tool for cutting holes in solid material where the cutting edges run in a general radial direction at one end of the tool and helical grooves extend from the cutting edges along the length of the tool to eject chips and sometimes admit coolant.

twisted pair 1. Two insulated wires (signal and return) which are twisted around each other. Since both wires have nearly equal exposure to any electrostatic or electromagnetic interference, the differential noise is slight. 2. A communications medium consisting of two insulated wires loosely twisted together.

two element feedwater control A control system whereby two process variables (steam flow and drum level) are used as inputs to the control loop that regulates feedwater flow to the drum to maintain the drum level at set point. The feedforward input is steam flow, with the output of the drum level controller as the primary control signal [S77.42].

two-out-of-three logic circuit (2/3 logic circuit) A logic circuit that employs three independent inputs. The output of the logic circuit is the same state as any two matching input states [S77.42].

two-phase A fluid state consisting of a mixture of liquid with gas or vapors [RP31.1].

two-piece-element clamp A two-piece-element clamp or pinch valve is a valve consisting of two flexible elastomeric elements or liners installed between a two-piece flanged body. The flexible elements or liners also extend over the flange faces and act as gaskets between the valve and the connecting piping [S75.08].

two-position action A type of control-system action that involves positioning the final control device in either of two fixed positions, without permitting it to stop at any intermediate position.

two-position controller See controller, two-position.

two's complement 1. A method of representing negative numbers in binary; formed by taking the radix complement of a positive number. 2. A form of binary arithmetic used in most computers to perform both addition and subtraction with the same circuitry where the representation of the numbers determines the operation to be performed.

two-sided sampling plan Any statistical quality control method whereby acceptability of a production lot is determined against both upper and lower limits.

two-stroke cycle An engine cycle for a reciprocating internal combustion engine that requires two strokes of the piston to complete.

two-way valve A valve with one inlet opening and one outlet opening [S75.05].

Tyndall effect A physical phenomenon first observed by Sir John Tyndall, who noted that particles suspended in a fluid could be seen readily if illuminated by strong light and viewed from the side, even though they could not be seen when viewed from the front in the same light beam; this effect is the basis for nephelometry, which involves measurement of the intensity of side-reflected light, and is commonly used in such applications as analyzing for trace amounts of silver in solution, determining the concentration of small amounts of calcium in titanium alloys, measuring bacterial growth rates, and controlling the clarity of beverages, potable water and effluent discharges.

U

U See velocity (ft/sec).

U bolt A rod threaded at both ends and bent into a U shape; it is most often used with a bar across the span of the U to provide a clamping force around a tubular or cylindrical object.

UL See Underwriters Laboratories.

ultimate analysis See analysis, ultimate.

ultimate cycle method See Ziegler-Nichols method.

ultimately controlled variable See variable, ultimately controlled.

ultimate strength See tensile strength.

ultrasonic Using frequencies above the audio-frequency range, i.e., above 20kHz [S37.1].

ultrasonically-assisted machining A machining method in which ultrasonic vibrations are imparted to a tool to make it cut with better quality or speed than would be possible using the same machining process but without vibrating the tool.

ultrasonic atomizer A type of atomizer that produces uniform droplets at low feed rates by flowing liquid over a surface which is vibrating at ultrasonic frequency.

ultrasonic bonding A method of joining two solid materials by subjecting a joint under moderate clamping pressure to vibratory shearing action at ultrasonic frequencies until a permanent bond is achieved; it may be used on both soft metals and thermoplastics.

ultrasonic cleaning Removing soil from a surface by the combined action of ultrasonic vibrations and a chemical solvent, usually with the part immersed.

ultrasonic coagulation A process that uses ultrasonic energy to bond small particles together, forming an aggregated mass.

ultrasonic delay line A constrained pathway for propagating sound so that the transit time along the pathway becomes a fixed time delay for another signal.

ultrasonic density sensor A device for determining density from the attenuation of ultrasound beams passing through a liquid or semisolid; a typical application involves immersing an ultrasonic transducer in fully agitated lime slurry, thus avoiding coating and clogging which occurs with other devices.

ultrasonic detector Any of several devices for detecting ultrasound waves and measuring one or more wave attributes.

ultrasonic drilling A method of producing holes of almost any desired shape in very hard materials, such as tungsten carbide or gemstones, by causing a suitable tool which is pressed against the workpiece to vibrate axially under the driving force of an ultrasonic transducer.

ultrasonic flowmeter A device for measuring flow rates across fluid streams by either Doppler-effect measurements or time-of-transit determination; in both types of flow measurement, displacement of the portion of the flowing stream carrying the sound waves is determined and flow rate calculated from the effect on soundwave characteristics.

ultrasonic frequency Any frequency for compression waves resembling sound where the frequency is above the audible range—that is, above about 15 kHz.

ultrasonic generator A device for producing compression waves of ultrasonic frequencies.

ultrasonic level detector Any of several devices that use either time-of-transit or intensity attenuation of an ultrasonic beam to determine the position of the upper surface in a body of confined liquid or bulk solids.

ultrasonic light diffraction Forming optical diffraction patterns when light passing through a longitudinal ultrasound field is refracted by interaction with the sound waves.

ultrasonic machining A machining method in which an abrasive slurry is driven against a workpiece by a tool vibrating axially at high frequency to cut an exact shape in the workpiece surface.

ultrasonic material dispersion Using ultrasound waves to break up one component of a mixture and disperse it in another to create a suspension or emulsion.

ultrasonics Technology associated with the production and utilization of sound having a frequency higher than about 15-20 kHz.

ultrasonic stroboscope A device for producing pulsed light by using ultrasound to modulate a light beam.

ultrasonic testing A nondestructive testing method in which high frequency sound waves are projected into a solid to detect and locate flaws, to measure thickness, or to detect structural differences.

ultrasonic thickness gage Any of several devices that use either resonance or pulse-echo techniques to determine the thickness of metal parts—sheet or plate thickness, or pipe-wall thickness, for example; the technique also may be used to determine coating thickness in applications where a suitable reflection can be obtained from the coating-substrate interface.

ultrasonic transducer A device for converting high-frequency electric impulses into mechanical vibrations, or vice versa, usually through the use of a magnetostrictive or piezoelectric material.

ultrasonic welding Same as ultrasonic bonding.

ultrasonoscope An instrument for displaying an echosonogram on an oscilloscope, and sometimes for providing auxiliary output to a chart recorder.

ultraviolet-erasable read-only memory (UVROM) A type of computer memory that can be erased or changed only by exposure to ultraviolet light.

ultraviolet radiation Electromagnetic radiation having wavelengths shorter than visible light and longer than low-frequency x-rays—that is, wavelengths of about 14 to 400 nanometers.

ultraviolet spectrophotometry Determination of the concentration of various compounds in a water solution or gas stream based on char-

acteristic absorption of ultraviolet rays; uv absorption patterns are not as distinctive "fingerprints" as their ir counterparts, but in many cases the former are more selective and sensitive for use in process control applications.

umbilical connection Any flexible grouping of electrical, mechanical or hydraulic connections between a machine, vehicle or robotics device and a source of power, control signals and data acquisition or auxiliary services.

unaccounted-for loss That portion of a boiler heat balance which represents the difference between 100 percent and the sum of the heat absorbed by the unit and all the classified losses expressed as percent.

unbalanced (to ground) Opposite of balanced (to ground); unbalanced (to ground) cable pairs can be susceptible to noise and crosstalk and can cause crosstalk to other pairs.

unbalance, dynamic The net force produced on the valve stem in any given open position by the fluid pressure acting on the closure member and stem within the pressure retaining boundary, with the closure member at a stated opening and with stated flowing conditions [S75.05].

unbalance, static The net force produced on the valve stem by the fluid pressure acting on the closure member and stem within the pressure retaining boundary with the fluid at rest and with stated pressure conditions [S75.05].

unbonded Stretched and unsupported between ends (usually refers to strain-sensitive wire) [S37.1].

unbonded strain gage A type of wire strain gage sometimes used in transducer applications where strain is determined from elastic tension developed across the gage between mechanical end connections.

unburned combustible The combustible portion of the fuel which is not completely oxidized.

unburned combustible loss See combustible loss.

uncertainty The interval within which the true value of a measured quantity is expected to lie with a stated probability.

unconditional branch See unconditional transfer.

unconditional jump See unconditional transfer.

unconditional transfer An instruction which switches the sequence of control to some specified location. Synonymous with unconditional branch and unconditional jump. Loosely, jump.

uncouple To disengage a screwed, pinned or latched connection.

undamped frequency See frequency, undamped.

underbead crack A crack in the heat-affected zone of a weldment that does not extend to the base metal surface.

undercut 1. An unfilled groove in the base metal along the toe of a weld. 2. A groove or recess along the transition zone from one cross-section to another, such as from a hub to a fillet, that leaves a portion of one cross-section undersized.

underdamped See damping.

underfill A condition whereby the face of a weld is lower than the position of an adjacent base metal surface.

underflow Pertaining to the condition that arises when a machine computation yields a nonzero result that is smaller than the smallest nonzero quantity that the intended unit of storage is capable of storing.

understressing Repeatedly stressing a part at a level below the fatigue limit or below the maximum service stress to improve fatigue properties.

Underwriters Laboratories (UL) An independent testing and certifying organization.

undocumented In data processing, the absence of instructions on how to use a program or computer.

unfired pressure vessel A vessel designed to withstand internal pressure, neither subjected to heat from products of combustion nor an integral part of a fired pressure vessel system.

unformatted ASCII A mode of data transfer in which the low-order seven bits of each byte are transferred; no special formatting of the data occurs or is recognized.

unformatted binary A mode of data transfer in which all bits of a byte are transferred without regard to their contents.

ungrounded Refers to the presence or absence of an electrical connection between the "low" side of the transducer element and the portion of the transducer intended to be in contact with the test structure. Method of ungrounding should be stated as "internally ungrounded" or "by means of separate stud." [RP37.2]

UNIBUS DEC's common communications bus to the CPU, memory, and all peripheral devices along which address, data, and control are trans-ferred on fifty-six lines; the form of communication is identical for all devices on this bus.

unidirectional pulse A wave pulse in which intended deviations from the normally constant values occur in only one direction.

unified screw thread A system of standard 60° V threads that are classified coarse, fine and extra-fine (UNC, UNF and UNEF) to provide different levels of strength and clamping power.

uniform corrosion Chemical reaction or dissolution of a metal characterized by uniform receding of the surface.

unilateral tolerance A method of dimensioning in which either the upper or lower limit of the allowable range is given as the stated size or location, and the permissible variation is given as a positive or negative tolerance from that size, but not both; the decision as to whether the upper or lower limit is given as the stated dimension depends on the critical value for the dimension, and should be chosen so that the tolerance is always away from the critical value and toward a less critical condition.

unilateral transducer A transducer that produces output waves related to input waves when connected in one direction, but that cannot produce such waves when input and output connections are reversed.

uninterruptible power source The use of resident batteries in a device to phase in when external power is interrupted.

uninterruptible power supply (UPS) A type of power supply that can provide electrical power even when line power is lost.

union A threaded assembly for joining the ends of two pipes or tubes where neither can be rotated to complete the joint; it usually consists of flanged members that are threaded or soldered onto the pipe ends and a ring member that surrounds the flange edges and makes a leak-tight seal.

unit 1. A collection of associated elements, loops, devices, and/or equipment modules that perform a coordinated function and which operates relatively independently. 2. A device having a special function. 3. A basic element. See arithmetic unit, binary unit, central processing unit, and control unit.

unit, central processing (CPU) The unit of a computing system that includes the circuits controlling the interpretation and execution of instructions [RP55.1].

unit sensitivity The specific amount that a measured quantity must rise or fall to cause a pointer or other indicating element to move one scale division on a specific instrument.

universal development system A development system that, by means of personality modules, can be used to develop the software and hardware for a range of microcomputers.

universal instrument See altazimuth.

universal joint A linkage for transmitting rotational motion and power between two shafts whose axes do not coincide, especially when the axis of one must be allowed to pivot through a small angle with respect to the other during operation.

universal output transformer An output transformer for an electronic device having several taps on its secondary winding so that it can be connected to almost any loudspeaker system by choosing the proper tap.

universal ratio set Abbreviated URS. An arrangement of variable resistors used as a highly accurate continuously adjustable arm of a Wheatstone bridge with a resolution of 0.001 ohm and a total range of 2111.110 ohms; a URS is particularly well suited for measuring unknown resistances in terms of highly accurate fixed-decade standards within a 10:1 ratio of the unknown resistances.

unlined body A body without a lining [S75.05].

unscheduled maintenance Any urgent need for repair or upkeep that was unpredicted or not previously planned, and must be added to or substituted for previously planned work.

unsprung weight That portion of a vehicle's gross weight that is comprised of the wheels, axles and various other components not supported by its springs.

unsteady flow A flow in which the flow rate fluctuates randomly with time and for which the mean value is not constant.

unterminated ramp A ramp that starts at the variable's initial value, becomes linear, and continues to a higher or lower value beyond the setpoint of interest, such that the instrument's or channel's desired output signal is obtained while the input ramp is still linear [S67.06].

update 1. To put into a master file changes required by current information or transactions. 2. To modify an instruction so that the address numbers it contains are increased by a stated amount each time the instruction is performed.

upgrade To increase the value or quality of an operating system or commercial product by incorporating changes in design or manufacture without changing its basic function.

upgrading Major changes to older control systems.

upper limit 1. The upper limit of the signal current is the current corresponding to the maximum value of the dc current signal [S50.1]. 2. The pneumatic signal corresponding to the maximum value of the transmitted input [S7.4].

upper range-limit See range-limit, upper.

upper range-value See range-value, upper.

UPS See uninterruptible power supply.

upset To cause a local increase in diameter or other cross-sectional dimension by applying an axial deforming force to a piece of rod or wire, such as is used to produce heads on nails or screws.

upsetting See cold heading.

upstream seating A seat on the upstream side of the ball, designed so that the pressure of the controlled fluid causes the seat to move toward the ball [S75.05]. See also seating, upstream.

uptake A conduit for exhaust gases connecting the outlet of a furnace or firebox to a chimney or stack.

up time The time during which equipment is either producing work or is available for productive work. Contrasted with downtime.

upward compatibility In data processing, the ability of a computer device or program to function on newer models.

USASCII U.S. Standard Code for Information Exchange. The standard code, using a coded charter set consisting of 7-bit coded characters (8 bits including parity check), used for information exchange among data processing systems, communications systems, and associated equipment. The USASCII set consists of control characters and graphic characters. See also ASCII.

use factor The ratio of hours in operation to the total hours in that period.

user In data processing, the person or client who makes use of a computer system.

user-friendly In data processing, a general term to describe programs that do not require extensive learning or technical skill to use successfully.

user interface The way a program communicates with an operator.

utility 1. Any general-purpose computer program included in an operating system to perform common functions. 2. Any of the systems in a process plant, manufacturing facility not directly involved in production; may include any or all of the following—steam, water, refrigeration, heating, compressed air, electric power, instrumentation, waste treatment and effluent systems.

utility program See utility routine.

utility routine A standard routine used to assist in the operation of the computer, e.g., a conversion routine, a sorting routine, a printout routine, or a tracing routine. Synonymous with utility program.

utility software A library of programs for general use in a computer system.

U-tube manometer A device for measuring gage pressure or differential pressure by means of a U-shaped transparent tube partly filled with a liquid, commonly water; a small pressure above or below atmospheric is measured by connecting one leg of the U to the pressurized space and observing the height of liquid while the other leg is open to the atmosphere; similarly, a small differential pressure is measured by connecting both legs to pressurized space—for example, high- and low-pressure regions across an orifice or venturi.

UV erasable PROM (EPROM) Memory whose contents can be erased by a period of intense exposure to UV radiation.

UVROM See ultraviolet-erasable read-only memory.

V

v See volt.

vacuum A low-pressure gaseous environment having an absolute pressure lower than ambient atmospheric pressure.

vacuum brake A type of power-assisted vehicle brake whose released position is maintained by maintaining a pressure below atmospheric in the actuating cylinder, and whose actuated position is obtained by admitting air at atmospheric pressure to one side of the cylinder.

vacuum breaker A device used in a water supply line to relieve a vacuum and prevent backflow. Also known as backflow preventer.

vacuum degassing Removing dissolved or trapped gases in a metal by melting or heating it under high vacuum.

vacuum deposition A process for coating a substrate with a thin film of metal by condensing it on the substrate in an evacuated chamber. See also vacuum plating.

vacuum filtration A process for separating solids from a suspension or slurry by admitting the mixture to a filter at atmospheric pressure (or higher) and drawing a vacuum on the outlet side to assist the liquid in passing through the filter element.

vacuum forming A method of forming sheet plastics by clamping the sheet to a stationary frame, then heating it and drawing it into a mold by pulling a vacuum in the space between the sheet and mold.

vacuum fusion A laboratory technique for determining dissolved gas content of metals by melting them in vacuum and measuring the amount of hydrogen, oxygen and sometimes nitrogen released during melting; the process can be used on most metals except reactive elements such as alkali and alkaline-earth metals.

vacuum gage Any of several devices for measuring pressures below ambient atmospheric.

vacuum-gage control circuit An electric circuit which energizes the tube of an electrically operated vacuum gage, controls and measures gage currents or voltages, and sometimes supplies and regulates power that degases tube elements.

vacuum-gage tube An enclosed portion of a pressure-measuring system connected to an evacuated chamber or system; its essential component is the pressure-sensing element, but it also includes the envelope and any support structure, plus the means for connecting the gage to the evacuated space.

vacuum jacketed valve See jacketed valve.

vacuum photodiode A vacuum tube in which light incident on a photoemissive surface (cathode) frees electrons, which are collected by the positively biased anode.

vacuum plating A process for producing a thin film of metal on a solid substrate by depositing a vaporized compound on the work surface, or by reacting a vapor with the surface, in an evacuated chamber. Also known as vapor deposition.

vacuum pump A device similar to a compressor whose inlet is attached to a chamber to remove noncondensible gases such as air and maintain the chamber at a pressure below atmospheric.

vacuum system A system consisting of one or more chambers that can withstand atmospheric pressure without completely collapsing, and having an opening for pumping gas out of the enclosed space.

vacuum tube A device for use in an electronic circuit to amplify d-c, audio, or microwave frequencies or rectify radio-frequency signals; it consists of an arrangement of metal emitters, grids and plates enclosed in a thin, evacuated glass envelope with a molded plastic base containing pin connectors that are attached to the tube internals.

validity The correctness, especially the degree of the closeness by which iterated results approach the correct result.

validity check A check based upon known limits or upon given information or computer results, e.g., a calendar month will not be numbered greater than 12, and a week does not have more than 168 hours. See also check, validity.

value, desired In process instrumentation, the value of the controlled variable wanted or chosen. The desired value equals the ideal value in an idealized system [S51.1].

value engineering The systematic use of engineering principles to identify the functions of a product or service, and to provide these functions reliably at lowest cost. Also known as value analysis; value control.

value, ideal In process instrumentation, the value of the indication, output or ultimately controlled variable of an idealized device or system. It is assumed that an ideal value can always be defined even though it may be impossible to achieve [S51.1].

value, measured The numerical quantity resulting, at the instant under consideration, from the information obtained by a measuring device [S51.1].

value referred to the input The value obtained by dividing an output value by the nominal gain of the subsystem [RP55.1].

value, rms (root-mean-square value) The square root of the average of the square of the instantaneous values [S51.1].

valve 1. A valve is a device used for the control of fluid flow. It consists of a fluid retaining assembly, one or more ports between end openings and a movable closure member which opens, restricts or closes the port(s) [S75.05]. 2. An in-line device in a fluid-flow system that can interrupt flow, regulate the rate of flow, or divert flow to another branch of the system. 3. A term used by the British to denote a vacuum tube since valve action is the way a vacuum tube operates with a stream of electrons.

valve, ball A valve with a rotary motion closure component consisting of a full ball or a segmented ball.

valve diaphragm A flexible member which is moved into the fluid flow passageway of a body to modify the rate of flow through the valve [S75.05].

valve, diaphragm type A valve with a flexible linear motion closure component which is moved into the fluid flow passageway of the body to modify the rate of flow through the valve by the actuator.

valve, floating ball A valve with a full ball positioned within the valve that contacts either of two seat rings and is free to move toward the seat ring opposite the pressure source when in the closed position to effect shutoff.

valve follower A linkage that transmits motion from a cam to the push rod of a valve, especially in an internal combustion engine.

valve, fuel control An automatically or manually operated device consisting essentially of a regulating valve and an operating mechanism. It is used to regulate fuel flow and is usually in addition to the safety shut-off valve. Such valve may be of the automatic or manually opened type.

valve, globe A valve with a linear motion closure component, one or more ports and a body distinguished by a globular shaped cavity around the port region.

valve, manual gas shutoff A manually operated valve in a gas line for the purpose of completely turning on or shutting off the gas supply.

valve, manual oil shutoff A manually operated valve in the oil line for the purpose of completely turning on or shutting off the oil supply to the burner.

valve, manual reset safety shut-off A manually opened, electrically latched, electrically operated safety shut-off valve designed to automatically shut off fuel when de-energized.

valve, motor driven reset safety shut-off An electrically operated safety shut-off valve designed to automatically shut off fuel flow upon being de-energized. The valve is opened and reset automatically by integral motor device only.

valve plug An obsolete term. See closure component.

Van de Graaf generator An electrostatic device that uses a system of belts to generate electric charges and carry them to an insulated electrode, which becomes charged to a high potential.

vane 1. A flat or curved machine element attached to a hub or rotor that is acted upon by a flowing stream of fluid to produce rotary motion. 2. A fixed or adjustable plate inserted in a gas or air stream used to change the direction of flow.

vane control A set of movable vanes in the inlet of a fan to provide regulation of air flow.

vane guide A set of stationary vanes to govern direction, velocity and distribution of air or gas flow.

vane type valve A fluid powered device in which fluid acts upon a movable pivoted member, the vane, to provide rotary motion to the actuator stem [S75.05].

Van Stone nipples (flanges) A pipe nipple made with one enlarged integral end held against another face with a loose flange around the nipple.

vapor The gaseous product of evaporation.

vapor barrier A sheet or coating of low gas permeability that is applied to a structural wall to prevent condensation and absorption of moisture.

vapor deposition See vacuum plating.

vapor-filled thermometer A type of filled-system thermometer in which temperature is determined from the vapor pressure developed from partial vaporization of a volatile liquid contained within the system.

vapor generator A container of liquid, other than water, which is vaporized by the absorption of heat.

vaporimeter 1. An apparatus in which the volatility of oils are estimated by heating them in a current of air. 2. An instrument used to determine alcohol content by measuring the vapor pressure of the substance.

vaporization The change from liquid or solid phase to the vapor phase.

vaporization cooling A method of cooling hot electronic equipment by spraying it with a volatile, nonflammable liquid of high dielectric strength; the liquid absorbs heat from the electronic equipment, vaporizes, and carries the heat to enclosure walls or to a radiator or heat exchanger. Also known as evaporative cooling.

vapor pressure 1. The pressure of a vapor corresponding to a given temperature at which the liquid and vapor are in equilibrium. Vapor pressure increases with temperature [RP31.1]. 2. The pressure (for a given temperature) at which a liquid is in equilibrium with its vapor. As a liquid is heated, its vapor pressure will increase until it equals the pressure above the liquid; at this point the liquid will begin to vaporize.

vapor pressure, Reid The vapor pressure of a liquid at 100 °F (311 K) as determined by ASTM Designation D 323-58, "Standard Method of Test for Vapor Pressure of Petroleum Products (Reid Method)." [RP31.1]

vapor pressure thermometer A temperature transducer for which the pressure of vapor in a closed system of gas and liquid is a function of temperature.

var A unit of measure for reactive power; it is calculated by taking the product of voltage, current, and the sine of the phase angle.

variable The symbolic representation of a logical storage location that can contain a value that changes during a discrete processing operation. See also measurand.

variable address See indexed address.

variable-area track A motion-picture sound track divided laterally into transparent and opaque areas, where the line of demarcation is an oscillographic trace that corresponds to the wave shape of the recorded sound.

variable-density track A motion-picture sound track of constant width, where the photographic density varies along the length of the track in accordance with a defined wave parameter of the recorded sound; the track is usually, but not always, of uniform density in the transverse direction at any point along the direction of travel.

variable, directly controlled In a control loop, the variable the value of which is sensed to originate a feedback signal [S51.1].

variable indirectly controlled A variable which does not originate a feedback signal, but which is related to, and influenced by, the directly controlled variable [S51.1].

variable-inductance accelerometer An instrument for measuring instantaneous acceleration of a body; it consists of a differential transformer with a center coil excited from an external a-c signal whose magnitude is proportional to displacement of a ferromagnetic core mass suspended on springs in the center of the three coils.

variable-inductance pickup A transducer that converts mechanical oscillations into audio-frequency electrical signals by varying the inductance of an internal coil.

variable-length record format A file format in which records are not necessarily the same length.

variable, manipulated A quantity or condition which is varied as a function of the actuating error signal so as to change the value of the di-

405

rectly controlled variable [S51.1].

variable, measured A quantity, property, or condition which is measured. It is sometimes referred to as the measurand. Common measured variables are temperature, pressure, rate of flow, thickness, speed, etc. [S51.1]

variable-reluctance pickup A transducer that converts mechanical oscillations into audio-frequency electrical signals by varying the reluctance of an internal magnetic circuit.

variable reluctance proximity sensor A device that senses the position (presence) of an actuating object by means of the voltage generated across the terminals of a coil surrounding a pole piece that extends from one end of a permanent magnet; coil voltage is proportional to the rate of change of magnetic flux as the object passes through the field near the pole piece.

variable reluctance tachometer A type of tachometer designed to measure rotational speeds of 10,000 to 50,000 rpm by detecting electrical pulses generated as an actuating element integral with the rotating body repeatedly passes through the magnetic field of a variable-reluctance sensor; the pulses are amplified and rectified, then used to control direct current to a milliammeter, which is calibrated directly in rpm.

variable-resistance accelerometer An instrument that measures acceleration by determining the change in electrical resistance in a measuring element such as a strain gage or slide wire whose dimensions are changed mechanically under the influence of acceleration.

variable-resistance pickup A transducer that converts mechanical oscillations into audio-frequency electrical signals by varying the electrical resistance of an internal circuit.

variable, ultimately controlled The variable whose control is the end purpose of the automatic control system [S51.1].

variable word-length Having the property that a machine word may have a variable number of characters. It may be applied either to a single entry whose information content may be changed from time to time or to a group of functionally similar entries whose corresponding components are of different lengths.

variometer A form of variable inductance, consisting of two coils connected in a series and arranged one inside the other, the inner equipped to rotate and, thereby, vary the mutual inductance between coils. This device was principally used in the early days of radio communications, but has found continued usefulness in electronics.

varmeter An instrument for measuring the electric power drawn by a reactive circuit. Also known as reactive volt-ampere meter.

varnish A transparent coating material consisting of a resinous substance dissolved in an organic liquid vehicle.

VAX 11-780 DEC's most powerful computer; uses thirty-two-bit words and has virtual memory.

VCO See voltage-controlled oscillator.

VDU See video display unit.

vectopluviometer A rain gage, or a circular array of four or more rain gages, that measures the direction and inclination of falling rain.

vector 1. A quantity having magnitude and direction, as contrasted with a scalar which has quantity only. 2. A one dimensional matrix. See matrix.

vectored interrupt An interrupt which carries the address of its service routine.

vector quantity A property or characteristic which is completely defined only when both magnitude and direction are given.

vector voltmeter A two-channel, high-frequency sampling voltmeter that can be connected to two input signals of the same frequency to measure not only their voltages but also the phase angle between them.

vee orifice "V" shaped flow control orifice which allows a characterized flow control as the gate moves in relation to the fixed Vee opening [S75.05].

vehicle 1. A body such as an aircraft or rocket designed to carry a payload aloft. 2. A self-propelled machine for transporting goods or personnel. 3. A solvent or other carrier for the resins and pigments in paint, lacquer, shellac or varnish.

velocimeter An instrument for measuring the speed of sound in gases, liquids or solids.

velocity The rate of change of a position vector with respect to time at any given point in space; the first derivative of distance with respect to time.

velocity head The pressure, measured in height of fluid column, needed to create a fluid velocity. Numerically, velocity head is the square of the velocity divided by twice the acceleration of gravity ($U2/2g$).

velocity limit A limit which the rate of change of a specified variable may not exceed [S51.1].

velocity limiting control See control, velocity limiting.

velocity meter A flowmeter that measures rate of flow of a fluid by determining the rotational speed of a vaned rotor inserted into the flowing stream; the vanes may or may not occupy the entire cross section of the flowpath.

velocity of approach A factor (F) determined by the ratio (m) of the valve orifice area to the inlet pipe area.

velocity pressure The measure of the kinetic energy of a fluid.

velocity-type flowmeter A flow-measurement device in which the fluid flow causes a wheel or turbine impeller to turn, producing a volume-time readout. Also known as current meter; rotating meter.

vena contracta The location where cross-sectional area of the flowstream is at its minimum. The vena contracta normally occurs just downstream of the actual physical restriction in a control valve [S75.05].

Venn diagram A graphical representation in which sets are represented by closed areas. The closed regions may bear all kinds of relations to one another, such as be partially overlapped, be completely separated from one another, or be contained totally one within another. All members of a set are considered to lie within or be contained within the closed region representing the set. The diagram is used to facilitate the determination of whether several sets include or exclude the same members.

vent Any opening or passage that allows gases to escape from a confined space to prevent the buildup of pressure or the accumulation of hazardous or unwanted vapors.

venturi A constriction in a pipe, tube or flume consisting of a tapered inlet, a short straight constricted throat and a gradually tapered outlet; fluid velocity is greater and pressure is lower in the throat area than in the main conduit upstream or downstream of the venturi; it can be used to measure flow rate, or to draw another fluid from a branch into the main fluid stream.

venturi meter A type of flowmeter that measures flow rate by determining the pressure drop through a venturi constriction.

venturi tube A primary differential-pressure producing device having a cone section approach to a throat and a longer cone discharge section. Used for high volume flow at low pressure loss.

verify 1. To determine whether a transcription of data or other operation has been accomplished accurately. 2. To check the results of key-punching.

vernier A short auxiliary scale which slides along a main instrument scale and permits accurate interpolation of fractional parts of the least division on the main scale.

vertical boiler A fire-tube boiler consisting of a cylindrical shell, with tubes connected between the top head and the tube sheet forms the top of the internal furnace. The products of combustion pass from the furnace directly through the vertical tubes.

vertical firing An arrangement of a burner such that air and fuel are discharged into the furnace, in practically a vertical direction.

vertical orifice installation, vertical orifice run, vertical meter run An orifice plate used in a vertical pipeline. See basic definitions.

vessel A container or structural enclosure in which materials—especially liquids, gases and slurries—are processed, stored or treated.

V/F (boilup-to-feed ratio) A quantity used to analyze the operation of a distillation column.

vibrating density sensor Any of several devices in which a change in natural oscillating frequency of a device element—cylinder, single tube, twin tube, U-tube or vane—is detected and related to density of process fluid flowing through the system.

vibrating quartz-crystal moisture sensor A device for detecting the presence of moisture in a sample gas stream by dividing the stream into two portions, one of which is dried, then alternately passing the two streams across the face of a hygroscopically sensitized quartz crystal whose wet and dry vibrational frequencies are continuously monitored and compared to the frequency of an uncoated sealed reference crystal.

vibrating-reed electrometer An instrument that uses a vibrating capacitor to measure small electrical charges, often in combination with an ionization chamber.

vibrating-reed tachometer A device consisting of an extended series of reeds of various lengths mounted on the same base; the device is placed on a vibrating surface, such as the enclosure of rotating equipment, and the frequency determined by observing which of the reeds is vibrating at its natural frequency.

vibration　A periodic motion or oscillation of an element, device, or system. Note 1: Vibration is caused by any excitation which displaces some or all of a particular mass from its position of equilibrium. The resulting vibration is the attempt of the forces, acting on and within the mass, to equalize. 2: The amplitude and duration of any vibration is dependent on the period and amplitude of the excitation and is limited by the amount of damping present [S51.1].

vibration damping　Any method of converting mechanical vibrational energy into heat.

vibration error　The maximum change in output at any measurand value within the specified range, when vibration levels of specified amplitude and range of frequencies are applied to the transducer along specified axes [S37.1].

vibration machine　A device for determining the effects of mechanical vibrations on the structural integrity or function of a component or system—especially electronic equipment. Also known as shake table.

vibration meter　A device for measuring vibrational displacement, velocity and acceleration; it consists of a suitable pickup, electronic amplification circuits, and an output meter.

vibration sensitivity　See vibration error.

vibration-type level detector　A device for detecting the level of solids in a bin or hopper, in which a tuning fork driven by a piezoelectric crystal vibrates freely when the level is below the sensor position and is inhibited from vibrating when bulk material surrounds the sensor.

vibratory separation　A technique for separating or classifying particulate solids using screens that are subjected to vibratory or oscillating motion.

vibrograph　An instrument for making an oscillographic recording of the amplitude and frequency of a mechanical vibration, such as by producing a trace on paper or film using a moving stylus.

vibrometer　A device for measuring the amplitude of a mechanical vibration. Also known as vibration meter.

vibronic isolation　Systems which minimize the transfer of vibrations from the floor and surrounding environment to the surface of an optical table or other equipment mounted on them.

vibronic transition　A simultaneous change in both vibrational and electronic energy state of a molecule, with the amount of energy involved similar to that for electronic transitions.

Vickers hardness　See diamond-pyramid hardness.

video　In radio telemetry, this is the term generally applied to a telemetry multiplex output from a radio receiver.

video card　In data processing, a plug-in circuit board that controls the display of data on the monitor.

video display unit (VDU)　Any one of several types of shared human interface devices that use digital video technology.

video receiver　The data output of a telemetry receiver; the multiplex of telemetry measurements.

video terminal　An operator terminal with a cathode-ray-tube (CRT) display instead of a printer; see CRT display.

virtual address space　A set of memory addresses that are mapped into physical memory addresses by the paging or relocation hardware where a program is executed.

virtual block　One of a collection of blocks comprising a file (or the memory image of that file). The block is virtual only in that its block number refers to its position relative to other blocks on the volume; that is, the virtual blocks of a file are numbered sequentially beginning with one, while their corresponding logical block numbers can be any random list of valid volume-relative block numbers.

virtual leak　A gradual release of gas by desorption from the interior walls of a vacuum system in a manner that cannot be accurately predicted; its effect on system operation resembles that of an irregularly variable physical leak.

virtual memory　The set of storage locations in physical memory, and on disk, that are referred to by virtual addresses. From the programmer's viewpoint, the secondary storage locations appear to be locations in physical memory. The size of virtual memory in any system depends on the amount of physical memory available and the amount of disk storage used for nonresident virtual memory.

virtual page number　The virtual address of a page of virtual memory.

virtual processor　Software which allows an individual user to consider a computer's resources to be entirely dedicated to him. A computer can simulate several virtual processors simultaneously.

viscometer An instrument that measures the viscosity of a fluid.

viscometer gage An instrument that determines pressure in a vacuum system by measuring the viscosity of residual gases.

viscosity Measure of the internal friction of a fluid or its resistance to flow.

viscosity, absolute The property by which a fluid in motion offers resistance to shear. Usually expressed as newton-seconds/meter^2 [RP31.1].

viscosity, kinematic The ratio of absolute viscosity to density. The SI unit is the meter^2/s [RP31.1].

viscous damping A method of converting mechanical vibration energy into heat by means of a piston attached to the vibrating object which moves against the resistance of a fluid—usually a liquid or air—confined in a cylinder or bellows attached to a stationary support.

viscous-drag-type density meter A type of meter for determining gas density by comparing the drag force on linked impellers driven by flow of a standard gas and the test gas; the balance point is a function of gas density, and the instrument can be calibrated to read directly in density units.

viscous flow See laminar flow.

visibility meter An instrument for directly or indirectly determining visual range in the earth's atmosphere.

visual display That part of an annunciator or lamp cabinet that indicates the sequence state. Usually consists of an enclosure containing lamps behind a translucent window. The lamps can be off, flashing, or on [S18.1].

visual display unit (VDU) A generic term used for display units based on technologies such as cathode ray tubes (CRTs), plasma discharge panels (PDPs), electroluminescent devices (ELs), liquid crystal displays (LCDs), etc. [S5.5].

Viton A A fluorocarbon rubber by E.I. du Pont de Nemours Co.

vitreous enamel A coating applied to metal by covering the surface with powdered alkaliborosilicate glass frit and fusing it onto the surface by firing at a temperature of 800 to 1600 °F (425 to 875 °C). Also known as porcelain enamel.

vitreous silica See silica glass.

vitreous slag Glassy slag.

vitrified wheel A grinding wheel made by compacting a mixture of abrasive particles and glass frit, then firing it to produce a bonded mass.

vocabulary A list of operating codes or instructions available to the programmer for writing the program for a given problem, for a specific computer, or for a specific language.

voice print An acoustic spectrograph that can be used to analyze sound patterns, especially the harmonic patterns that distinguish one person's voice from another's.

vol See volume.

volatile 1. Of a liquid, having appreciable vapor pressure at room or slightly elevated temperature. 2. Of a computer, having memory devices that do not retain information if the power is interrupted.

volatile matter Those products given off by a material as gas or vapor, determined by definite prescribed methods.

volatile memory Memory whose contents are lost when the power is switched off.

volatile storage 1. A storage device in which stored data are lost when the applied power is removed, for example, an acoustic delay line. 2. A storage area for information subject to dynamic change.

volatization See vaporization.

volt A unit of electromotive force which when steadily applied to a conductor whose resistance is one ohm will produce a current of one ampere.

voltage amplification The ratio of voltage of an output signal to voltage of the corresponding input signal.

voltage, common mode (CMV) 1. That amount of voltage of the same polarity and phase common to both input lines. Common mode voltage can be caused by magnetic induction, capacitive coupling, and resistive coupling [RP55.1]. 2. A voltage of the same polarity on both sides of a deferential input relative to ground.

voltage-controlled oscillator (VCO) An oscillator in which the output frequency is dependent upon the input voltage [S51.1].

voltage divider An electronic network that consists of multiple impedance elements connected in series; for a given voltage impressed across the network, one or more lower output voltages can be obtained by tapping across one or more node pairs in the network.

voltage, normal mode 1. The actual voltage

difference between input signal lines [RP55.1]. 2. A voltage induced across the input terminals of a device [S51.1].

voltage-range multiplier A separate device installed externally to an instrument so its voltage range can be extended beyond the upper limit of the instrument scale; it consists principally of a special type of series resistance or impedance element.

voltage ratio For potentiometric transducers, the ratio of output voltage to excitation voltage, usually expressed in percent [S37.1].

voltage standing-wave ratio In a waveguide, the ratio of the amplitude of the electric field at a voltage minimum to the amplitude at an adjacent voltage maximum.

voltage-type telemeter A system for transmitting information to a remote location using the amplitude of a single voltage as the telemeter signal.

volt-ampere meter An instrument for measuring apparent power—the product of voltage and current—in an a-c power circuit; in high-power applications the scale is usually graduated in kilovolt-amperes.

voltmeter An instrument for determining the magnitude of an electrical potential; it generally is constructed as a moving-coil instrument having high internal series resistance; if the high internal resistance is replaced with a low-resistance shunt connected in parallel with the instrument terminals, it can function as an ammeter.

voltmeter-ammeter An instrument consisting of a voltmeter and an ammeter in the same housing, but with separate electrical connections.

volt-ohm-milliammeter A test instrument having different ranges for measuring voltage, resistance and current flow (in the milliampere range) in electrical or electronic circuits. Also known as circuit analyzer; multimeter; multiple-purpose meter.

volume (vol) 1. The magnitude of a complex audio-frequency current measured in standard volume units on a graduated scale. 2. The three-dimensional space occupied by an object. 3. A measure of capacity for a tank or other container in standard units. 4. A mass storage media that can be treated as file-structured data storage.

volume control A device or system that regulates or varies the output-signal amplitude of an electronic circuit, such as for varying the loudness of reproduced sound.

volume flow rate Calculated using the area of the full closed conduit and the average velocity in the form, Q=VxA, to arrive at the total quantity of flow.

volume indicator A standard instrument for indicating the magnitude of a complex wave such as an electronic signal for reproducing speech or music; the magnitude in volume units equals the number of decibels above a reference level established by connecting the instrument across a 600-ohm resistor that is dissipating 1 mW of power at 100 Hz.

volume meter Any flowmeter in which actual flow of a fluid is determined by measuring a characteristic associated with the flow.

volume of air The number of cubic feet of air per minute expressed at fan outlet conditions.

volumetric efficiency For a reciprocating engine or gas compressor, the ratio of volume of working fluid (at a specified temperature and pressure) admitted divided by piston displacement.

volumetric flow rate (q) The volumes of fluid moving through a pipe or channel within a given period of time.

volute A spiral casing for a centrifugal pump or fan; it allows the speed developed at the rotor vanes to be converted to pressure without hydraulic shock.

vortex 1. The swirling motion of a liquid in a vessel at the entrance to a discharge nozzle. 2. The point in a cyclonic gas path where the outer spiral converges to form an inner spiral and where the two spirals change general direction by 180°.

vortex shedding A phenomenon that occurs when fluid flows past an obstruction; the shear layer near the obstruction has a high velocity gradient, which makes it inherently unstable; at some point downstream of the immediate vicinity of the obstruction, the shear layer breaks down into well-defined vortices, which are captured by the flowing stream and carried further downstream.

vortex-type flowmeter A device that uses differential-pressure variations associated with formation and shedding of vortices in a stream of fluid flowing past a standard flow obstruction—usually a circular element with a T-shaped cross section—to actuate a sealed detector at a frequency proportional to vortex shedding which, in turn, provides an output signal directly related to flow rate.

votator A device used principally in food-processing industries for simultaneously chilling and mechanically working a continuous emul-sified stream, such as in the production of margarine.

vulcanizing Producing a hard, durable, flexible rubber product by steam curing a plasticized mixture of natural rubber, synthetic elastomers and certain chemicals.

W

w   See flow rate (lb/hr).

W   See watt.

wafer   1. A thin disc of a solid substance. 2. A thin part or component, such as a filter element.

wafer body   A body whose end surfaces mate with the pipeline flanges. It is located and clamped between the piping flanges by long bolts extending from flange to flange. A wafer body is also called a flangeless body [S75.05]. See also body, Wafer.

wafer-type temperature detector   A type of resistance-thermometer element designed with fine insulated wire of copper, nickel or platinum sandwiched between protective sheets of insulating material inside a sealed wafer-type enclosure; theses combine small mass with good thermal contact to give fast response times.

waist   The center portion of a vessel, tank or container that is smaller in cross section than adjacent sections.

wait condition   As applied to tasks, the condition of a task such that it is dependent on an event or events in order to enter the ready condition.

wait state   In data processing, the elapsed time between the request for data from memory, and when the memory chip responds.

wall box   A structure in a wall of a steam generator through which apparatus, such as sootblowers, extend into the setting.

wandering sequence   A welding technique in which increments of a weld bead are deposited along the seam randomly in both increment length and location.

warm-up period   1. The time required after energizing a device before its rated performance characteristics apply [S51.1]. 2. The period of time, starting with the application of excitation to the transducer, required to assure that the transducer will perform within all specified tolerances [S37.1].

wash   1. A stream of air or other fluid sent back along the axis of a propeller or jet engine. 2. A surface defect in castings caused by heat from the metal rising in the mold which induces expansion and shear of interface sand in the cope cavity. 3. A coating applied to the face of a mold prior to casting. Also known as mold wash. 4. To remove soil from parts, especially using a detergent or soap solution. 5. To remove cuttings or debris from a hole during drilling by introducing a liquid stream into the borehole and flushing it out.

washer   1. A ring-shaped component used to distribute a fastener's holding force, insulate or cushion a nut or bolthead from its bearing surface, lock a nut in place, or improve tightness of a bolted joint. 2. A machine for mechanically agitating parts or materials in a detergent solution. Also known as washing machine.

waste   1. Rubbish from a building, or refuse from a manufacturing or process plant. 2. Dirty water from domestic or industrial uses. 3. The amount of excavated material remaining after the hole has been backfilled. 4. A relatively loose mass of threads or yarns used for wiping up spilled oil or other liquids.

waste fuel   Any by-product fuel that is waste from a manufacturing process.

waste heat   Sensible heat in non-combustible gases.

waste lubrication   A method of delivering oil or other lubricant to a bearing surface by wicking action using cloth waste to absorb and transfer the lubricant.

watchdog   In control systems, a combination of hardware and software which acts as an interlock scheme, disconnecting the system's output from the process in event of system malfunction.

watchdog timer   An electronic internal timer

413

which will generate priority interrupt unless periodically recycled by a computer. It is used to detect program stall or hardware failure conditions. See also timer, watchdog.

water A liquid composed of two parts of hydrogen and sixteen parts oxygen by weight.

water calorimeter A device for measuring radio-frequency power by determining the rise in temperature of a known volume of water in which the radio-frequency power is absorbed.

water column (w.c.) A vertical tubular member connected at its top and bottom to the steam and water space respectively of a boiler, to which the water gage, gage cocks, high and low level alarms and fuel cutoff may be connected.

water cooling Using a stationary or flowing volume of water to absorb heat and disperse it or carry it away.

water-flow pyrheliometer A device for determining intensity of solar radiation in which the radiation sensor is a blackened water calorimeter; radiation intensity is calculated from the rise in temperature of water flowing through the calorimeter at a constant rate.

water gage The gage glass and its fittings for attachment.

water gas Gaseous fuel consisting primarily of carbon monoxide and hydrogen made by the interaction of steam and incandescent carbon.

water hammer 1. A sudden increase in pressure of water due to an instantaneous conversion of momentum to pressure. 2. A series of shocks, sounding like hammer blows, caused by suddenly reducing fluid-flow velocity in a pipe.

water jacket A casing around a pipe, process vessel or operating mechanism for circulating cooling water.

water level The elevation of the surface of the water in a boiler.

water path In ultrasonic testing, the distance from the search unit to the workpiece in a water column or immersion test setup.

waterproof Impervious to water. Compare with water resistant.

waterproof grease A viscous lubricant that does not dissolve in water and that resists being washed out of bearings or other moving parts.

waterproofing agent A substance used to treat textiles, paper, wood and other porous or absorbent materials to make them shed water rather than allow it to penetrate.

water resistant Slow to absorb water or to allow water to penetrate, often expressed as a maximum allowable immersion time. Compare with waterproof.

water tube A tube in a boiler having the water and steam on the inside and heat applied to the outside.

water tube boiler A boiler in which the tubes contain water and steam, the heat being applied to the outside surface.

water vapor A synonym for steam, usually used to denote steam of low absolute pressure.

watt (W) Metric unit of power. The rate of doing work or the power expended equal to 107 ergs/second, 3.4192 Btu/hour or 44.27 footpounds/minute.

watt-hour meter An integrating meter that automatically registers the integral of active power in a circuit with respect to time, usually providing a readout in kW-h.

wattmeter An instrument for directly measuring average electric power in a circuit.

wave Variation of a physical attribute of a solid, liquid or gaseous medium in such a manner that some of its parameters vary with time at any position in the medium, while at any instant of time the parameters vary with position.

wave analyzer An electronic instrument for measuring magnitude and frequency of the various sinusoidal components of a complex electrical signal.

wave filter A transducer that separates waves by introducing relatively small insertion loss into waves of one or more frequency bands while introducing relatively large insertion loss into waves of other frequencies.

waveform digitizer A device which generates a digital signal corresponding to an analog waveform which it receives.

wave front 1. Of a wave propagating in a bulk medium, any continuous surface where the wave has the same phase at any given instant in time. 2. Of a wave propagating along a continuous surface, any continuous line where the wave has the same phase at any given instant in time.

wave gage A device for measuring the height of waves on the ocean or a large lake and for measuring the period between successive waves.

waveguide An elongated volume of air or other dielectric used in guided transmission of electromagnetic waves; it usually consists of a circular or rectangular tube of dimensions chosen for efficient propagation of a specific frequency or frequencies; the tube walls may be electrically conductive, or they may constitute surfaces where permittivity or permeability, or both, are discontinuous.

wave impedance   In an electromagnetic wave, the ratio of the transverse electric field to the transverse magnetic field.

wave interference   A pattern of varying wave amplitude caused by superimposing one wave on another in the same medium.

wavelength   In any periodic wave, the distance from any point on the wave to a point having the same phases on the next succeeding cycle; the wavelength, λ, equals the phase velocity, v, divided by the frequency, f.

wavelength-division multiplexing   Combination of two or more signals so they can be transmitted over a common optical path, usually through a single optical fiber, by a technique in which the signals are generated by light sources having different wavelengths.

wavelength meter   An instrument which measures the wavelength of a laser beam, or other monochromatic source of light.

wavemeter   An instrument for determining the wavelength of an a-c or high frequency signal, either directly or indirectly. The usual method is by measuring the frequency and converting to wavelength. Also called frequency meter.

wave motor   A power conversion device for producing mechanical power from the lifting power of sea waves.

wave normal   A unit vector perpendicular to a wave front and having a positive component coincident with the direction of propagation; in an isotropic medium, the wave normal lies along the direction of the propagation.

wave soldering   A soldering technique used extensively to bond electronic components to printed circuit boards; soldering is precisely controlled by moving the assemblies across a flowing wave of solder in a molten soldering bath; the process also minimizes heating of the assemblies, and thus avoids one of the causes of early failure in electronic components. Also known as flow soldering.

wave tail   The trailing portion of a signal-wave envelope, in time or distance, where the rms wave amplitude decreases from its steady-state value to the end of the wave train.

Wb   See weber.

w.c.   See water column.

wear   Progressive deterioration of a solid surface due to abrasive or adhesive action resulting from relative motion between the surface and another part or a loose solid substance.

wear oxidation   See fretting.

weatherometer   A test apparatus used to estimate the resistance of materials and finishes to deterioration when exposed to climatic conditions; it subjects test surfaces to accelerated weathering conditions such as concentrated ultraviolet light, humidity, water spray and slat fog.

weatherproof   Capable of being exposed to an outdoors environment without substantial degradation for an extended period of time.

weather resistance   The relative ability of a material or coating to withstand the effects of wind, rain, snow and sun on its color, luster, and integrity.

web   1. The vertical plate connecting upper and lower flanges of a rail or girder. 2. The central portion of the tool body in a twist drill or reamer. 3. A thin section of a casting or forging connecting two regions of substantially greater cross section.

weber (Wb)   Metric unit for magnetic flux.

weep   A term usually applied to a minute leak in a boiler joint which forms droplets (or tears) of water very slowly.

weighbridge-type belt scale   A scale mounted above or below a belt conveyor that supports a section of the conveyor belt via a structural suspension system (weigh carriage) and weigh idlers [RP74.01].

weigh carriageEM]A structure supporting the weigh idlers, which in turn transmits weight to the load reactor [RP74.01].

weigh idlers   Idlers positioned in the weigh carriage assembly so that they sense the weight of the material on the conveyor belt and transmit the weight through the carriage to the load reactor [RP74.01].

weighing   The process of determining either the mass or weight of a body depending upon the apparatus and the procedure employed [RP31.1].

weigh-scale   A device for determining either the mass or the weight of a body depending upon the apparatus and procedure employed [RP31.1].

weight (wt)   1. The force with which a body is attracted by gravity. The newton is the unit force in this Standard [RP31.1]. 2. As in place weight. The multiplier value associated with a digit be-

415

cause of its position in a set of digits. The second place from the right has a place weight of 10 in decimal and 16 in hexadecimal, so a 2 in this location is worth 2x10 in decimal and 2x16 in hexadecimal. See also significance.

weighting Artificial adjustment of a measurement to account for factors peculiar to conditions prevailing at the time the measurement was taken.

weights Reference units of mass such as counterpoise "weights" used with lever balances and dead "weights" used in calibrating balances, scales, and pressure gauges [RP31.1].

weir An open-channel flow measurement device analogous to the orifice plate-flow constriction.

weir pond See stilling basin.

weir type valve A body having a raised contour contacted by a diaphragm to shut off fluid flow [S75.05]. See also body, weir type.

welded strain gage A type of foil strain gage especially designed to be attached to a metal substrate by spot welding; used almost exclusively in stress analysis.

weld ends Valve end connections which have been prepared for welding to the line pipe or other fittings. May be butt weld (BWE), or socket weld (SWE) [S75.05]. See also end connections, welded.

welder A person or machine that makes a welded joint.

welding Producing a coherent bond between two similar or dissimilar metals by heating the joint, with or without pressure, and with or without filler metal, to a temperature at or above their melting point.

welding force See electrode force.

welding ground See work lead.

weldment A structure or assembly whose parts are joined together by welding.

weld metal The metal in the fusion zone of a welded joint.

well 1. A pressure-tight tube or similarly shaped chamber, closed at one end, and usually having external threads so it can be screwed into a tapped hole in a pressure vessel to form a pressure-tight means of inserting a thermocouple and other temperature-measuring element. Also known as thermowell. 2. A chamber at the bottom of a condenser, vacuum filter or other vessel where liquid droplets collect. Also known as hot well.

well-type manometer A type of double-leg, glass-tube manometer in which one leg is substantially smaller than the other; the large-diameter leg acts as a reservoir whose liquid level does not change appreciably with changes in pressure.

Westphal balance An instrument for determining the density of solids and liquids by direct reading, using a balance with movable weights; to measure liquid density, a plummet of known weight and volume is suspended in the liquid, whereas to measure solid density, a sample of the solid is suspended in a liquid of known density.

wet assay Determining the amount of recoverable mineral in an ore or metallurgical residue, or the amount of specific elements in an alloy, using flotation, dissolution and other wet-chemistry techniques.

wet back Baffle provided in a firetube boiler joining the furnace to the second pass to direct the products of combustion, that is completely water cooled.

wet-back boiler A baffle provided in a firetube boiler or water leg construction covering the rear end of the furnace and tubes, and is completely water cooled. The products of combustion leaving the furnace are turned in this area and enter the tube bank.

wet basis The more common basis for expressing moisture content in industrial measurement, in which moisture is determined as the quantity present per unit weight or volume of wet material; by contrast, the textile industry uses dry basis or regain moisture content as the measurement standard.

wet-bulb temperature The lowest temperature which a water wetted body will attain when exposed to an air current. This is the temperature of adiabatic saturation.

wet-bulb thermometer A thermometer whose bulb is covered with a piece of fabric such as muslin or cambric that is saturated with water; it is most often used as an element in a psychrometer.

wet classifier A device for separating solids in a liquid-solid mixture into fractions by making use of the difference in settling rates between small and large particles. The classifications can take place on a moving stream with appropriate spiral trays or in a tank or pond that is not agitated.

wet grinding Any technique for removing surface layers of a metal part by abrasive action in the presence of a fluid, such as water or soluble oil, which cools and lubricates the work surface and carries away grinding debris.

wet leg The liquid-filled low-side impulse line in a differential pressure level measuring system.

wetness A term used to designate the percentage of water in steam. Also used to describe the presence of a water film on heating surface interiors.

wet steam Steam containing moisture.

wetting The process of supplying a water film to the water side of a heating surface.

wet-type differential pressure meter A design of differential-pressure instrument in which pressure difference is determined across an intermediate liquid in the instrument, by means of either an inverted-bell or float-type indicating mechanism.

Wheatstone bridge A four-arm resistance bridge, usually having three fixed resistances and one variable resistance.

wheel A simple machine consisting of a circular rim connected by a web or by spokes to a central hub or axle about which it revolves; in most applications, a wheel is used to support a load that rests on the axle while allowing the load to be moved easily from location to location.

wheel bearing A device that allows a wheel to revolve about a stationary axle with low friction while supporting a heavy load; the chief types include journal, roller and ball bearings.

wheel diagram A method of representing a time-multiplexed format in the manner of a mechanical commutator or rotary switch.

wheel dresser A tool for refacing a grinding wheel to restore its dimensional accuracy and ability to cut work metal.

white light A mixture of colors of visible light that appears white to the eye. A mixture of the three primary colors is sufficient to produce white light.

white noise 1. Random noise that has a constant energy per unit bandwidth at every frequency in the range of interest [RP74.01]. 2. A noise whose power is distributed uniformly over all frequencies and has a mean noise power-per-unit bandwidth; since idealistic white noise is an impossibility, bandwidth restrictions have to be applied.

white radiation See Bremsstrahlung.

Whitworth screw thread A British standard screw threaded characterized by a 55° V form with rounded crests and roots.

wicking 1. Distribution of a liquid by capillary action into the pores of a porous solid or between the fibers of a material such as cloth. 2. Flow of solder under the insulation on wire, especially stranded wire.

wide-area network In data processing, the inter-connection of computers that can be miles apart in contrast to computers interconnected within one building.

wideband radiation thermometer A low-cost pyrometer that responds to a wide spectrum of the total radiation emitted by a target object; depending on the lens or window material used, the instrument responds to wavelengths from 0.3 μm to between 2.5 and 20 μm, which usually represents a significant fraction of the total radiation emitted. Also known as broadband pyrometer; total radiation pyrometer.

wideband recording (instrumentation tape) A mode of recording or playback where the frequency response at a given tape speed is "wide."

wide range mechanical atomizing oil burner A burner having an oil atomizer with a range of flow rates greater than that obtainable with the usual mechanical atomizers.

wide range metering The measurement of flow rates over a wide range of values at a defined accuracy.

Wien bridge A type of a-c bridge circuit that uses one leg containing a variable resistance and a variable capacitance to achieve balance with a leg containing a resistance and capacitance in parallel; although a Wien bridge can be used for measurements of this type, its more usual application is in determining frequency of an unknown a-c excitation signal.

wild card In MS-DOS, a symbol used to search for any other symbol in a file.

winch A machine, usually power driven, consisting chiefly of a horizontal drum on which to wind cable, rope or chain and with which to apply a pulling force for hauling or hoisting; a winch with a vertical drum is usually known as a capstan.

Winchester disk A high-storage-capacity, small, moderately priced magnetic recording medium with a nonremovable storage element.

windbox A chamber below the grate or surrounding a burner, through which air under pressure is supplied for combustion of the fuel.

windbox pressure The static pressure in the windbox of a burner or stoker.

window 1. An opening in the wall of a building or the sidewall of a vehicle body, usually covered with transparent material such as glass, which admits light and permits occupants to see out. 2. An interval of time when conditions are favorable for taking some predetermined action, such as launching a space vehicle. 3. A defect in thermoplastic sheet or film, similar to a fisheye but generally larger. 4. An aperture for passage of magnetic or particulate radiation. 5. An energy range or frequency range that is relatively transparent to the passage of waves. 6. A span of time when specific events can be detected, or when specific events can be initiated to produce a desired result. 7. In data processing, an area of a computer screen that can be temporarily opened to run a second program without disturbing the original program.

window (nameplate) A component of a visual display made from a translucent material that is illuminated from the rear and labeled to identify the monitored variable [S18.1].

wind tunnel An apparatus consisting of a duct, fans or other air-handling equipment, and instrumentation; it is used to study effects of airflow past solid objects.

windup Saturation of the integral mode of a controller developing during times when control cannot be achieved, which causes the controlled variable to overshoot its setpoint when the obstacle to control is removed.

wing nut An internally threaded fastener having radial projections that allow it to be tightened or loosened by finger pressure.

winning Recovering a metal from an ore or chemical compound.

winterize To prepare a measurement station for local winter conditions.

wiped joint A type of soldered joint in which molten filler metal is applied, then distributed between the faying surfaces by sliding mechanical motion.

wiper (movable contact) That portion of the potentiometric assembly which slides on the resistance element. It is connected to a terminal and provides an electrical output as a function of the shaft position relative to the body [S37.12].

wire 1. A very long, thin length of metal, usually of circular cross section, and usually made by drawing through a die. 2. A continuous length of metallic electrical conductor; it may be single-strand, stranded or tinsel construction, and may be bare or covered with flexible insulating material; its size is usually standardized to provide a predetermined maximum current-carrying capacity, and if insulated it may be rated for maximum circuit voltage as well.

wire cloth Screening made of wires crimped or woven together.

wired program computer A computer in which the instructions that specify the operations to be performed are specified by the placement and interconnection of wires. The wires are usually held by a removable control panel, allowing flexibility of operation, but the term is also applied to permanently wired machines which are then called fixed program computers. Related to fixed program computer.

wiredrawing Reducing the cross section of wire or rod by pulling it through a die.

wire gage 1. A device for measuring the diameter of wire or thickness of sheet metal. 2. A standard series of sizes for wire diameter or sheet metal thickness, usually indicated by size codes consisting of a series of consecutive numbers.

wire insulation A flexible covering for electrical wire, which has relatively high dielectric strength to prevent inadvertently grounding the conductor, and which may have specific mechanical or chemical attributes that protect the conductor from damage or environmental effects in service.

wire rope A flexible length of stranded metal wire suitable for carrying substantial axial tension loads; it is most often used to haul or hoist loads or to act as a tensioned stationary support member in a structure, framework or truss.

wire stripper A tool for removing insulation from the end of insulated wire to prepare the wire for termination.

wiring closet The room or location where the telecommunication wiring for a building, or section of a building, comes together to be interconnected.

Wobbe index The ratio of the heat of combustion of a gas to its specific gravity. For light hydrocarbon gases the Wobbe index is almost a linear function of the gas' specific gravity.

wobble switch Any of several designs of momentary or limit switches which are actuated by physical contact of an object with an extended wire, rod or cable projecting from the switch body; in most instances, wobble switches can be actuated by an object approaching from any direction in a plane perpendicular to the axis of the actuator.

Wollaston wire   Extremely fine platinum wire used in electroscopes, microfuses and hot wire instruments; it is made by coating platinum wire with silver, drawing the sheathed wire, and dissolving the silver away with acid.

Wood's glass   A type of glass that is relatively opaque to visible light but relatively transparent to ultraviolet rays.

word   1. A group of bits that contain a measurement, command, tag or other information—typically sixteen bits. 2. A collection of bits (8, 12, 16, or 32) that represents the basic instruction or data in the computer. 3. In PCM telemetry terminology, one group of bits which represents a specific measurement; typically, a binary number. See also alphabetic word, computer word, machine word, and numeric word.

word length   The number of characters or bits in a machine word. In a given computer, this number may be constant or variable.

word processing   Software and computer hardware that can manipulate text in a variety of ways, such as formatting, moving, replacing, etc.

word selector   A hardware device that can select a given word or words whenever they occur in the format, and present each to the user as a display and/or analog output.

word time   In a storage device that provides serial access to storage locations, the time interval between the appearance of corresponding parts of successive words.

word wrap   The ability of word processing programs to automatically start another line when a designated line-length is reached.

work-around   An action required to complete the process run, even though all equipment is not working satisfactorily. You may have to run part of the process on manual, or you might jump out of an interlock until the maintenance can be scheduled.

work function   An electrical potential difference corresponding to the amount of work that must be done to remove an electron from the surface of a metal.

work hardening   See strain hardening.

working fluid   Fluid that does the work for a system.

working load   The maximum service load that an individual structural member, or an entire structure, is designed to carry.

working pressure   The maximum allowable operating pressure for an internally pressurized vessel, tank or piping system, usually defined by applying the ASME Boiler and Pressure Vessel Code or the API piping code.

working space   See working storage.

working storage   A portion of the internal storage reserved for the data upon which operations are being performed. Synonymous with working space and contrasted with program storage.

working temperature   The temperature of the fluid immediately upstream of a primary device.

work lead   The electrical conductor that connects the workpiece to a welding machine or other source of welding power. Also known as ground lead; welding ground.

World Federation   The joining together of three international regions: a) The Americas (Canadian MAP Interest Group and U.S. MAP/TOP Users Group) and Western Pacific (Australian MAP Interest Group), b) Asia (Japan MAP Users Group), and c) Europe (European MAP Users Group).

worm   A shaft having at least one complete spiral tooth around the pitch surface, and used as the driving member for a worm gear or worm wheel.

worm gear   A gear with teeth cut on an angle so it can be driven by a worm; it is used to transmit power and motion between two nonparallel, nonintersecting shafts.

worm wheel   A gear wheel with curved teeth that mesh with a worm; it is usually used to transmit power and motion from the worm shaft to a nonintersecting shaft whose axis is at right angles to the worm shaft.

worst resolution   The magnitude of the largest of all output steps over the unit range expressed as a percentage of VR [S37.6].

wow   In instrumentation tape recording and playback, high-frequency tape speed variations. See flutter.

wrap-around liner   In a butterfly valve body, a liner extending around the end faces of the wafer body to form a gasket seal with the pipe flanges. The liner may cover all or part of the flange contact area of the wafer body [S75.05].

wrap forming   See stretch forming.

wrapper sheet   The outside plate enclosing the firebox in a firebox or locomotive boiler. Also the thinner sheet in the shell of a two thickness boiler drum.

wringing fit   A type of interference fit having zero to slightly negative allowance.

write   To transfer information and store it. The information may be written in memory devices like RAM, or on storage media such as magnetic discs or tapes. Sometimes the storage element is absent, as in writing to a display.

write-protect notch   A cut-out in the diskette envelope that prevents the computer from writing on the diskette, but does not prevent the computer from reading the diskette.

write time   The amount of time it takes to record information. Related to access time.

wrought alloy   A metallic material that has been plastically deformed, hot or cold, after casting to produce its final shape or an intermediate semifinished product.

wt   See weight.

wye   A pipe fitting similar to a tee, but in which the branch is at a 45° angle to the run.

wye-delta bridge   A d-c bridge arrangement where resistors no larger than one megohm are used in Wye configurations to simulate high-accuracy, high-stability resistors of one gigohm (10^9 ohms) or greater; these arrangements are used to accurately determine the value of very large resistances or to calibrate a bridge.

X

xerography   A dry copying process involving the photoelectric discharge of an electrostatically charged plate. The copy is made by tumbling a resinous powder over the plate, the remaining electrostatic charge discharged and the resin transferred to paper or an offset printing master.

x-ray diffraction analyzer   Any of several devices for detecting the positions of monochromatic x-rays diffracted from characteristic scattering planes of a crystalline material; used primarily in detecting and characterizing phases in crystalline solids.

x-ray diffractometer   An instrument used in x-ray crystallography to measure the diffracted angle and intensity of x-radiation reflected from a powdered, polycrystalline or single-crystal specimen.

x-ray emission analyzer   An apparatus for determining the elements present in an unknown sample (usually a solid) by bombarding it with electrons and using x-ray diffraction techniques to determine the wavelengths of characteristic x-rays emitted from the sample; wavelength is used to identify the specific atomic species responsible for the emission, and relative intensity at each strong emission line can be used to quantitatively or semiquantitatively determine composition.

x-ray fluorescence analyzer   An apparatus for analyzing the composition of materials (solid, liquid or gas) by exciting them with strong x-rays and determining the wavelengths and intensities of secondary x-ray emissions.

x-ray goniometer   An instrument for measuring the angle between incident and refracted beams of radiation in x-ray analysis.

x-ray microscope   An apparatus for producing greatly enlarged images by projection using x-rays from a special ultra-fine-focus x-ray tube, which acts essentially as a point source of radiation.

x-ray monochromator   A device for producing an x-ray beam having a narrow range of wavelengths; it usually consists of a single crystal of a selected substance mounted in a holder that can be adjusted to give proper orientation.

x-rays   Short-wavelength electromagnetic radiation, having a wavelength shorter than about 15 nanometers, usually produced by bombarding a metal target with a stream of high-energy electrons; wavelengths are in the same range as gamma rays, longer than cosmic rays but shorter than ultraviolet; like gamma rays, x-rays are very penetrating and can damage human tissues, induce ionization, and expose photographic films.

x-ray thickness gage   A device used to continually measure the thickness of moving cold-rolled sheet or strip during the rolling process; it consists of an x-ray source on one side of the strip and a detector on the other—thickness is proportional to the loss in intensity as the x-ray beam passes through the moving material.

X-value   A term sometimes used to designate the inductive or capacitative reactance of an a-c electrical device or circuit.

XY plotter   A device used in conjunction with a computer to plot coordinate points in the form of a graph.

XY recorder   A recorder for automatically drawing a graph of the relationship between two experimental variables; the position of a pen or stylus at any given instant is determined by signals from two different transducers that drive the pen-positioning mechanism in two directions at right angles to each other.

Y

Y See expansion factor.

yield The quantity of a substance produced in a chemical reaction or other process from a specific amount of incoming material.

yield stress The force per unit area at the onset of plastic deformation, as determined in a standard mechanical-property test such as a uniaxial tension test.

yoke 1. The structure which rigidly connects the actuator power unit to the control valve [S75.05]. 2. A clamping device to embrace and hold two other parts. 3. A slotted crosshead used in some steam engines instead of a connecting rod. 4. The framework surrounding the rotor of a d-c generator or motor which supports the field coils and provides magnetic linkage between them.

Z

Z See compressibility factor.

zap In data processing, a slang word meaning to erase or wipe-out data.

zero a device To erase all the data stored on a volume and reinitialize the format of the volume.

zero adjuster A mechanism for repositioning the pointer on an instrument so that the instrument reading is zero when the value of the measured quantity is zero.

zero adjustment See adjustment, zero.

zero-based conformity See conformity, zero-based.

zero-based linearity See linearity, zero-based.

zero bias A positive or negative adjustment to instrument zero to cause the measurement to read as desired.

zero code error A measure of the difference between the ideal (0.5 LSB) and the actual differential analog input level required to produce the first positive LSB code to transition (00...00 to 00...01).

zero defects A management program that encourages perfect performance in a manufacturing operation, and usually provides workers with rewards and incentives for achieving perfection.

zero elevation 1. For an elevated-zero range, the amount the measured variable zero is above the lower range-value. It may be expressed either in units of the measured variable or in percent of span [S51.1]. 2. Biasing the zero output signal to raise the zero to a higher starting point. Usually used in liquid level measurement for starting measurement above the vessel connection point.

zero error See error, zero.

zero frequency gain See gain, zero frequency.

zero governor A regulating device which is normally adjusted to deliver gas at atmospheric pressure within its flow rating.

zero level A reference level used for comparing signal intensities in electronic or sound-reproduction systems—in electronics, the zero level is usually taken as 0.006 W of power; in sound reproduction, it is usually taken as the threshold of hearing.

zero-measured output The output of a transducer, under room conditions unless otherwise specified, with nominal excitation and zero measurand applied [S37.1].

zero shift 1. In process instrumentation, any parallel shift of the input-output curve [S51.1]. 2. A change in the zero-measurand output over a specified period of time and at room conditions. NOTE: This error is characterized by a parallel displacement of the entire calibration curve [S37.1]. 3. A shift in the instrument calibrated span evidenced by a change in the zero value. Usually caused by temperature changes, overrange, or vibration of the instrument.

zero suppression 1. For a suppressed-zero range, the amount the measured variable zero is below the lower range-value. It may be expressed either in units of the measured variable or in percent of span [S51.1]. 2. The elimination of nonsignificant zeros in a numeral. 3. Biasing the zero output signal to produce the desired measurement. Used in level measurement to counteract the zero elevation caused by a wet-leg.

zero, zero out The procedure of adjusting the measuring instrument to the proper output value for a zero-measurement signal.

Ziegler-Nichols method A method of determination of optimum controller settings when tuning a process-control loop (also called the ultimate cycle method). It is based on finding the proportional gain which causes instability in a closed loop.

zinc plating An electroplating coating of zinc on a steel surface which provides corrosion protection in a manner similar to galvanizing.

zinc rich coating A single component zinc-rich coating which can be applied by brush, spray or dip and dries to a gray matte finish. ZRC is accepted by Underwriters Laboratories, Inc., as the equivalent to hot dip galvanizing.

zone 1. A portion of internal storage allocated for a particular function or purpose. 2. On a multi-position controller, the range of input values between selected switching points or any switching point and range-limit [S51.1]. 3. The international method for specifying the probability that a location is made hazardous by the presence, or potential presence, of flammable concentrations of gases or vapors.

zone 0 A location in which a concentration of a flammable gas or vapor mixture is continuously present, or is present for long periods of time.

zone 1 A location in which a concentration of a flammable gas or vapor mixture is likely to occur in normal operation.

zone 2 A location in which a concentration of a flammable gas or vapor mixture is unlikely to occur in normal operation and if it does occur, will exist only for a short period of time.

zone bit 1. One of the two leftmost bits in a binary coding system in which six bits are used to define each character. 2. Any bit in a group of bit positions used to indicate the classification of the group—numeral, alpha character, special sign or command, for instance.

zone control A method of controlling temperature or some other process characteristic by dividing a physical area or process flowpath into several regions, or zones, and independently controlling the process characteristic in each zone.

zone, dead 1. For a multi-position controller, a zone of input in which no value of output exists. It is usually intentional and adjustable [S51.1]. 2. A predetermined rage of input through which the output remains unchanged, irrespective of the direction of change of the input signal. There is but one input-output relationship. Dead zone produces no phase lag between input and output [S51.1].

zone, intermediate Any zone not bounded by a range-limit [S51.1].

zone, live A zone in which a value of the output exists [S51.1].

zone, neutral A predetermined range of input values in which the previously existing output value is not changed [S51.1].

zones Formerly called divisions. A zone is an area of similar probability of the presence and concentration of the potentially explosive mixture. It is part of the area classification. (The other part being the gas group.) Three zones are recognized in the UK: Zone 0—In which an explosive gas-air mixture is continuously present or present for long periods; Zone 1—In which an explosive gas-air mixture is likely to occur in normal operation; Zone 2—In which an explosive gas-air mixture is not likely to occur in normal operation, and if it occurs will exist only for a short time.

Z-value A term sometimes used to designate the impedance of a device or circuit, which is the vector sum of resistance and reactance (X-value).

Zyglo method A technique for liquid-penetrant testing to detect surface flaws in a metal using a special penetrant that fluoresces when viewed under ultraviolet radiation.

Abbreviations & Acronyms

A

a Ampere and Area

A Anode; Ångstrom; Area of pipe; Area; Acceleration (in general); Ampere

AA Auxiliary Building; Arithmetical Average; Aluminum Association

AB Air Blast

ABBR Abbreviate

ABC Automatic Brightness Control

ABCB Air Blast Circuit Breaker

ABN Airborne

ABND Abandoned

ABNL Abnormal

ABR Absorber

ABS Absolute; Air Break Switch; Acrylonitrile Butadiene Styrene

ABSV Absorptive

ABT About; Air Blast Transformer

ABUT Abutment

ABV Above

Ac Actinium

a-c Alternating Current

A/C Air Condition

AC Acre; Alternating Current

ACB Air Circuit Breaker

ACC Accumulator

ACCEL Accelerate

ACCESS Accessory

ACCT Account

ACCUM Accumulate

ACET Acetylene

ACFM Actual Cubic Foot (per min.)

AC-FT Acre-Foot

ACK ACKnowledge

A/CM/°/F Amperes per Centimeter per °F

ACO Analog Control Output

ACP Auxiliary Control Panel

ACR Auxiliary Control Room

A/CS Air Conditioning (Cooling-Heating System)

ACST Acoustic

ACTE Actuate

ACTG Actuating

ACTL Actual

ACTN Activation

ACTR Actuator

A/CU Air Conditioning Unit

ACV Automatic Check Valve; Alarm Check Valve

A/D Analog/Digital; Analog-to-Digital

ADC Analog-to-Digital Converter

ADCCP Advanced Data Communications Control Procedures

ADD Addition

ADDT Additive

ADH Adhesive

ADJ Adjust

ADLC Advanced Data Link Control

ADP Ammonium Dihydrogen Phosphate

ADPCM Adaptive Differential Probe Code Modulation

ADPT Adapter

ADS Automatic Dispatch System; Automatic Door Seal

ADV Advance

AF Audio Frequency

AFC Automatic Frequency Control

AFIPS American Federation of Information Processing Societies

AFM Abrasive Flow Machinery

AFMDC Air Force Machinability Data Center

AFT After

AFW Auxiliary Feedwater

AFWP Auxiliary Feedwater Pump

Ag Silver

AGA American Gas Association

AGC Automatic Gain Control

AGGR Aggregate

AGTR Agitator

AH Ampere-hour

AHU Air Handling Unit

AIChE American Institute of Chemical Engineers

AIR CLD Air Cooled

AIR COND Air Condition

AIT AutoIgnition Temperature

Al Aluminum

Al$_2$O$_3$ Aluminum Oxide (alumina)

ALC Alcohol

ALGOL ALGorithmic-Oriented Language

ALK Alkaline

ALM Alarm

ALT Altitude

ALT. Alternate

ALTNTR Alternator

ALTRN Alteration

ALU Arithmetic and Logical Unit

ALY Alloy

Am Americium

AM Forenoon; Amplitude; Amplitude Modulation; Ammeter

A/M Automatic/Manual
AMB Amber; Ambient
AMIG Australian MAP Interest Group
AMP Ampere
AMPL Amplifier
AMT Amount
ANAL Analysis
ANALZ Analyzer
ANHYD Anhydrous
ANK Alphanumeric Keyboard
ANL Automatic Noise Limiter
ANN Annunciator
ANNS Annulus
ANSI American National Standards Institute (of ASME)
ANT Antenna
AO Access Opening
AOP Auxiliary Oil Pump
AP Access Panel; Acidproof
APD Avalanche Photodiode
APHA American Public Health Association
API American Petroleum Institute
APL A Programming Language; Airplane
APP Apparatus
APRM Average Power Range Monitoring
APT Automatically Programmed Tools; Apartment
Ar Argon
AR Auxiliary Relay
ARM Armature
ARMD Armored
ARR Arrange; Arrester
ARRGT Arrangement
ART Artificial
As Arsenic
A/S Air Supply
AS Air Supply
ASA American Standards Association
ASC Automatic Sensitivity Control; Accredited Standard Committee
ASI Analog Status Input
ASCII American Standard Code for Information Interchange
ASME American Society of Mechanical Engineers
ASO Analog Status Output
ASR Automatic Send/Receive
ASSOC Associate
ASST Assistant
ASSY Assembly
ASTM American Society for Testing and Materials

At Astatine
AT Airtight; Ampere Turn; Atomic
ATE Automatic Test Equipment
atm Atmosphere
ATT Attach
ATTEN Attenuator
ATTN Attention
at.wt. Atomic Weight
Au Gold
AUD Audible
AUI Access Unit Interface
AUTC Automatic
AUTH Authorized
AUTOXFMR Autotransformer
AUTRAN Automatic Utility Translator
AUX Auxiliary
AUXBLDG Auxiliary Building
AVC Automatic Volume Control
AVDP Avoirdupois
AVE Automatic Volume Expansion
AVG Average
AVI Aviation
AWG American Wire Gauge
AWS American Welding Society
AZ Azimuth

B

b Barn
B Bandwidth; Boron
Ba barium
BACT Bacteriological
BAF Baffle
BAL Balance
BAR Barometer
BARR Barrier
BASIC Beginner's All-purpose Symbolic Instruction Code
BAS NET Basic Network
BASW Bell Alarm Switch
BAT Battery
BBL barrel
BC Belt Conveyor
BCD Binary-Coded Decimal
BCH Bose Chaudhuri-Hocquenghem
BCN Beacon
BCOMP Buffer COMPlete
BCT Bushing Current Transformer
BD Board
BDC buffered data channel

BDL Bundle
BDPLT Bedplate
BDR Breeder
BDY Boundary
Be beryllium
Bé Baumé specific gravity scale
BEL Below
BER bit error rate
BESS Bessemer
BET Between
bEv One billion electron volts; also written Bev, BEv, or BEV
BF Boiler Feed
BFBP Boiler Feed Booster Pump
BFO Beat Frequency Oscillator
BFP Boiler Feed Pump
BFPT Boiler Feed Pump Turbine
BFW Boiler Feed Water
BG Base Group; Back Gear
BH Boiler House
BHD Bulkhead
BHN Brinell hardness
BHP Boiler Horsepower; Brake Horsepower
BHP-HR Brake Horsepower-Hour
BHS Building Heating System
Bi bismuth
BIM Binary Input Multiplexer
BISYNC BInary SYNchronous Communications
BIT Bituminous
Bk berkelium
BK Bank; Black; Brake
BKR Breaker
BLDG Building
BLDR Boulder
BLK Black; Blank; Block
BL Blue; Bottom Layer
BLO Blower
BLR Boiler
BLSTGPWD Blasting Powder
BLT Borrowed Light
BLU Blue
BLWDN Blow Down
BM Beam
BMEP Brake Mean Effective Pressure
BNI Bureau d'Orientation de la Normalisatin en Informatique
BO Blowoff
BOC Blowout Coil
BOM Binary Output Multiplexer
BOP Balance of Plant
BOR Borrow

BOT Bottom
BP Blueprint; Back Pressure; Boiler Pressure; Boiling Point
BPC Back Pressure Control
BPD Barrels Per Day
BPH Barrels Per Hour
BPHM By-pass Handwheel Motor
BPHS By-pass Handwheel Switch
BPS Bits per second
BPV Back Pressure Valve
BPVLV Back Flow Prevention Valve
Br Bromine
BR Branch; Brown; Brush
BRDG Bridge
BRG Bearing
BRK Break
BRKT Bracket
BRN Brown
BRT Brightness
BSI The British Standard Institution
BSMT Basement
BSTR Booster
BT Bus Tie
BTFLY VLV Butterfly Valve
BTL Bottled
BTU British Thermal Units
BTUH British Thermal Units per Hour
BUBLR Bubbler
BUF Buffer
BUT Button
BUZ Buzzer
BW Back Wash
BWR Boiler Water Reactor
BWV Back Water Valve
BYP By-pass

C

C Hundred; Celsius; Centigrade (obsolete, use Celsius); Coulomb; Carbon; Coefficient of Discharge, dimensionless; Cast (used with other materials); Cycle
Ca Calcium
CA Cable; Constant Amplitude; Compressed Air
CAB Cabinet
CAD Computer Aided Design
CAD/CAM Computer-Aided Design/Computer-Aided Manufacturing
CAE Computer Aided Engineering
CAL Caliber; Calibrate; Calorie; Computer-Aided

Learning; Conversional Algebraic Language

CALC Calculate

CAL/cm³ Calorie per Cubic Centimeter

CAL/in.³ Calories per Cubic Inch

CAM Camber; Computer Aided Manufacturing

CAMAC Computer Automated Measurement And Control

CAMR Camera

CAP Capacitor; Capacity; Capital; Code Alarm Paging

CARR Carrier

CARR CUR Carrier Current

CASA/SME Computer and Automated Systems Association of the Society of Manufacturing Engineers

CASE Common Applications Service Elements

CAT Catalogue; Chemical Addition Tank

CATV Community Antenna Television

CAUS Caustic

CAV Cavity

CAVIT Cavitation

CB Common Battery; Circuit Breaker; Center of Buoyancy; Catch Basin

CBAL Counterbalance

CBEMA Computer and Business Equipment Manufacturers Association

CBW Constant Bandwidth

CC Cubic Centimeter; Cooling Coil; Code Call; Component Cooling

CCAP Code Call Alarm and Paging

CCITT Consultative Committee on International Telephony and Telegraphy

CCR Control Complexity Ratio

CCS Component Cooling System

CCT Constant Current Transformer

CCTV Closed-Circuit Television

CCW Counterclockwise; Condensor Circulating Water

CCWP Condensor Circulating Water Pump

CCWPS Condensor Circulating Water Pump Station

CCWS Condensor Circulating Water System

C_d Relative Capacity Factor ($C_d = C_v / d^2$)

cd Candela

Cd Cadmium

CD Cut Down

CD PL Cadmium Plate

CDR Current Directional Relay

Ce Cerium

CE Commutator End

CEA Control Element Assembly

CEM Cement

CER Ceramic

C_f Liquid Pressure Recovery Factor; same as F_L.

Cf Californium

CF Center of Floatation; Centrifugal Force

CFGN Configuration

cfm Cubic Feet per Minute

cfs Cubic Feet per Second

CG Center of Gravity; Centigram; Chain Grate

CGS Centimetre-Gram-Second

CH Chain

CHAN Channel

CHAR Character

CHBR Chamber

CHCS Chemical Cleanin System

CHEM Chemical

CHFR Chamfer

CHG Change; Charge

CHGR Charger

CHK Check

CHW Chilled Water

Ci Curie

CI Cast Iron

CID Computer Interface Device

CIM Computer Integrated Manufacturing

CIMS Computer Integrated Manufacturing System

CIP Cast Iron Pipe

CIR Circle; Circular

CISP Cast Iron Soil Pipe

CKT Circuit

Cl Chlorine

CL Closing; Clearance; Class; Center Line; Cutter Location; Carload

CLD Cooled

CLG Cooling; Ceiling

CLK Clock

CLN Chlorination

CLOS Central Lubricating Oil System

CLP Clamp

CLR Cooler; Clear

CLSD Closed

CLV Clevis

cm Centimetre

Cm Curium

CM Center Matched

cm² Square Centimetre

cm³ Cubic Centimetre

cm³/min Cubic Centimetre per Minute

cm³/s Cubic Centimetre Per Second

CMIG Canadian MAP Interest Group

CMIL Circular Mil

CMOS Complementary Metal Oxide Semicon-

ductor
CMPD Compound
CMPNT Component
CMPR Compare
CMRR Common-Mode Rejection Ratio
CMV Common Mode Voltage
cm/s Centimetre per Second
CN Change Notice
CNC Computer Numerical Control
CNDCT Conductor; Conductivity
CNDS Condensate
CNTFGL Centrifugal
CNTMT Containment
CNTNR Container
CNTOR Contactor
CNTR Counter
CNVR Conveyor
Co Cobalt
CO Cut Out; Company; Carbon Monoxide; Cleanout; Compliance Officer; Change Order
CO_2 Carbon Dioxide
COAX Coaxial
COBOL COmmon Business-Oriented Language
CODAB COnfiguration DAta Block
CODIL COntrol DIagram Language
COEF Coefficient
COL Column
COLL Collector
COM Common; Computer Output Microfilm
COMB Combine; Combination; Combustion
COML Commercial
COMM Communication; Commutator
COMP Computation; Computer; Composition; Compensate
COMPL Complete
COMPT Compartment
CONC Concrete; Concentrate
COND Condition; Condensor; Conductivity
CONN Connector
CONST Constant
CONSTR Construction
CONT Control; Controller; Continue; Continuous; Contact
CONTAM Contaminated
CONT BLDG Control Building
CONTR Contractor; Contract
CONT STA Control Station
CONV Converter
COOL Coolant
COR Corner
CORR Correct; Correction
CO_2S Carbon Dioxide Storage, Fire Protection
and Purging System
COS Corporation for Open Systems
COT Cotton
COV Cover
CP Candlepower; Circular Pitch; Center of Pressure; Chemically Pure
CPLG Coupling
CPM Cycles per Minute; Critical Path Method
CPRS Compress
CPRSR Compressor
CPS Characters per Second, Cycles per Second, or Conversions per Second depending on content.
CPU Central Processing Unit
Cr Chromium
CR Crushed; Control Relay; Carriage Return Character
CRC Cyclic Redundancy Check
CRD Control Rod Drive
CRDS Control Rod Drive System
CRG Carriage
CRK Crank
CRKC Crankcase
CRSP Correspond
CRT Cathode Ray Tube
CRUIS Cruising
Cs Cesium
CS Cast Steel; Condensate System
CSA Canadian Standards Association
CSG Casing
CSMA/CD Carrier Sense Multiple Access with Collision Detect
CSP Computer Set Point
CSTG Casting
CT Current Transformer; Cooling Towers; Center Tap
CTD Coated
CTI Comparative Tracking Index
CTN Carton
CTR Center
CTRS Contrast
CTWT Counterweight
cu Cubic
Cu Copper
cu cm Cubic Centimeter
cu in Cubic Inch
cu ft Cubic Feet
cu m Cubic Meter
cu mm Cubic Millimeter
cu mu Cubic Micrometer
CUR Current
CUST Customer
cu yd Cubic Yard

C_v Value-Sizing Coefficient
CV Control Valve; Check Valve
CVD Chemical Vapor Deposition
CW Clockwise; Cold Water; Cooling Water
CWA Current Word Address
CWP Circulating Water Pump
CWR Chilled Water Return
CWS Chilled Water Supply
CWT Hundredweight
CX Composite
CYL Cylinder

D

d Diameter of a Valve Inlet in Inches
D Drop; Density; Diameter of Pipe
DA Double Acting
D/A Digital-to-Analog
DAC Digital-to-Analog Converter
DAS Data Acquisition System
DASD Direct Access Storage Device
DAVC Delayed Automatic Volume Control
dB Decibel
DB Dry Bulb; Decibel
DBL Double
DBLR Doubler
DBT Dry Bulb Temperature
d-c Direct Current
DC Direct Current; Data Communication
DCA Decay
DCE Data Communications Equipment
DCS Distributed Control System
DDA Digitally-Directed Analog
DDC Direct Digital Control
DDCS Distributed Digital Control Systems
DDCMP Digital Data Communications Message Protocol
DE District Engineer
DEC Decimal; Digital Equipment Corporation
DECR Decrease
DEFL Deflect
DEG Degree
DEL Delineation
DELIV Deliver
DEM Demodulator
DESCR Describe
DESIG Designation
DEV Develop; Development
DEVN Deviation
DF Double Feeder; Drive FIT

DFT Deaerating Food Tank; Diagnostic Function Test
DG Double Glass; Double Groove (insulators); Display Generator; Diesel Generator
DGB Diesel Generator Building
DHMR Dehumidifier
DI Ductile Iron; Demand Indicator
dia. Diameter
DIA Diameter
DIAG Diagram; Diagonal
diam Diameter
DIAPH Diaphragm
DIFF Differential
DIFF T Differential Temperature
DIFF TR Differential Time Relay
DIM Dimension
DIN Standards Institution of the Federal Republic of Germany (West Germany)
DI/OU Data Input/Output Unit
DIP Dual In-line Package
DIR Direction; Director
DIR CONN Direct Connected
DIS Draft International Standard
DISC Disconnect
DISCH Discharge
DISCR Discriminator
DISP Dispatch
DIST District; Distance
DISTR Distribute
DIV Divide; Division; Diverter
DK Deck
DL Drawing List
DM Demand Meter; Decimeter
DMA Direct Memory Access
DMH Drop Manhole
DMNRLZR Demineralizer
DMPR Damper
DMT Dead Man Timer
DN Down
DNS Downscale
DNSTR Downstream
DO Ditto; Dissolved Oxygen; Diesel Oil
DOLM Dolomite
DOS Disk Operating System
DP Door Post; Differential Pressure; Dew Point; Dash Pot; Draft Proposal
DPDT Double Pole Double Throw
DPDT SW Double Pole Double Throw Switch
DPSL Disposal
DPST Double Pole Single Throw
DPST SW Double Pole Single Throw Switch
DP SW Double Pole Switch

DPT   Dew Point Temperature
DPV   Dry Pipe Valve
DR   Drive; Drain; Drainage; Door
DRWN   Drawn
DS   Downspout
DSL   Diesel
DSTL   Distill
DSTLT   Distillate
DTL   Diode-Transistor Logic
DUP   Duplicate
DV   Device
DW   Distilled Water
DWG   Drawing
DWT   Deadweight
DX   Duplex
DXD   Differential Expansion Detector
DYN   Dynamo; Dynamic
DYNM   Dynamotor
DYNMT   Dynamometer
DYN S   Dynamic Snubber

E

E   East; Modulus of Elasticity
EA   Each
EAROM   Electrically Alterable Read-Only Memory
EBCDIC   Extended Binary Coded Decimal Interchange Code
EBOP   Emergency Bearing Oil Pump
ECC   Eccentric
ECI   Essential Control Instrumentation
ECM   Electrochemical Machining or Electrochemical Milling
ECMA   European Computer Manufacturers Association
ECMCH   Electromechanical
ECON   Economizer
ECR   Electric Control Room
ECSA   Exchange Carriers Standard Association
EC SW   End Cell Switch
ED   Equipment Drain; Eccentricity Detector
EDM   Electrical Discharge Machining
EDP   Electronic Data Processing
EDT   Equipment Drain Tank
EDUC   Eductor
EEPROM   Electrically Erasable and Programmable Read-Only Memory
EFF   Effective; Efficiency
EFL   Effluent

EFS   Emergency Feedwater System
EG   For Example
EGTS   Emergency Gas Treatment System
EHC   Electrical Hydraulic Controls
EHP   Effective Horsepower
EHT   Exhaust Hood Thermostat
EHV   Extra High Voltage
EHW   Extreme High Water
E/I   Voltage to Current
EIA   Electronics Industry Association
EIS   Extended Instruction Set
EJCTR   Ejector
EL   Elevation; Elastic Limit
ELEC   Electric
ELECT   Electrolyte
ELECTC   Electrolytic
ELEM   Element; Elementary
ELEV   Elevate; Elevator
ELL   Elbow
ELONG   Elongation
ELW   Extreme Low Water
ELYHD   Electrohydraulic
EMER   Emergency
EMF   Electromotive Force
EMI   Electromagnetic Interference
EMO   Electric Motor Operated
EMP   Electromagnetic Pulse
EMUG   European MAP Users Group
ENCL   Enclose
ENG   Engine
ENGR   Engineer
ENGRG   Engineering
ENTM   Entrance
ENV   Envelope
EOF   End of File
EOP   Emergency Oil Pump
EOT   End of Tape
EP   Explosion ProofEmergency Power
E/P   Voltage to Pneumatic
EPA   Enhanced Performance Architecture or Environmental Protection Agency
EPROM   Electrically Programmable Read-Only Memory
EQ   Equal; Equation
EQL   Equalizer
EQPT   Equipment
EQUIV   Equivalent
Er   Erbium
EREC   Erection
Es   Einsteinium
ES   Environmental System; Electrostatic
ESC   Escape

ESS Extraction Steam System
ESSN Essential
EST Estimate; Estimated
ETC And So Forth
Eu Europium
eV Electron Volt
EVAC Evacuation
EVAP Evaporator
EWICS European Workshop on Industrial Computer Systems
EXC Excitation; Exciter; Excavate; Excessive
EXCAV Excavation
EXCH Exchange
EXCL Exclusive
EXEC Executive
EXH Exhaust
EXIST Existing
EXP Expand; Expansion; Experiment; Expose; Expulsion
EXPL Exploration
EXP V Expansion Valve
EXT Extension; Exterior; External; Extinguish
EXTR Extraction; Extrude

F

2F Two Frequency
f Fuel; Friction Factor; Force, Pounds; Velocity of Approach
F Flat; Fire; Farad; Fahrenheit; Fluorine
FACIL Facility
FAI Fail As Is
FAIL Failure
F/A RATIO Fuel-Air Ratio
FB Fuse Block; Fuel Building; Flat Bar
FBR Fiber
FC Front Connected; Footcandle; Fail Close; Fire Control
FCC Frame Code Complement
FCT Filament Center Tap
FCV Flow Control Valve
F_d Valve Style Modifier
FD Forced Draft; Flow Diagram; Floor Drain; Feed
FDB Forced Draft Blower
FDC Fire Deparment Connection
FDDI Fiber Distributed Data Interface
FDM Frequency-Division Multiplex
FDN Foundation
FDR Feeder

F DR Fire Door
FDRY Foundry
Fe Iron
FEA Feature
FEC Fire Extinguisher Cabinet
$FeCl_3$ Ferric Chloride
FED Federal
FEM Female
FEN Fence
FFT Fast-Fourier Transform
FGR Finger
FH Fire Hydrant; Fire Hose
FHC Fire Hose Cabinet
FHP Fractional Horsepower
FHR Fire Hose Rack
FIFO First-In, First-Out
FIG Figure
FIL Fillet; Filament
FILL Filling
FIX Fixture
F_K Ratio of Specific Heats Factor
FK Fork
F_L Liquid Pressure—Recovery Factor
FL Footlambert; Flush; Fluid; Floor; Flashing
FLD Field
FLDNG Flooding
FLEX Flexible
FLG Flange
FLHLS Flashless
F_{LP} Combined Liquid Pressure Recovery Factor and Piping Geometry Factor of a Valve with Reducers
fl oz Fluid Ounce
FLSF Failsafe
FLT Float
FLTR Filter
FLUOR Fluorescent
Fm Fermium
FM Flowmeter; Fire Main; Field Multiplex; Frequency Modulation
FMG Framing
FMS Flexible Manufacturing System
FNP Fusion Point
FO Fuel Oil; Fail Open
FOA Forced Oil Air
FOB Free on Board
FOC Focus
FOIRL Fiber Optic Inter-Repeater Link
FORG Forging
FORTRAN FORmula TRANslating System
FOS Fuel Oil System
FP Flame Proof; Freezing Point

FP Piping-Geometry Factor
FPD Feet per Day
fpm Foot per Minute
FPRF Fireproof
FPRFG Fireproofing
fps Feet per Second
FPS Foot-Pound Second; Fire Protection System
FPY Feet per Year
Fr Francium
FR Reynolds-Number Factor; Front; Frame
FRAC Fractional
FRATE Frame Rate
FR BEL From Below
FREEBD Freeboard
FREQ Frequency
FREQ CH Frequency Changer
FRT Freight
FRWK Framework
F_S Laminar-Flow Coefficient
FS Frequency Shift; Float Switch
FSBL Fusible
FSK Frequency Shift Keying
FSR Full Scale Range
FST Forged Steel
FSY Frame Synchronizer
ft Foot; Feet
FT Flametight; Foot; Feet
ft² Square Feet
ft³ Cubic Feet
ft/D Feet per Day
ft³/hr Cubic Feet per Hour
ft³/min Cubic Feet per Minute
ft³/s Cubic Feet per Second
FTAM File Transfer Access and Management
ft lb Foot Pound
ft/min Feet per Minute
ft/s Feet per Second
ft/sec Feet per Second
ft/yr Feet per Year
FU Fuse
FURN Furnish
FUT Future
FV Front View
FW Fresh Water; Feedwater
FWA First Word Address
FWCS Feedwater Control System
FWD Forward
FWS Flow Switch

G

g Gram
G Acceleration (due to gravity); Green; Grid; Girder; Giga; Gas; Specific Gravity; Gram
Ga Gallium
GA Gas Analyzer; Gage; Gaugue
gal Gallon
GALL Gallery
gal/min Gallons per Minute
gal/s Gallons per Second
GALV Galvanize
GAR Garage
GASO Gasoline
G-CAL Gram Calorie
GCB Gas Circuit Breaker
g/cm² Grams per Square Centimetre
g/cm³ Grams per Cubic Centimetre
GCS Generator Cooling System
Gd Gadolinium
GD Guard
Ge Germanium
GEN General; Generator
GHZ Giga Hertz
GI Galvanized Iron
GIC Gate Input Card
GIL Green Indicating Lamp
g/L Grams per Litre
GL Glass; Glaze
GLCM Governor Load Change Motor
GLV Globe Valve
G-M Gram-meter
GMAW Gas Metal-Arc Welding
GOR Gate Output Register
GPAD Gallons per Acre per Day
GPG Grains per Gallon; Gas Pressure Gage
gph Gallons per Hour (flow)
GPH Graphite
GPR General Purpose Register
gps Gallons per Second
GPSS General-Purpose Simulation System
GR Grade; Group
GRAD Graduation
GRAPH Graphic
GRD Ground; Grind
GRG Grading
GRIN Graded Index Fiber (optical fiber)
GRN Green
GROM Grommet
GRV Groove

GRWT Gross Weight
GS Gas Stripper; Galvanized Steel
GSWR Galvanized Steel Wire Rope
GT Grease Trap
GTAW Gas Tungsten-Arc Welding
GTRB Gas Turbine
GTV Gate Valve
GVA Gigavolt-Ampere
GVAR Gigavar
GVL Gravel
GW Gigawatt
Gy Gray
GYP Gypsum
GYRO Gryoscope

H

h Hour; Valve-Travel Function (fluid head feet)
H High; Hard; Henry; Hydrogen; Valve-Travel in Inches (fluid head)
H_2 Hydrogen
HBS Heating Boiler System
H_2CrO_4 Chromic Acid
H_2O Water
H_2S Hydrogen Sulphide
H_2S_4 Sulphuric Acid
HC Holding Coil; Heating Cabinet; Heating Coil; Hand Control
HCl Hydrochloric Acid
HD Head
HDDR High Density Digital Recording
HDL Handle
HDLC High-Level Data Link Control
HDLS Headless
HDR Header
HDVS Heater Drain & Vent System
HD WHL Hand Wheel
HEX Hexagon
Hf Hafnium
HF High Frequency
Hg Mercury
HG Hand Generator
HGT Height
HH High-High; Hand Hole
HL High-Low
HLCD High-Level Computing Device
HLDG Holding
HLHI High-Level Human Interference
HLOI High-Level Operator Interface
Ho Holmium

HOL Hollow
HORIZ Horizontal
hp Horsepower
HP High Pressure; Horsepower
HPFPS High Pressure Fire Protection System
hp-h Horsepower-Hour
HPI High Pressure Injection
hp/in. Horse Power per Inch
hp/in.^{3}/min Horse Power per Cubic Inch per Minute
HPIS High Pressure Injection System
HQ Headquarters
hr Hour
HR Hand Reset; Hour
HRS Heat Rejection System
HS High Speed; Hand Switch
HSE House
HSG Housing
HSTAT Humidistat
HT High Tension; High Temperature; Heat
HTG Heating
HTR Heater
HTW High Temperature Water
HUP Holdup Pump
HUT Holdup Tank
HV High Voltage
HVAC Heating, Ventilation and Airconditioning
HVY Heavy
HW Hot Well; Hot Water; Head Water
HWC Hot Water, Circulating
HWGW Head Water Gage Well
HWY Highway
HX Heat Exchanger
HYB Hybrid
HYDR Hydraulic
HYDRELC Hydroelectric
HYDRO Hydrostatic
HYPCL Hypochlorite
Hz Hertz

I

I Island; Current or Moment of Inertia; Iodine
IA Instrument Air
IACS International Annealed Copper Standard
IAE Integral Absolute Error
IAR Instruction Address Register
IB Instruction Book
IC Interior Communication; Integrated Circuit
I&C Instrument and Control

ICP Integrated Circuit Piezoelectric
ICS Interphone Control Station; Integrated Control System
ID Induced Draft
IDENT Identical; Identify
IDF Intermediate Distributing Frame
IDLH Immediate Danger to Life and Health
I/E Current to Voltage
IEC International Electrotechnical Commission
IEE Institution of Electrical Engineers (England)
IEEE Institute of Electrical and Electronics Engineers
IF Intermediate Frequency
IFAC International Federation of Automatic Control
IFF If and Only If
IFIP International Federation for Information Processing
IGN Ignition
IHP Indicated Horsepower
I/I Current to Current
IL Indicating Lamp
ILLUM Illuminate
IMC Institute of Measurement and Control
IMP Impact; Impedance; Imperial; Impulse
In Indium
in. or ″ Inch
in.² Square Inches
INBD Inboard
INC Incoming
INCAND Incandescent
INCIN Incinerator
INCL Include
INCOLR Intercooler
INCR Increase
IND Inductance; Induction; Indicate; Indicator
INDEP Independent
INFO Information
in.²/h Square Inch per Hour
in. Hg Inches of Mercury
in. H₂O Inches of Water
INJ Injection
in.²/min Square Inch per Minute
in.³ Cubic Inches
in.³/h Cubic Inch per Hour
in.-in. Inch-Inch
in./in. Inch per Inch
INL Inlet
in.³/min Cubic Inch per Minute
in.-lb Inch-Pound
in./min Inch per Minute
in./rev Inches per Revolution

in./s Inches per Second
INSP Inspect
INST Instantaneous
INSTL Install
INSTM Instrumentation
INSTR Instruction; Instrument
INT Intersect; Internal; Interior; Integral; Integrating
INTCHG Interchangeable
INTCP Intercept
INTCPR Interceptor
INTER Interrupt
INTERCOM Intercommunication
INTK Intake
INTLK Interlock
INTMD Intermediate
INTMT Intermittent
INV Invert
INVR Inverter
INVS Inverse
I/O Input/Output
IOS Insulating Oil System
I/P Current to Pressure Transducer; Current/Pneumatic; Current/Pressure
IP Intermediate Pressure
IPA Intermediate Power Amplifier
IPM Interruptions per Minute
IPS Interruptions per Second; Inches per Second
IPTS International Practical Temperature Scale
ir Infrared
Ir Iridium
IR Instantaneous Relay
IRIG Inter-Range Instrumentation Group
IRM Intermediate Range Monitoring
IRREG Irregular
IS Inside; International Standard
ISA Instrument Society of America
ISDN Integrated Systems Digital Network
ISLN Isolation
ISO International Standards Organization
ISOL Isolate
ISS Issue
IT Insulting Transformer
ITAE Integral Time Absolute Error
ITI Industrial Technology Institute
ITR Inverse Time Relay
IVD Integrated Voice Data LAN
IWS Injection Water System
IX Ion Exchanger

J

J Jack; Joule
JB Junction Box
JCL Job Control Language
J/cm² Joule per Square Centimetre
JCT Junction
JEIDA Japan Electronic Industry Development Association
JIS Japanese Industrial Standards
JNL Journal
JOVIAL Jules'Own Version of International Algorithmic Language
JP Jacket Pump
JT Joint
JW Jacket Water

K

K Kilo; Kelvin; Key; Kip (1,000 lb); Ratio of Specific Heats; Kilobyte
K$_a$ Air Spring Rate (lb/in)
K$_B$ Bernoulli Coefficient
KBX Keybox
kc Kilocycles
K$_c$ Cavitation Index
kcal Kilocalorie
KCIL Circular Mils, Thousands
KCMIL Thousand Circular Mills
KD Knock Down
K-FT Kip-Feet
kg Kilogram
kg per cu m Kilograms per Cubic Meter
kg-m Kilogram Meter
kg/m² Kilograms per Square Metre
kg/m³ Kilograms per Cubic Metre
kg/mm² Kilograms per Square Millimetre
kg/mm³ Kilograms per Cubic Millimetre
KGPS Kilograms per Second
kg/s Kilograms per Second
K$_h$ Fluid Force Coefficient
kHz Kilohertz
KIN Kinescope
KIP Thousand Pound
KIP-FT Thousand Foot Pound
kl Kilolitre
km Kilometre
KMPH Kilometer per Hour
KMPS Kilometer per Second

km/s Kilometre per Second
KN SW Knife Switch
KO Knock-Out
KP Kick Plate
kPa Kilopasca
KP&D Kick Plate and Drip
Kr Krypton
KS Kilocycles per Second
ksi 1,000 Pounds per Square Inch
KST Keyseat
kV Kilovolt
kVA Kilovolt-Ampere
KVAH Kilovolt-Ampere/Hour
KVAR Kilovar; Reactive Kilovolt Ampere
KVARH Kilovarhour
KVVD Kelvin-Varney Voltage Divider
kW Kilowatt
KWH Kilowatt-Hour
KWY Keyway

L

L Low; Line; Left; Lamp; Lambert; Litre; Long; Length
La Lanthanum
LA Lightning Arrester
LAB Laboratory
LAD Ladder
LAM Laminate
LAN Local Area Network
LAT Lateral; Latitude
LAV Lavatory
lb Pound
LB/FT² Pounds per Square Feet
lb/ft³ Pound Mass per Cubic Foot
LB-FT Pound-Foot
lb/gal Pound Mass per Gallon
LB-HR Pounds per Hour
lb/in² Pound Mass per Square Inch
lb/in³ Pound Mass per Cubic Inch
LB-IN Pound-Inch
LBS Load Bearing Switch
LC Locked Closed; Lighting Cabinet; Level Controller
LCD Liquid Crystal Display
LCU Local Control Unit
LDG Landing
LDS Load Dispatch System
LED Light Emitting Diode
LF Low Frequency

LG   Length
LH   Left Hand
Li   Lithium
LIFO   Last In, First Out
LIM   Limit; Limiter
LIM SW   Limit Switch
LIN   Linear
LIN FT   Linear Foot
LIQ   Liquid
LIS   Latch Indicator Switch
LISP   List Processing Language
LK   Factor in Sound-Prediction Formula; Link
LKG   Leakage
LL   Low-Low; Live Load
LLC   Logical Link Control
LLEI   Low-Level Engineering Interface
LLHI   Low-Level Human Interface
LLM   Load Limit Motor
LLOI   Low-Level Operator Interface
LLR   Load Limiting Resistor
lm   Lumen
L/min   Litre per Minute
LMST   Limestone
LO   Lubricating Oil; Locked Open
LOFF   Leakoff
LOG   Logarithmic
LONG   Longitude
LOX   Liquid Oxygen
L_p Sound Pressure Level, dBA
LP   Liquid Petroleum; Low Pass; Low Point; Low Pressure; Linear Programming
LPI   Low Pressure Injection
LPIS   Low Pressure Injection System
LPM   Lines per Minute
L_{po} Sound Pressure Level at a Point Four Feet Downstream of a Valve and Three Feet from the Surface of the Pipe
LPOF   Low-Pass Output Filter
LPS   Low Speed
LPW   Lumens per Watt
Lr   Lawrencium
LR   Load Ratio; Load Ratio Control
LS   Level Switch
LSAP   Link Service Access Point. See Sap
LSB   Least Significant Bit
LSI   Large-Scale Integration
LSR   Load Shifting Resistor
LT   Low Temperature; Low Tension; Low Torque; Light
LTD   Limited
LTDN   Letdown
LTG   Lighting

Lu   Lutetium
LV   Low Voltage
LVA   Low Vacuum Alarm
LVDT   Linear Variable Differential Transformer
LVL   Level
LVRT   Linear Variable Reluctance Transducer
LWR   Lower
Lx   Factor in Sound-Prediction Formula
L_x Lux

M

μ Micro
μA Microampere
μF Microfarad
μH Microhenry
μM Micrometer
μMHO Micromho
μSEC Microseconds
μV Microvolt
$\mu V/M$ Microvolts per Meter
μW Microwatt
m   Milli; Mega; Metre; Area Ratio
M   Thousand; Bending Moment
$M\Omega$ Megohm
m^2 Square Metre
m^3 Cubic Metre
m^3/h Cubic Metre per Hour
mA   Milliamperes (electrical)
MA   Master
MAC   Media Access Control
MACH   Machine
MAG   Magnet; Magnetic; Magneto
MAINT   Maintenance
MANF   Manifold
MAP   Manufacturing Automation Protocol
MAR   Marine
mA-s   Milliampere Second
MASER   Microwave Amplification by the Stimulated Emission of Radiation
MATL   Material
MAX   Maximum
MAX DIF P   Maximum Differential Pressure
MAX DIF T   Maximum Differential Temperature
MB   Millibars
MBH   British Termal Units per Hour (thousand)
MBM   Magnetic Bubble Memory
MC   Megacycles
MCC   Motor Control Center
MCF   Thousand Cubic Feet
MCR   Main Control Room

MDF   Main Distributing Frame
Md   Mendelevium
MDS   Megawatt Demand Setter
MECH   Mechanism; Mechanical
MED   Mean Effective Difference; Medical; Medium
MEP   Mean Effective Pressure
MET   Meteorological; Metal
MF   Medium Frequency; Motor Field
MFD   Manufactured
MFG   Manufacturing
MFP   Main Feedwater Pump
MFPT   Main Feedwater Pump Turbine
MFR   Manufacture
MFRR   Manufacturer
Mg   Magnesium
MG   Motor Generator; Milligram
MGD   Million Gallons per Day
MgO   Magnesium Oxide
MH   Millihenry; Manhole
MHz   Megahertz
MI   Miles
MICR   Magnetic Ink Character Recognition
MIMO   Multiple-Input-Multiple-Output Control System
MIN   Minute; Minimum
MIN DIF PRESS   Minimum Differential Pressure
MIN DIF TEMP   Minimum Differential Temperature
mips   Million Instructions per Second
MIS   Management Information System
MISC   Miscellaneous
MIX   Mixture
MKUP   Makeup
mL   Millilitre
ML   Millilambert
MLDG   Molding
mm   Millimetre (length)
mm²   Square Millimetres
mm Hg   Millimeters of Mercury
mm H₂O   Millimeters of Water
mm²/min   Square Millimetres per Minute
mm²/s   Square Millimetres per Second
mm³   Cubic Millimetres
mm³/h   Millimetres per Hour
mm³/min   Cubic Millimetres per Minute
mm³/s   Cubic Millimetres per Second
MMA   Maximum-Minimum Algorithm
MMFS   Manufacturing Messaging Format Standard
m/min   Metres per Minute

mm/m   Millimetres per Linear Metre
mm/min   Millimetres per Minute
mm/rev   Millimetres per Revolution
mm/s   Millimetres per Second
Mn   Manganese
MN   Main
MNL   Manual
MNOS   Metal-Nitride-Oxide Semiconductor
Mo   Molybdenum
MO   Motor Operated; Month; Master Oscillator
MOCS   Multiple-Output Control System
MOD   Motor Operated Disconnect; Modulator
MODEM   Modulator and Demodulator
mol wt   Molecular Weight
MON   Monitor
MOS   Metal-Oxide-Semiconductor
MoS₂   Molybdenum Sulfide
MOT   Motor
MOV   Motor Operated Valve
MP   Melting Point
MPG   Miles per Gallon
MPH   Miles per Hour
MPS   Meters per Second
MPU   MicroProcessor Unit
MPX   Multiplexer
ms   Milliseconds
MSB   Most Significant Bit
MSI   Medium Scale Integration
MSIS   Main Steam Isolation Signal
MSL   Mean Sea Level
MSS   Main Steam Systems; Manufacturers Standardization Society of the Valves and Fittings Industry Inc.
MSTRE   Moisture
MSV   Main Steam Vault
MSW   Masterswitch
MTBF **Mean-Time-Between-Failures**
MTD   Mounted
MTG   Mounting
MTR   Meter (instrument); Metering
MTS   Master Trip Solenoid
MTTF   Mean-Time-to-Failure
MTTR   Mean Time to Repair
mu   micron (0.001 mm)
MULT   Multiple
MUX   Multiplexer
mV   Millivolt
MVA   Megavoltz Ampere
MVC   Manual Volume Control
MW   Microwave; Milliwatt; Megawatt
MWH   Megawatt Hour
MWP   Maximum Working Pressure

441

N

n Power Law Index
N North; Newton; Nitrogen
N₂ Nitrogen
Na Sodium
NACE National Association of Corrosion Engineers
Na₂SO₄ Sodium Sulphate
NaCl Sodium Chloride
NaClO₃ Sodium Chlorate
NaNO₂ Sodium Nitrite
NaNO₃ Sodium Nitrate
NaOH Sodium Hydroxide
NAS National Aerospace Standards
NASA National Aeronautics and Space Administration
Nb Columbium (niobium)
NBS National Bureau of Standards
NC Normally Closed
NE Northeast
NEC National Electric Code
NEG Negative
NEMA National Electrical Manufacturers Association
NET Network
NEUT Neutral
NFPA National Fire Protection Association
Ni Nickel
NICET National Institute for Certification in Engineering Technologies
NIOSH National Institute for Occupational Safety and Health
NIST National Institute of Standards and Technology
N-m Newton Metre
NMRR Normal-Mode Rejection Ratio
NNI Netherlands Standards Institution
NNS Non-Nuclear Safety
NO Number; Normally Open
NOM Nominal
NOR Normal
NORP Normal Power
NOR POOL EL Normal Pool Elevation
NOZ Nozzle
NP Nameplate
NPS National Pipe Straight Thread; Nominal Pipe Size (diameter)
NPSH Net Positive Suction Head
NPT National Pipe Taper Thread
NRC Nuclear Regulatory Commission

NRZ Non-Return-to-Zero
NSE nth Sequential Algorithm
NSEC Nano Second
NSR Nuclear Safety Related
NUC Nuclear
NW Northwest

O

Ω Ohm
O Oxygen
OA Overall
OBE Operating Basis Earthquake
OC Overcurrent
OCB Oil Circuit Breaker
O-C-O Open-Close-Open
OCR Optical Character Recognition
OD Outside Diameter
OEM Original Equipment Manufacturer
OG Off-Gas
OGR Outgoing Repeater
OGS Off-Gas System
OGT Outgoing Trunk
OI Oil Insulated
OLG Oil Level Gage
OPER Operator
OPNG Opening
OPR Operate
OPT Optical
ORF Orifice
ORNT Orientation
OS Oil Switch; Operating System
OSC Oscilloscope; Oscillograph
OSHA Occupational Safety and Health Administration
OSI Open System Interconnection
OSTB Oscillograph Test Block; Oscilloscope Test Block
OS&Y Outside Screw and Yoke
OTM Overtempered Martensite
OUT Outlet; Output; Outside; Outgoing
OUTBD Outboard
OVFL Overflow
OVHD Overhead
OVLD Overload
OVSP Overspeed
OVV Overvoltage
OXD Oxidized
OXY Oxygen
oz Ounce

oz avdp Ounce Avoirdupois
OZ-FT Ounce-Foot
oz/ft² Ounces per Square Foot
OZ-IN Ounce-Inch
oz/in² Ounces per Square Inch

P

3PDT SW Triple Pole, Double Throw Switch
3PST SW Triple Pole, Single Throw Switch
3P SW Triple Pole Switch
4PDT SW 4 Pole Double Throw Switch
4PST SW 4 Pole Single Throw Switch
4P SW 4 Pole Switch
p Pressure, psia; Page
P Pole; Pico; Phosphorous; Poise
P&ID Piping and Instrumentation Drawing
Pa Pascal
PA Public-Address System; Power Amplifier; Plant Air
PADS Plant Alarm and Display System
PAM Pulse Amplitude Modulated
PAM/FM Frequency Modulation of a Carrier by Pulse Amplitude Modulated Information
PAM/FM/FM Frequency Modulation by Frequency Modulated Subcarriers by Pulse Amplitude Modulated Information
PAR Parallel
PART Partial; Particulate
PASS Passage
PASSWY Passageway
PAT Patent
PATT Pattern
PAW Plasma Arc Welding
PAX Private Automatic Exchange
Pb Lead
PB Push Button; Pull Box
PB STA Push Button Station
PBW Proportional Bandwidth
PBX Private Branch Exchange
P$_c$ Thermodynamic Critical Pressure, psia
PC Pulsating Current; Power Cabinet; Piece; Pico Coulomb; Plant Computer; Personal Computer or Programmable Controller
PCB Printed Circuit Board; Power Circuit Breaker
PCF Pounds per Cubic Foot
PCH Punch
PCI Protocol Control Information
PCIU Process Control Interface Unit

PCM Pulse Code Modulation
PCT Percent
PD Product Detector; Potential Difference; Pitch Diameter; Proportional Derivative
PD² Plastically Deformed Debris
PDD Programmable Data Distributor
PDM Pulse Duration Modulation
PDP Plasma Discharge Panel
PDU Protocol Data Unit
PEN Penetration
PERC Percission
PERF Perforate
PERM Permanent
PERP Perpendicular
PERS Personnel
PERT Program Evaluation and Review Technique
PF Power Factor; Pico Farad
PFD Preferred; Process Flow Diagram
PFM Power Factor Meter
pH Hydrogen (Ion Concentration)
PH Phase
PHOTO Photograph
PHYS Physical
P/I Pneumatic to Current Coverter
PI Proportional-Integral
PIA Peripheral Interface Adaptor
PID Proportional Integral Derivative
P&ID Piping and Instrumentation Drawing; Process and Instrument Drawing (diagram)
PIEZ Piezometer
PIO Programmed Input/Output
PIP Peripheral Interchange Program
PIX Picture
PK Peck
PKG Packing
PL Pile
PL/1 Programming Language/1
PLATF Platform
PLC Power Line Carrier; Programmable Logic Controller
PLOT Plotting
PLSTC Plastic
PLT Plant; Pilot
PM Afternoon; Power Metallurgy
P/M Powder Metallurgy
PMP Pump
PMS Plant Monitoring System
PMWS Primary Makeup Water System
PNEU Pneumatic
PNL Panel
PO Pneumatic Operated

P1/O2   Part 1 of 2
POCV   Piston-Operated Check Valve
POL   Polarized; Problem-Oriented Language
PORT   Portable
pos   Positive
POSN   Position
POT   Potential; Potentiometer
POT W   Potable Water
pp   Pages
P-P   Push-Pull
ppb   Parts per Billion
PPD   Pour Point Depressant
PPH   Pounds per Hour
PPI   Programmed Peripheral Interface
ppm   Parts per Million (1/106)
PPS   Pulses per Second; Plant Protection System
PPTR   Precipitator
PR   Purple; Pair; Pilot Relay
PRCS   Process
PREAMP   Preamplifier
PREP   Prepare
PRESS   Pressure
PRF   Proof
PRFCN   Purification
PRI   Primary
PRMG   Priming
PRMLD   Premolded
PRNG   Purging
PROJ   Project
PROM   Programmable Read Only Memory
PROPNL   Proportional
PROT   Protection
PRP   Purple
PRS   Press
PRT   Pressurizer Relief Tank
PRV   Pressure Reducing Valve
PS   Pressure Switch
PSD   Power Spectral Density
PSF   Pounds per Square Foot
PSFA   Pounds per Square Foot Absolute
PSFG   Pounds per Square Foot Gage
psi   Pounds per Square Inch
psia   Pounds per Square Inch, Absolute
psig   Pounds per Square Inch, Gage
PSK   Phase Shift Keying
PSS   Packet Switching System
pt   Part
PT   Point; Potential Transformer
PTM   Pulse Time Modulation
PTN   Partition
PTR   Printer; Paper Tape Reader
PTY   Party

PU   Pickup
PULL B S   Pull Button Switch
PULV   Pulversizer
PV   Process Variable
PW   Power
PWR   Pressurized Water Reactor
PWR SPLY   Power Supply
PYR   Pyrometer
PZR   Pressurizer

Q

q   Volumetric Flow Rate
QCB   Queue Control Block
QT   Quart
QTR   Quarter
QTY   Quantity
QTZ   Quartz
QUAD   Quadrant

R

r   Radius
R   Right; Red; Rankin; Ratio; Rankline; Gas Content, consistent with units
$R_{A,B \text{ or } C}$ Rockwell Hardness; A, B or C scales
RAD   Radian
RADN   Radiation
radar   RAdar Detection And Ranging
RAM   Random Access Memory
RB   Roller Bearing; Reactor Building
RBM   Rod Block Monitoring
RC   Remote Control
RCD   Record
RCDG   Recording
RCDR   Recorder
RCPB   Reactor Coolant Pressure Boundary
RCPN   Reception
RCPT   Receptacle
RCTL   Resistor-Capacitor-Transistor Logic
RCV   Receive
RCVD   Received
RCVG   Receiving
RCVR   Receiver
RCW   Raw Cooling Water
RD   Round; Root Diameter; Ratio Detector
RDL   Radial
Re   Reynolds Number
REAC   Reactor; Reactive

RECIP Reciprocal; Reciprocate
RECIR Recirculate
RECIRCN Recirculation
RECL Reclosing
RECOG Recognition
RECT Rectifier
REDUC Reduction
REF Refinery; Reference
REFR Refrigerate; Refractory
REFRIG Regrigerant
REG Register; Regulate; Regular
REGEN Regenerative
REM Roentgen Equivalent Man
REP Roentgen Equivalent Physical
REQD Required
RES Resistance
RESID Residual
RET Retainer; Retard
RETR Retractable
Rev Value-Reynolds Number
REV Reverse; Revolution
RF Roof; Radio Frequency
RFI Radio Frequency Interference
RFM Reactive Factor Meter
RGA Relative Gain Array
RH Relative Humidity; Reheat; Right Hand Thread
RHEO Rheostat
Rhm Roentgens per Hour at One Metre
RHR Roughness Height Rating; Residual Heat Removal; Reheater
RHRP Residual Heat Removal Pump
RHRS Residual Heat REmoval System
RIL Red Indicating Lamp
RIP Real-Time Interrupt Process
RLF Relief
RLSE Release
RLY Relay
RM Room
RMS Root Mean Square
RNG Range
RNRZ Randomized Non-Return-to-Zero
ROM Read-Only Memory
ROT Rotate; Rotary
RPM Revolutions per Minute
rps Revolutions per Second
RPU Remote Processing Unit
RSV Reserve; Reheat Stop Valve
RSVR Reservoir
RSW Raw Service Water
RTC Real Time Clock
RTD Resistance Temperature Device; Resistance

Temperature Detector
RTL Real-Time Language; Resistor-Transistor Logic
RTN Return
RUB Rubber
RUPT Rupture
RV Relief Valve
RVM Reactive Voltmeter
RW Raw Water
RWCLS Raw Water Chlorination System

S

s Second
S South; Side; Single; Siemens
SAE Society of Automative Engineers
SAF Safety
SAL Salinometer
SALV Salvage
SAMA Scientific Apparatus Makers Association
SAN Sanitary
SAP Service Access Point
SAS Service Air System
SAT Saturate
SAW Surface Acoustic Wave
Sb Antimony
SB Stuffing Box; Soot Blower; Sleeve Bearing; Service Building
SC Scale; Standing Committee
SCADA Supervisory Control and Data Acquisition
SCFH Standard Cubic Feet per Hour (flow)
SCFM Standard Cubic Feet per Minute
SCHEM Schematic
SCIO Serial Channel Input/Output
s/cm² Second per Square Centimetre
SCP Spherical Candle Power
SCR Short Circuit Ratio; Silicon-Controlled Rectifier
SCRB Scrubber
SCS Supervisory Control System
S/D Synchro-to-Digital Converter
SD Shutdown
SDGS Standby Diesel Generator System
SDLC Synchronous Data Link Control
SDR Sender
SDS Station Drainage System
SDU Service Data Unit
SE Southeast
SEC Second; Secondary

SECT Section
SEG Segment
SEL Select; Selector
SEMI AUTO Semi-Automatic
SEP Separator
SEQ Sequence
SEQL Sequential
SER Series; Serial
SERV Service
SET Settling
SEW Sewer
SF Single Feeder; Single Frequency
sfm Surface Feet per Minute
SG Steam Generator
SH Shunt
SHF Super High Frequency
SHLD Shield
SHP Shaft Horsepower
SHR CKT Short Circuit
SHTDN Shutdown
Si Silicon
SI Systeme Internationale d'Unites
SIG Signal
SIL Silence
SiO$_2$ Silicon Dioxide
SJAE Steam Jet Air Ejector
SK Sink
SKIM Skimmer
SL Slide; Sea Level
SLD Sealed
SLV Sleeve
SM Small
SMAW Shielded Metal Arc Welding
SME Society of Manufacturing Engineers
SMI Serial Model Interface
SMK Smoke
SMPL Sample
SMPLG Sampling
Sn Tin
SNAP Sub-Network Access Protocol
SND Sound
SNL Speed No Load
SOBU Seal Oil Backup
SOC Socket
SOL Solenoid
SOV Solenoid Operated Valve; Shut-Off Valve
SP Static Pressure; Standpipe; Speed; Space; Spare; Specific; Set Point; Standards and Practices
SPCR Spacer
SPDL Spindle
SPDT Single Pole, Double Throw

SPDT SW Single Pole, Double Throw Switch
SP GR Specific Gravity
SPHER Spherical
SP HT Specific Heat
SPKR Speaker
SPL Special; Sound Pressure Level
SP PH Split Phase
SPST Single Pole; Single Throw
SPST SW Single Pole, Single Throw Switch
SP SW Single Pole Switch
SQ Square
SQ CG Squirrel Cage
sq cm Square Centimeter
sq ft Square Feet
sq in Square Inch
sq m Square Meter
sq mm Square Millimeter
SQRT Square Root
sr Steradian
SRF Side Relief Angle
SS Station Service; Solid State; Sampling System
SSB Single Sideband
SSF Saybold Seconds Furol (Oil Viscosity)
SSU Seconds Saybolt Universal (Oil Viscosity)
S+T Speech-Plus-Tone
ST Steam
STA Station; Stationary
STAT Stator
STBY Standby
STD Standard
STEER Steering
STG Starting
STGEN Steam Generator
STK Stock
STOR Storage
STR Strainer; Stream; Strip; Structural; Structure
STUP Startup
ST W Storm Water
STWY Stairway
SU Startup
SUB Submerged
SUBSTA Substation
SUC Suction
SUM Summary
SUMR Summer
SUP Supply; Support
SUPP Supplement
SUPPR Suppression
SUPV Supervise; Supervisor
SUSP Suspend; Suspended
SV Safety Valve

SW Switch; Southwest; Short Wave; Salt Water; Stud Arc Welding
SWBD Switchboard
SWG Swing
SWGR Switchgear
SWR Switcher; Switch Register
SWYD Switchyard
SX Simplex
SYNSCP Synchroscope
SYS System
SYSGEN System Generation

T

t Metric Ton (tonne); temperature, °F
T Truss; Toll; Time; Tee; Teeth; Tesla; Temperature Absolute; Temperature, °Rankin
Ta Tantalum
TAB Tabulate
TACH Tachometer
TAGS Technical Assistance Groups
TAN Tangent
TAPPI Technical Association of the Pulp and Paper Industry
TARP Tarpaulin
Tb Terbium
TB Turbine Building; Test Block; Terminal Board; Terminal Block
T$_c$ Thermodynamic Critical Temperature, °Rankin
Tc Technetium
TC Trip Coil; Thermocouple; Technical Committee
T/C Thermocouple
TCO Trip Cut Out
TD Turbine Drive; Time Delay
TDC Time Delay Closing
TDM Tandem; Time Division Multiplex
TDMA Time Division Multiplex Access
TDO Time Delay Opening
Te Tellurium
TE Thermal Element
TECH Technical
TEL Telephone
TEMP Temporary; Temperature
TEMP IND Temperature Indicator
TENS Tension
TENT Tentative
TER Tertiary
TERM Terminal
TERR Territory

TETDS Turbine Extraction Taps and Drain System
TFE Telemetry Front End; Tetraflouroethylene
T-G Turbine Generator
TGCS Turbogenerator Control System
TGL Toggle
Th Thorium
TH Total Head, Feet
THD Total Dynamic Head
THERM Thermometer
THERMO Thermostat
THK Thick
THR Thrust; (Throughput) Algorithm
THRM Thermal
THROT Throttle
THRU Through
THU Thursday
Ti Titanium
TIC Temperature Indicator Control
TIR Total Indicator Reading
TJ Telephone Jack
TK Tank
TKT Ticket
Tl Thallium
TL Tie Line; Test Link
TLC Telephone Line Carrier; Thin-Layer Chromatography
TLG Telegraph
TLM Telemeter
TLV Threshold Limit Value
TLV-STEL Threshold Limit Value-Short Term Exposure Limit
TLV-TWA Threshold Limit Value-Time Weighted Average
Tm Thulium
TM Temperture Modifier; Temperature Meter
TMBR Timber
TMS Temperature Monitoring System
TN Train; Tennessee
TNG Training
TNL Tunnel
TO Turnout
TOL Tolerance
TOP Technical and Office Protocol
TON-MI Ton Mile
TOR Torque
TOT Total
TP Total Pressure
TPF Taper per Foot
TPH Tons per Hour
tpi Threads per Inch
TPI Turns per Inch; Teeth per Inch; Taper per

Inch
TPR Taper
TR Transmit-Receive; Tons of Refrigeration
TRANS Transfer; Transportation
TRANSV Transverse
TRD Tread
TRF Tuned Radio Frequency
TRI Triode
TRIM Trimmer
TRK Trunk
TRS Transverse Rupture Strength
TRTMT Treatment
TS Thermal Switch; Tensile Strength; Taper Shank; Tangent Spiral
TSCA Toxic Substance Control Act
TSF Tons per Square Foot
TT Transfer Trip
TTL Transistor/Transistor Logic
TTP Telephone Twisted Pair
TTR Transfer Trip Relay
TUB Tubing
TUE Tuesday
TUR Turret
TURB Turbine
TURBOGEN Turbine Generator
TV Television
TW Twisted
TWP Township
TWR Tower
TWX Teletypewriter Exchange
TX Texas

U

U Unit; Uranium; Velocity (ft/sec)
UART Universal Asynchronous Receiver Transmitter
UC Unit Cooler
UH Unit Heater
UHF Ultra High Frequency
UL Underwriters Laboratories
ULT Ultimate
UNC Unified Coarse Thread
UNDG Underground
UNDW Underwater
UNEF Unified Extra Fine Thread
UNF Unified Fine Thread
UNIV Universal
UO Unit Operator
UPR Upper

UPS Upscale; Uninterruptible Power Supply
UPSTR Upstream
USART Universal Synchronous/Asynchronous Receiver Transmitter
USASCII U.S. Standard Code for Information Exchange
USRT Universal Synchronous Receiver Transmitter
UT Utah
uv Ultraviolet
UV Under Voltage
UVD Under Voltage Device
UVROM Ultraviolet-Erasable Read-Only Memory

V

v Volt; Specific Volume, ft^3/lb
V Violet; Volt; Valve; Vanadium; Volume, $in.^3$
vA Volt Ampere
VA Voltampere
VAC Vacuum
VAR Variable; Reactive Volt Ampere
VB Valve Box
VCO Voltage-Controlled Oscillator
VCT Volume Control Tank
VDF Video-Frequency
VDU Visual Display Unit; Video Display Unit
VEG Vegetable
VEL Velocity
VENT Ventilation; Ventilate
VERT Vertical
VFO Variable Frequency Oscillator
VG Voice Frequency
VHF Very-High Frequency
VIB Vibration; Vibrate
VID Video
VIO Violet
VISC Viscosity
VIT Vitreous
VLF Very-Low Frequency
VLSI Very Large Scale Integration
VM Voltmeter
V/MIL Volts per Mil
VOL Volume
VP Vent Pipe; Velocity Pressure
VPD Valve Position Detector
VPS Vacuum Priming System
VR Voltage Relay
VRS Varies

VS Versus; Ventilating System; Vent Stack; Variable Spring Support
VT Vibration Transmitter; Vermont; Vacuum Tube
VTI Venturi

W

2W Two Wires
2WT 2-Way Trunk
4W Four-Wire
w Flow Rate (lb/hr)
W White; Wide; Width; West; Wall; Watt; Tungsten
W/ With
WA Washington
WAN Wide Area Network
WASH Washer
Wb Weber
WB Wet Bulb
WBT Wet Bulb Thermometer
w.c. Watercolumn (hydrostatic head)
WC Word Count
WD Wind
WDG Winding
WDS Waste Disposal System
WDW Window
WED Wednesday
WG Water Gage; Working Group
WH Watthour; Water Heater
WHM Watthour Meter
WHSE Warehouse
WHT White
WI Wisconsin
WIL White Indicating Lamp
WIR Wiring
WK Week
WL Water Line
WM Wattmeter
WN Weld Neck
W/O Without
WOG Water, Oil or Gas
WP Weatherproof
WPC Watts per Candle
WPF Waterproof
WPR Working Pressure
WR² Flywheel Effect
W S EL Water Surface Elevation
WS Wetted Surface
WSP Working Steam Pressure
WT Weight; Watertight; Water Tank

WTR Water
WTS Water Treatment System
WV West Virginia
WW Waterworks
WWO Waste Water & Oil
WY Wyoming

X

x Pressure Drop Ratio
XARM Cross Arm
XBAR Cross Bar
XBEAM Cross Beam
XCONN Cross Connection
XCVR Transceiver
XDCR Transducer
Xe Xenon
XFMR Transformer
XLTR Translator
XMSN Transmission
XMTG Transmitting
XMTR Transmitter
XPL External Party Line
XRD Cross Road
XS Transfer Switch
XST External Start
x_t Pressure Drop Ratio Factor
XTAL Crystal
XTIE Crosstie
x_{tp} Value of x, for a Valve Fitting Assembly

Y

Y Yellow; Yttrium; Expansion Factor
YD Yard
yd³ Cubic Yard
YEL Yellow
YIL Yellow Indicating Lamp
YR Year

Z

Z Compressibility Factor
ZFN Zero-Order Fixed-Aperture Non-Redundant Sample Algorithm
Zn Zinc
Zr Zirconium
ZrO₂ Zirconium Oxide
ZS Zone Switch
ZSS Zero Speed Switch

449

Symbols

Symbols Used in Engineering

μ   Symbol for coefficient of friction (Poisson's ratio).
μA   Symbol for microampere.
μF   Symbol for microfarad.
μm   Symbol for micrometre.
μs   Symbol for microsecond.
μV   Symbol for microvolt.
μW   Symbol for microwatt.

Greek Letters Used in Engineering

Capital Letter	Lower Case Letter	Common Engineering Usage Capital Letter	Lower Case Letter
A	α		Angles, Attenuation Constant, Coefficients, Adsorption Factor, Area
B	β		Angles, Phase Constant, Coefficients
Γ	γ	Complex Propagation Constant	Angles, Specific Gravity, Propagation Constant, Electrical Conductivity
Δ	δ	Increment, Decrement, Determinant, Permittivity	Increment, Decrement, Angles, Density
E	ϵ		Dielectric Constant, Permittivity, Base of Natural Logarithms, Electric Intensity
Z	ζ		Coefficients, Coordinates
H	η		Efficiency, Surface Charge Density, Intrinsic Impedance, Coordinates, Hysteresis
Θ	θ		Angles, Angular Phase Displacement, Time Constant, Reluctance
I	ι		Unit Vector
K	κ		Coupling Coefficient, Susceptibility

452

Greek Letters Used in Engineering

Capital Letter	Lower Case Letter	Common Engineering Usage Capital Letter	Lower Case Letter
Λ	λ	Permeance	Wave Length, Attenuation Constant
M	μ		Prefix Micro, Amplification Factor, Permeability
N	ν		Frequency, Reluctance
Ξ	ξ		Coordinates
Π	π		3.1416
P	ρ		Coordinates, Resistivity, Volume Charge Density
Σ	σ	Summation	Electrical Conductivity, Surface Charge Density, Leakage Coefficient, Complex Propagation Constant
T	τ		Density, Time Constant, Time-Phase Displacement, Volume Resitivity, Transmission Factor
Φ	φ	Scaler Potential	Angles, Magnetic Flux
X	χ		Angles, Electrical Susceptibility
Ψ	ψ		Angles, Coordinates, Phase Difference, Dielectric Flux
Ω	ω	Ohms, Resistance in Ohms, Solid Angle	Angular Velocity

Recommended International Symbols

SYMBOL		SYMBOL DEFINITION
IEC417		PROTECTIVE GROUNDING TERMINAL: A terminal which must be connected to earth ground prior to making any other connections to the equipment.
GREEN COLOR American National Standard National Electrical Code ANSI Cl-1975		SUPPLY CIRCUIT PROTECTIVE GROUNDING TERMINAL: (1) A green-colored terminal screw with a hexagonal head; (2) a green-colored pressure wire connector.
IEC417		GROUNDED TERMINAL: A grounded terminal which, as far as the operator is concerned, is already grounded by means of an earth grounding system.
IEC4l7		A terminal to which or from which an alternating (sine wave) current or voltage may be applied or supplied.
IEC4l7		A terminal to which or from which a direct current or voltage may be applied or supplied.
IEC4l7		A terminal to which or from which an alternating and direct current or voltage may be applied or supplied.
IEC417		EXPLANATION: This marking indicates that the operator must refer to an explanation in the operating instructions.
RED COLOR IEC417		HIGH VOLTAGE TERMINAL: A terminal at which a voltage with respect to another terminal or part exists or may be adjusted to 1000 volts or more.
IEC4l7		ON: Power: connection to the principal supply circuit.
IEC4l7		OFF: Power: disconnection from the principal supply circuit.
IEC335.1	3 $\sim$	A terminal to which or from which a three-phase alternating (sine-wave) current or voltage may be applied or supplied.
IEC335.1	3N $\sim$	A terminal to which or from which a three-phase alternating (sine-wave) current or voltage and neutral conductor may be applied or supplied.

454

Symbols for Actuators and Valves

Actuator (ACTR)

Represents the final control element that determines the state of a two-state device.

Desired Device State is
CLOSED

Desired Device State is
OPEN

The use of a letter in the symbol to designate the type of actuator is optional. Other choices include:

Character Designation

M = Electrical Motor
S = Solenoid
H = Hydraulic
A = Air Motor

Throttling Actuator (TACT)

Represents a diaphragm actuator that can affect multiple positions of the control device.

Manual Actuator (MATR)

Represents a manually-operated valve actuator.

Valve (VLVE)

Represents GLOBE, GATE, BALL, and NEEDLE valves used to regulate fluid flow through piping systems. Can be used with various combinations of actuators to convey multiple manipulation schemes.

Actual State is
CLOSED

Actual State is
OPEN

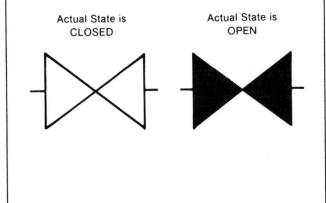

Symbols for Valves

3-Way Valve (VLV3)

Represents a valve used in piping systems to select flow paths or regulate between flow paths. Can be used with various combinations of actuators to convey multiple manipulation schemes.

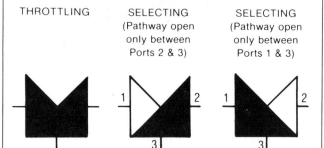

THROTTLING   SELECTING   SELECTING
                 (Pathway open   (Pathway open
                 only between   only between
                 Ports 2 & 3)   Ports 1 & 3)

Note: Port numbers are not part of symbol.

Butterfly Valve (BVLV)

Represents a butterfly valve, damper, or vane used to throttle (modulate) fluid flow through a pipe, duct, or stack

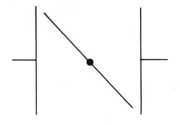

Check Valve (CVLV)

Represents a device that mechanically limits fluid flow to only one direction in a piping system—typically a check valve or back-draft damper.

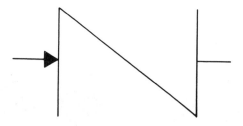

Arrow shows direction of allowable flow and is part of the symbol.

Relief Valve (RVLV)

Represents a one-way mechanically actuated pressure relief valve. While these valves are normally closed, two symbols are shown to accommodate those situations where feedback signals are provided to indicate actual status.

Normally closed valve   Normally closed valve
that is actually   that is actually
CLOSED   OPEN

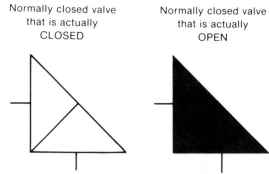

Symbols for Separators and Dryers

Cyclone Separator (CSEP)

A device used for solid, liquid, or vapor separation.

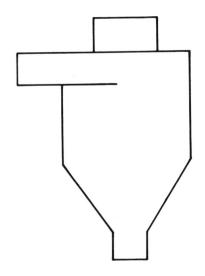

Rotary Separator (RSEP)

A rotary device for separating solids from liquids.

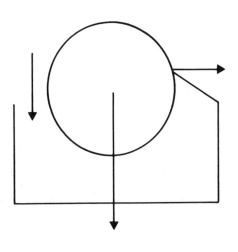

Spray Dryer (SDRY)

A device used for evaporation of liquids from mixtures of solids and liquids.

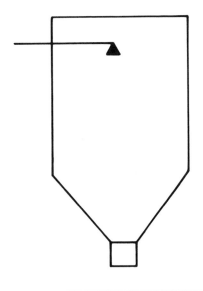

Symbols for Precipitators and Scrubbers

Electrostatic Precipitator (EPCP)

A device used to separate solid particles from a gas (e.g., in a smoke stack) by means of an electrostatically charged grid.

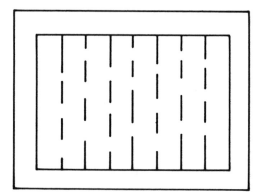

Scrubber (SCBR)

A device that uses a liquid spray to scrub gas.

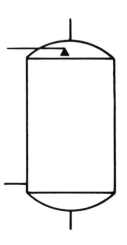

Symbols for Rotating Equipment

Blower (BLWR)

A device used to convey a gas under slight pressure.

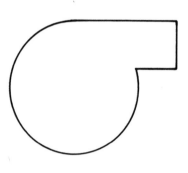

Compressor (CMPRO)

A device used to convey a gas under high pressure.

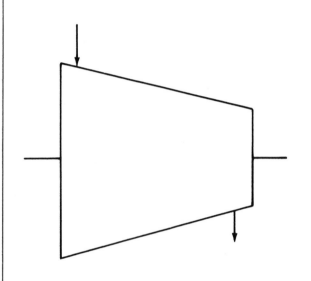

Pump (PUMP)

Represents that class of equipment used to transport slurries or liquids by internal rotary action. Examples are centrifugal, gear, lobe, etc.

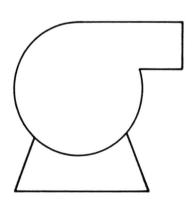

Turbine (TURB)

A device using the force of expanding gas to propel rotating equipment.

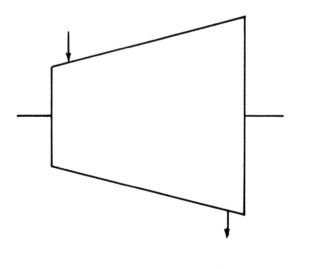

Symbols for Containment Vessels

Distillation Tower (DTWR)

A packaged or trayed distillation tower used for separation. Packing or trays may be shown to indicate type of distillation tower.

Jacketed Vessel (JVSL)

A vessel with a heating or cooling jacket. Jacket may be on straight shell, on bottom head, on top head, or any combination, as required to match the actual process vessel.

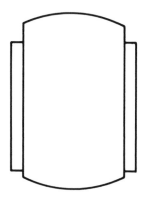

Reactor (RCTR)

A chemical reactor. Internal details may be shown to indicate type of reactor.

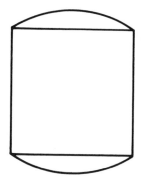

Vessel (VSSL)

A vessel or separator. Internal details may be shown to indicate type of vessel. Can also be used as a pressurized vessel in either a vertical or horizontal arrangement.

Atmospheric Tank (ATNK)

A tank for materials stored under atmospheric pressure.

Bin (BINN)

A container used to store solid or granular material that is discharged from the bottom.

Floating Roof Tank (FTNK)

A tank for liquids with roof of vessel moving up and down with a change in stored volume.

Gas Holder (GHDR)

A tank for gases with roof of vessel moving up and down with a change in stored volume.

Symbols for Containment Vessels

Pressure Storage Vessel (PVSL) A pressurized spherical vessel for storage of gases and liquids.	**Weigh Hopper (WHPR)** A vessel used for weighing material.

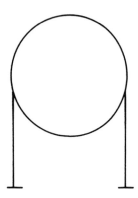

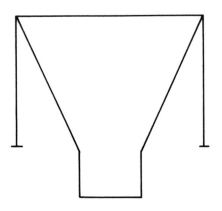

Symbols for Electrical Devices

Circuit Breaker (CBRK)

Representation of a circuit breaker for electrical systems. See STATE INDICATOR symbol for alternative use.

Manual Contactor (MCTR)

A power distribution switch used for device isolation.

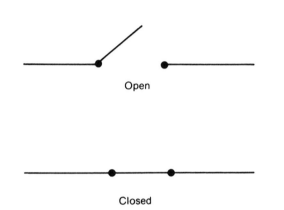

Delta Connection (DLTA)

Representation of a 3-phase delta connection.

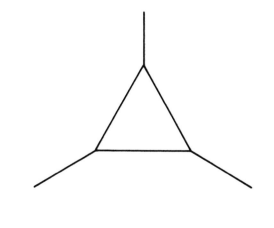

Fuse (FUSE)

Representation of a fuse as an over-current protection device.

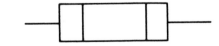

Symbols for Electrical Devices

Motor (MOTR)

An AC or DC motor.

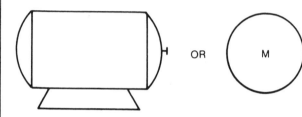 OR

This is the preferred symbol for process diagrams (base optional).

This is the preferred symbol for electrical diagrams.

State Indicator (STAT)

Used to represent binary states. For example: closed circuit/circuit open, etc.

Circuit Closed   Circuit Open

Transformer (XFMR)

A universal transformer.

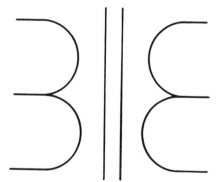

WYE Connection (WYEC)

Representation of a 3-phase wye (star) connection.

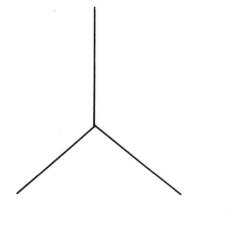

Symbols for Filters

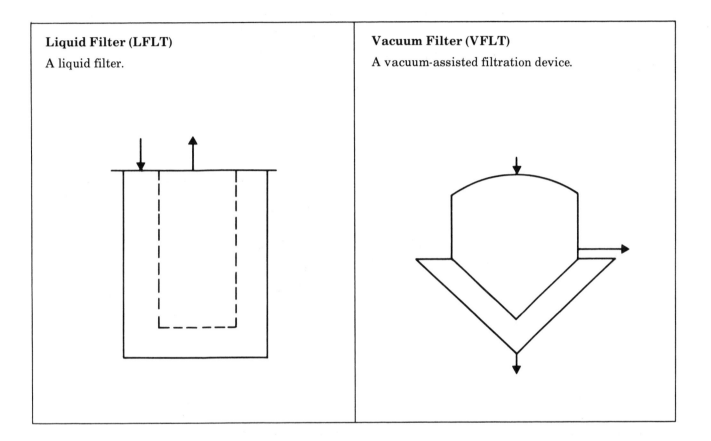

Liquid Filter (LFLT)

A liquid filter.

Vacuum Filter (VFLT)

A vacuum-assisted filtration device.

Symbols for Heat Transfer Devices

Exchanger (XCHG) Heat transferral equipment. An alternative symbol is depicted.	**Forced Air Exchanger (FAXR)** A forced-air heat exchanger.

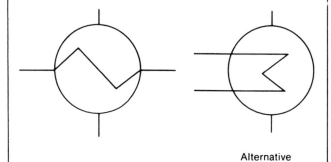

Alternative

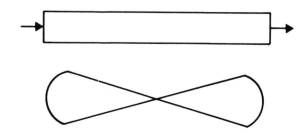

Furnace (FURN) Process heater or furnace. Internal details may be shown as needed.	**Rotary Kiln (KILN)** Typical gas, oil, coal- or coke-fired kiln.

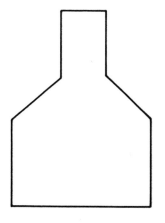

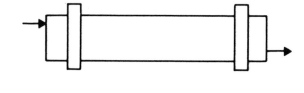

Symbols for HVAC Devices

Cooling Tower (CTWR)

A device for use in HVAC or other processes indicating the atmospheric cooling of water by forced evaporation.

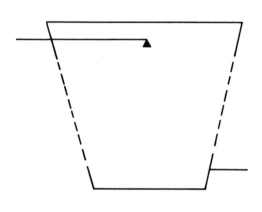

Evaporator (EVPR)

An HVAC device used to represent the exchange of heat between a liquid or gas and a refrigerant.

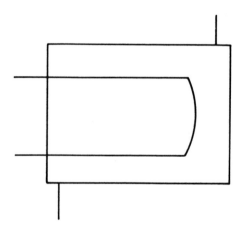

Finned Exchanger (FNXR)

A high surface transfer device used to exchange heat between a liquid or gas and air.

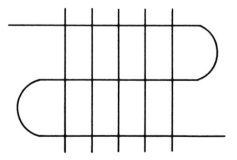

Symbols for Material Handling and Mixing

Conveyor (CNVR)

Belt conveyors, chain conveyors, and roller conveyors used in association with other symbols to represent more complex equipment such as a paper machine.

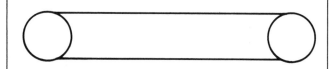

Mill (MILL)

Rotating rod, ball, autogenous, or semiautogenous mill used for size reduction of solids.

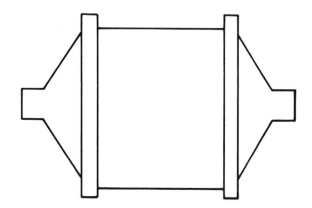

Roll Stand (RSTD)

Roll stand used in metal, paper, rubber, plastic, and glass industries.

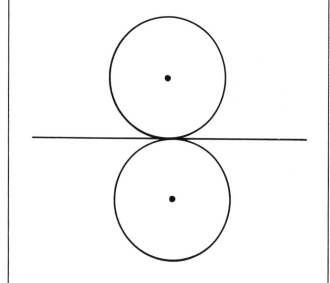

Rotary Feeder (RFDR)

A rotary feeder used to convey material in dry powder form from one location to another.

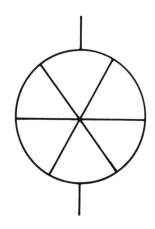

Screw Conveyor (SCMV)

A typical screw conveyor or screw pump.

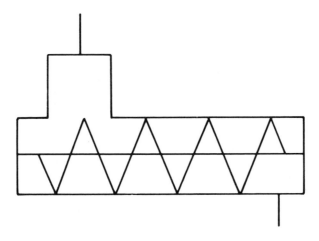

Agitator (AGIT)

A blade, propeller or paddle-type agitator.

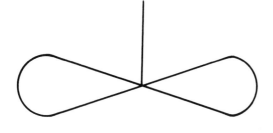

Inline Mixer (IMIX)

A mixing device used to continuously blend materials.

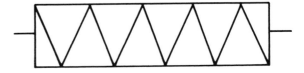

Symbols for Compressors

Reciprocating Compressor (RECP)

A reciprocating compressor or pump represents that class of equipment used to transport slurries or liquids by reciprocating action. Examples are pistons, diaphragms, plungers, etc.

Symbols for Function Blocks and Function Designations

The function designations associated with controllers, computing devices, converters and relays may be used individually or in combination. The use of a box avoids confusion by setting off the symbol from other markings on a diagram and permits the function to be used as a stand-alone block on conceptual designs.

FUNCTION	SYMBOL	MATH EQUATION	GRAPHIC REPRESENTATION	DEFINITION
SUMMING	Σ	$M = x_1 + x_2 + \ldots + x_n$		THE OUTPUT EQUALS THE ALGEBRAIC SUM OF THE INPUTS. (THE INPUTS MAY BE LABELED WITH POSITIVE OR NEGATIVE SIGNS).
AVERAGING	Σ/n	$M = \dfrac{x_1 + x_2 + \ldots + x_n}{n}$		THE OUTPUT EQUALS THE ALGEBRAIC SUM OF THE INPUTS DIVIDED BY THE NUMBER OF INPUTS.
DIFFERENCE	$\triangle$	$M = x_1 - x_2$		THE OUTPUT EQUALS THE ALGEBRAIC DIFFERENCE OF THE TWO INPUTS.
PROPORTIONAL	K, 1:1, 2:1	$M = Kx$		THE OUTPUT IS DIRECTLY PROPORTIONAL TO THE INPUT. IN THE CASE OF A VOLUME BOOSTER, "K" MAY BE REPLACED BY 1:1, FOR INTEGER GAINS, 2:1, 3:1, ETC., MAY BE SUBSTITUTED FOR K.
INTEGRAL	$\int$	$M = \dfrac{1}{T_I} \int x\, dt$		THE OUTPUT VARIES IN ACCORDANCE WITH BOTH MAGNITUDE AND DURATION OF THE INPUT. THE OUTPUT IS PROPORTIONAL TO THE TIME INTEGRAL OF THE INPUT.
DERIVATIVE	d/dt	$M = T_D \dfrac{dx}{dt}$		THE OUTPUT IS PROPORTIONAL TO THE RATE OF CHANGE (DERIVATIVE) OF THE INPUT.

Symbols for Function Blocks and Function Designations

FUNCTION	SYMBOL	MATH EQUATION	GRAPHIC REPRESENTATION	DEFINITION
MULTIPLYING	$\boxed{\times}$	$M = x_1 x_2$		THE OUTPUT EQUALS THE PRODUCT OF THE TWO INPUTS.
DIVIDING	$\boxed{\div}$	$M = \dfrac{x_1}{x_2}$		THE OUTPUT EQUALS THE QUOTIENT OF THE TWO INPUTS.
ROOT EXTRACTION	$\boxed{\sqrt[n]{\ }}$	$M = \sqrt[n]{x}$		THE OUTPUT EQUALS THE ROOT (I.E., CUBE ROOT, FOURTH ROOT, 3/2 ROOT, ETC.) OF THE INPUT. IF n IS OMITTED, A SQUARE ROOT IS ASSUMED.
EXPONENTIAL	$\boxed{x^n}$	$M = x^n$		THE OUTPUT EQUALS THE INPUT RAISED TO A POWER (I.E., SECOND, THIRD, FOURTH, ETC.).
NONLINEAR OR UNSPECIFIED FUNCTION	$\boxed{f(x)}$	$M = f(x)$		THE OUTPUT EQUALS SOME NONLINEAR OR UNSPECIFIED FUNCTION OF THE INPUT.
TIME FUNCTION	$\boxed{f(t)}$	$M = Xf(t)$ $M = f(t)$		THE OUTPUT EQUALS THE INPUT TIMES SOME FUNCTION OF TIME OR EQUALS SOME FUNCTION OF TIME ALONE.
HIGH SELECTING	$\boxed{>}$	$M = \begin{cases} x_1 \text{ FOR } x_1 \geq x_2 \\ x_2 \text{ FOR } x_1 \leq x_2 \end{cases}$		THE OUTPUT IS EQUAL TO THE GREATER OF THE INPUTS.

Symbols for Function Blocks and Function Designations

FUNCTION	SYMBOL	MATH EQUATION	GRAPHIC REPRESENTATION	DEFINITION
LOW SELECTING	$\vee$	$M = \begin{cases} X_1 \text{ FOR } X_1 \leq X_2 \\ X_2 \text{ FOR } X_1 \geq X_2 \end{cases}$		THE OUTPUT IS EQUAL TO THE LESSER OF THE INPUTS.
HIGH LIMITING	$\boxed{A}$	$M = \begin{cases} X \text{ FOR } X \leq H \\ H \text{ FOR } X \geq H \end{cases}$		THE OUTPUT EQUALS THE INPUT OR THE HIGH LIMIT VALUE WHICHEVER IS LOWER.
LOW LIMITING	$\boxed{\forall}$	$M = \begin{cases} X \text{ FOR } X \geq L \\ L \text{ FOR } X \leq L \end{cases}$		THE OUTPUT EQUALS THE INPUT OR THE LOW LIMIT VALUE WHICHEVER IS HIGHER.
REVERSE PROPORTIONAL	$-k$	$M = -KX$		THE OUTPUT IS REVERSELY PROPORTIONAL TO THE INPUT.
VELOCITY LIMITER	$\boxed{A}$	$\dfrac{dM}{dt} = \dfrac{dX}{dt} \begin{cases} \dfrac{dX}{dt} \leq H \text{ AND} \\ M = X \end{cases}$ $\dfrac{dM}{dt} = H \begin{cases} \dfrac{dX}{dt} \geq H \text{ OR} \\ M \neq X \end{cases}$		THE OUTPUT EQUALS THE INPUT AS LONG AS THE RATE OF CHANGE OF THE INPUT DOES NOT EXCEED A LIMIT VALUE. THE OUTPUT WILL CHANGE AT THE RATE ESTABLISHED BY THIS LIMIT UNTIL THE OUTPUT AGAIN EQUALS THE INPUT.
BIAS	$+$ $-$ $\pm$	$M = X \pm b$		THE OUTPUT EQUALS THE INPUT PLUS (OR MINUS) SOME ARBITRARY VALUE (BIAS).
CONVERT	$\boxed{*/*}$	OUTPUT = f(INPUT)	NONE	THE FORM OF THE OUTPUT SIGNAL IS DIFFERENT FROM THAT OF THE INPUT. * E - VOLTAGE I - CURRENT P - PNEUMATIC A - ANALOG B - BINARY * H - HYDRAULIC O - ELECTROMAGNETIC, SONIC R - RESISTANCE(ELECT.) D - DIGITAL

Symbols for Function Blocks and Functions Designations

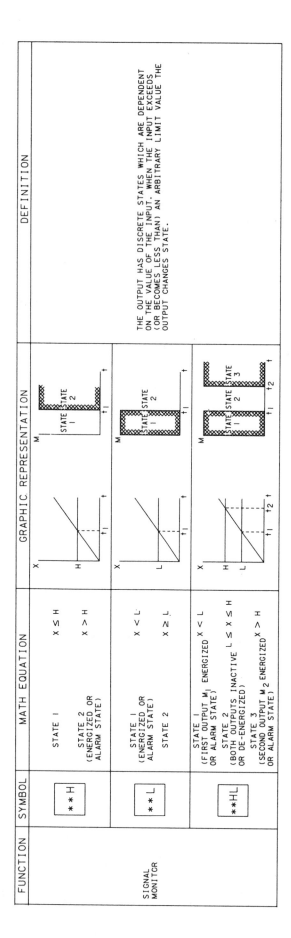

FUNCTION	SYMBOL	MATH EQUATION	GRAPHIC REPRESENTATION	DEFINITION
SIGNAL MONITOR	**H	STATE 1 $\quad X \leq H$ STATE 2 $\quad X > H$ (ENERGIZED OR ALARM STATE)		THE OUTPUT HAS DISCRETE STATES WHICH ARE DEPENDENT ON THE VALUE OF THE INPUT. WHEN THE INPUT EXCEEDS (OR BECOMES LESS THAN) AN ARBITRARY LIMIT VALUE THE OUTPUT CHANGES STATE.
	**L	STATE 1 $\quad X < L$ (ENERGIZED OR ALARM STATE) STATE 2 $\quad X \geq L$		
	**HL	STATE 1 $\quad X < L$ (FIRST OUTPUT M_1 ENERGIZED OR ALARM STATE) STATE 2 $\quad L \leq X \leq H$ (BOTH OUTPUTS INACTIVE OR DE-ENERGIZED) STATE 3 $\quad X > H$ (SECOND OUTPUT M_2 ENERGIZED OR ALARM STATE)		

THE VARIABLES USED IN THE TABLE ARE:

b - ANALOG BIAS VALUE.

$\dfrac{d}{dt}$ - DERIVATIVE WITH RESPECT TO TIME.

H - AN ARBITRARY ANALOG HIGH LIMIT VALUE.

$\dfrac{1}{T_i}$ - INTEGRATING RATE.

L - AN ARBITRARY ANALOG LOW LIMIT VALUE.

M - ANALOG OUTPUT VARIABLE.

n - NUMBER OF ANALOG INPUTS OR VALUE OF EXPONENT.

t - TIME.

T_D - DERIVATIVE TIME.

X - ANALOG INPUT VARIABLE.

$X_1, X_2, X_3, \ldots, X_n$ - ANALOG INPUT VARIABLE (1 TO N IN NUMBER).

474

Instrument Line Symbols

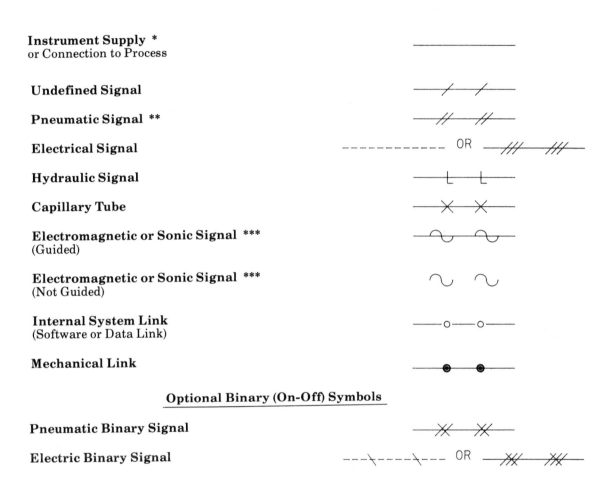

Instrument Supply *
or Connection to Process

Undefined Signal

Pneumatic Signal **

Electrical Signal   OR

Hydraulic Signal

Capillary Tube

Electromagnetic or Sonic Signal ***
(Guided)

Electromagnetic or Sonic Signal ***
(Not Guided)

Internal System Link
(Software or Data Link)

Mechanical Link

Optional Binary (On-Off) Symbols

Pneumatic Binary Signal

Electric Binary Signal   OR

NOTE: "Or" means user's choise. Consistency is recommended.

* The following abbreviations are suggested to denote the types of power supply. These designations may also be applied to purge fluid supplies.

    AS —Air Supply
        IA—Instrument Air
        PA—Plant Air
    ES —Electrical Supply
    GS —Gas Supply
    HS —Hydraulic Supply
    NS —Nitrogen Supply
    SS —Steam Supply
    WS—Water Supply

The supply level may be added to the instrument supply line, e.g., AS-100, a 100-psig air supply; ES-24DC, a 24-volt direct current power supply.

** The pneumatic signal symbol applies to a signal using any gas as the signal medium. If a gas other than air is used, the gas may be identified by a note on the signal symbol or otherwise.

*** Electromagnetic phenomena including heat, radio waves, nuclear radiation, and light.

Symbols for General Instruments or Functions

	PRIMARY LOCATION *** NORMALLY ACCESSIBLE TO OPERATOR	FIELD MOUNTED	AUXILIARY LOCATION *** NORMALLY ACCESSIBLE TO OPERATOR
DISCRETE INSTRUMENTS	1 * ⊖ IP1 **	2 ◯	3 ⊖
SHARED DISPLAY, SHARED CONTROL	4	5	6
COMPUTER FUNCTION	7	8	9
PROGRAMMABLE LOGIC CONTROL	10	11	12

* Symbol size may vary according to the user's needs and the type of document. A suggested square and circle size for large diagrams is shown above. Consistency is recommended.

** Abbreviations of the user's choise such as IP1 (Instrument Panel #1), IC2 (Instrument Console #2), CC3 (Computer Console #3), etc., may be used when it is necessary to specify instrument or function location.

*** Normally inaccessible or behind-the-panel devices or functions may be depicted by using the same symbols but with dashed horizontal bars, i.e.:

13	14	15
	6TE 2584-23 INSTRUMENT WITH LONG TAG NUMBER	INSTRUMENTS SHARING COMMON HOUSING *
16	17	18
PILOT LIGHT	C 12 PANEL MOUNTED PATCHBOARD POINT 12	P ** PURGE OR FLUSHING DEVICE
19	20	21
R ** RESET FOR LATCH-TYPE ACTUATOR	DIAPHRAGM SEAL	I ** *** UNDEFINED INTERLOCK LOGIC

* It is not mandatory to show a common housing.

** These diamonds are approximately half the size of the larger ones.

*** For specific logic symbols, see ANSI/ISA standard S5.2.

Symbols for Control Valve Bodies and Dampers

1 GENERAL SYMBOL	2 ANGLE	3 BUTTERFLY	4 ROTARY VALVE
5 THREE-WAY	6 FOUR-WAY	7 GLOBE	8
9 DIAPHRAGM	10	11 DAMPER OR LOUVER	12

Further information may be added adjacent to the body symbol either by note or code number.

Symbols for Actuators

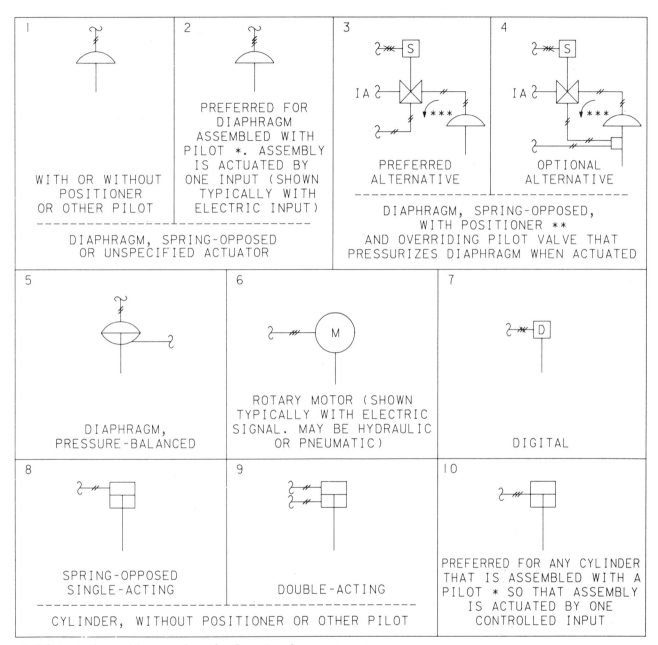

1 — WITH OR WITHOUT POSITIONER OR OTHER PILOT	2 — PREFERRED FOR DIAPHRAGM ASSEMBLED WITH PILOT *. ASSEMBLY IS ACTUATED BY ONE INPUT (SHOWN TYPICALLY WITH ELECTRIC INPUT)

DIAPHRAGM, SPRING-OPPOSED OR UNSPECIFIED ACTUATOR

3 — PREFERRED ALTERNATIVE

4 — OPTIONAL ALTERNATIVE

DIAPHRAGM, SPRING-OPPOSED, WITH POSITIONER **
AND OVERRIDING PILOT VALVE THAT
PRESSURIZES DIAPHRAGM WHEN ACTUATED

5 — DIAPHRAGM, PRESSURE-BALANCED

6 — ROTARY MOTOR (SHOWN TYPICALLY WITH ELECTRIC SIGNAL. MAY BE HYDRAULIC OR PNEUMATIC)

7 — DIGITAL

8 — SPRING-OPPOSED SINGLE-ACTING

9 — DOUBLE-ACTING

CYLINDER, WITHOUT POSITIONER OR OTHER PILOT

10 — PREFERRED FOR ANY CYLINDER THAT IS ASSEMBLED WITH A PILOT * SO THAT ASSEMBLY IS ACTUATED BY ONE CONTROLLED INPUT

* Pilot may be positioner, solenoid valve, signal converter, etc.

** The positioner need not be shown unless an intermediate device is on its output. The positioner tagging, ZC, need not be used even if the positioner is shown. The positioner symbol, a box drawn on the actuator shaft, is the same for all types of actuators. When the symbol is used, the type of instrument signal, i.e., pneumatic, electrical, etc., is drawn as appropriate. If the positioner symbol is used and there is no intermediate device on its output, then the positioner output signal need not be shown.

*** The arrow represents the path from a common to a fail-open port. It does not correspond necessarily to the direction of fluid flow.

Symbols for Symbols for Actuators

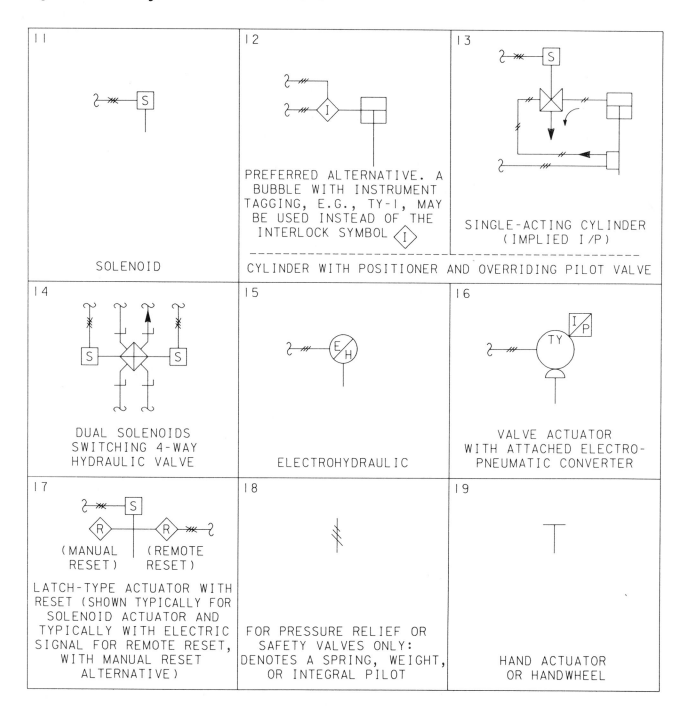

11	12	13
	PREFERRED ALTERNATIVE. A BUBBLE WITH INSTRUMENT TAGGING, E.G., TY-1, MAY BE USED INSTEAD OF THE INTERLOCK SYMBOL	SINGLE-ACTING CYLINDER (IMPLIED I/P)
SOLENOID	CYLINDER WITH POSITIONER AND OVERRIDING PILOT VALVE	

14	15	16
DUAL SOLENOIDS SWITCHING 4-WAY HYDRAULIC VALVE	ELECTROHYDRAULIC	VALVE ACTUATOR WITH ATTACHED ELECTRO-PNEUMATIC CONVERTER

17	18	19
(MANUAL RESET) (REMOTE RESET)		
LATCH-TYPE ACTUATOR WITH RESET (SHOWN TYPICALLY FOR SOLENOID ACTUATOR AND TYPICALLY WITH ELECTRIC SIGNAL FOR REMOTE RESET, WITH MANUAL RESET ALTERNATIVE)	FOR PRESSURE RELIEF OR SAFETY VALVES ONLY: DENOTES A SPRING, WEIGHT, OR INTEGRAL PILOT	HAND ACTUATOR OR HANDWHEEL

Symbols for Self-Actuated Regulators, Valves, and Other Devices

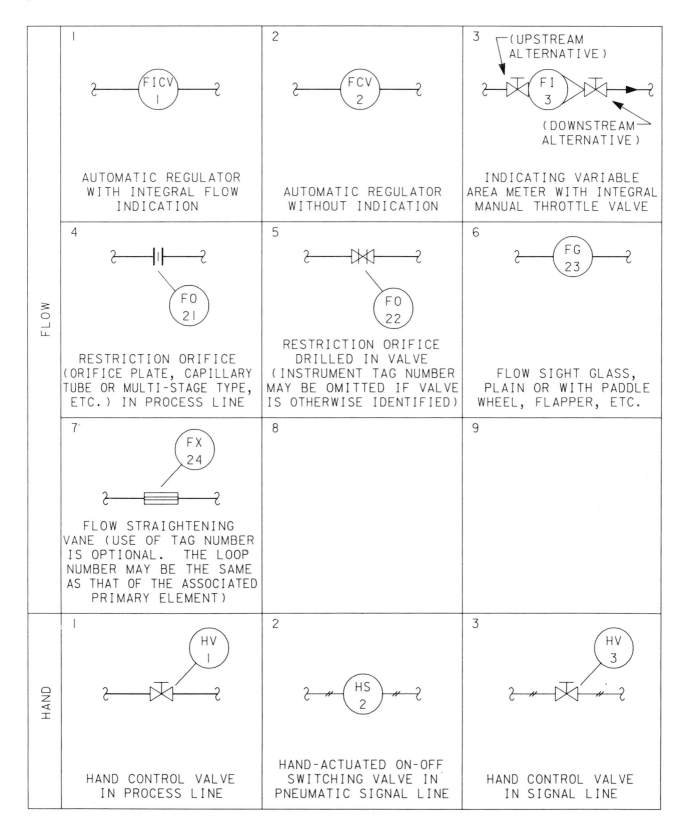

1 FICV 1 — AUTOMATIC REGULATOR WITH INTEGRAL FLOW INDICATION	**2** FCV 2 — AUTOMATIC REGULATOR WITHOUT INDICATION	**3** (UPSTREAM ALTERNATIVE) FI 3 (DOWNSTREAM ALTERNATIVE) — INDICATING VARIABLE AREA METER WITH INTEGRAL MANUAL THROTTLE VALVE
4 FO 21 — RESTRICTION ORIFICE (ORIFICE PLATE, CAPILLARY TUBE OR MULTI-STAGE TYPE, ETC.) IN PROCESS LINE	**5** FO 22 — RESTRICTION ORIFICE DRILLED IN VALVE (INSTRUMENT TAG NUMBER MAY BE OMITTED IF VALVE IS OTHERWISE IDENTIFIED)	**6** FG 23 — FLOW SIGHT GLASS, PLAIN OR WITH PADDLE WHEEL, FLAPPER, ETC.
7 FX 24 — FLOW STRAIGHTENING VANE (USE OF TAG NUMBER IS OPTIONAL. THE LOOP NUMBER MAY BE THE SAME AS THAT OF THE ASSOCIATED PRIMARY ELEMENT)	**8**	**9**
1 HV 1 — HAND CONTROL VALVE IN PROCESS LINE	**2** HS 2 — HAND-ACTUATED ON-OFF SWITCHING VALVE IN PNEUMATIC SIGNAL LINE	**3** HV 3 — HAND CONTROL VALVE IN SIGNAL LINE

FLOW

HAND

481

Symbols for Self-Actuated Regulators, Valves, and Other Devices

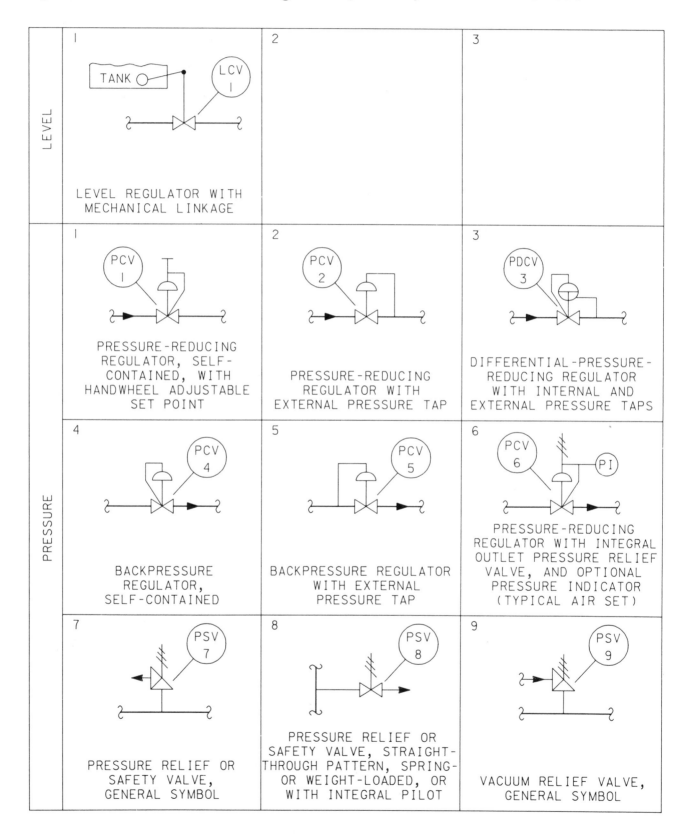

LEVEL

1 — LEVEL REGULATOR WITH MECHANICAL LINKAGE

PRESSURE

1 — PRESSURE-REDUCING REGULATOR, SELF-CONTAINED, WITH HANDWHEEL ADJUSTABLE SET POINT

2 — PRESSURE-REDUCING REGULATOR WITH EXTERNAL PRESSURE TAP

3 — DIFFERENTIAL-PRESSURE-REDUCING REGULATOR WITH INTERNAL AND EXTERNAL PRESSURE TAPS

4 — BACKPRESSURE REGULATOR, SELF-CONTAINED

5 — BACKPRESSURE REGULATOR WITH EXTERNAL PRESSURE TAP

6 — PRESSURE-REDUCING REGULATOR WITH INTEGRAL OUTLET PRESSURE RELIEF VALVE, AND OPTIONAL PRESSURE INDICATOR (TYPICAL AIR SET)

7 — PRESSURE RELIEF OR SAFETY VALVE, GENERAL SYMBOL

8 — PRESSURE RELIEF OR SAFETY VALVE, STRAIGHT-THROUGH PATTERN, SPRING- OR WEIGHT-LOADED, OR WITH INTEGRAL PILOT

9 — VACUUM RELIEF VALVE, GENERAL SYMBOL

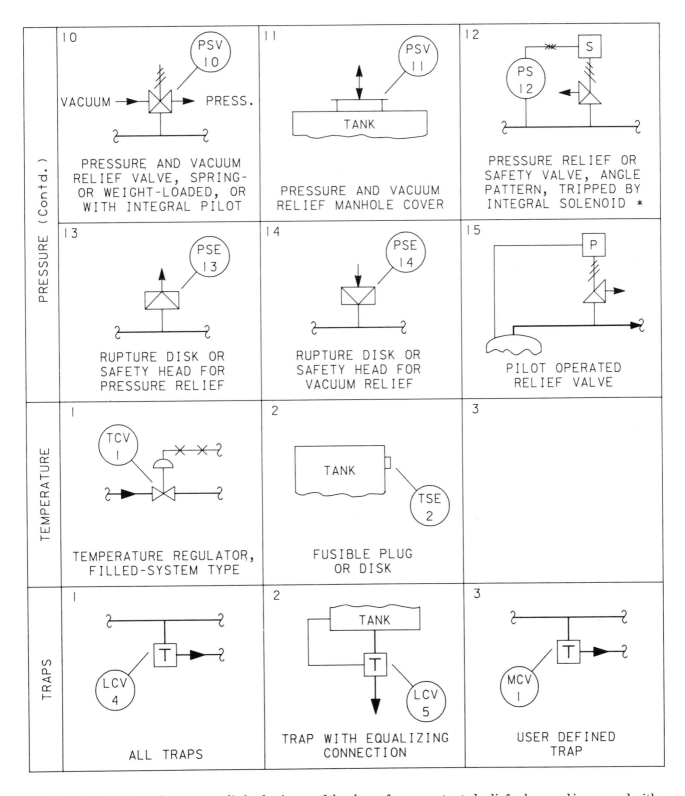

PRESSURE (Cont'd.)

10 PRESSURE AND VACUUM RELIEF VALVE, SPRING- OR WEIGHT-LOADED, OR WITH INTEGRAL PILOT

11 PRESSURE AND VACUUM RELIEF MANHOLE COVER

12 PRESSURE RELIEF OR SAFETY VALVE, ANGLE PATTERN, TRIPPED BY INTEGRAL SOLENOID *

13 RUPTURE DISK OR SAFETY HEAD FOR PRESSURE RELIEF

14 RUPTURE DISK OR SAFETY HEAD FOR VACUUM RELIEF

15 PILOT OPERATED RELIEF VALVE

TEMPERATURE

1 TEMPERATURE REGULATOR, FILLED-SYSTEM TYPE

2 FUSIBLE PLUG OR DISK

3

TRAPS

1 ALL TRAPS

2 TRAP WITH EQUALIZING CONNECTION

3 USER DEFINED TRAP

* The solenoid-tripped pressure relief valve is one of the class of power-actuated relief valves and is grouped with the other types of relief valves even though it is not entirely a self-actuated device.

Symbols for Actuator Action in Event of Actuator Power Failure

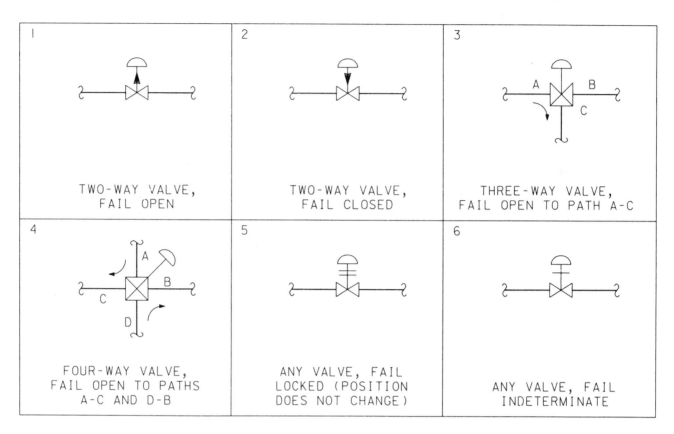

The failure modes indicated are those commonly defined by the term, "self-position". As an alternative to the arrows and bars, the following abbreviations may be employed:

FO—Fail Open
FC—Fail Closed
FL—Fail Locked (Last Position)
FI— Fail Intermediate

Symbols for Primary Elements

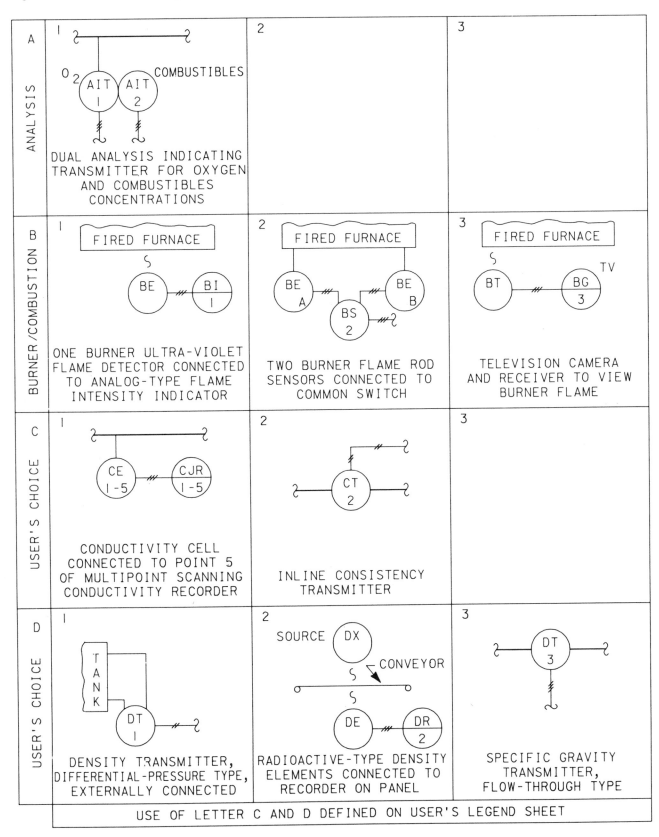

	1	2	3
A ANALYSIS	DUAL ANALYSIS INDICATING TRANSMITTER FOR OXYGEN AND COMBUSTIBLES CONCENTRATIONS		
B BURNER/COMBUSTION	ONE BURNER ULTRA-VIOLET FLAME DETECTOR CONNECTED TO ANALOG-TYPE FLAME INTENSITY INDICATOR	TWO BURNER FLAME ROD SENSORS CONNECTED TO COMMON SWITCH	TELEVISION CAMERA AND RECEIVER TO VIEW BURNER FLAME
C USER'S CHOICE	CONDUCTIVITY CELL CONNECTED TO POINT 5 OF MULTIPOINT SCANNING CONDUCTIVITY RECORDER	INLINE CONSISTENCY TRANSMITTER	
D USER'S CHOICE	DENSITY TRANSMITTER, DIFFERENTIAL-PRESSURE TYPE, EXTERNALLY CONNECTED	RADIOACTIVE-TYPE DENSITY ELEMENTS CONNECTED TO RECORDER ON PANEL	SPECIFIC GRAVITY TRANSMITTER, FLOW-THROUGH TYPE

USE OF LETTER C AND D DEFINED ON USER'S LEGEND SHEET

Symbols for Primary Elements

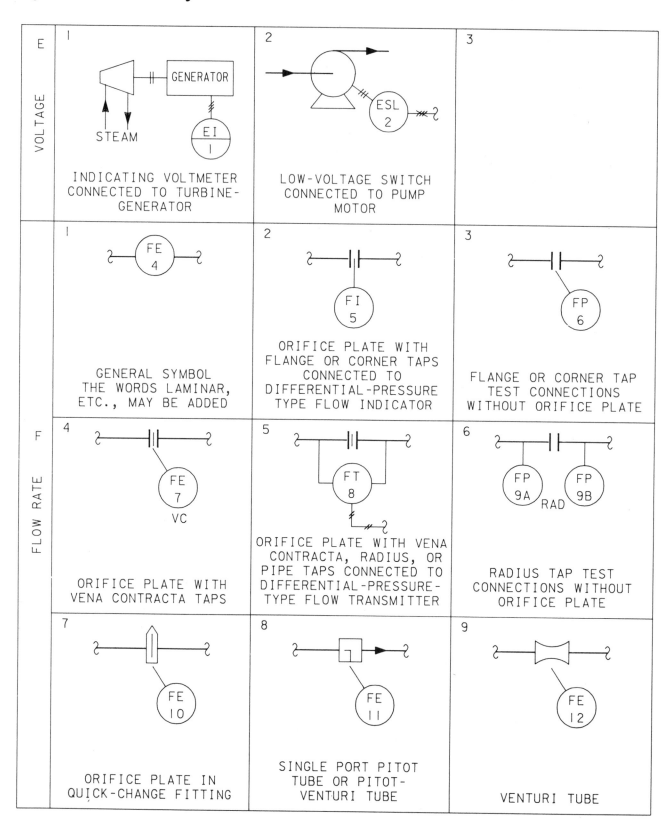

	1 INDICATING VOLTMETER CONNECTED TO TURBINE-GENERATOR	2 LOW-VOLTAGE SWITCH CONNECTED TO PUMP MOTOR	3

E — VOLTAGE

1 GENERAL SYMBOL THE WORDS LAMINAR, ETC., MAY BE ADDED	2 ORIFICE PLATE WITH FLANGE OR CORNER TAPS CONNECTED TO DIFFERENTIAL-PRESSURE TYPE FLOW INDICATOR	3 FLANGE OR CORNER TAP TEST CONNECTIONS WITHOUT ORIFICE PLATE
4 ORIFICE PLATE WITH VENA CONTRACTA TAPS	5 ORIFICE PLATE WITH VENA CONTRACTA, RADIUS, OR PIPE TAPS CONNECTED TO DIFFERENTIAL-PRESSURE-TYPE FLOW TRANSMITTER	6 RADIUS TAP TEST CONNECTIONS WITHOUT ORIFICE PLATE
7 ORIFICE PLATE IN QUICK-CHANGE FITTING	8 SINGLE PORT PITOT TUBE OR PITOT-VENTURI TUBE	9 VENTURI TUBE

F — FLOW RATE

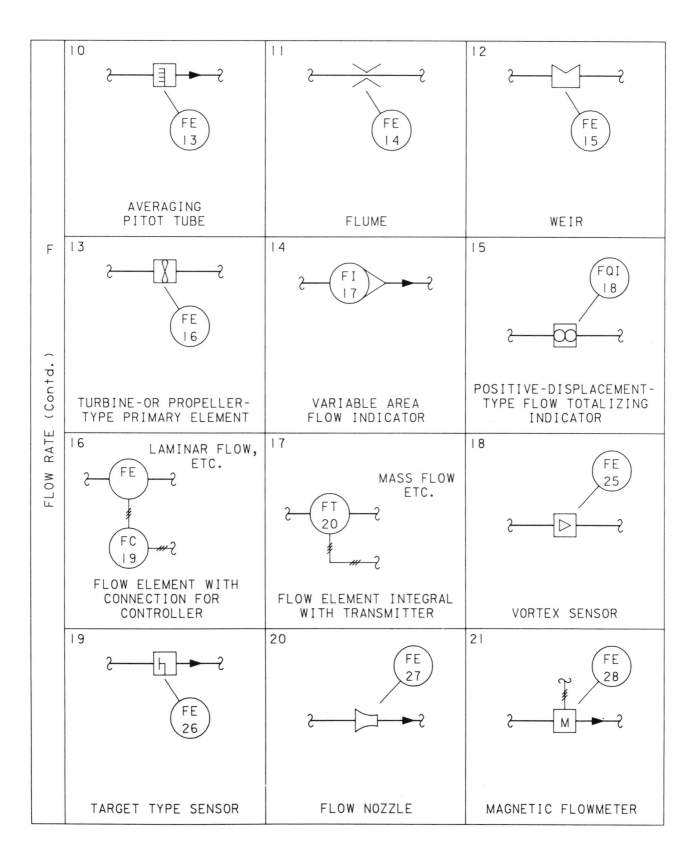

FLOW RATE (Cont'd.)

10 AVERAGING PITOT TUBE	11 FLUME	12 WEIR
13 TURBINE-OR PROPELLER-TYPE PRIMARY ELEMENT	14 VARIABLE AREA FLOW INDICATOR	15 POSITIVE-DISPLACEMENT-TYPE FLOW TOTALIZING INDICATOR
16 LAMINAR FLOW, ETC. FLOW ELEMENT WITH CONNECTION FOR CONTROLLER	17 MASS FLOW ETC. FLOW ELEMENT INTEGRAL WITH TRANSMITTER	18 VORTEX SENSOR
19 TARGET TYPE SENSOR	20 FLOW NOZZLE	21 MAGNETIC FLOWMETER

Symbols for Primary Elements

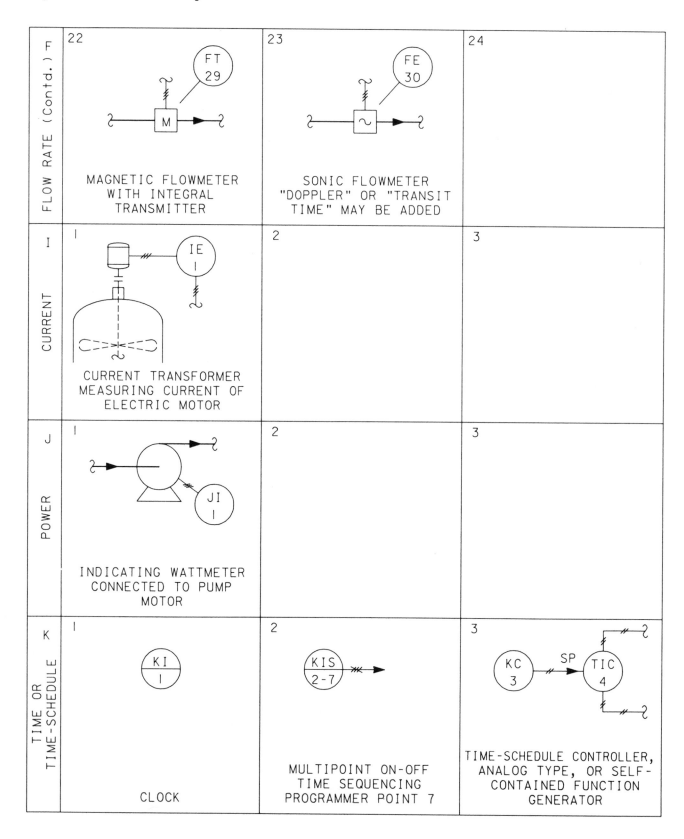

	22 MAGNETIC FLOWMETER WITH INTEGRAL TRANSMITTER	23 SONIC FLOWMETER "DOPPLER" OR "TRANSIT TIME" MAY BE ADDED	24
F FLOW RATE (Cont'd.)			
I CURRENT	1 CURRENT TRANSFORMER MEASURING CURRENT OF ELECTRIC MOTOR	2	3
J POWER	1 INDICATING WATTMETER CONNECTED TO PUMP MOTOR	2	3
K TIME OR TIME-SCHEDULE	1 CLOCK	2 MULTIPOINT ON-OFF TIME SEQUENCING PROGRAMMER POINT 7	3 TIME-SCHEDULE CONTROLLER, ANALOG TYPE, OR SELF-CONTAINED FUNCTION GENERATOR

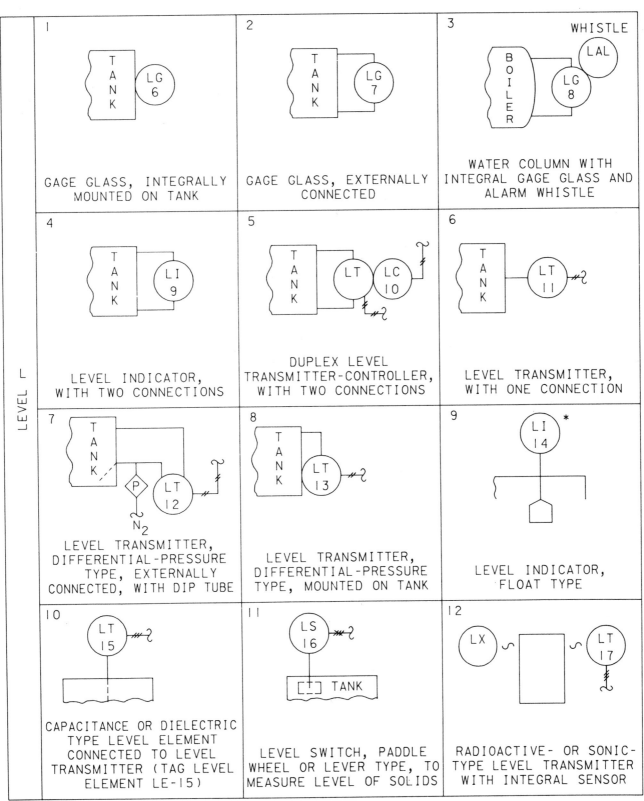

1	2	3
TANK LG 6	TANK LG 7	WHISTLE — BOILER LG 8 LAL
GAGE GLASS, INTEGRALLY MOUNTED ON TANK	GAGE GLASS, EXTERNALLY CONNECTED	WATER COLUMN WITH INTEGRAL GAGE GLASS AND ALARM WHISTLE

4	5	6
TANK LI 9	TANK LT LC 10	TANK LT 11
LEVEL INDICATOR, WITH TWO CONNECTIONS	DUPLEX LEVEL TRANSMITTER-CONTROLLER, WITH TWO CONNECTIONS	LEVEL TRANSMITTER, WITH ONE CONNECTION

7	8	9
TANK P LT 12 N2	TANK LT 13	LI 14 *
LEVEL TRANSMITTER, DIFFERENTIAL-PRESSURE TYPE, EXTERNALLY CONNECTED, WITH DIP TUBE	LEVEL TRANSMITTER, DIFFERENTIAL-PRESSURE TYPE, MOUNTED ON TANK	LEVEL INDICATOR, FLOAT TYPE

10	11	12
LT 15	LS 16 [·] TANK	LX LT 17
CAPACITANCE OR DIELECTRIC TYPE LEVEL ELEMENT CONNECTED TO LEVEL TRANSMITTER (TAG LEVEL ELEMENT LE-15)	LEVEL SWITCH, PADDLE WHEEL OR LEVER TYPE, TO MEASURE LEVEL OF SOLIDS	RADIOACTIVE- OR SONIC- TYPE LEVEL TRANSMITTER WITH INTEGRAL SENSOR

LEVEL L

* Notations such as "mounted at grade" may be added.

Symbols for Primary Elements

L **LEVEL (Cont'd.)**	13 REMOTE VIEWING OF GAGE GLASS BY USE OF TELEVISION	14 LEVEL GLASS WITH ILLUMINATOR	15

M **USER'S CHOICE**	1 MOISTURE RECORDER (IF THERE IS A SEPARATE PRIMARY ELEMENT, IT SHOULD BE TAGGED ME-2	2 SELF-CONTAINED HUMIDITY CONTROLLER IN ROOM	

USE OF LETTER M TO BE DEFINED IN USER'S LEGEND

P **PRESSURE OR VACUUM**	1 PRESSURE INDICATOR, DIRECT-CONNECTED	2 WITH PRESSURE LEAD LINE	3 LINE-MOUNTED
		PRESSURE INDICATOR CONNECTED TO DIAPHRAGM SEAL WITH FILLED SYSTEM	
	4 PRESSURE ELEMENT, STRAIN-GAGE TYPE, CONNECTED TO PRESSURE INDICATING TRANSMITTER (TAG STRAIN GAGE PE-19)	5	6

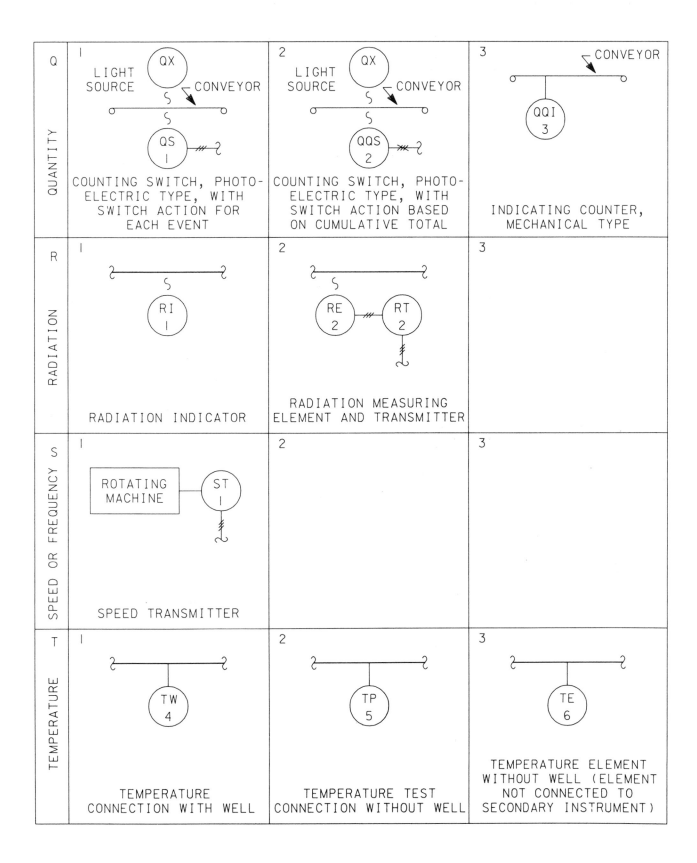

Q QUANTITY	1 LIGHT SOURCE QX CONVEYOR QS 1 COUNTING SWITCH, PHOTO-ELECTRIC TYPE, WITH SWITCH ACTION FOR EACH EVENT	2 LIGHT SOURCE QX CONVEYOR QQS 2 COUNTING SWITCH, PHOTO-ELECTRIC TYPE, WITH SWITCH ACTION BASED ON CUMULATIVE TOTAL	3 CONVEYOR QQI 3 INDICATING COUNTER, MECHANICAL TYPE
R RADIATION	1 RI 1 RADIATION INDICATOR	2 RE 2 RT 2 RADIATION MEASURING ELEMENT AND TRANSMITTER	3
S SPEED OR FREQUENCY	1 ROTATING MACHINE ST 1 SPEED TRANSMITTER	2	3
T TEMPERATURE	1 TW 4 TEMPERATURE CONNECTION WITH WELL	2 TP 5 TEMPERATURE TEST CONNECTION WITHOUT WELL	3 TE 6 TEMPERATURE ELEMENT WITHOUT WELL (ELEMENT NOT CONNECTED TO SECONDARY INSTRUMENT)

Symbols for Primary Elements

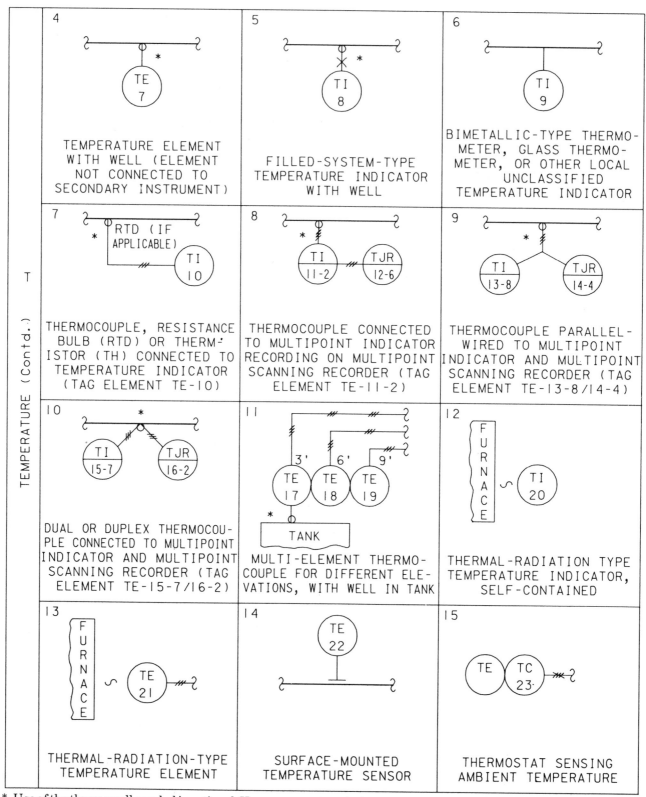

4 TE 7	5 TI 8	6 TI 9
TEMPERATURE ELEMENT WITH WELL (ELEMENT NOT CONNECTED TO SECONDARY INSTRUMENT)	FILLED-SYSTEM-TYPE TEMPERATURE INDICATOR WITH WELL	BIMETALLIC-TYPE THERMO-METER, GLASS THERMO-METER, OR OTHER LOCAL UNCLASSIFIED TEMPERATURE INDICATOR
7 RTD (IF APPLICABLE) TI 10	8 TI 11-2 TJR 12-6	9 TI 13-8 TJR 14-4
THERMOCOUPLE, RESISTANCE BULB (RTD) OR THERM-ISTOR (TH) CONNECTED TO TEMPERATURE INDICATOR (TAG ELEMENT TE-10)	THERMOCOUPLE CONNECTED TO MULTIPOINT INDICATOR RECORDING ON MULTIPOINT SCANNING RECORDER (TAG ELEMENT TE-11-2)	THERMOCOUPLE PARALLEL-WIRED TO MULTIPOINT INDICATOR AND MULTIPOINT SCANNING RECORDER (TAG ELEMENT TE-13-8/14-4)
10 TI 15-7 TJR 16-2	11 3' 6' 9' TE 17 TE 18 TE 19 TANK	12 FURNACE TI 20
DUAL OR DUPLEX THERMOCOU-PLE CONNECTED TO MULTIPOINT INDICATOR AND MULTIPOINT SCANNING RECORDER (TAG ELEMENT TE-15-7/16-2)	MULTI-ELEMENT THERMO-COUPLE FOR DIFFERENT ELE-VATIONS, WITH WELL IN TANK	THERMAL-RADIATION TYPE TEMPERATURE INDICATOR, SELF-CONTAINED
13 FURNACE TE 21	14 TE 22	15 TE TC 23
THERMAL-RADIATION-TYPE TEMPERATURE ELEMENT	SURFACE-MOUNTED TEMPERATURE SENSOR	THERMOSTAT SENSING AMBIENT TEMPERATURE

TEMPERATURE (Contd.)   T

* Use of the thermowell symbol is optional. However, use or omission of the symbol should be consistent throughout a project.

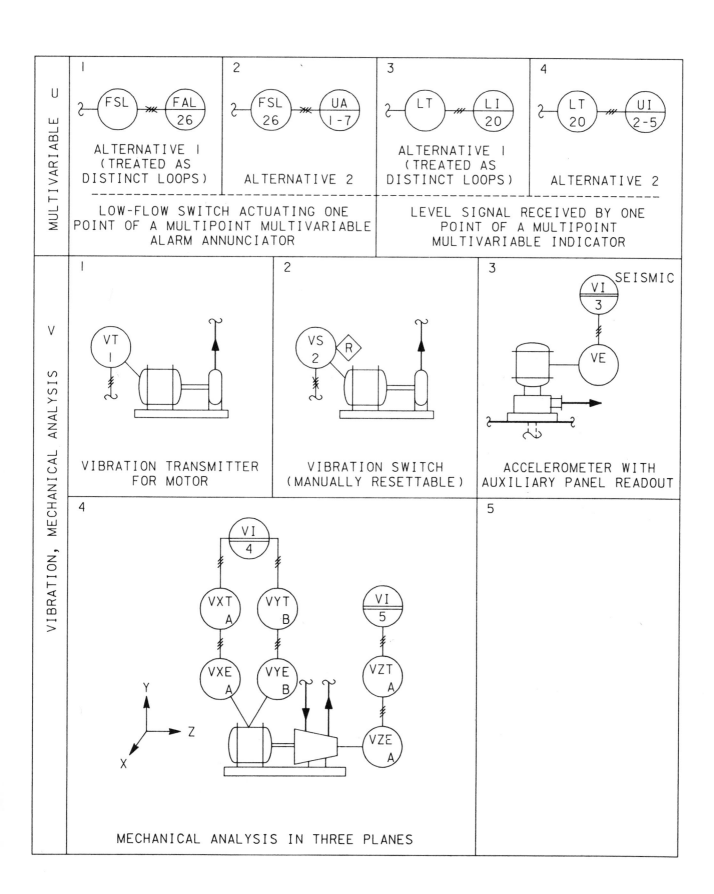

MULTIVARIABLE U

1. ALTERNATIVE I
(TREATED AS
DISTINCT LOOPS)

2. ALTERNATIVE 2

LOW-FLOW SWITCH ACTUATING ONE
POINT OF A MULTIPOINT MULTIVARIABLE
ALARM ANNUNCIATOR

3. ALTERNATIVE I
(TREATED AS
DISTINCT LOOPS)

4. ALTERNATIVE 2

LEVEL SIGNAL RECEIVED BY ONE
POINT OF A MULTIPOINT
MULTIVARIABLE INDICATOR

VIBRATION, MECHANICAL ANALYSIS V

1. VIBRATION TRANSMITTER
FOR MOTOR

2. VIBRATION SWITCH
(MANUALLY RESETTABLE)

3. ACCELEROMETER WITH
AUXILIARY PANEL READOUT

4. MECHANICAL ANALYSIS IN THREE PLANES

5.

Symbols for Primary Elements

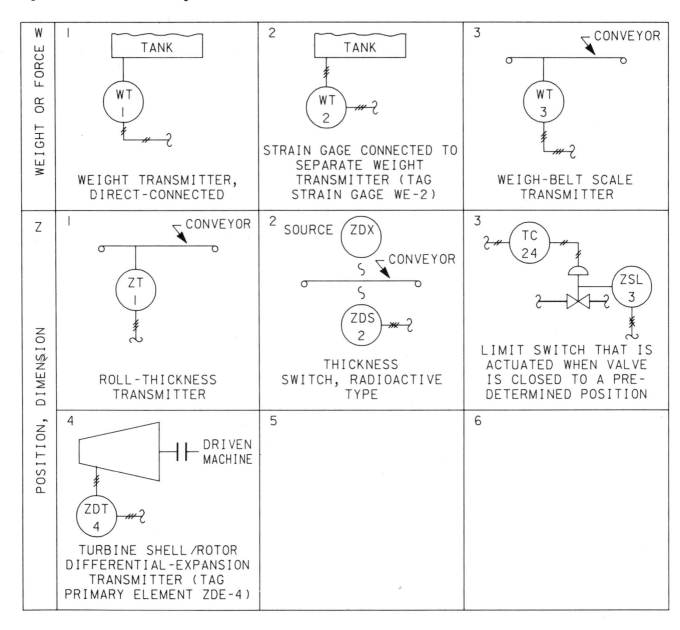

	1	2	3
W WEIGHT OR FORCE	TANK WT 1 WEIGHT TRANSMITTER, DIRECT-CONNECTED	TANK WT 2 STRAIN GAGE CONNECTED TO SEPARATE WEIGHT TRANSMITTER (TAG STRAIN GAGE WE-2)	CONVEYOR WT 3 WEIGH-BELT SCALE TRANSMITTER

	1	2	3
Z POSITION, DIMENSION	CONVEYOR ZT 1 ROLL-THICKNESS TRANSMITTER	SOURCE ZDX CONVEYOR ZDS 2 THICKNESS SWITCH, RADIOACTIVE TYPE	TC 24 ZSL 3 LIMIT SWITCH THAT IS ACTUATED WHEN VALVE IS CLOSED TO A PRE-DETERMINED POSITION
	4	5	6
	DRIVEN MACHINE ZDT 4 TURBINE SHELL/ROTOR DIFFERENTIAL-EXPANSION TRANSMITTER (TAG PRIMARY ELEMENT ZDE-4)		

Examples of Function Symbols

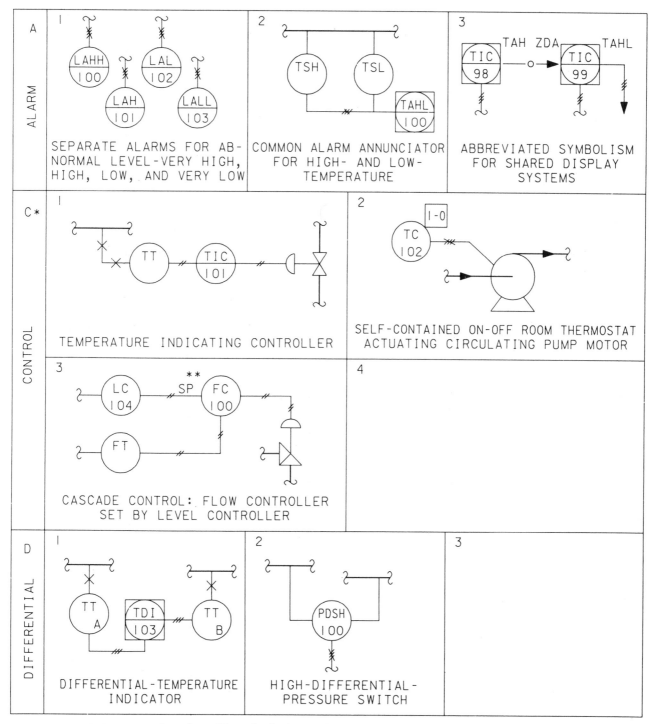

A ALARM	1 SEPARATE ALARMS FOR ABNORMAL LEVEL-VERY HIGH, HIGH, LOW, AND VERY LOW	2 COMMON ALARM ANNUNCIATOR FOR HIGH- AND LOW-TEMPERATURE	3 ABBREVIATED SYMBOLISM FOR SHARED DISPLAY SYSTEMS

LAHH 100   LAL 102   LAH 101   LALL 103

TSH   TSL   TAHL 100

TIC 98   TAH ZDA   TAHL   TIC 99

C* CONTROL

1 TEMPERATURE INDICATING CONTROLLER

TT   TIC 101

2 SELF-CONTAINED ON-OFF ROOM THERMOSTAT ACTUATING CIRCULATING PUMP MOTOR

TC 102   1-0

3 CASCADE CONTROL: FLOW CONTROLLER SET BY LEVEL CONTROLLER

LC 104   ** SP   FC 100   FT

4

D DIFFERENTIAL

1 DIFFERENTIAL-TEMPERATURE INDICATOR

TT A   TDI 103   TT B

2 HIGH-DIFFERENTIAL-PRESSURE SWITCH

PDSH 100

3

* It is expected that control modes will not be designated on a diagram. However, designations may be used outside of the controller symbol, if desired in combinations such as [%], [∫], [1-0].

** A controller is understood to have integral manual set-point adjustment unless means of remote adjustment is indicated. The remote set-point designation is SP.

495

Symbols for Diagramming Binary Logic

FUNCTION	SYMBOL	DEFINITION	EXAMPLE
4.1 INPUT	Statement of Input —| |— Alternatively: (circle with arrow) Statement of Input Initiating instrument or device number, if known	An input to the logic sequence	The start position of a hand switch HS-1, is actuated to provide an input to start a conveyor. Alternative diagrams: a) HS-1 Start Conveyor Manually —| |— b) (HS 1) Start Conveyor Manually —| |—
4.2 OUTPUT	—| Statement of Output Alternatively: —| Statement of Output (circle with arrow) Operated instrument or device number, if known	An output from the logic sequence.	An output from the logic sequence commands valve HV-2 to open. Alternative diagrams: a) —| Open Valve HV-2 b) —| Open Valve (HV 2)
4.3 AND **BASIC**	A| B|— [A] —|D C|	Logic output D exists if and only if all logic inputs A, B, and C exist.	Operate pump if suction tank level is high and discharge valve is open. Tank Level High |— Valve Open |— [A] —| Operate Pump
4.4 OR **BASIC**	A| B|— [OR] —|D C|	Logic output D exists if and only if one or more of logic inputs A, B, and C exist.	Stop compressor if cooling water pressure is low or bearing temperature is high. Water Pressure Low |— Bearing Temperature High |— [OR] —| Stop Compressor

FUNCTION	SYMBOL	DEFINITION	EXAMPLE
4.5 QUALIFIED OR	*Internal details represent numerical quantities (see "Definition").	Logic output D exists if and only if a specified number of logic inputs A, B, and C exist. Mathematical symbols, including the following, shall be used, as appropriate, in specifying the number: a. $=$ equal to b. $\neq$ not equal to c. $<$ less than d. $>$ greater than e. $\not<$ not less than f. $\not>$ not greater than g. $\leq$ less than or equal to [equivalent to f] h. $\geq$ greater than or equal to [equivalent to e]	a) Operate mixer if two, and only two, bins are in service. Red Bin In Service — Blue Bin In Service — White Bin In Service — Yellow Bin In Service — $=2$ — Operate Mixer b) Stop reaction if at least two safety devices call for stop. Device #1 Actuated — Device #2 Actuated — Device #3 Actuated — Device #4 Actuated — Device #5 Actuated — $\not<2$ — Stop Reaction c) Operate materials feeder if at least one and no more than two mills are in service. Mill #1 In Service — Mill #2 In Service — Mill #3 In Service — ≥1 $\not>2$ — Operate Feeder
4.6 NOT	**The NOT symbol may be drawn tangent to an adjacent logic symbol.**	Logic output B exists if and only if logic input A does not exist.	Shut off fuel gas if burners no. 1 and no. 2 are not on. Burner No. 1 On — Burner No. 2 On — A — Shut Off Fuel Gas Some Alternatives: Burner No. 1 On — Burner No. 2 On — A — Shut Off Fuel Gas Burner No. 1 On — Burner No. 2 On — OR — Shut Off Fuel Gas

BASIC

FUNCTION	SYMBOL	DEFINITION	EXAMPLE
4.7 MEMORY (Flip-Flop) BASIC	**a)** A ─ **S** ─ C B ─ **R** ─ D* *Output D shall not be shown if it is not used.	**S represents *set memory* and R represents *reset memory*.** **Logic output C exists as soon as logic input A exists. C continues to exist, regardless of the subsequent state of A, until the memory is reset, i.e., terminated by logic input B existing. C remains terminated regardless of the subsequent state of B, until A causes the memory to be set.** **Logic output D, if used, exists when C does not exist, and D does not exist when C exists.**	If tank pressure becomes high, vent tank and continue venting, regardless of pressure, until venting is stopped by manual actuation of hand switch, HS-1, provided that the pressure is not high. If the venting is stopped, a compressor may be started. 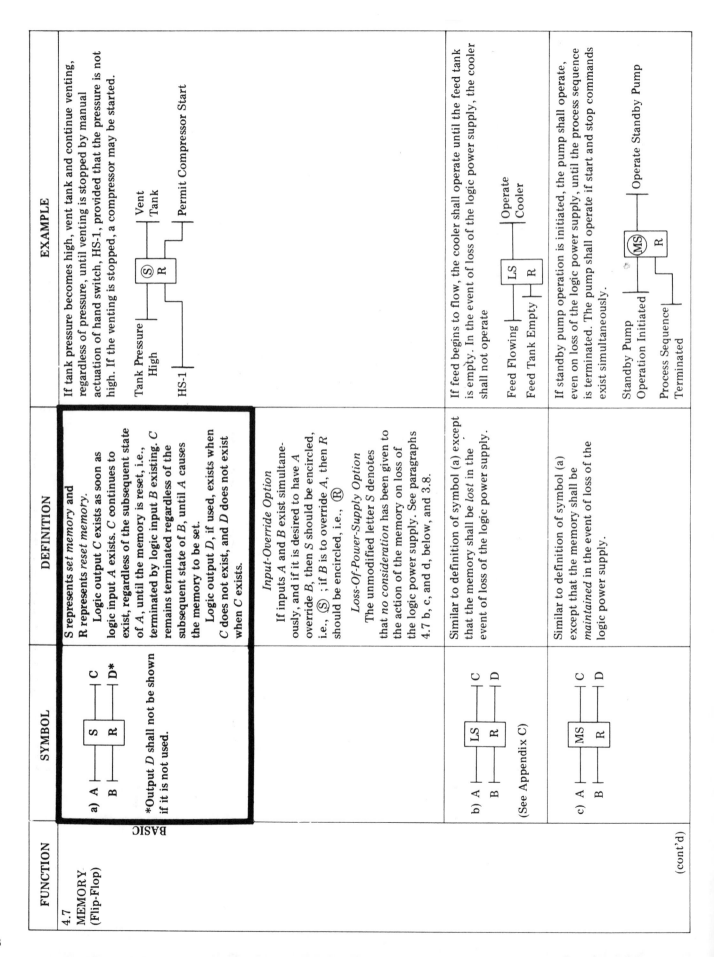 Tank Pressure High ─ S ─ Vent Tank HS-1 ─ R ─ Permit Compressor Start
		Input-Override Option If inputs A and B exist simultaneously, and if it is desired to have A override B, then S should be encircled, i.e., Ⓢ ; if B is to override A, then R should be encircled, i.e., Ⓡ	
		Loss-Of-Power-Supply Option The unmodified letter S denotes that *no consideration* has been given to the action of the memory on loss of the logic power supply. See paragraphs 4.7 b, c, and d, below, and 3.8.	
	b) A ─ **LS** ─ C B ─ **R** ─ D (See Appendix C)	Similar to definition of symbol (a) except that the memory shall be *lost* in the event of loss of the logic power supply.	If feed begins to flow, the cooler shall operate until the feed tank is empty. In the event of loss of the logic power supply, the cooler shall not operate Feed Flowing ─ LS ─ Operate Cooler Feed Tank Empty ─ R ─
	c) A ─ **MS** ─ C B ─ **R** ─ D	Similar to definition of symbol (a) except that the memory shall be *maintained* in the event of loss of the logic power supply.	If standby pump operation is initiated, the pump shall operate, even on loss of the logic power supply, until the process sequence is terminated. The pump shall operate if start and stop commands exist simultaneously. Standby Pump Operation Initiated ─ MS ─ Operate Standby Pump Process Sequence Terminated ─ R ─

(cont'd)

498

FUNCTION	SYMBOL	DEFINITION	EXAMPLE
4.7 (cont'd)	d) A —[NS]— C B —[R]— D	Similar to definition of symbol (a) except that *after consideration* it is deemed *not significant*, so far as the process is concerned, whether the memory is maintained or lost in the event of loss of power supply.	If reservoir level is low, operate fill pump until either level is high or water quality is unsatisfactory. It is not significant to the process what happens to the pump on loss of the logic supply. If start and stop commands are simultaneous, the pump shall stop. Reservoir Level Low —[NS]—[®]— Operate Fill Pump Reservoir Level High —[OR] Water Quality Unsatisfactory —
4.8 TIME ELEMENT	a) A —[*]— B *For functional details, see the following (also see Section 3.9):	Logic output *B* exists with a time relationship to logic input *A* as specified.	
BASIC	b) A —[DI t]— B (<u>D</u>elay <u>I</u>nitiation of output)	**The continuous existence of logic input *A* for time *t* causes logic output *B* to exist when *t* expires. *B* terminates when *A* terminates.**	If reactor temperature exceeds a high limit continuously for 10 seconds, block catalyst flow. Resume flow when temperature does not exceed the limit. Reactor Temp. High —[DI 10 s]— Block Catalyst Flow
BASIC	c) A —[DT t]— B (<u>D</u>elay <u>T</u>ermination of output)	**The existence of logic input *A* causes logic output *B* to exist immediately. *B* terminates when *A* has terminated and has not again existed for time *t*.**	If system pressure falls below a low limit, operate compressor at once. Stop the compressor when pressure is not low continuously for one minute. System Press. Low —[DT 1 min.]— Operate Compressor
BASIC	d) A —[PO t]— B (<u>P</u>ulse <u>O</u>utput)	**The existence of logic input *A*, regardless of its subsequent state, causes logic output *B* to exist immediately. *B* exists for time *t* and then terminates.**	If vessel purge fails for any period of time, operate evacuation pump for 3 minutes and then stop the pump. Vessel Purge Fails —[PO 3 min.]— Operate Evacuation Pump

(cont'd)

FUNCTION	SYMBOL	DEFINITION	EXAMPLE
4.8 (cont'd)	A generalized method for diagramming all time functions is outlined as follows. The symbols that are defined are intended to be illustrative but are not all-inclusive.		
	e1) A ⊢ ... ⊢ B	Input logic state exists. Input logic state does not exist. Output logic state exists. Output logic state does not exist. The time at which the logic input A is initiated is represented by the left-hand edge of the box. Passage of time is from left to right and is usually shown unscaled. The logic output B always begins and ends in the same state within the time-element box. More than one output may be shown, if required.	
	e2) A ⊢ ... ⊢ B	The timing of logic may be applied to either the existence state or the non-existence state, as applicable. Output logic state exists. Output logic state does not exist.	
	f1) A ⊢ t_1 ⊢ B	The continuous existence of logic input A for time t_1 causes logic output B to exist when t_1 expires. B terminates when A terminates.	Avoid nuisance alarms on high level by actuating alarm only if level remains high continuously for 0.5 second. The alarm signal terminates when there is no high level. Level Is High — $\boxed{s\ 0.5}$ — Actuate Alarm
	f2) A ⊢ t_1 — t_2 ⊢ B	The continuous existence of logic input A for time t_1 causes logic output B to exist when t_1 expires. B terminates when A has been terminated continuously for time t_2.	Purge immediately with inert gas when combustibles concentration is high. Stop the purge when concentration is not high continuously for 5 minutes. Combustibles Concentration Is High — $\boxed{0\ \ 5\ min.}$ — Purge With Inert Gas

(cont'd)

FUNCTION	SYMBOL	DEFINITION	EXAMPLE
4.8 (cont'd)	f3) A ⊣ $\|t_3$ t_4 ⊢ B	The termination of logic input A and its continuous non-existence for time t_3 cause logic output B to exist when t_3 expires. B terminates when either (1) B has existed for time t_4, or (2) A again exists, whichever occurs first.	Steam is turned on for 15 minutes beginning 6 minutes after agitator has stopped except that the steam shall be turned off if the agitator restarts. Agitator Operating ⊣ 6 min. 15 min. ⊢ Steam On
	f4) A ⊣ t_1 t_4 ⊢ B	The existence of logic input A, regardless of its subsequent state, causes logic output B to exist when time t_1 expires. B exists for time t_4 and then terminates. *	If pressure dips to low value momentarily, block modulating control of turbine immediately, maintain for 1½ minutes, then release turbine to modulating control. Pressure Low ⊣ 0 min. 1½ min. ⊢ Turbine Modulating Control Blocked
	f5) A ⊣ t_1 t_4 ⊢ B	The continuous existence of logic input A for time t_1 causes logic output B to exist when t_1 expires. B exists for time t_4, regardless of the state of A, and then terminates. *	If pH is low continuously for ½ minute, add caustic for 3 minutes. pH Low ⊣ ½ min. 3 min. ⊢ Add Caustic
	f6) A ⊣ t_1 t_4 ⊢ B	The continuous existence of logic input A for time t_1 causes logic output B to exist when t_1 expires. B terminates when either (1) B has existed for time t_4, or (2) A terminates, whichever occurs first. *	If temperature is normal continuously for 5 minutes, add reagent for 2 minutes except that reagent shall not be added if temperature is abnormal. Temperature Normal ⊣ 5 min. 2 min. ⊢ Add Reagant
	*For symbols f4, f5, and f6, the action of logic output B depends on how long logic input A is in continuous existence, up to the line break for A. Beyond the break in A, the state of A is not significant to the completion of the B sequence. If it is desired to have a B time segment, e.g., t_1, go to completion only if A exists continuously, then A must be drawn beyond that segment. If A is drawn past the beginning but not beyond the end of a time segment, then the segment will be initiated and go to completion regardless of whether A exists only momentarily or longer.		
4.9 SPECIAL	A ⊣ ⟋ ⊢ B Statement of Special Requirements	Logic output B exists with a relationship to logic input A as specified in the statement of special requirements. The statement may cover a logic function not otherwise specified in this standard or a logic system that is further defined elsewhere.	

Symbols for Fittings

1 PURPOSE

The purpose of this recommended practice is to aid in the proper specification and application of instrument tube fittings by standardizing nomenclature. Proper use of this recommended practice should result in data that: More accurately describes a particular fitting for all interested parties; permits better communications between manufacturer, vendor, purchaser, and ultimate user; provides standardized nomenclature to be used by drafting, billing, purchasing, and stores groups; makes more clear the use of standardized drafting symbols and abbreviations for the more commonly used instrument tube fittings.

2 SCOPE

This recommended practice defines nomenclature for tube fittings most commonly used in instrumentation. It is not intended as a substitute for manufacturers' catalog numbers, nor does it apply to special fittings.

This recommended practice is intended to apply to mechanical flared and flareless tube fittings as commonly used in instrument tubing systems.

SPECIFICATIONS

3.1 Each tube fitting in a bill of material should be specified in the following order:

  a. Fitting material
  b. Tube outside diameter (OD)

  c. Pipe thread (NPT* or NPTF**)
  d. Standard name (see Section 4)

Example: Brass 1/4″ OD x 1/4″ NPT Male Connector.

3.2 Special tube fittings not covered by this recommended practice should be specified as listed in manufacturers' catalogs. Manufacturer and catalog number should be given after size and other pertinent data.

3.3 Each tube fitting in a purchase order should be specified according to paragraphs 3.1 or 3.2 with the manufacturer's catalog number, if desired for added clarity.

3.4 The ports for each tube tee or cross shown in paragraphs 4.8 to 4.13 should be specified in the order indicated by the port numbers in the detailed drawings.

Example using 4.9: Brass 1/4″ tube x 1.4″ NPT x 1/4″ tube Male Run Tee.

3.5 Standard abbreviations may be used on drawings and bills of material as shown in Section 4.

3.6 Symbols used on drawings should be as shown in Section 4.

3.7 When metric tubing sizes are in use, the specification should use metric dimensions.

---

* ANSI Standard Taper Pipe Thread from ANSI B2.1-1968
** ANSI Standard Taper Pipe Thread (Dryseal) from ANSI B1.20.3-1976

Detailed	Abbreviation	Schematic Symbol
4.1 Male Connector	**MC**	
4.2 Female Connector	**FC**	

Detailed	Abbreviation	Schematic Symbol
4.3 Male Elbow	ME	
4.4 Female Elbow	FE	
4.5 Union Elbow	UE	
4.6 Tubing Union	UC	
4.7 Bulkhead Union	BU	

Detailed	Abbreviation	Schematic Symbol
4.8 Union Tee 1 —————— 2 3	UT	
4.9 Male Run Tee 1 —————— 2 3	MRT	
4.10 Male Branch Tee 1 —————— 2 3	MBT	
4.11 Female Run Tee 1 —————— 2 3	FRT	
4.12 Female Branch Tee 1 —————— 2 3	FBT	

Detailed	Abbreviation	Schematic Symbol
4.13 Union Cross 	UCR	
4.14 Bulkhead Male Connector 	BMC	
4.15 Bulkhead Female Connector 	BFC	
4.16 Reducing Union 	RU	
4.17 Male Adapter 	MA	

Detailed	Abbreviation	Schematic Symbol
4.18 Female Adapter	FA	
4.19 Tubing Cap	CA	
4.20 Plug	PL	
4.21 Reducer	RE	
4.22 Thermocouple Style Male Connector	TMC	